W9-BJN-472

J. Sosnowiak

STRUKTUR UND EIGENSCHAFTEN DER MATERIE
IN EINZELDARSTELLUNGEN
BEGRÜNDET VON M. BORN UND J. FRANCK
HERAUSGEGEBEN VON S. FLÜGGE
XXV

ATOMIC AND IONIC IMPACT PHENOMENA ON METAL SURFACES

BY

DR. MANFRED KAMINSKY
ARGONNE NATIONAL LABORATORY
ARGONNE, ILLINOIS

1965

NEW YORK
ACADEMIC PRESS INC., PUBLISHERS

BERLIN · HEIDELBERG · NEW YORK
SPRINGER-VERLAG

SPRINGER-VERLAG
BERLIN · HEIDELBERG · NEW YORK

Published in the U.S.A. and Canada by
ACADEMIC PRESS INC., PUBLISHERS
111 Fifth Avenue, New York, New York 10003

Library of Congress Catalog Card Number 64–24914

Printed in Germany

Preface

The collisions of neutral or charged gaseous particles with solid surfaces govern many physical and chemical phenomena, as has been recognized for a long time. The gas/solid phenomena in turn depend on a great variety of processes such as the charge transfer of the gas/solid interface, adsorption and desorption, the energy transfer between an incident particle and the surface, etc.

Our knowledge of these processes, however, is only fragmentary. This is partly due to the difficulty in adequately controlling the experimental conditions. Consequently, until recently the data were usually so complex that reliable information about a particular elementary process could not be deduced. Within the last five to ten years, however, the techniques of ultra-high vacuum and surface preparation have developed rapidly and there has been a booming and widespread interest in the role of gas/solid interactions in such diverse fields as plasma physics, thermonuclear reactions, thermionic energy conversion, ion propulsion, sputtering corrosion of the surface of satellites and ion engines, ion getter pumps, deposition of thin films, etc. This led to extensive investigations of numerous gas/solid phenomena, such as surface ionization, sputtering, emission of secondary electrons and ions from surfaces under atom and/or ion impact, ion neutralization, and the thermal accomodation of gaseous particles on surfaces. As a result, it has become possible to gather a variety of valuable information. An unfortunate result, however, has been that the studies of different impact phenomena on surfaces have advanced so rapidly that often data have been considered as isolated problems and contact between different fields has tended to be lost.

The leading thought in the preparation of the manuscript has been to give a compact synopsis of certain atomic and ionic impact phenomena on surfaces and thus to facilitate interdisciplinary communication between the different fields—as Massey and Burhop [*461*] did so successfully on a similar subject in chapter IX of their book published a decade ago.

The large amount of information available today on atomic and ionic impact phenomena on solid surfaces obviously necessitates a selection of topics. In the following, only impact phenomena on metal surfaces will be considered; and even here only certain types of impact phenomena have been selected. The author wishes to point out explicitly that his selection by no means indicates the relative importance of the subjects chosen but rather reflects his own interests and activities in the field.

The intention has been to give the reader a balanced view of the subject, from both the experimental and theoretical standpoints, and to emphasize the well established principles, which emerge and the many peculiarities, obscurities, and uncertainties which still remain to be resolved. In organizing the manuscript, consideration was given to the sensitivity with which the atomic and ionic impact phenomena under discussion depend on the nature of the metal surface and on the forces of interaction between incident particles and the metal surface. These topics have been treated in separate chapters at the beginning. In the subsequent discussion of particular atomic and ionic impact phenomena, each capter begins with a definition of the quantity investigated and then presents a brief critical discussion of the experimental techniques employed in the study of a particular collision phenomenon. This is followed by a description of the experimental results.

The data have been considered against the available theoretical background. However, in order to stay within the planned size of the book, it often was not feasible to provide a detailed mathematical theory. Instead, the emphasis has been laid on a presentation of the more important concepts of the theories available and on a discussion of the physical phenomena. Yet in most cases a physical account of different theories has been given, and in some important cases even a more detailed description of a particular theory has been provided.

An extensive bibliography enables the reader to go back to the original literature for more detailed studies. Some pains have been taken to formulate adequate definitions of the quantities discussed and, whenever possible, to use the symbols consistently throughout the book. The somewhat complicated designations of formulas, tables, and figures were adopted to facilitate their identification with the appropriate section.

The original manuscript was prepared as an article in the "Ergebnisse der Exakten Naturwissenschaften" on the invitation of one of the editors, Professor S. Flügge. However, when it was completed in the fall of 1962, Professor Flügge offered the alternative possibility of publishing it as a monograph in the series „Struktur und Eigenschaften der Materie". In pursuance of this suggestion, the author expanded the very concise original manuscript in several places to clarify and strengthen some of the points made and to include some of the more important new contributions to the literature of the field. The references added in bringing the survey up to the fall of 1963, when the type was actually set, are the ones whose numbers are followed by a letter (to avoid renumbering all the references).

It is a pleasure to acknowledge the generous assistance received from many people in preparing this monograph. I wish to express my sincere gratitude to Dr. MORTON HAMMERMESH, Associate Director of the Argonne National Laboratory and former Director of the Physics Division, for his constant interest in the progress of the manuscript and for giving me the opportunity and the facilities to undertake this task. In addition, I am most heavily indebted to him for several stimulating discussions concerning theoretical aspects of ionic impact phenomena and especially for his kindness in translating a considerable portion of the manuscript from German into English. I am also very grateful to Dr. LOWELL M. BOLLINGER, Director of the Physics Division of the Argonne National Laboratory, for his permission to carry this work to its completion.

For communications concerning the original concept of the article, the author is indebted to Professor W. WALCHER, University of Marburg, Germany.

It is a pleasure to acknowledge the editorial assistance of Dr. FRANCIS THROW, whose contributions to the article have been invaluable. His great interest in the field reported here resulted in many helpful discussions concerning the preparation of the article. In his careful reading of the manuscript he directed the author's attention to numerous errors and made many suggestions which helped to clarify the expression.

I am indebted to many colleagues for their permission to use their drawings and figures. For assistance in the search for journals and publications from the U.S.S.R., I acknowledge gratefully the help of Mr. NORMAN P. ZAICHICK, from Argonne's library service.

I am very grateful to Mrs. CATHERINE YACK for the careful technical preparation of the original German text. For their patience and careful attention to detail in the accurate preparation of the final English text, I wish to express special thanks to Mrs. JUDITH Z. MILLER, Mrs. KAY PEMBLE, Miss JOANN METES, and in particular to Miss JOSEPHINE MILLER, on whom fell the principal responsibility.

I also wish to acknowledge most gratefully the careful work of Mr. RICHARD W. SNYDER and especially of Mr. BERNARD GREENBERG in preparation of the figures. I sincerely appreciate the technical assistance of Mr. GORDON GOODWIN in preparing some of the material used in the manuscript. I should also like to thank my wife, Mrs. ELISABETH KAMINSKY, for her invaluable help in correcting the proofs and for her great understanding and encouragement in bringing the task to its completion. The excellent cooperation of the Springer-Verlag in the preparation of the book is gratefully acknowledged.

Argonne, Illinois, September, 1963 MANFRED KAMINSKY

Contents

Correction

On page 87, line 3, paragraph 2: Instead of "proposed" read "used".

1. The Nature of the Metal Surface

1.1. The Heterogeneous Surface

1.1.1. Physical Heterogeneity

1.1.1.1. Lattice Defects and Interstitial Positions

In investigating the interaction of an atom, molecule, or ion beam with a metal surface, one must consider the actual structure of the metal. An ideal crystal can be pictured as an infinitely extended three-dimensional array of points whose homogeneity requires that the elementary cells of the space lattice be precisely placed relative to one another. The arrangement of the lattice points in space must have the symmetry of one of the 230 space groups. In describing the deviations of an imperfect crystal from the strict periodic arrangement of the ideal crystal we can, according to CORRENS [*126*], distinguish between defects and dislocations. The defects include vacancies [SCHOTTKY defects, shown schematically in Fig. 1.1.1.1.a, A] as well as interstitial sites [FRENKEL defects, shown schematically in Fig. 1.1.1.1.a, B]. When ions move to interstitial positions, their previous lattice positions are left vacant. Although in the description of defects it is usually assumed that the individual elementary cells are arranged strictly parallel in the crystal, deviations from such an arrangement can occur in the imperfect crystal. Such deviations, which are collectively referred to as "dislocations", consist in slight tilts or displacements of one region of the crystal relative to another. The crudest models of the

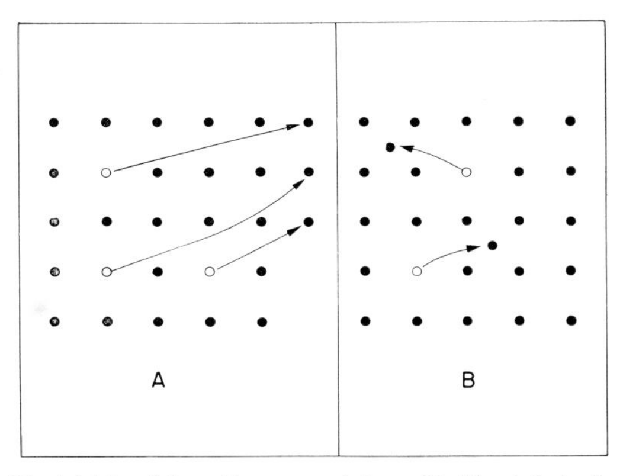

Fig. 1.1.1.1.a. Schematic representation of lattice defects. In SCHOTTKY defects (A), lattice atoms diffuse to the crystal surface, leaving unoccupied lattice sites (open circles). In FRENKEL defects (B), lattice atoms move to interstitial positions leaving previous lattice positions vacant. Interstitial atoms and vacancies have equal concentrations

actual structure are slightly canted "mosaic blocks" or dendrite structures (BUERGER [*103*]) or, in some cases, spiral structures.

Such defect and dislocation phenomena can undoubtedly be expected on the crystal surface as well, although conclusions about the situation on the surface should not be drawn from the behavior in the crystal interior without further consideration. According to GIBBS [*234*] the equilibrium between the surface and the phase in the crystal interior is determined by whether the perturbations lead to an increase or decrease of surface energy.

On an atomic scale, the surface of a real crystal is never completely smooth. But even the surface of an ideal crystal, as BURTON, CABRERA and FRANK [*106*] have shown, may roughen and develop steps because of thermal motions. They find that planes with low indices remain smooth at temperatures below the melting point, while surfaces with high indices begin to roughen. According to KOSSEL [*406, 407*] and STRANSKI et al. [*686, 387*], the possible characteristic sites on a surface of a cubic lattice (Fig. 1.1.1.1.b) are in a completed edge (1), at a vertex (2), in a completed surface (3), on the incomplete edge (4), at an incomplete corner (5), at a step (6), and on the surface (7). Vacant lattice positions may also occur on the surface (8). An especially important position is the half-crystal position (1/2), which occurs most frequently in gradual buildup of the crystal. The binding forces which are exerted on a given particle by its neighbors naturally depend very strongly on its precise location. For homopolar crystals and also (as STRAUMANIS [*688*] and MÜLLER [*509*] have shown) for metal crystals, the binding forces exerted on a given particle by its neighbors may be assumed to add approximately linearly. If the binding energies of nearest and next nearest neighbors are denoted by E_1 and E_2 (where $E_2 \approx 0.1\, E_1$), then the total binding energy of an interior constituent (for example in a cubic crystal) would be represented by $E = 6E_1 + 12E_2$. In a similar manner, following KNACKE and STRANSKI [*387*], the binding energies may be computed for surface elements in the above mentioned positions (1)—(7) and (1/2)*.

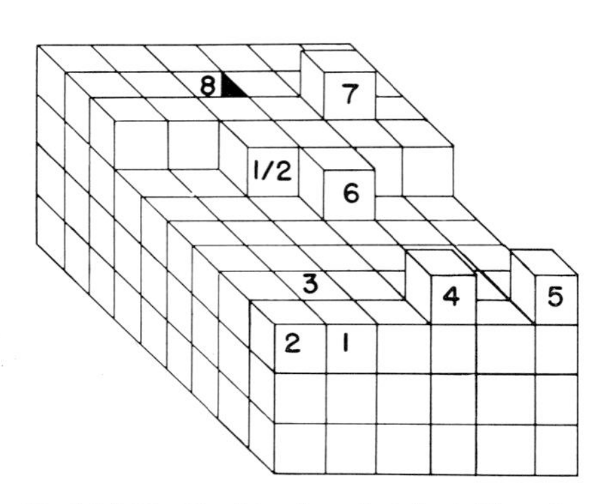

Fig. 1.1.1.1.b. Possible characteristic lattice sites on the surface of a cubic lattice. (After KOSSEL and STRANSKI)

* Thus, for the positions (1)—(7) and (1/2), the binding energies are:
$E(1) = 4E_1 + 4E_2$; $E(2) = 3E_1 + 3E_2$; $E(3) = 5E_1 + 8E_2$; $E(4) = E_1 + 3E_2$; $E(5) = E_1 + 2E_2$; $E(6) = 2E_1 + 6E_2$; $\mathrm{E}(7) = E_1 + 4E_2$; $E(1/2) = 3E_1 + 6E_2$.

The values of the binding energies for the different positions decrease in the order

$$E(3) > E(1) > E(1/2) > E(2) > E(6) > E(7) > E(5).$$

These considerations, of course, apply only very roughly in calculations of the adsorption energies of foreign atoms at the various lattice positions since the equilibrium separation between the adatom and the surface atom is commonly known only inadequately. However, the effect of surface structure on adsorption energies depends on the nature of the binding forces between the adatom and the surface atom. Finally the effect of surface heterogeneity has been invoked in interpreting the results obtained from various experiments, such as the investigation of catalytic reactions (SCHWAB et al. [*644*]), inelastic reflection of atoms at surfaces (ROBERTS [*593, 594*]), and desorption rates (STRANSKI [*387*]).

1.1.1.2. Occurrence of Different Crystal Planes on the Surface

The presence of different crystal planes on a metal surface results in a heterogeneous distribution of adsorption forces on the face. This

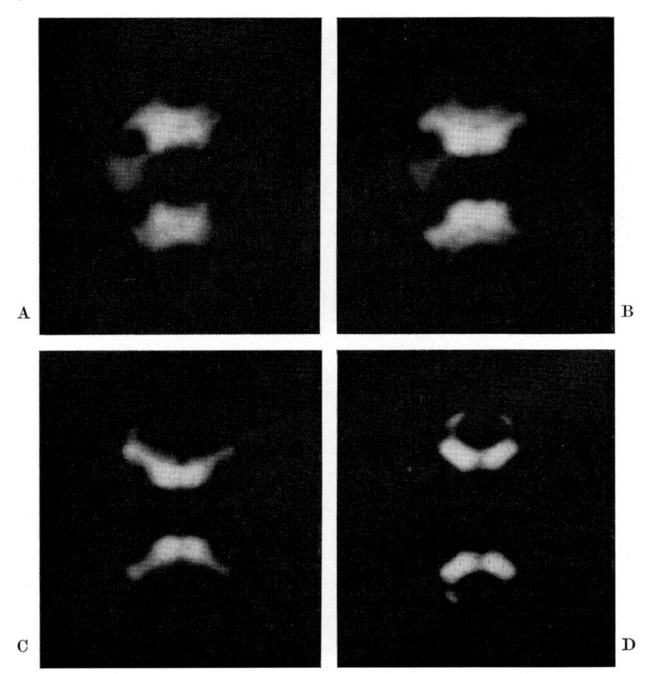

Fig. 1.1.1.2.a. Field-emission-microscope photographs of the adsorption of Xe on a wolfram emitter tip held at 79° K. All photographs are oriented to show a (110) plane at the center and (100) planes at top and bottom. To maintain an emission current of 5×10^{-9} A, the necessary voltage was 4770 V at $t = 0$ (clean point) but fell as the adsorbed layer formed. Photograph A: 4720 V at $t = 1.75$ min; B: 4600 V at $t = 4$ min; C: 4280 V at $t = 8$ min; D: 2900 V at $t = 47$ min. Final pressure $\approx 10^{-7}$mm (EHRLICH and HUDDA [*172*])

heterogeneity of metal surfaces is apparent from many experimental results such as, for example, measurements of field emission, thermionic emission, thermal accommodation, physical and chemical adsorption, and catalytic activity. Thus, for example, EHRLICH and HUDDA [*173, 176*] have recently used the field-emission method to study the adsorption of noble gases on wolfram and molybdenum, both of which are body-centered cubic. When xenon atoms with a gas temperature of 300° K impinged on a wolfram needle at 79° K, they found that the adsorbed

Fig. 1.1.1.2.b. Hard-sphere model of Xe adsorbed on a wolfram emitter. Scale: 1 Å = 1 cm. Most favorable adsorption sites are on the (310) and (611) planes. (EHRLICH [*174*])

atoms arranged themselves around the (100) pole as shown in the ring of bright emission in Fig. 1.1.1.2.a. This ring-shaped region, which results from the increase in electron emission as a consequence of xenon adsorption, is bounded on its inner side by the edges of the (100) plane and on its outer side by the edges of the (100) and (111) planes (see model shown in Fig. 1.1.1.2.b). Fig. 1.1.1.2.c shows a similar behavior in the adsorption of xenon on molybdenum. The preferential adsorption on the (120), (130), and (116) planes can be understood theoretically, since the number of lattice atoms surrounding the adatom is a maximum on these surfaces.

Investigations of the work functions of metal surfaces also show a marked dependence on the crystal planes that constitute the surface.

Table 1.1.1.2. A lists the work functions of individual crystal planes in wolfram and silver, as obtained by four different authors by the methods stated.

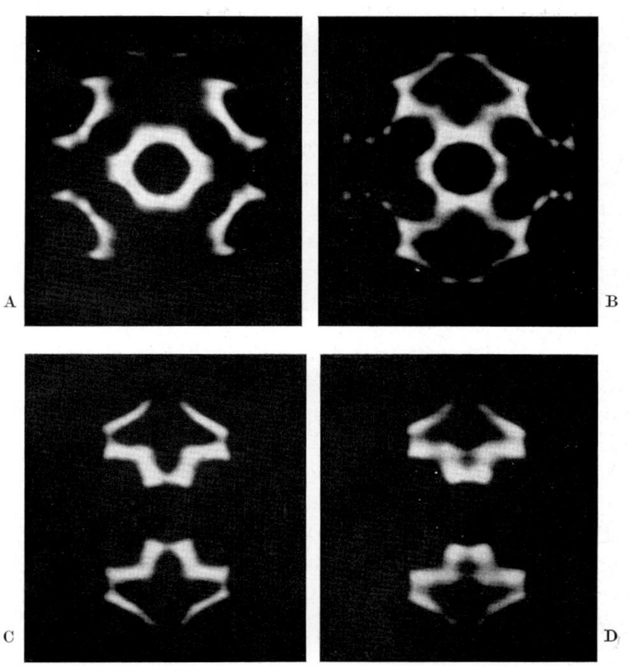

Fig. 1.1.1.2.c. Field-emission-microscope photographs of the adsorption of Xe on a molybdenum emitter tip held at 79° K. The adsorption occurs preferentially around the [*100*] poles, although it is enhanced along the edges of the (110) planes. Voltage for 5×10^{-9} A emission: clean point, case A: $V = 3800$ V at $t = 0$; case B: $V = 3380$ V at $t = 18$ min; case C: $V = 3050$ V at $t = 29$ min; case D: $V = 2780$ V at $t = 121$ min. (Ehrlich and Hudda [*174, 176*])

Table 1.1.1.2.A. *Experimental data for the work functions of different crystal planes for wolfram and silver surfaces*

Metal	Plane	Smith [*660*] Thermionic emission (eV)	Hutson [*334*] Thermionic emission (eV)	Müller [*511*] Field emission (eV)	Farnsworth [*201*] Photo emission (eV)
W	111	4.38	4.3	4.39	—
	001	4.56	4.44	—	—
	011	5.26	5.09	5.70—5.99	—
	116	4.29	4.20	4.30	—
Ag	111	—	—	—	4.75
	100	—	—	—	4.81

The results seem to indicate that the value of the work function depends on the density of packing in the particular crystal plane. The greater the number of nearest neighbors and the shorter the distance to them, the greater is the work function. It should be mentioned in this connection that SACHTLER [*622*] has suggested that the work function may be represented by the product of metal density and ionization potential of the individual metal atom plus an additive constant.

For polycrystalline metal surfaces, the experimentally determined value of the work function is an average of the work functions of all the crystal planes that are present on the surface. It is made up of the contributions $e\varphi_\nu$ from the individual crystal planes, i.e.,

$$e\bar{\varphi} = e \sum c_\nu \varphi_\nu ,$$

where c_ν is the fraction of the metal surface that is occupied by the ν th crystal plane. We shall see in the following chapters that the work function of a surface is one of the important quantities determining the interaction energy for the atom, molecule, and ion beams impinging on the metal surface. The use of polycrystalline surfaces for investigating such interaction energies presents an essential difficulty because, as a result of the inadequate knowledge of both φ_ν and c_ν, the effect of the work function on the interaction energy is very poorly known. From this it follows that only atomically clean faces of single crystals should be used in such investigations.

The difference in adsorption from one crystallographic plane to another in the same crystal has been confirmed by numerous other experiments. Thus, for example, sodium is preferentially adsorbed on the (211) planes of wolfram and molybdenum surfaces. Barium is preferentially adsorbed on the (111) planes of wolfram, while oxygen atoms are preferentially adsorbed on the (100) planes of wolfram and molybdenum ([*69*, *70*, *507*]). The adsorption process on different crystallographic planes has been considered theoretically by LENEL [*430*], BARRER [*50*], ORR [*538*], and YOUNG [*788*]. In this connection we should mention the idea of VOLKENSHTEIN [*744*], who explains chemisorption phenomena on surfaces in terms of their inhomogeneities (which he calls "microdefects"), which are not localized at given positions on the surface but are mobile and interact with one another.

Different crystallographic planes of a given crystal can show marked differences in catalytic activity. Thus, the experiments of GWATHMEY et al. [*132*, *749*], show that, in the case of copper single-crystal spheres, those positions that are parallel to the most densely packed (111) planes show the highest catalytic activity. Other authors (SOSNOVSKY [*668*],

Beeck et al. [*67*], and Westrick et al. [*779*]) have shown that catalytic activity depends markedly on the nature of the crystal surface.

1.1.2. Chemical Heterogeneity

By "chemical heterogeneity" we mean the imperfections in a crystal as the result of the presence of foreign atoms in the host lattice. Such inclusion of foreign atoms into the lattice can occur intentionally (e.g., in preparing crystals of substitutional alloys such as Ag-Au) or accidentally (e.g., in the process of fabrication). Even when present in very low concentration, the foreign atoms on the surface of the lattice can serve as "active centers"—i.e., they may be active in adsorption and catalytic processes and thus have effect on the heats of activation. (Further information can be obtained from the review articles of Schwab et al. [*644*] and Constable [*123*]). Such impurities in the body of the metal can wander through the crystal lattice and thus also arrive at the surface. Lazarus [*426*] recently discussed the dependence of such diffusion rates on electrostatic interactions between the foreign atom and lattice atoms, and on the effects which depend on the sizes of foreign and lattice atoms. In this connection, one may call attention to the interesting mass spectrometric investigations of Datz et al. [*139*], who observed a spontaneous emission of positive alkali ions from a heated wolfram wire on which single-crystal domains were present. In this process the pulses were aperiodic and consisted principally of the potassium ions which occurred as impurities in the wolfram lattice. The individual pulses contained up to 10^6 alkali ions in a pulse time of less than 100 μ sec. (The rise times of the pulses varied from 2 to 50 μsec, depending on the surface temperature.) From the large number of alkali ions emitted per unit time, the authors concluded that these pulses were due to impurities concentrated primarily in microcracks* and cavities rather than at dislocations at the edges and vertices of the crystal—the otherwise preferred positions for emission processes (see also C. Kittel, Introduction to Solid State Physics, Wiley & Sons, New York, 1953). The results of Reynolds and Michel [*588*], who actually found an emission of even such polymeric species as K_2^+, K_3^+, K_{15}^+ from heated wolfram filaments, seem to support the thesis of Datz et al. [*139*] that there are impurity-filled pockets in the filament, since such polymers apparently can form only under fairly high pressures.

Ahearn [*3*] recently studied the impurities of metal crystals and metal surfaces by use of a spark-discharge source and a double-focusing mass spectrometer. He was able to detect extremely small concentrations of impurity (one foreign atom per 10^7 lattice atoms).

* Such microcracks are assumed to be large compared to the atomic spacing of the individual lattice components.

1.1.3. Induced Heterogeneity

By "induced heterogeneity" we mean that as a result of an adsorption process the metal surface is covered by a layer of foreign atoms from the surrounding gas. The rate (particles/cm²-sec) at which gas molecules impinge on the surface in the equilibrium state is given by $n = \frac{1}{4} N\overline{v}$, where N is the number of molecules per cc and $\overline{v}$ is the mean thermal speed. Thus if it is assumed that the background pressure is 10^{-6} mm Hg, that the condensation coefficient (sticking probability) is unity, and that 1.15×10^{15} adsorption positions are available in a monatomic layer of (say) wolfram, then one estimates that such a monatomic layer of foreign atoms would be adsorbed onto the surface within 1 sec. This is in agreement with the experimental results of BEKKER and HARTMAN [*60*]. At a background pressure of 10^{-9} mm Hg, the time to build up a monolayer would be 1000 sec, in agreement with the experimental results of HAGSTRUM [*279, 283*].

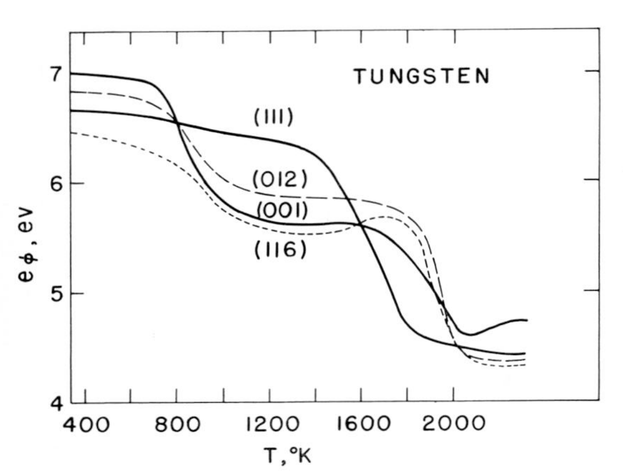

Fig. 1.1.3.a. Changes of the work function of different wolfram single-crystal planes with increasing oxygen desorption, measured with a field-electron microscope after heating the wolfram crystal for periods of 1 min in steps of 100° K. (E. W. MÜLLER [*512*])

However, as will be seen for example in the inelastic scattering of atoms from surfaces (Chap. 6), such adsorption layers can be drastically affected by the characteristics of the gas-solid interaction.

For "clean" crystal surfaces, it appears that the change in work function when a layer of gas is adsorbed on a surface is different for different crystal planes. For example, the experiments of MÜLLER show that the work function of wolfram single-crystal planes covered by oxygen decreases with increasing desorption (increasing surface temperature), i.e., with decreasing degree of foreign gas covering (Fig. 1.1.3.a). The investigation of the same system by BECKER and BRANDES [*62*] shows similar relations. In photoelectric measurements of the work function of wolfram, EISINGER [*177—179*] found that covering the (113) plane with hydrogen, nitrogen, or carbon monoxide changed the work function of this plane by the following amounts.

$$\text{Hydrogen-covered W}: \Delta(e\varphi) = -0.45 \text{ eV}$$

$$\text{Nitrogen-covered W:} \quad \Delta(e\varphi) = -0.35\ \text{eV}$$
$$\text{CO-covered W:} \quad \Delta(e\varphi) = -0.86\ \text{eV}$$

In accordance with the usual convention, the negative sign here indicates merely that a dipole layer is formed with the negative charge away from the metal, although the change in work function occurs in the opposite sense (i.e., the work function is increased). In addition to the measurements on clean metal surfaces, there is also a large amount of experimental material available on the changes in work function of polycrystalline metallic surfaces when covered with foreign atoms. Recently EBERHAGEN [*169*] has written a comprehensive survey, to which the reader is referred.

From a classical electrostatic argument, it has been suggested (for example, by CULVER and TOMPKINS [*131*]) that the change in the work function should depend on the degree of covering through the relation

$$\Delta(e\varphi) = 4\pi\, \delta_m \theta \mu_0 ,$$

where δ_m is the number of available adsorption sites per cm² on the surface, θ is the fraction of the surface positions which are already covered, and μ_0 is the dipole moment of the adatom-surface system. The μ_0 is assumed to be independent of φ.

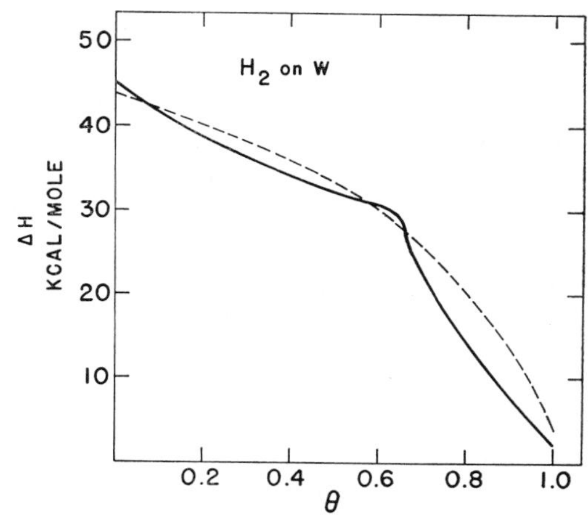

Fig. 1.1.3.b. Changes of the heat of adsorption ΔH with the degree of surface coverage θ for hydrogen adsorption on wolfram films (solid line, BEECK [*68*]) and a wolfram wire (dotted line, ROBERTS [*595*])

This change in work function as a result of surface coverage has a corresponding effect on heats of chemisorption, which will be discussed in more detail in Chap. 5. Here we shall only, as an example, give Fig. 1.1.3.b, in which the dependence of the heats of chemisorption of hydrogen on (a) tungsten layers, or (b) tungsten wires are shown. The heat of chemisorption is seen to decrease markedly as the degree of covering increases.

From this marked effect of surface coverage on the quantities which are characteristic for the interaction of the gas atom (or ion) and the metal surface, it is evident that unknown layers of foreign gas atoms on the surface (for example, because of inadequate vacuum) can seriously distort such results. For this reasón it is necessary to carry out such investigations at extremely high vacuum and to use various techniques to clean the surfaces, as will be discussed in more detail in Chap. 3.

1.2. Theoretical Description of Metal Surfaces The Work Function

Here we shall assume a familiarity with the features of the first, essentially classical, theory of metals developed by Drude and Lorentz, which describes the motion of electrons in metals as an electron gas, in analogy to the kinetic theory of gases. Pauli and Sommerfeld showed that this theory leads to contradictions in the values of the specific heats of metals. These contradictions are removed when it is realized that the large number and small mass of the electrons in a metal makes the Maxwell distribution inapplicable; Fermi statistics hold instead. We shall also assume that the reader is familiar with Sommerfeld's quantum theory of the free electron, which started from the concepts that (1) the valence electrons in the metal are "free", as assumed by Drude and Lorentz, but the energies of the valence electrons are discrete, and (2) in accordance with the Pauli exclusion principle, each quantum state can be occupied by only two electrons with antiparallel spins. We do not discuss the extension of these concepts to such problems as the motion of electrons in a periodic field (Bloch's theorem), three-dimensional motion of the electron, the band theory of metals, and the theory of Brillouin zones.

Even though the present state of the theory of metals contains many extensions and further developments (see for example the review article by H. Jones [*359*]), the older concept of a freely moving electron gas confined within the metal by a surface potential barrier can still be retained for the purpose of the following discussion. With this concept, we can use thermodynamic arguments to treat those phenomena that are related to the potential barrier at the surface, e.g., the extraction of electrons from a metal surface (the work function). On the other hand, we may use quantum mechanical considerations based on particular models (as noted above).

The following is a somewhat more detailed consideration of the several divergent definitions of the work function of a metal, since this quantity is of great importance for the interaction of gas atoms, molecules, and ions impinging on the metal surface.

In analogy to the work function of a neutral atom, one can define an isothermal work function in the following way. In a system of N_i ions and N_e electrons in the interior of a metal at temperature $T = 0$, let $N_i = N_e$ so that the interaction energy of the electrons be exactly compensated by that of the ions, so that each "cell" is assumed to be electrically neutral. The total energy of the system may be represented by $E(N_i, N_e)_{T=0}$. The work function is the change in the total energy of the system when one electron is removed, i.e.,

$$e\varphi = [E(N_i, N_e - 1) - E(N_i, N_e)]_{T=0}.$$

If T is constant and greater than zero, the analogous expression for the isothermal work function is

$$e\varphi_{\text{isoth}}(T) = [E(N_i, N_e - 1) - E(N_i, N_e)]_{V, T=\text{const}} = \left(\frac{\partial E}{\partial N_e}\right)_{V, T=\text{const}}. \tag{1.2–1}$$

Frequently one uses the "adiabatic" work function (for example, in the RICHARDSON equation for thermionic emission), which is defined as

$$e\varphi_{\text{adiab}} = -\eta_0 = \left(\frac{\partial E}{\partial N_e}\right)_{V, S=\text{const}}. \tag{1.2–2}$$

Here S is the (constant) entropy of the system and η_0 denotes the level of the electrochemical potential with respect to the energy of the electron at rest at infinity.

The difference between φ_{isoth} and φ_{adiab} is usually very small, as will be briefly shown in the following. From thermodynamics we have

$$\eta_0 = \left(\frac{\partial E}{\partial N_e}\right)_{S, V=\text{const}} = \left(\frac{\partial F}{\partial N_e}\right)_{T, V=\text{const}}, \tag{1.2–3}$$

where $F = E - TS$ is the free energy,

$$\left(\frac{\partial E}{\partial N_e}\right)_{T, V=\text{const}} = \left(\frac{\partial F}{\partial N_e}\right)_{T, V=\text{const}} + T\left(\frac{\partial S}{\partial N_e}\right)_{T, V=\text{const}}, \tag{1.2–4}$$

and $S = -(\partial F/\partial T)_{V, N_e=\text{const}}$ is the entropy. Then Eq. (1.2–4) can be written

$$\begin{aligned}\left(\frac{\partial E}{\partial N_e}\right)_{T, V=\text{const}} &= \left(\frac{\partial E}{\partial N_e}\right)_{S, V=\text{const}} - T\,\frac{\partial^2 F}{\partial N_e\,\partial T}\\ &= \left(\frac{\partial E}{\partial N_e}\right)_{S, V=\text{const}} - T\left(\frac{\partial \eta_0}{\partial T}\right)_{V, N_e=\text{const}}.\end{aligned}$$

This leads to

$$\varphi_{\text{isoth}} = \varphi_{\text{adiab}} + T\left(\frac{\partial \eta_0}{\partial T}\right)_{V, N_e=\text{const}}. \tag{1.2–5}$$

To estimate the magnitude of $T(\partial\eta_0/\partial T)_{V, N_e=\text{const}}$, we will use some of the results of the SOMMERFELD theory of metals. From Eq. (1.2–3) it follows that

$$T\left(\frac{\partial \eta_0}{\partial T}\right)_{V, N_e=\text{const}} = T\left(\frac{\partial E_i}{\partial T}\right) = -\frac{\pi^2}{6}\,E_{i0}\left(\frac{T}{T_{cr}}\right)^2. \tag{1.2–6}$$

Here the characteristic kinetic energy E_i of the electron (Fig. 1.2.a) is given by

$$E_i = E_{i0}\left[1 - \frac{\pi^2}{12}\left(\frac{T}{T_{cr}}\right)^2\right], \tag{1.2–7}$$

with

$$E_{i0} = \frac{\hbar^2}{2 m_e}\left(\frac{3n}{8\pi}\right)^{2/3}, \tag{1.2–8}$$

and the critical degeneracy temperature T_{cr} is

$$T_{cr} = \frac{\hbar^2}{2 m_e k}\left(\frac{3n}{8\pi}\right)^{2/3}. \tag{1.2–9}$$

Here $n = N_e/V$ is the total number of electrons in volume V.

The degeneracy temperature T_{cr} is of the order of $70000°$ K, $E_{i0} \approx 10$ eV, and for $T = 1000°$ K, the difference between the isothermal and adiabatic work functions is

$$\varphi_{\text{isoth}} - \varphi_{\text{adiab}} \approx 0.003 \text{ eV}.$$

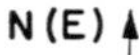

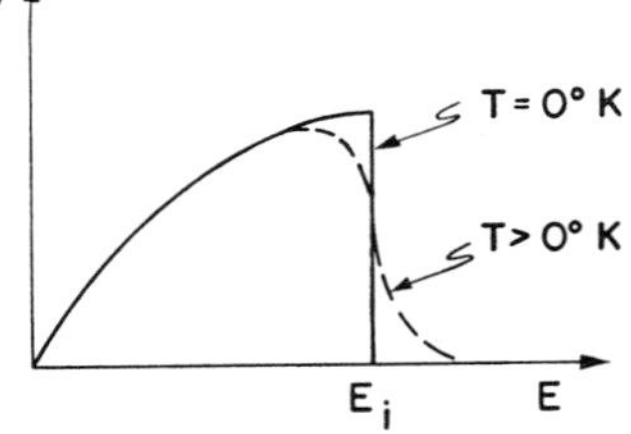

Fig. 1.2.a. Schematic diagram of the distribution function $N(E)$, the number of electrons which have energies in the range between E and $E + dE$, plotted against their kinetic energy E for the case $T = 0°$ K and $T > 0°$K. The characteristic electron energy E_i is given by Eq. (1.2–7)

After these general thermodynamic considerations, let us go on briefly to some particular models.

In the classical theory of metals, the work function was equated to the jump in potential energy at the boundary of the metal. It is equal to the minimum energy that must be given to a free electron in the electron gas at $T = 0$ to enable it to escape from the metal. If $f(r)$ is a force field at the metal boundary, which is directed toward the interior of the metal and hinders the exit of free electrons into the outside space, the work to make possible the emergence of the electron from a point $(-i)$ in the interior of the metal is

$$e\varphi = \int_{-i}^{\infty} f(r)\,dr. \tag{1.2–10}$$

This idea of the work function is revised in SOMMERFELD's model of the metal. According to SOMMERFELD, the integral in (1.2–10) determines only the "outer work function" W_a which is equal to the total depth of the potential well (Fig. 1.2.b). In contrast to the classical theory, it is assumed that at $T = 0$ not all the N_e electrons have zero kinetic energy, but that their energies are distributed from zero up to W_i, the FERMI level. The work function $e\varphi$ is thus the minimum energy necessary to bring a conduction electron at $T = 0$ into the energy continuum, i.e.,

$$e\varphi = W_a - W_i. \tag{1.2–11}$$

It should be noted that at normal temperatures the energy levels above the FERMI energy are so sparsely occupied that they may be neglected in extending these considerations to $T > 0$.

This potential well is still a very crude model of the variation of the potential energy of an electron at a metal surface; it must be modified to agree better with reality. Thus we must remember that the potential energy associated with the lattice structure is a periodic function of the coordinates. Furthermore, modern theories (OLDEKOP and SAUTER [*534*], CUTLER and GIBBONS [*134*], BARDEEN [*47*] and LEWIS [*438*]) take account of the polarization part of the work function, and consequently consider how the polarizing effect of the individual electron in the metal affects the distribution of the other conduction electrons. Furthermore account is taken of a "double-layer contribution to the work function" which is ascribed to the following mechanism. The free conduction electrons can emerge from the boundary of the metal and build up a negatively charged layer near the surface. This layer, together with the excess positive charge of the ions below the surface, forms an electrical double layer whose field slows down electrons emerging from the metal surface (OLDEKOP and SAUTER [*534*]).

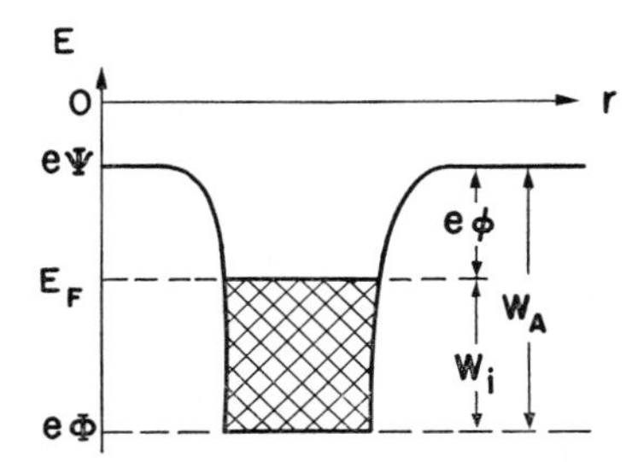

Fig. 1.2.b. Potential energy diagram for the free metal electrons at a metal surface (potential-well model)

The temperature dependence of the work function should also be considered briefly. This has not been considered carefully enough in investigations of, for example, thermionic emission, positive, surface ionization, and desorption processes (Chap. 8). Thus most of the investigations mentioned in these chapters have been carried out at various high surface temperatures, while in evaluating the results (for example, by using the RICHARDSON equation or the SAHA-LANGMUIR equation) φ was assumed to be constant and independent of temperature. Actually however, the charge density on the metal surface decreases with increasing surface temperature or, in other words, the FERMI level

$$W_i = \frac{\hbar^2}{2\,m_e}\left(\frac{3\,n}{8\,\pi}\right)^{3/2} \qquad (1.2\text{–}12)$$

decreases with increase in temperature, since the electron density n in the metal decreases because of thermal expansion of the metal. But the "external work function" W_a also decreases with increasing temperature because of the decrease in the surface charge responsible for the potential jump at the metal boundary. The magnitude and sign of the temperature coefficient $d\varphi/dT$ depends on how fast W_a and W_i change with T. For example, the experimental results (POTTER [*576*], LANGMUIR [*424*], REIMAN [*585*], and NOTTINGHAM [*525*] on some metals (such as wolfram) show increases in the range from 10^{-4}—10^{-5} eV/degree. This means that over a small temperature interval, the temperature dependence of φ can be represented as

$$\varphi(T) = \varphi(T_0) + a(T - T_0). \tag{1.2–13}$$

If one inserts this equation into the RICHARDSON equation, one obtains

$$\begin{aligned} j &= A_0 \exp\left(-\frac{ea}{k}\right) T^2 \exp\left[-\frac{e(\varphi(T_0) - aT_0)}{kT}\right] \\ &= AT^2 \exp\left(-\frac{e\psi}{kT}\right). \end{aligned} \tag{1.2–14}$$

If one plots $(\ln j)/T^2$ against $1/T$, the RICHARDSON line no longer gives φ at $T = 0$ but rather $\psi = \varphi(T_0) - aT_0$. The experimental results must be correspondingly corrected.

Finally, we want to consider how the work function changes as a result of adsorption layers on the metal surface. (Such layers were mentioned briefly in Section 1.1.3.) As a free atom approaches the metal surface, there is a perturbation of the discrete energies of its outer electrons. In the subsequent adsorption process we must distinguish between the following three limiting cases of possible electronic interactions (cf. schematic diagram, Fig. 1.2.c).

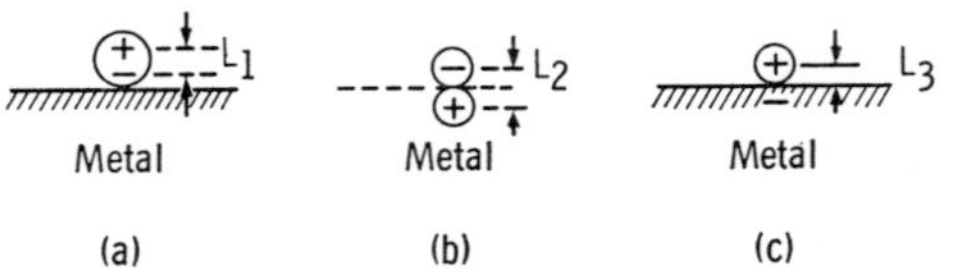

Metal (a) L_1 Metal (b) L_2 Metal (c) L_3

(a) $E_W = \dfrac{m_e c^2 \chi}{2N_A r^3}$, [Lennard-Jones]

(b) $E_{hom} = E_{coulomb} + E_{exchg.}$, [Heitler-London]

(c) $E_{het} = e(\phi - I) + \dfrac{e^2}{4L}$, [De Boer]

Fig. 1.2.c. Schematic diagram of three limiting cases of electronic interactions in the adsorption process: (a) physical adsorption (VAN DER WAALS forces), (b) weak chemisorption (homopolar forces), (c) strong chemisorption (heteropolar forces)

In the first case, the electron shell of the adatom is filled (as for example, in a noble gas). There is no exchange of electrons between the metal surface and the noble gas atom, but a slight polarization of the noble gas atom may occur if the electron affinity of the metal surface is sufficiently large (the case of "physical adsorption", discussed in Chapter 4). In the second case, exchange forces as described by HEITLER and LONDON may cause a weak chemisorptive bond of the covalent type (chapters 4 and 5) between the adsorbent and adsorber—as in the case of H_2 on Ni (ELEY [*180*])—and thereby change the work function. In the third case, two possibilities must be distinguished. The work function of the metal surface may be small in comparison with the ionization energy of the adatom. Then none of the split electron states of the adatom lie below the FERMI energy W_i, and consequently a valence electron of the adatom is transferred to the metal surface. As a result, the surface layer becomes positively charged and consequently the work function of the metal is reduced (cf. Fig. 1.2.d). In the other possibility the work function of the metal surface is large in comparison with the ionization energy of the adatom. Then an electron is transferred from the metal surface to the adatom and the surface layer becomes negatively charged which consequently leads to an increase of the work function (cf. Fig. 1.2.e).

Naturally various intermediate cases are possible between these limits. Thus, for example, "physical adsorption" may be due not only

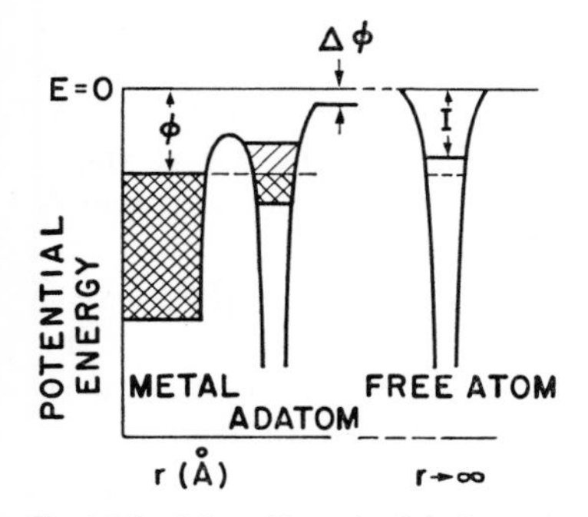

Fig. 1.2.d. Schematic potential diagram for the adsorption of an electropositive adatom (e.g., Cs on W)

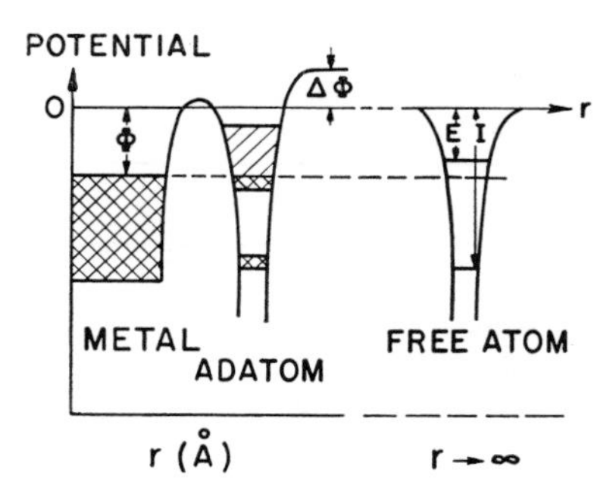

Fig. 1.2.e. Schematic potential diagram for the adsorption of an electronegative adatom (e.g., O_2 on W)

to the polarization of the electron core of the adatom but also to a partial overlapping of the electron states of the adatom-metal system. The resultant dipole layer reduces the work function for the metallic electrons and, according to MIGNOLET [*480*], the change in work function is proportional to $[(\varphi/I) - (S/\varphi)]$, where I is the ionization potential and S is the electron affinity of the adsorbed substance.

Attempts have been made to express these changes in work function in terms of the surface coverage θ of the surface. A classical electrostatic argument (potential jump at a double layer) shows that the work function φ of a clean metal surface increases (or decreases) by the amount $4\pi\, e\, \sigma_1 \mu_\theta \theta$ if the surface layer of adatoms is charged negatively (positively), i.e.,

$$e\varphi = e\varphi_0 \pm 4\pi\, e \sigma_1 \mu_\theta \theta, \tag{1.2–15}$$

where σ_1 is the number of surface atoms per cm². Here it is assumed that the interaction of the adatoms can be neglected. However, as $\sigma_1 \theta$ increases one must take account of such an interaction. Thus MILLER [*481*] has shown that the strength of the mutual depolarization of a plane arrangement of dipoles is

$$\mu_\theta = \frac{\mu_0}{(1 + 9\,\alpha\theta/a_1^3)}\,, \tag{1.2–16}$$

where a_1 is the separation of the dipoles at $\theta = 1$ and α is the polarizability of the adsorbed complex. Thus Eq. (1.2–15) becomes

$$\Delta(e\varphi) = \pm \frac{4\pi e \mu_0 \sigma_1}{1 + \varkappa\theta}\,\theta\,, \tag{1.2–17}$$

where $\varkappa = 9\alpha/a_1^3$ and $\sigma_1\theta$ is the number of dipoles per cm². According to BOUDART [*87*], Eq. (1.2–17) is valid for the adsorption of Cs on wolfram for $\theta < 0.4$. The experimental values are given in Table 1.2.A. The agreement between experimental and calculated values is satisfactory.

Table 1.2.A. *Change in work function with varying degree of coverage of Cs on wolfram*

θ	$\Delta(e\varphi)_{calc}$ [*87*] eV	$\Delta(e\varphi)_{exper}$ [*705*] eV
0.167	1.28	1.31
0.263	1.82	1.82
0.345	2.22	2.22
0.392	2.44	2.41

SACHTLER et al. [*96*] give a different expression for the change in work function with varying degree of coverage θ. They derived this relation by use of the PAULING model of the homopolar bond. In the derivation they superposed the eigenfunctions of three limiting structures—(1) the pure homopolar form M-X, where M = metal and X = adsorbed material, (2) the ionized form M^-X^+, and (3) the inverse structure M^+X^-—and applied ELEY's method [*181*] to determine the heats of adsorption of systems with homopolar binding by use of ELEY's method, namely,

$$\Delta(e\varphi) = \frac{4\pi(300 \times 10^{-18})\,\sigma_1\theta}{3.15}\,e(\varphi - \psi)\,,$$

where $\sigma_1\theta$ is the number of surface dipoles per cm², φ is the electron work function of the metal, and ψ is the "absolute electronegativity" of the adsorbed material, as defined by MULLIKEN [*515*]. It is the arithmetic mean of the ionization potential I and the electron affinity S of the adsorbed substance, i.e., $\psi = \frac{1}{2}(I + S)$. SACHTLER et al. show on the basis of experimental material (see their Table 2) that their relation gives at least the correct sign for the change in work function for both acceptor-adsorption and donor-adsorption systems.

2. Determination of the Work Function of Metal Surfaces

2.1. Thermal Emission of Electrons

The significance of the work function as an important quantity in the interaction phenomena of atoms and molecules impinging on metal surfaces is evident from the preceeding chapter. We shall therefore describe some of the techniques for measuring the work function of clean or gas-covered metal surfaces.

One of the commonest methods uses the thermal emission of electrons from the heated metal surface under investigation. Such measurements using single-crystal wires have been made for instance with a method due to SMITH [660], shown in Fig. 2.1.a. (Similar arrangements have been used by NICHOLS [523].) In this method, electrons emitted from the hot cathode wire A are attracted toward concentric anode B. Those that pass through collimating slits S and S' are collected on the positive collector D. The voltage on electrode C is negative with respect to D in order to suppress the secondary electrons emitted from the collector and from slits S and S'. In SMITH's arrangement, the wolfram single-crystal wire A can be rotated in order to investigate the dependence of the work function on the crystal plane. If the anode voltage is chosen large enough that the electron current has attained its saturation value (i.e., is not space-charge limited), then the electron current density j depends on the temperature T and the work function φ in accordance with the RICHARDSON equation

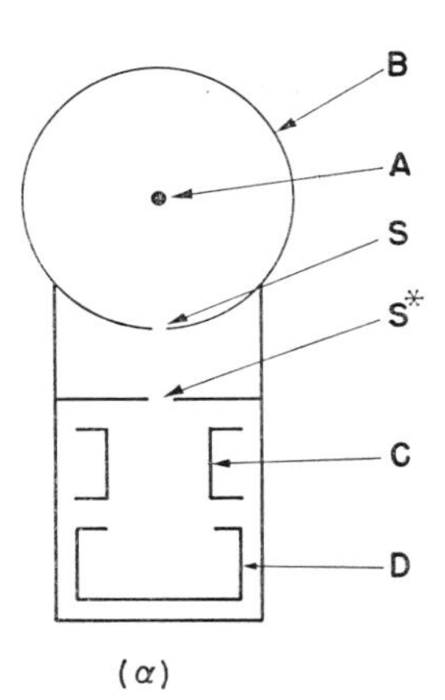

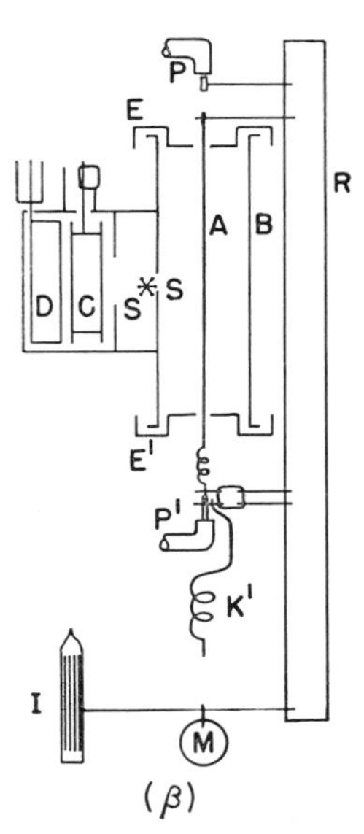

Fig. 2.1.a. α and β. Top view (diagram α) and cross section (diagram β) of the experimental arrangement used to measure the thermoelectric emission constant for different single-crystal planes. The single-crystal wire A is mounted in the rotor R so that it can be rotated in front of the slit S cut in the cylindrical anode B. Cup D is the electron collector, and C is a secondary-electron suppressor electrode maintained at a negative voltage with respect to D (SMITH [660])

$$j_0 = AT^2 \exp\left[-(W_a - W_i)/kT\right] = AT^2 \exp(-e\varphi/kT), \quad (2.1\text{–}1)$$

where

$$A = \frac{4\pi m_e k^2 e}{h^3} = 120 \frac{\text{amp sec}}{\text{cm}^2\,\text{degree}}, \quad (2.1\text{–}2)$$

in which m_e is the electron mass, k the BOLTZMANN constant, and e the charge on the electron. Plotting $\ln(j_0/T^2)$ against $1/T$ gives the RICHARDSON line, from whose slope the work function can be determined.

However, many factors which effect the results have not been taken into account in deriving this equation. Thus, as has been shown by HUTSON [334] and by DOBREZOW et al. [151], the observed energy distribution of the emitted electrons deviates from the MAXWELL velocity distribution in that it has fewer electrons in the low-energy region. According to NOTTINGHAM [526], this can be explained by reflection at the metal boundary. The RICHARDSON equation should therefore be corrected by a reflection factor r, which represents the fraction of the electrons that have sufficient energy to escape from the surface but are reflected at the surface potential barrier of the cathode. The reflection coefficient r can amount to several per cent for pure metals. Furthermore, the work function φ should not be assumed to be independent of temperature (see Chap. 1) and the effect of the surface heterogeneity (cf. Chap. 1) must be taken into account. Finally the RICHARDSON equation must also be corrected for the fact that applying an anode potential may raise the field intensity in front of the emitter to such a large value that the potential barrier at the metal surface is distorted and the work function is consequently reduced (SCHOTTKY effect). For experiments of this sort, the reader is referred to the work of SMITH [660] and the summary of DOBREZOW [153].

An extended RICHARDSON equation in which some of the above effects are included is

$$j = (1 - r)\exp\left(-\frac{e a}{k}\right)\left(\frac{4\pi m_e k^2}{h^3}\right) T^2 \exp\left\{\frac{-e[\overline{\varphi}(T_0) - a T_0]}{k T}\right\} \exp\left(\frac{e\sqrt{e\,\mathcal{E}}}{k T}\right). \tag{2.1–3}$$

Here the term $\overline{\varphi}(T) = \overline{\varphi}(T_0) + a(T - T_0)$ takes account of the temperature dependence of the average work function of a heterogeneous surface, and $\mathcal{E}$ is the electric field intensity at the surface. Naturally one must also correct for the errors which occur as a result of the experimental procedure itself, as for example the temperature drop along the heated wire or the SKOUPY effect*.

In the case of a heterogeneous surface, this method gives a minimum value for the average work function $e\overline{\varphi}_{\min}$, since the electron emission from patches with $e\varphi_{\min}$ is enhanced. As can be seen from the experimental work-function data compiled by EBERHAGEN [169] and MICHAEL-

* SKOUPY [655], JOHNSON [357, 358], and SCHMIDT [639] found that when a wolfram wire is heated with direct current, steps develop on the surface. This effect is attributed to migration of wolfram ions on the surface in the external electric field.

SON [473], the method of thermal electron emission has been successfully applied to the measurement of the work function of about 57 clean metals and numerous gas-covered surfaces.

2.2. Thermal Emission of Ions

In addition to the method of thermelectron emission, there is also a possibility of using thermal emission of ions for determining the work function of metal surfaces. If an incandescent metal surface is struck by gas atoms or molecules, some of them may desorb from the surface as neutral particles and some as positive or negative ions. This method has recently been applied very successfully by REYNOLDS [587], who determined the work funktion of the (110) and (111) crystal plane of a tungsten monocrystal filament.

The process of ionization of atoms or molecules impinging on a hot metal surface (surface ionization) can, for example, be described by the degree of ionization α, which gives the ratio of the number of ions N_i which are emitted per unit time from the surface of the metal to the number of atoms N_0 which are emitted in the same time from the surface. The SAHA-LANGMUIR equation gives the degree of ionization α in terms of the work function $e\varphi$ of the metal, the ionization energy eI of the colliding atom, and the surface temperature T. A modified form, which takes into account the ion (atom) reflection and the temperature dependence of φ is

$$\alpha_{\mathcal{E}=0} = \frac{\dot{N}_i}{\dot{N}_0} = \frac{P_i N_i}{P_0 N_0} = \frac{g_+}{g_0}\left[\frac{1-r_i}{1-r_0}\right] \exp\left\{\frac{e[\varphi(T)-I]}{kT}\right\}, \qquad (2.2\text{–}1)$$

where $r_{i,0}$ is the reflection coefficient for the ion or atom, $P_{i,0}$ the desorption probability for the ion or atom, $N_{i,0}$ the number of adsorbed ions or atoms, and $g_{+,0}$ the statistical weights of ionic and atomic states. (For more details see Chap. 8.)

This relation is derived under the simplifying assumption that the atoms and molecules which impinge on the surface come into thermodynamic equilibrium with the surface, where the field $\mathcal{E}$ accelerating the ions is assumed to be zero. For the case in which the ions are accelerated by an externally applied field $\mathcal{E}$, Eq. (2.2–1) must be modified in the following way. The image force with which the adion is attracted to the metal is reduced in an external field by an amount $e\mathcal{E}$ (normal SCHOTTKY effect). The resulting expression for the field dependence of the degree of ionization α is

$$\alpha(\mathcal{E}) = \alpha_{\mathcal{E}=0} \exp\left(e\sqrt{e\mathcal{E}}\,/kT\right). \qquad (2.2\text{–}2)$$

This relation is confirmed by experiment, as is described in detail in Chapter 8. One can show that if $\alpha \ll 1$ and $N = N_i + N_0$ is constant then the surface ionization current $i = eN_i$ may be expressed by a relation analogous to Eq. (2.2–1), namely,

$$i_{\mathcal{E}=0} = A \exp\left\{e[\varphi(T) - I]/kT\right\}. \qquad (2.2\text{–}3)$$

If one plots ln $i_{\mathcal{E}=0}$ against $1/kT$, the difference $e[\varphi(T) - I]$ can be determined from the slope of the line. If one uses atoms or molecules with known ionization energy eI and makes sure that the condition $\alpha \ll 1$ is satisfied (e.g., as for the system Na on wolfram), the work function $e\varphi$ can be determined in this way. The use of relative measurements can considerably extend the domain of applicability of the above relations. One must further note that the SAHA-LANGMUIR equation (2.2–1) is valid only for clean surfaces. As will be further discussed in Chap. 8, gas covering of the surface introduces discrepancies between experimental results and values calculated from Eq. (2.2–1).

Finally, it should be mentioned that measurements of the thermal emission of ions do not yield the same average value of the work function of the polycrystalline surface as would be obtained from measurements of thermoelectron emission. Ion emission is dominated by the patches having $e\varphi_{\max}$ so that $e\overline{\varphi}$ is large; but a low value of $e\overline{\varphi}$ is obtained in electron emission, which occurs mainly from the patches with work function $e\varphi_{\min}$. (See remarks under 2.1.)

2.3. Photoelectric Method

The photoelectric method has also been applied successfully to the determination of the work function of clean and gas-covered metal surfaces. As shown by EINSTEIN, the maximum kinetic energy $(E_{\text{kin}})_{\max}$ which an electron can attain (at $T = 0°$ K) after absorption of a photon and overcoming the potential barrier W_a is

$$(E_{\text{kin}})_{\max} = h\nu - (e\varphi)_{T=0}. \qquad (2.3\text{–}1)$$

Thus no photoemission is possible for $h\nu < e\varphi$, and $\nu_0 = e\varphi/h$ gives the "red limit", the lowest frequency at which the effect occurs. One can thus write Eq. (2.3–1) as

$$(E_{\text{kin}})_{\max} = h(\nu - \nu_0). \qquad (2.3\text{–}2)$$

This relation was used for determining the work function. In particular if the voltage V_s applied between the photocathode and the anode is just sufficient to prevent the emission of electrons from the photocathode, it follows that

$$eV_s = h(\nu - \nu_0) = h\nu - e\varphi. \qquad (2.3\text{–}3)$$

If the frequency ν is known, the work function $e\varphi$ can be determined from this. However, since such measurements are made at $T > 0°$ K, the kinetic energies of the electrons may be greater than $(E_{\text{kin}})_{\text{max}}$, so this method does not lead to precise values for $e\varphi$.

More exact methods involve measurements of the photoelectron current density j_φ as a function of the irradiation frequency ν or the cathode temperature T. The dependence of j_φ on frequency can be represented by two different types of characteristic curves, which are shown in Fig. 2.3.a. Curve α of this figure shows a "normal" characteristic curve in which the ratio i_φ/J of photoelectron current i_φ to flux of radiant energy (ordinate) increases monotonically with the frequency ν (abscissa). This is the case for most clean metal surfaces and is called the "normal" photoeffect. Curve β shows a typical deviation from the normal characteristic; i.e., in certain frequency ranges, maxima can now occur in the photocurrent curve. The existence of maxima in the sensitivity of the photocathode is called "spectral selectivity" so this effect is called the "selective photoeffect". It occurs, for example, in photocathodes in which thin films of alkali metal are put on metallic backings, but no reaction occurs between the backing and the film. Such a reaction frequently can be avoided only by inserting an interstitial isolating layer.

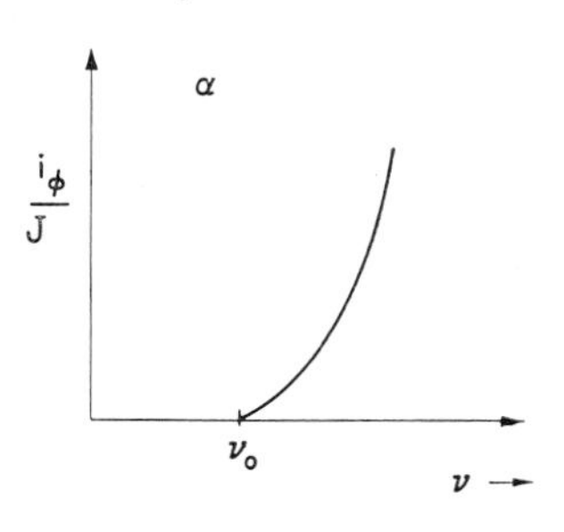

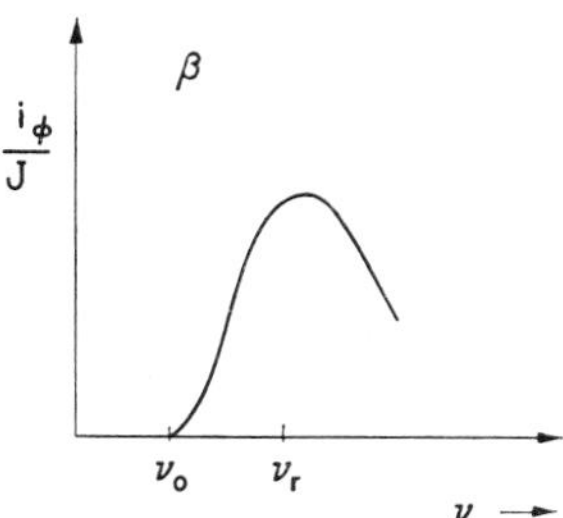

Fig. 2.3.a. The ratio of the photoelectron current i_φ to the flux of radiant energy J as a function of the photon frequency ν. Curve α is characteristic of the "normal" photoeffect, occuring most commonly at clean metal surfaces. Curve β is characteristic of the "selective" photoeffect, showing the existence of a maximum in the electron emissivity for a certain frequency range. This process may occur for example in photocathodes in which films of alkali metal are put on metallic backings

According to FOWLER [221], the photocurrent density j_φ is given by

$$j_\varphi = \alpha e \frac{4\pi m_e k T}{h^3} \int\limits_{W_a - h\nu}^{\infty} \ln\left[1 + \exp\left(\frac{W_i - W_x}{kT}\right)\right] dW_x, \qquad (2.3\text{–}4)$$

in which the expression

$$\frac{4\pi m_e k T}{h^3} \ln\left[1 + \exp\left(\frac{W_i - W_x}{kT}\right)\right] dW_x \qquad (2.3\text{–}5)$$

represents the rate (electrons per sec per cm²) at which the metal boundary is struck by electrons of the SOMMERFELD electron gas with energies in the range between W_x and $W_x + dW_x$. The coefficient α is the factor by which the density of excited electrons at the emitter boundary is lower than the density of the normal electron gas, and is proportional to the number of photons absorbed in the surface layer. If one introduces the new variables

$$t = \frac{W_i - W_x}{kT} \quad \text{and} \quad \delta = \frac{h(\nu - \nu_0)}{kT} = \frac{W_i - (W_a - h\nu)}{kT},$$

one can write Eq. (2.3–4) more simply as

$$j_\varphi = \alpha e \frac{4\pi m_e k^2 T^2}{h^3} \int_{-\infty}^{\delta} \ln(1 + e^t)\, dt\,. \tag{2.3–6}$$

The function

$$f(\delta) = \int_{-\infty}^{\delta} \ln(1 + e^t)\, dt \tag{2.3–7}$$

can be represented by one of two series expansions:

For $\delta \leqslant 0,\ h\nu \leqslant e\varphi, \quad f(\delta) = e^\delta - \frac{e^{2\delta}}{2^2} + \frac{e^{3\delta}}{3^2} \mp \cdots.$

For $\delta \geqslant 0,\ h\nu \geqslant e\varphi, \quad f(\delta) = \frac{\pi^2}{6} + \frac{\delta^2}{2} - \left[e^{-\delta} - \frac{e^{-2\delta}}{2^2} + \frac{e^{-3\delta}}{3^2} \mp \cdots\right].$

Equation (2.3–6) can also be written as

$$j_\varphi = \alpha A T^2 f(\delta) = \alpha A T^2 f\left[\frac{h(\nu - \nu_0)}{kT}\right] \tag{2.3–8}$$

or

$$\ln\left(\frac{j_\varphi}{T^2}\right) = B + \Phi\left[\frac{h(\nu - \nu_0)}{kT}\right], \tag{2.3–9}$$

where $B = \ln \alpha A$ and $\Phi(\delta) = \ln f(\delta)$ is the Fowler function.

To evaluate the results of the measurements, one can use the Fowler isotherm method. In this method the photocurrents i_φ are measured for various frequencies ν at a given temperature T and referred to the same number of photons of absorbed light $J/h\nu$; i.e., after

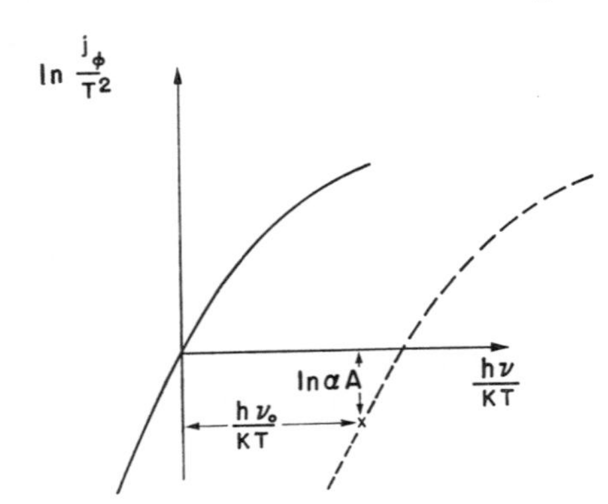

Fig. 2.3.b. Illustration of Fowler's [221] isotherm method of determining the work function from the spectral distribution curve of photoelectron emission from a metal surface at a given temperature T. The photocurrents i_φ are measured for various frequencies ν at a given temperature T and referred to the same number of photons of adsorbed light $(h\nu/J)$; i.e., after computing $j_\varphi = (i_\varphi/J)h\nu$ near the threshold the quantity $\ln(j_\varphi/T^2)$ is plotted vs. $h\nu/kT$ (solid line in the figure). Using the same graphical units the function $\Phi(\delta)$ (Eq. 2.3–9) is plotted vs. δ (dotted line). Superimpose the two curves by a simple translation of axes. The shift along the $\log(j_\varphi/T^2)$ axis represents the constant $B = \ln \alpha A$. The shift along the $h\nu/kT$ axis gives $\nu_0/kT = e\varphi/kT$

computing $j_\varphi = (i_\varphi/J)h\nu$, one plots $\ln(j_\varphi/T^2)$ as a function of $h\nu/kT$ (Fig. 2.3.b). By measuring the intercept of the FOWLER curve on the abscissa, one determines $h\nu_0/kT$, from which the work function of the metal can be determined. The intercept on the ordinate gives the quantity $\ln \alpha A$.

DU BRIDGE [*164*] gave another method which uses the curves for constant ν. He measures the photocurrent density at constant frequency but with a variable photocathode temperature T. If the variable δ in Eq. (2.3–6) is replaced by the new variable

$$\varepsilon = \ln\left[\frac{h(\nu - \nu_0)}{k}\right] - \ln T,$$

the result is

$$j_\varphi = \alpha A T^2 F(\varepsilon) \qquad (2.3\text{–}10)$$

so that

$$\ln\left(\frac{j_\varphi}{T^2}\right) = \ln \alpha A + \ln F(\varepsilon) = B + \chi(\varepsilon). \qquad (2.3\text{–}11)$$

If the experimental values of $\ln(j_\varphi/T^2)$ are plotted against $\ln T$, the horizontal intercept $\ln[h(\nu - \nu_0)/k]$ of the curve at a given frequency ν

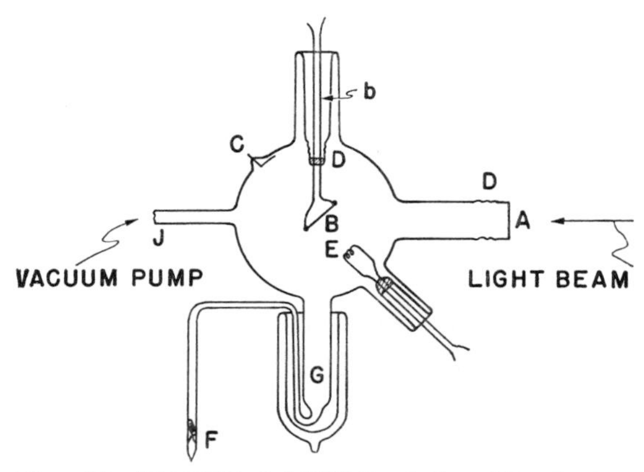

Fig. 2.3.c. Photocell used by SUHRMANN et al. [*693*] for studies of the chemisorption of gases on metal surfaces and corresponding work function changes

can be used to determine the critical frequency ν_0, from which the work function of the metal can be found.

Both methods allow a rather exact determination of φ. Still, in the photoelectric method the measured work function of a polycrystalline material is always less than the mean work function $\overline{\varphi}$, since patches of high work function tend to be excluded from the emission process. Fig. 2.3.c shows a photocell which was recently used by SUHRMANN and SACHTLER [*693*] for measurements of the work function. The metal foil B serves as the photocathode and the anode is a metallic coating C on the inner surface of the photocell. Monochromatic light passes through the quartz window A to the photocathode B. A hot filament E permits electron bombardment of cathode B (to clean the surface as well as to evaporate the metal under investigation).

The photoelectric method for determining work functions has been applied in many modern experiments, for instance, by SUHRMANN [*697*], TOYE [*726*], SACHTLER et al. [*623*] and PERRY et al. [*556*].

2.4. Method Based in the Field Emission of Electrons

Applying an intense electric field ($\approx 10^7$ V/cm) at a metal surface will distort the potential barrier at the surface so that when electrons from the FERMI sea strike this barrier they will have a finite probability of tunneling through and being emitted. According to FOWLER and NORDHEIM [*220*], the electron current density j is given by

$$j = \frac{4\sqrt{W_i/\varphi}}{W_i + \varphi}\,\frac{e^3\,\varepsilon^2}{8\pi h\varphi}\,\exp\left(\frac{-8\pi\sqrt{2m}\,\varphi^{3/2}}{3he\,\varepsilon}\right), \tag{2.4–1}$$

where W_i is the FERMI edge, $e\varphi$ the thermal work function, and ε the field intensity at the emitter. NORDHEIM [*524*] has corrected this formula so that it takes account of the SCHOTTKY effect (see Section 2.1) which somewhat reduces the field intensity for a given current density. He found that if $e\varphi$ is expressed in eV and ε in V/cm, then the current density in amp/cm² is given by

$$j = \frac{1.57\times 10^{-6}\,\varepsilon^2}{\varphi\, t^2(x)}\,\exp\left[\frac{-6.83\times 10^7\,\varphi^{3/2} f(x)}{\varepsilon}\right]. \tag{2.4–2}$$

The functions $t(x)$ and $f(x)$ with the argument $x = 3.79\times 10^{-4}\sqrt{\varepsilon}/\varphi$ can be obtained from the tabulated values of GOOD and MÜLLER [*250*].

The work function $e\varphi$ is the only constant of the material which appears in the above Eqs. (2.4–1) and (2.4–2). If one makes a plot of $\ln(j/\varepsilon^2)$ against $1/\varepsilon$, the slope is determined principally by the work function. The slope is given by

$$\frac{d\log_{10}(j/\varepsilon^2)}{d(1/\varepsilon)} = -\,2.96\times 10^7\,\varphi^{3/2}\,S(x)\,. \tag{2.4–3}$$

The function $S(x)$ with the argument $x = 3.79\times 10^{-4}\sqrt{\varepsilon}/\varphi$ has also been tabulated by GOOD and MÜLLER [*250*].

Fig. 2.4.a shows a schematic view of a field electron microscope (MÜLLER [*508, 510*]) which consists of a cathode point (radius of curvature $\approx 500-100$ Å), an anode ring, an evaporating coil, and a luminescent screen. When a voltage of about 10 kV is applied between anode and cathode, the electrons are emitted from the tip and fly onto the luminescent screen.

Müller and Drechsler [*160*] and Smirnow and Schuppe [*659*] have used this method to determine the work functions of individual

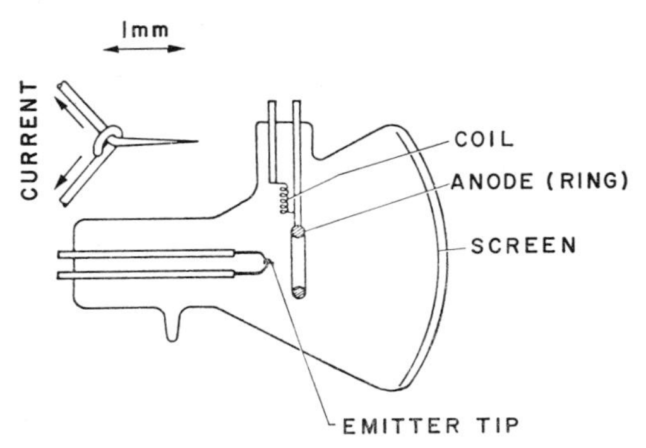

Fig. 2.4.a. Schematic diagram of Müller's [*510*] field-emission microscope

crystal planes by measuring the emission currents. With this same method, the change in work function as a result of covering the surface with gas has been investigated by numerous authors (Ehrlich et al. [*172*], Gomer et al. [*244*], Hayward et al. [*299*], Klein [*386*], and Lewis [*438*]).

2.5. Measurements of Contact Potential

As one can see from Fig. 2.5.a, the contact potential $V_{1,2}$ of two metals which are in electrical contact and have work functions $e\varphi_1$ and $e\varphi_2$ is given by the difference

$$V_{1,2} = -(\varphi_1 - \varphi_2). \qquad (2.5\text{–}1)$$

If the state of surface 2 is kept unchanged, one can determine the change in the contact potential as a result of gas absorption on surface 1, and thus find the corresponding change in work function $\Delta\varphi$ since

$$\Delta V_{1,2} = V_{1,2}{}^* - V_{1,2} = -(\varphi_1{}^* - \varphi_1) \text{ if } \varphi_2 = \text{const}. \qquad (2.5\text{–}2)$$

A negative value of $\Delta V_{1,2}$ indicates that the electrons of the metal are displaced toward the adatom; and conversely, positive values indicate an electron displacement toward the metal. There are many techniques for determining the contact potential. One is based on the fact that the resulting contact potential difference is compensated by an opposing voltage. This procedure has been used in several different ways by Zismann [*806*], Potter [*576*], Mignolet [*478*], Kolm [*403*], Schaafs [*628*], and MacFadyen and Holbeche [*449*].

Another measuring technique is the so-called magnetron method, which was first applied by OATLEY [528] and then by WEISSLER and WILSON [776] to determine how gas absorption changes the contact potential. Here an axial magnetic field causes the electrons emitted by a hot filament to travel in spiral paths to an anode which is coaxial with

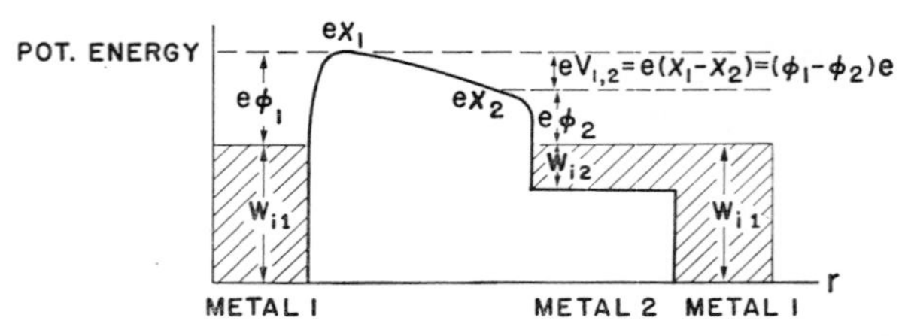

Fig. 2.5.a. Potential diagram at the junction between two metal plates with work functions $e\varphi_1$ and $e\varphi_2$, respectively. Electrons flow from metal 2 (which originally had the higher FERMI level) to metal 1 until the two FERMI levels reach the same energy. The added negative charge on metal 1 raises the external electrostatic energy to $e\chi_1$ while the loss of electrons by metal 2 lowers its external electrostatic energy to $e\chi_2$. The contact potential $V_{1,2}$ is given by $V_{1,2} = \chi_1 - \chi_2 = \varphi_1 - \varphi_2$ (EBERHAGEN [169])

the filament. It has been shown that if $\mathfrak{H}_0$ is the field strength necessary to reduce the anode current to half its original value, the total potential difference between anode and cathode is

$$V = V_a + V_{A,K} + V_{\text{kin}} = 8\,\mathfrak{H}_0^2 r^2 (e/m_e) = K I_0^2, \qquad (2.5\text{–}3)$$

where V_a is the applied anode voltage, $V_{A,K}$ is the contact potential difference between the anode and cathode, V_{kin} is a temperature-dependent term which depends on the kinetic energy of the emitted electrons, r is the anode radius, e/m_e is the charge-to-mass ratio of the electrons, I_0 is the anode current, and K is a constant of proportionality. If I_0 is measured for different values of V_a then the quantity $(V_{A,K} + V_{\text{kin}})$ can be determined from the graph of I_0^2 against V_a. Then since V_{kin} is known with sufficient accuracy, $V_{A,K}$ is also determined.

A disadvantage of the above methods of measuring contact potential difference is that the work function cannot be determined directly, but only through differences.

2.6. Electron Reflection Method

Another method which can be used for determining the work function of metal surfaces, is the use of an "electron mirror" microscope (see for example HENNEBERG and RECKNAGEL [305], HOTTENROTH [324], WISKOTT [783, 784], and WEISSENBERG et al. [51]).

As one can see from the schematic Fig. 2.6.a taken from WEISSENBERG, the beam of electrons after being bunched by magnetic condensing lenses strike the "electron mirror" (objective). The reflected electrons are then separated from the incident ones by traversing a magnetic deflecting field, after which they come to a projection lens which throws

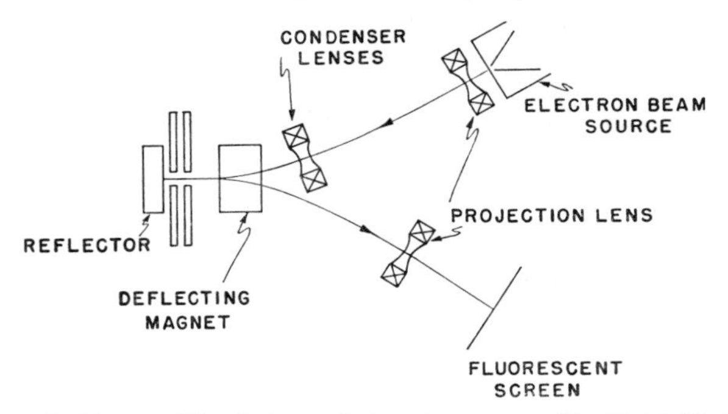

Fig. 2.6.a. Schematic diagram of the electron reflector microscope used by BARTZ, WEISSENBERG and WISKOTT [51]

an enlarged image on the fluorescent screen. If the surface of the electron mirror consists, for example, of two different metals with work function $e\varphi_1$ and $e\varphi_2$, the equipotential surfaces in front of the metal surface will bulge toward the metal with smaller work function.

The electrons in the parallel beam which impinge on the negatively charged "mirror" should have sufficiently uniform kinetic energy that

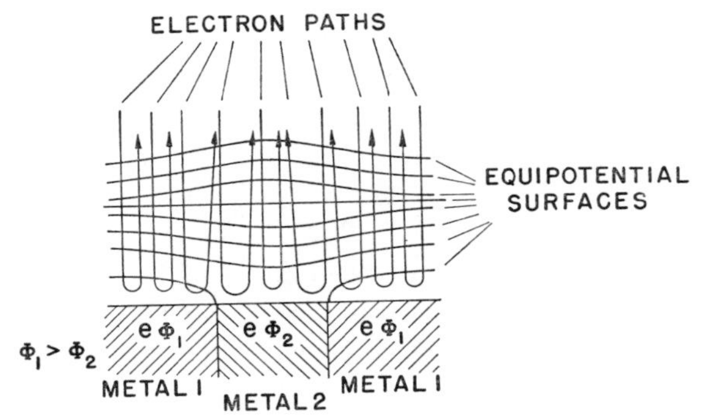

Fig. 2.6.b. Schematic diagram of the electron trajectories near a metal reflector surface which consists of regions with different work functions ($e\varphi_1$ and $e\varphi_2$). (After EBERHAGEN [169])

they come into the immediate neighborhood of the surface without touching the mirror surface. As a result of the distortion of the equipotential surfaces immediately in front of the mirror (see Fig. 2.6.b), the reflected electrons are deflected toward the surface region (metal 2) with the lower work function $e\varphi_2$. If these reflected electron beams are made visible on a fluorescent screen, the brighter regions in the image correspond to zones of lower work function. One thus obtains a faithful and enlarged reproduction of the contact potential differences as they occur on the surface. According to WISKOTT [*783, 784*], the resolution of such an arrangement is already 80 Å at a field intensity ε of about 10^5 V/cm in front of the mirror.

A disadvantage of this method for determining contact potential differences is that surface heterogeneities produce effects similar to those of contact potential difference.

3. Preparation of Metal Surfaces

3.1. Thermal Desorption; the Flash-Filament Method

The importance of a clean surface for investigations of the interaction of atom, molecule, and ion beams with surfaces is evident from the foregoing. The methods for preparation of surfaces with a minimum degree of physical, chemical, and induced heterogeneity (see Sections 1.1.1, 1.1.2, and 1.1.3) can be divided into those in which surface layers are removed (e.g., through thermal desorption, electrolysis, or ion bombardment) and those in which metal films are deposited (as, for example, by an evaporation process or electroplating*).

A frequently used method is that of thermal desorption, in which the metal surface is heated under vacuum to such a high temperature that physically and chemically adsorbed molecules are desorbed. Unfortunately this method requires the use of high temperatures and is therefore not generally applicable. According to WHEELER [*780*], the peak temperature T_h required to guarantee an adequate cleaning of the surface can be estimated roughly as

$$T_h \approx 20\,\Delta H\,,$$

where ΔH is the binding energy (kcal/mole) of the layer of impurity atoms on the metal surface. Such values of ΔH can be obtained from Tables 29 and 64 of TRAPNELL's monograph [*728*]. Table 3.1.A is a compilation of such values of T_h together with the melting points T_s of the metals for various metal-adatom systems. One can see from this table that the desorption method fails for some systems, since the material evaporates rapidly or even melts at the required temperature T_h.

An important application of the desorption procedure is the flash-filament method. APKER [*19*] used the following method in his first investigations. He introduced an additional hot filament (for example W)

* For modern summaries of such cleaning processes, the reader is referred to the discussion by MOESTA [*487*], EBERHAGEN [*169*], CULVER and TOMPKINS [*131*], H. MAYER [*464*], ALPERT [*11*] and BASSETT, MENTER and PASHLEY [*53*].

into an ion gauge and heated it in vacuum to so high a temperature that it was freed of adsorbed gas. The filaments which had been baked out

Table 3.1.A. *List of temperatures T_h necessary to outgas metal surfaces covered with different gases. The temperature T_s is the melting point of the metal*

System	H (kcal/mole)	T_h (°K)	T_s (°K)	System	H (kcal/mole)	T_h (°K)	T_s (°K)
O_2 on W	155	3100	3653	H_2 on Fe	32	640	1808
N_2 on W	95	1900	3653	C_2H_4 on Fe	68	1360	1808
H_2 on W	45	900	3653	NH_3 on Fe	45	900	1808
C_2H_4 on W	102	2040	3653	CO on Fe	32	640	1808
NH_3 on W	66	1320	3653	O_2 on Ni	130	2600	1726
N_2 on Ta	140	2800	3303	CO on Ni	35	700	1726
H_2 on Ta	45	900	3303	H_2 on Ni	31	620	1726
C_2H_4 on Ta	138	2760	3303	C_2H_4 on Ni	58	1160	1726
O_2 on Fe	75	1500	1808	NH_3 on Ni	37	740	1726
N_2 on Fe	40	800	1808				

in this fashion were then cooled for a time t_c and then reheated. Thereupon a pressure pulse, detectable by the ionization gauge, developed as the gas particles, adsorbed during the cooling time t_c, were again desorbed upon heating.

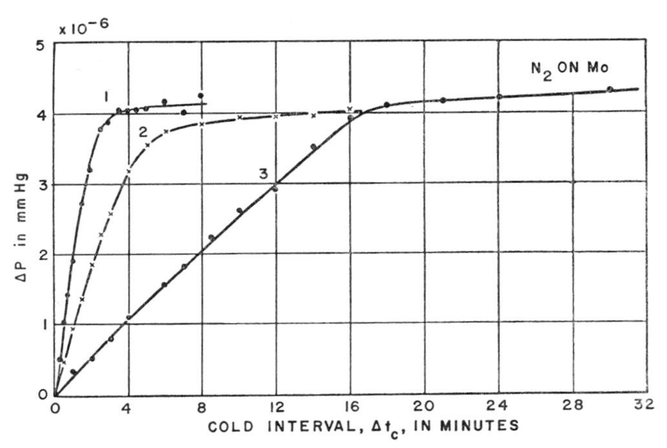

Fig. 3.1.a. The maxima Δp of pressure pulses obtained by flashing a molybdenum filament, plotted as a function of the cold time interval Δt_c between such flashes. The total pressure p of the vacuum system is the parameter. Curve *1*: $p = 1.5\times10^{-8}$ mm Hg; curve *2*: $p = 5\times10^{-9}$ mm Hg; curve *3*: $p = 1.8\times10^{-9}$ mm Hg (HAGSTRUM [*279, 281*])

If the maxima of such pressure pulses are studied as a function of the cooling time t_c and the results are plotted with the pressure of the vacuum system as a parameter, the pressure curves obtained are found to read saturation for cooling times beyond a certain value t_{cm}, as can be seen in Fig. 3.1.a taken from HAGSTRUM [279]. The critical cooling time t_{cm} may be thought of as the time necessary for building up a monatomic layer of foreign atoms on the surface. These times are in order-of-magnitude agreement with our estimates in Section 1.1.3 for a vacuum of about 10^{-9} mm Hg, especially when one substitutes the correct accommodation coefficients for the various systems.

Therefore, at least for wolfram and molybdenum surfaces studied in the pressure range of 10^{-9} mm Hg, one may regard the flash-filament method as a suitable procedure for eliminating most of the surface impurity layers. A disadvantage of the heating process is that any foreign atoms that may be imbedded in the lattice are enabled to diffuse more easily to the surface. FARNSWORTH et al. [205] have observed a contamination of nickel surfaces which were being degassed by such a heat treatment.

BECKER and HARTMAN [60] modified the flash-filament method somewhat in order to investigate adsorption processes on surfaces (nitrogen on wolfram).

3.2. Ion Bombardment of the Surface

Another method of cleaning is to bombard the metal surface with ions having energies of a few hundred to several thousand electron volts in an ultrahigh vacuum. The familiar process of cathode sputtering (discussed in more detail in Chap. 10) can be employed to remove the layers of foreign atoms from the surface as well as layers of cathode material. FARNSWORTH et al., for example, have been able to prepare clean nickel surfaces by bombardment with argon ions and subsequent thermal desorption. EGGLETON and TOMPKINS [170] have employed a similar combination of thermal desorption and ion bombardment for cleaning oxidized iron wires. After outgassing and then bombarding the iron wire with oxygen (in order to stop the diffusion of hydrogen to the metal surface), they were able to obtain oxide-free surfaces by subsequent bombardment by nitrogen and neon ions.

Unfortunately the ion bombardment produces some undesirable effects on the surface. For example, the bombarding ions may be captured for a certain length of time in the target lattice (ALMEN et al. [9, 10]). Then they can later arrive at the surface via diffusion or "burst" processes and there once again produce impurity (contamination) layers and either desorb or result in chemical heterogeneities on the surface.

Thus BROWN and LECK [97], VARNERIN and CARMICHAEL [734], and CARMICHAEL and TRENDELENBURG [108] have recently made a detailed investigation of the re-emission of such ions shot into a target. CARMICHAEL and TRENDELENBURG were able to observe the re-emission of noble gas atoms which were originally shot into a nickel target as ions and then were driven out of the lattice as a result of bombardment with another type of noble gas ion. Their results lead to the conclusion that, at the ion energy employed (about 100 eV), a nickel target can absorb up to the equivalent of 15 atomic layers of helium before saturation begins, whereas the absorption of other noble gases is equivalent to only about one layer of atoms. Their explanation is that the helium ions are more easily incorporated in the interstitial spaces than are the other noble gas atoms.

Another effect that can occur during the ion bombardment of the target is the development of physical heterogeneities on the surface. Thus, for example, TRILLAT [*729*] showed that bombarding a gold single-crystal layer with 12-keV ions converted it into a polycrystalline layer. OGILVIE [*532*] found a similar disturbing effect in the bombardment of silver single-crystal layers with argon ions in the energy range from 12 to 4000 eV. The original single-crystal layer is split up after the ion bombardment into a number of small crystallites (around 100 Å in length), the crystallites being oriented with respect to the original single-crystal layer [surfaces parallel to the (111), (110), (100) planes] as if a rotation had occurred around the [112] axis.

FARNSWORTH et al. [*202*] found that bombarding titanium, nickel, and platinum surfaces with argon ions likewise produced disturbances of the crystal lattice at the surface. GÜNTHERSCHULZE and TOLLMIEN [*265*] observed that bombardment of metals such as Pt, Al, Bi, and Mg with ions induced the development of submicroscopic cones on the surface (but Au, Pt, Fe, and Ni did not show the effect). The development of physical heterogeneities as a result of ion bombardment of surfaces can thus be regarded as well established for many systems. Finally we may mention the investigations of WEISSLER and WILSON [*776*] who observed changes in the work function of evaporated layers of wolfram and silver after they were bombarded with hydrogen, oxygen, freon 12, helium, and argon ions.

In conclusion, it seems clear that ion bombardment alone does not adequately clean the surface in general, but that a combination with other methods (as for example thermal desorption, through which the chemical and physical heterogeneities are at least partially annealed) can be successful.

3.3. Evaporation of Thin Layers

Among the methods for obtaining clean metal surfaces, the evaporation technique has been of outstanding importance because it enables one to deposit layers of various thicknesses under controlled conditions. Since the basis of this technique is very well known because of its wide industrial application (see also MAYER [*464*], POWELL,

CAMPBELL, and GONSER [*578*], HOLLAND [*318*] and WHEELER [*780*]), a very brief description will suffice here.

Whenever an evaporated layer of definite thickness is desired, the evaporation can be done with a Knudsen cell like those used for producing molecular beams. (See, for example, J. A. ALLEN [*7*].) The number of atoms or molecules emerging per unit time then follows the laws of molecular flow (cf. KNUDSEN [*391*]). If a welldefined layer thickness is not needed, then the material is most frequently evaporated by placing it on a hot ribbon or filament. Before starting the evaporation, these filaments or ribbons are outgassed under high vacuum by heating them almost to the boiling point of the material to be deposited. The evaporation process is then carried out under the best vacuum obtainable in order to make sure that the number of molecules of the residual gas which strike the target surface is small compared with the number of molecules to be evaporated. Recently BROWN [*98*] and FAUST [*208*] have fully described the application of such outgassing methods in their studies of the evaporation of beryllium, aluminum, wolfram, and platinum surfaces.

Furthermore, one must be careful that the vapor pressure of the ribbon used to heat the substance is negligible in comparison with that of the substance to be evaporated. Even when this requirement seems well satisfied, as in the case of evaporating germanium from a wolfram ribbon, HEAVENS [*303*] was able to show by a radioactive tracer method that traces of the ribbon material could be detected in the evaporated layer.

Finally the state and temperature of the target surface requires special attention. If the beam particles are to deposit on the surface to be plated, the temperature of the surface must be below the condensation point of the plating material. Numerous authors (COCKCROFT [*119*], ESTERMANN et al. [*193*], CRAMER [*130*], etc.) found that for many systems the target temperature must actually be far below the condensation temperature of the plating material if condensation is to occur. According to VOLLMER [*745*], condensation is possible only when nucleation occurs on the surface; and for this at least two molecules must collide during their residence time on the surface. The occurence of such processes depends on the matching of the beam density on the one hand and the mean adsorption lifetime and surface mobility of the adatom on the other. The presence of impurity layers on the underlying surface can influence this process decisively, so that a very careful cleaning of the target surface is essential. (An excellent discussion of the subject has been given by FRAUENFELDER [*225a*].) Similarly, the effects of physical heterogeneities (such as roughness or porosity of the surface) must be taken into account.

It should be mentioned in this connection that several authors (e.g., de Boer [*81*] and Pashley [*545*]) found that certain crystal planes were preferred in the film deposits. Thus Pashley [*545*] showed that the (111) planes predominated on the surfaces of deposited silver films about 1500—2000 Å thick, produced at about 270° C in a glow discharge. In evaporated layers of silver on a rock salt cleavage surface as substrate, the (100) planes predominated on the surface. Pashley et al. [*53*] observed a similar preference for certain crystal planes on the surfaces of evaporated films of gold on copper.

3.4. Galvanotechnic Procedures

The widely used engineering procedure of electrolysis, whose basis is well known from elementary physics, has as yet had little application in the preparation of thin clean surface films. The difficulties in applying this method for obtaining such layers have been described in more detail by Statescu [*681*] and Machu [*450*]. However, recently Brame and Evans [*92*] have reported success in producing pure gold films on an electrolytically polished silver surface by means of the galvanotechnic method. Erbacher [*183, 185*], also mentions success in electrochemical deposition of noble metal atoms on the surfaces of base metal electrodes, where he was actually able to prepare coatings down to a completely flawless monatomic layer.

Despite these individual successes, the galvanotechnic method for producing clean smooth surfaces appears not to be generally applicable.

4. Binding Forces Effective in the Collision of Atoms and Molecules with Metal Surfaces

This chapter, which surveys the possible binding forces involved when an atom or molecule collides with a metal surface, lays the foundation for the discussion of the energetics of the adsorption process (Chapter 5). As has been shown schematically in Fig. 2.1.c, three limiting types of binding forces must be distinguished: (a) van der Waals forces, (b) exchange (homopolar) forces, and (c) heteropolar (Coulomb) forces. These are discussed more specifically in the following sections.

4.1. van der Waals Forces—Physical Adsorption

4.1.1. Homogeneous Surfaces

A distinction usually is made between three types of van der Waals forces namely, those that depend on orienting electric dipoles (the directional effect), on electrostatic induction, and on interaction between transient dipoles (the dispersion effect). These various forces are given by expressions (4.1–1), (4.1–2), and (4.1–3), respectively.

The forces that depend on the directional effect arise from the fact that molecules with permanent electric dipole moments μ exert directional forces on one another as soon as their distance of separation r is sufficiently small. According to DEBYE [*141*] and KEESOM [*368, 369, 370*], the exchange energy is given by

$$E_W = -\frac{2}{3}\frac{\mu_1^2\mu_2^2}{r^6}\frac{1}{kT}. \tag{4.1–1}$$

If a molecule that has a permanent electric moment μ is approached by one that has none, an electric moment proportional to the polarizability α of the latter is induced. Then at a distance of separation r the energy of interaction between these dipoles is

$$E_I = \frac{2\,\alpha\mu^2}{r^6}. \tag{4.1–2}$$

The dispersion effect, which was first introduced by LONDON [*441, 442*], takes account of the fact that two atoms can exert dipole forces on one another by virtue of their transient dipole moments. Such a moment develops between the positive nuclear charge of the atom and the center of gravity of the electronic charge, so its magnitude and direction will vary periodically. However, since all directions are equally probable, the time average of the dipole moment will be equal to zero. If now such an atom with an instantaneous dipole enters the field of another atom with an instantaneous dipole moment, there results an additional induced dipole whose periodic variation follows the variations of the inducing dipole. These short-term perturbations of the electron vibrations may be treated in analogy to a dispersion effect. According to LONDON, the interaction energy which results from these instantaneous dispersion forces is inversely proportional to the sixth power of the distance of separation r of the atoms, i.e., $E_i = C/r^6$. LONDON finds

$$E_i = \frac{C}{r^6} \quad \text{with} \quad C = -\frac{3}{2}\alpha_1\alpha_2\frac{I_1 I_2}{I_1 + I_2}, \tag{4.1–3}$$

where $I_{1,2}$ are the ionization energies for atoms 1 and 2, and $\alpha_{1,2}$ are their polarizabilities. SLATER and KIRKWOOD [*656*], on the basis of a different assumption, find

$$C = -\frac{3}{4\pi}\frac{eh}{\sqrt{m_e}}\frac{\alpha_1\alpha_2}{(\alpha_1/n_1)^{1/2} + (\alpha_2/n_2)^{1/2}}, \tag{4.1–4}$$

where n_1 and n_2 are the numbers of electrons in the outermost shells of atoms 1 and 2, and m_e and e are the mass and charge of an electron. In treating the VAN DER WAALS forces operative in the interactions of atoms with metal surfaces, it is not clear how the incident atom (or molecule) interacts with a metal atom. The difficulty arises from the fact that the metal atom is not actually "free", but rather is coupled with its neighbors in the atomic lattice.

Lennard-Jones [*431*] used a highly simplified model. He regarded the metal as a completely polarizable body and treated the interaction between the metal surface and the fluctuating dipole of the impinging molecule by means of the classical method of electrical image forces. The molecule itself was treated as a quantum mechanical system on which the image potential acts as a perturbation. For the interaction energy he found

$$E_I = -\frac{m_e c^2 \chi}{2 N_A r^3}, \tag{4.1–5}$$

where c is the speed of light, χ is the magnetic susceptibility of the gas, N_A is Avogadro's number, and m_e is the mass of the electron.

Bardeen [*48*], who treated the adatom-metal system completely quantum mechanically, obtained

$$E_I = \frac{-m_e c^2 \chi}{2 N_A r^3} \; \frac{c_B e^2/(2 r_s h \nu_0)}{1 + [c_B e^2/(2 r_s h \nu_0)]}, \tag{4.1–6}$$

where c_B is a numerical constant, r_s is the radius of a sphere which on the average contains one electron, and $h\nu_0$ is a characteristic energy approximately equal to the ionization energy of the gas.

If the values of the individual quantities are substituted, the term $c_B e^2/2 r_s h \nu_0$ is found to be approximately equal to one. This means that values given by Eq. (4.1–6) are only about half as large as those given by Eq. (4.1–5).

Margenau and Pollard [*457*] used another model for the effective interaction involved. They divided the metal into small elements and considered the interaction between the adatom and the instantaneous dipole fields in each such element. Summation over the contributions of the individual elements then gives

$$E_I = E_1 + E_2 = \frac{e^2 \alpha}{16 r^3} \left(\frac{h n}{\pi m_e \nu_0} - \frac{c_B}{r_s} \right), \tag{4.1–7}$$

where α is the polarizability of the adatom, r is the shortest distance of the adatom from the metal, and n is the number of conduction electrons per cubic centimeter in the metal. The symbols m_e, ν_0, r_s, and c_B have the same meanings as in Eqs. (4.1–5) and (4.1–6).

The first term E_1 comes from the polarization of the metal by the continuously varying (fluctuating) dipole of the adatom; since it is positive, it represents a repulsive force. The second term E_2 comes from the polarization of the adatom by the motion of the electrons in the metal.

PROSEN, SACHS and TELLER [*579*] point out that the three equations (4.1–5), (4.1–6), and (4.1–7) are valid only for large values of r. For small values of r, the interaction energy computed on the basis of the SOMMERFELD theory of free electrons becomes proportional to $1/r$. For the interaction energy at small values of r they derived the relation

$$E_I = \frac{\pi \alpha^2 e^2 n}{2r}, \tag{4.1–8}$$

where the symbols have the same meanings as in Eq. (4.1–7). In Chap. 5 these various expressions for the interaction energy will be compared with the experimental data.

All of the proceeding derivations were carried out on the assumption that the surfaces were homogeneous. Heterogeneities in the surface, however, would affect the VAN DER WAALS binding energies, as discussed qualitatively in Section 4.1.2.

4.1.2. Heterogeneous Surfaces

As POLANYI [*575*], LENNARD-JONES [*431*], and DE BOER and CUSTERS [*79*] have shown, the VAN DER WAALS adsorption depends very strongly on the physical heterogeneity of the surface. They showed, for example, that the binding forces for adatoms adsorbed in small holes in the surface (schematic diagram α in Fig. 4.1.2.a) are much greater than those for adatoms on a flat surface. For adatoms in cracks (diagram β) the binding forces are still greater than those for atoms adsorbed in holes. On the other hand, the VAN DER WAALS forces are relatively very weak at protruding corners, edges, and peaks on the surface. The maximum binding forces occur, in agreement with the model of MARGENAU and POLLARD [*457*], in positions where the adatoms are surrounded by the maximum number of lattice atoms. A beautiful corroboration of this result is given by the recent investigation of EHRLICH and HUDDA [*172*], whose field-emission photographs reveal, for example, a preferential adsorption of xenon on the (310) and (611) planes of a wolfram surface (cf. Figs. 1.1.1.2.a and 1.1.1.2.b). These are the planes in which the atoms are most densely packed.

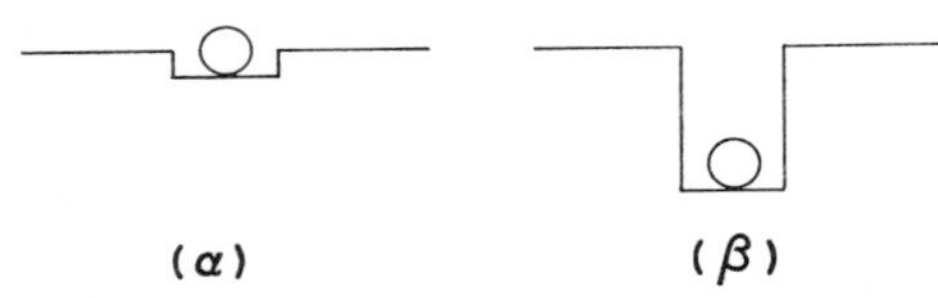

Fig. 4.1.2.a. Schematic diagram of the adsorption of adatoms in small holes in the surface (diagram α) and in deeper cracks (diagram β)

4.2. Exchange Forces—Weak Chemisorption on Homogeneous and Heterogeneous Surfaces

The potential energy diagram, shown in Fig. 4.2.a for the interaction of an H_2 molecule with a metal surface M (e.g., wolfram), elucidates the transition from the action of VAN DER WAALS binding forces to that of exchange forces. As an H_2 molecule approaches a metal surface, it can be "physically" adsorbed so that the potential energy curve ($H_2 + M$) is realized. If however, the molecule receives a certain activation energy* ΔH_{Act}, it can transfer to the potential curve ($M + 2H$) on which (after dissociation into H atoms) its atomic constituents become weakly chemisorbed. This weak chemisorption is characterized by the fact that, in contrast to the previously discussed VAN DER WAALS binding forces, the electron shell of the adatom may penetrate that of the metal. Both the metal and the adatom give up an electron with unpaired spin to a "binding orbital". The metal electron usually comes from the d band (the 3d, 4d, 5d band, depending on the kind of metal).

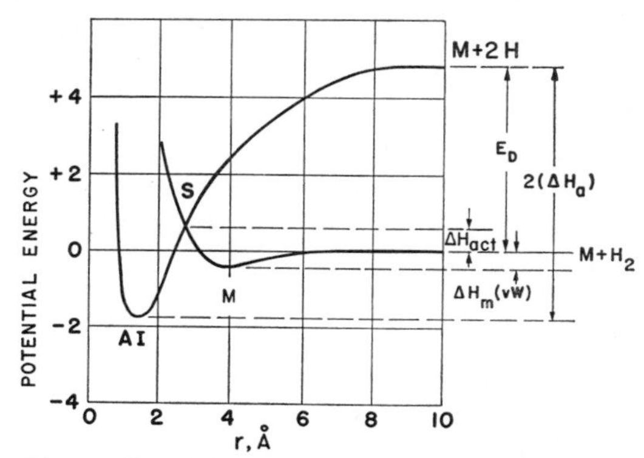

Fig. 4.2.a. Potential energy diagram for the adsorption of H_2 on a metal. In curve $M \rightarrow (M + H_2)$ the H_2 molecule is physically adsorbed with ΔH_m(vW) as the heat of adsorption (VAN DER WAALS forces acting). In curve $AI \rightarrow (M + 2H)$ the molecule, after acquiring an activation energy ΔH_{act}, can dissociate into two H atoms prior to chemisorption. The difference ΔH_a is the heat of chemiisorption for the hydrogen atom

This type of force, which is referred to as an exchange force, homopolar force, or covalent force, was described quantum mechanically by HEITLER and LONDON in the familiar example of the H_2 molecule. The semi-empirical relations for computing homopolar binding energies for gas-metal systems will be discussed in Chap. 5. These relations were derived by ELEY [*181*] and STEVENSON [*684*], who based their calculations on the PAULING method of linear combination of the eigenfunctions of the possible valence structures which contribute to the binding.

According to DOWDEN [*158*], covalent adsorption forces act preferentially when certain conditions are satisfied. First of all, the work function of the metal surface must be large. Moreover, the density of electrons in the *d* level must be large compared to that in the *s* and *p* levels inside the FERMI sea**. Finally the presence of unoccupied "atomic orbitals" (here in the sense of PAULING's [*546, 548*] definition) on the surface favors the occurrence of covalent binding.

* There are also systems in which the activation energy needed can be very small or even zero. Such systems have been pointed out by DOWDEN [*158*], COUPER and ELEY [*128*], and KEMBALL [*371*].

** The density of the *d* level is a maximum in pure nickel and in the corresponding metals of transition series.

The question whether or not the forces occurring in adsorption are covalent in nature can be answered on the basis of a combination of experimental methods, e.g., through measurements of surface potential, measurements of electrical resistance, and magnetic investigations of the adatom-metal system. In their studies of the adsorption of H_2 on nickel surfaces, BROEDER et al. [*96*] found that the saturation magnetization of ferromagnetic nickel is reduced by the adsorption process. This is explained by the fact that the metal electrons which participate in the covalent binding come from the 3d shell with initially unpaired spins. Additional information concerning the strength and direction of the charge displacement which occurs in covalent binding between the "shared electrons" and the positive charge centers of the adatom and surface atom can be obtained, for example, from measurements of surface potentials. From such measurements one can determine the magnitude and direction of the moment of the electric dipole formed by the adatom and the metal. (For further details, see MIGNOLET [*478*] and ELEY [*182*].) By measuring the changes of electrical resistance of thin metal films after adsorption of gases, R. SUHRMANN et al. [*694*] were able to draw conclusions concerning the nature of the adsorption forces. There is scant experimental information* about the influence of surface heterogeneities on covalent binding forces, largely because of the scarcity of systems in which covalent binding forces are the principal forces active in the adsorption process.

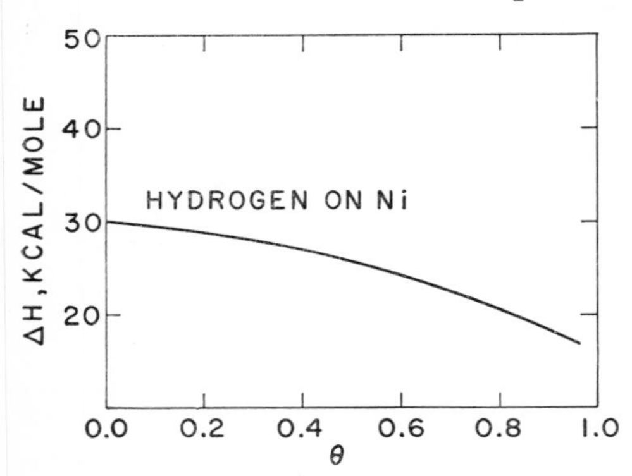

Fig. 4.2.b. Change of the heat of chemisorption ΔH with the surface coverage degree θ for the case of the adsorption of hydrogen on nickel films sintered at 23° C. (BEECK [*68*])

There is as yet no direct experimental proof of the effect of surface heterogeneities such as steps, cracks, holes, etc., on covalent binding forces. The effect of various crystallographic planes on the covalent adsorption process can be seen, for example, in the field-emission experiments of GOMER et al. [*243*]. He found that the diffusion rate of hydrogen atoms adsorbed on body-centered wolfram had an especially high value on the densely packed (110) planes.

How strongly chemical heterogeneities affect the covalent adsorption process can be seen, for example, from the discussion of MAXTED [*463*]. He showed that the activity for hydrogenation of transition metals

* Further details can be found, for example, in the reports of R. V. CULVER and F. C. TOMPKINS [*131*], S. S. ROGINSKI [*600*], and J. H. DE BOER [*83*].

decreases when they are contaminated by sulfides and thioles which easily give up electrons to the d band of the metal.

Finally, information concerning the adsorption of hydrogen on nickel films and the effect of surface covering on heats of chemisorption can be obtained from the investigations of BEECK [68]. As seen in Fig. 4.2.b, the heat of chemisorption decreases slightly with increasing hydrogen covering of the surface, in agreement with the results of SCHUIT and DE BOER [642]*. TRAPNELL [727] found that as the extent of coverage increased from $\theta = 0$ to $\theta = 1$ the heat of chemisorption of hydrogen on wolfram was reduced by about 42 kcal. He attributed this mainly to surface heterogeneities.

4.3. Heteropolar Binding Forces—Strong Chemisorption

4.3.1. Atoms on Homogeneous Surfaces

In contrast to the covalent forces, which involve only an interpenetration of the electron shells of the adatom and metal, the heteropolar binding forces are associated with the transfer of an electron from adatom to metal or vice versa. This electron transfer forms a layer of ions on the surface. This layer, in turn, induces an oppositely charged layer in the metal surface (Coulomb image force) so that a double layer results (see schematic diagram in Fig. 4.3.1.a). In addition to these heteropolar binding forces between the adion and its oppositely charged image in the surface, additional attractive forces may also occur as a result of interaction between the adion and the image forces of nearby adions if the surface is sufficiently covered.

Fig. 4.3.1.a. Schematic diagram of a double layer formed by adsorbed positive ions and their negative image charges at a metal surface

Provided that the coverage degree θ is small, that the distance d between adion and surface is so large that the image-force model is valid, and finally that the surface is homogeneous, the image-force energy E_{im} may be respresented by

$$E_{im}^{(+)} = -(I - \varphi)\, e + \frac{e^2}{4d} \qquad (4.3.1\text{–}1)$$

if the metal is an electron acceptor. But if the metal is an electron donor, the image-force energy is

$$E_{im}^{(-)} = (S - \varphi)\, e + \frac{e^2}{4d}, \qquad (4.3.1\text{–}2)$$

where S is the electron affinity of the adsorbed atom. Such heteropolar binding forces are effective, for example, in the adsorption of alkali and

* BOUDART [87] has pointed out the possibility (see also DE BOER [84]) that the hydrogen atoms may diffuse deep enough to be embedded under the topmost layer of nickel atoms.

alkaline earth metals on such metal surfaces as wolfram, molybdenum, and platinum*. If a sodium atom approaches a wolfram surface as shown in the energy diagram in Fig. 4.3.1.b, it will traverse the curve A^*SI and will be adsorbed as an ion, since the minimum potential energy I on the "ionic curve" is below A the minimum of the "atomic curve". At an infinite distance from the surface, the energy difference between the two curves is equal to the difference between the ionization energy eI of the sodium atom and the work function $e\varphi$ of the wolfram surface.

For the case in which Cs atoms strike a wolfram surface, the atomic curve and the ionic curve no longer intersect. The ionic curve is then always below the atomic curve, (see Fig. 8.2.a) this means that the Cs atoms are adsorbed on the surface as ions, as will be described in more detail in Chap. 8 on surface ionization. We note that the heats of adsorption $\Delta H = E_i + e(\varphi + I)$ are much higher for this present case of strong chemisorption than for weak chemisorption or physical adsorption.

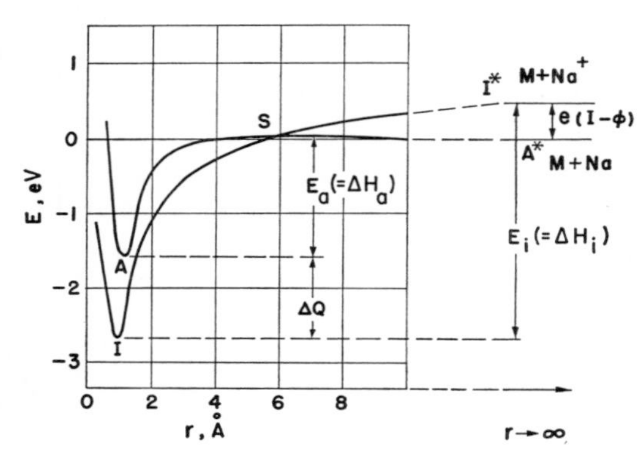

Fig. 4.3.1.b. Potential energy diagram for theadsorption of sodium atoms and ions on a metal surface

For the transfer of an electron from the adatom to a metal surface, the condition $e\varphi > eI$ must be satisfied. However, this condition does not apply to the case of adsorption of *ions* striking the surface. For this case the applicable relation is $E_i + e\varphi > eI + E_a$, where E_a is the adsorption energy of the adatom in atomic form, and E_i is its adsorption energy as an ion.

Metal surfaces may also act as electron donors. As may be concluded from the experiments of Lukirsky et al. [*447*] on the selective photoeffect for potassium surfaces covered with hydrogen, each adsorbed hydrogen atom takes an electron from the potassium surface. Thus they form a negatively charged layer of hydrogen ions on the metal surface.

4.3.2. Polar Molecules on Homogeneous Surfaces

If a polar molecule (e.g., CO) with a dipole moment μ approaches a metal surface, the attraction arising between the two is equivalent to the

* In this connection we should point out attempts to use dispersion forces in computing the interaction of alkali and alkaline earth atoms with metal surfaces. (See, for example, M. Drechsler [*162, 163*].) However, such treatments ignore the experimental fact (supported, for example, by measurements of contact potential differences) that there is electron transfer from the adatom to the metal.

interaction of two dipoles of opposite sign. The second dipole is located where the first would be imaged in the mirror surface. This corresponds to the remarks in the preceding section concerning the image force between an adatom and a metal surface. LORENZ and LANDE [*444*] and MAGNUS [*451*] have calculated the dipole-dipole interaction energy by use of this concept of image force, but have omitted any possible polarization of the molecule by the field of the metal surface. They found that when the dipole makes an angle β with the normal to the surface the interaction energy is

$$E = \frac{-\mu^2}{16\,r^3}\,(1 + \cos^2\beta)\,. \qquad (4.3.2\text{–}1)$$

From this it follows that when the dipole is perpendicular to the surface ($\beta = 0°$), the energy is

$$E = \frac{-\mu^2}{8\,r^3}\,, \qquad (4.3.2\text{–}2)$$

and that when the dipole is parallel to the surface ($\beta = 90°$), it is

$$E = \frac{-\mu^2}{16\,r^3}\,. \qquad (4.3.2\text{–}3)$$

In his calculations, JACQUET [*354*] took account of the possibility that the adsorbed dipole molecule becomes polarized by its own image in the metal. His expression for the interactions is

$$E = \frac{-\mu^2}{8\,r^3}\left[\frac{2}{1-(\alpha/4\,r^3)}\cos^2\beta + \frac{1}{1-(\alpha/8\,r^3)}\sin^2\beta\right]. \qquad (4.3.2\text{–}4)$$

Here α is the polarizability of the adsorbed molecule. The values calculated from Eq. (4.3.2–4) are approximately 30—40% greater than those from Eq. (4.3.2–1) and give a better fit to the experimental values.

4.3.3. Heterogeneous Surfaces

As seen from Eqs. (4.3.1–1) and (4.3.1–2), the magnitude of the value of the work function of the surface is one of the most important determining factors for the binding energy in the case of heteropolar binding forces. Furthermore, in accordance with what was said in Chap. 1, effects which are produced by physical, chemical, and induced heterogeneities influence the magnitude of the effective heteropolar binding forces*.

Thus, for example, the field-emission measurements of BENJAMINS and JENKINS [*69, 70*] show how the physical heterogeneity arising from the presence of different crystallographic planes on the metal surface affects the process of strong chemisorption. They found that sodium and

* A more detailed description of this type of influence on heteropolar binding forces is given, for example, by ROGINSKI [*600*], DE BOER [*83*], and TRAPNELL [*728*].

barium are preferentially adsorbed on the (211) planes of molybdenum. On wolfram, barium is preferentially adsorbed on the (111) planes. MARTIN [*458*] showed that Cs is preferentially adsorbed on wolfram on the (110) and (211) planes, as would be expected from Eqs. (4.3.1–1) and (4.3.1–2), since these surfaces correspond to the highest work function.

That chemical heterogeneity also affects heteropolar binding forces has been shown by DOWDEN [*159*] in his discussion of adsorption processes on nickel and nickel alloy surfaces. In the case of nickel, the addition of slight impurities of elements in the B subgroup of the periodic system (e.g., Cu, Ag, Au, Zn, etc.), whose weakly bound valence electrons are easily transferred to the 3d band of nickel, was found to increase the heat of chemisorption. Small additions of Pt or Pd to nickel, on the other hand, can reduce the heats of chemisorption. Further examples of the effects of chemical heterogeneity on the process of strong chemisorption have been obtained, for instance, from studies of the activity of catalysts (ROGINSKI [*600*]).

Finally the strong influence of induced heterogeneity on the process of strong chemisorption should be noted. (See also Section 1.1.3.) How strongly the heteropolar binding forces depend on the covering of the surface by foreign atoms is determined partly by the magnitude and direction of the dipole moments which are produced by adsorption of foreign atoms on the surface. In addition, the extent of the surface covering is of primary importance. Thus the change in the work function as well as the amount of free uncovered surface which is still available for adsorption, depends on the coverage degree. Likewise, with higher degrees of coverage the increasing interaction of the surface dipoles with one another will have an additional effect on the binding forces.

The example of the chemisorption of Cs on W illustrates how strongly the heats of chemisorption depend on the coverage degree θ. This is a highly simplified example of "surface heterogeneity" because here the Cs atoms themselves may be regarded as a sort of "foreign atom layer", which covers a certain fraction of the surface. In their experiments, TAYLOR and LANGMUIR [*705*] found that the heat of chemisorption decreased with increasing θ (Fig. 5.4.c in next chapter). Over the range $0.06 \leqslant \theta \leqslant 0.60$, their data were well fitted by the empirical relation

$$\Delta H(\theta) = \frac{64}{1 + 0.714\,\theta} \cdot (\text{kcal/mole})\,. \tag{4.3.3–1}$$

DE BOER [*83*] has derived an expression for the change in the heat of chemisorption with increasing θ. His relation predicts a decrease by an amount proportional to $\theta^{3/2}$, but is not able to reproduce the experimental results for coverage degrees less than 0.2.

5. Energetics of Surface Reactions

5.1. General Remarks and Definitions

After the preceeding discussion of the possible binding forces which can act on an atom or molecule impinging on a metal surface, the energetics of the adsorption and desorption processes will now be considered in more detail. The first step is to define various quantities such as heats of adsorption, heats of activation, and the like which will be used in the following. In this connection, the reader is referred to the excellent presentation by E. Hückel [*325*] and the work of Hill [*312*], from which one can obtain definitions which are more rigorously founded thermodynamically.

The integral heat of adsorption ΔH_{int} is the amount of heat Q that is liberated at equilibrium from an area F of the surface on which N molecules of the gas are adsorbed. If E_g is the energy per molecule in the gas phase and E_{ads} the energy per molecule in the adsorbed state, and if it is also assumed that no external work is done, then the expression for ΔH_{int} is

$$\Delta H_{\text{int}} = \frac{Q}{F} = \frac{N}{F}\,(E_g - E_{\text{ads}})\,. \qquad \left[\frac{\text{cal}}{\text{cm}^2}\right]. \qquad (5.1\text{–}1)$$

The differential heat of adsorption ΔH_{diff} is more often used. If an amount of heat dQ per cm² is liberated when a surface already covered with N molecules of gas adsorbs dN additional molecules at constant temperature without any external work, then the differential heat of adsorption ΔH_{diff} is defined as dQ/dN. It is given by

$$\Delta H_{\text{diff}} = \left(\frac{\partial \Delta H_{\text{int}}}{\partial N}\right)_T = \frac{E_g - E_{\text{ads}}}{F} - \frac{N}{F}\left(\frac{dE_{\text{ads}}}{dN}\right)_T. \qquad \left[\frac{\text{cal}}{\text{cm}^2}\right] \qquad (5.1\text{–}2)$$

(Note that for an ideal gas E_g depends only on the temperature.) Whenever the adsorption process cannot proceed without external work or without change of temperature, the heat of adsorption may be obtained from the above-defined adsorption equations if the thermal adsorption equations are known. Different values of heats of adsorption will be obtained, depending on whether the thermal adsorption equations ("adsorption isotherms") are based on the "single molecular adsorption" equations of Langmuir [*423*], Freundlich [*229*], and Temkin [*711, 712*] or on the "multimolecular adsorption" equations of Brunauer, Emmett, and Teller [*100*].

An isothermal heat of adsorption can be defined for the case in which the adsorption process takes place under isothermal conditions. For a fixed total number of gas molecules, the increase in the number of adsorbed particles is associated with the work done in the volume change dV accompanying the adsorption. HÜCKEL defines this isothermal heat of adsorption as

$$\Delta H_{\text{isoth}} = \Delta H_{\text{diff}} + RT . \qquad (5.1\text{–}3)$$

Definitions of other heats of adsorption (such as isosteric*, isopycnic, and isobaric), may be found in HÜCKEL's monograph [*325*]. A discussion of various experimental methods of determining heats of adsorption can be found in the more recent presentations of TRAPNELL [*728*], HONIG [*321*], and BEEBE [*63*]. It turns out (see also the comment of KINGTON and ASTON [*384*]) that the calorimetrically determined values of the heat of adsorption can lie between the values of the differential and isothermal heats of adsorption, i.e.,

$$\Delta H_{\text{diff}} < \Delta H_{\text{calorim}} < \Delta H_{\text{isoth}} = \Delta H_{\text{diff}} + RT . \qquad (5.1\text{–}4)$$

However, since in many cases RT is small in comparison with the values of ΔH_{diff} and ΔH_{isoth}, the experimental values often agree (within the limits of error) with the differential heats of adsorption. The meaning of the heat of adsorption or desorption and of the true and apparent activation energies ΔH_{tr} and ΔH_{Act} can best be seen from the potential energy curve in Fig. 5.1.a. If a diatomic molecule (such as N_2) approaches a metal surface (such as Fe), it can follow the potential curve (*1*) and be physically adsorbed at the potential minimum E_M. But if the molecule can take up an energy which is equal to or greater than the apparent activation energy ΔH_{Act} it can go over into curve (*2*) and its atomic constituents are chemisorbed. (For further details, see P. ZWIETERING and T. ROUKENS [*807*].) The energy difference $(P - E_a)$ is called the true activation energy** ΔH_{tr}.

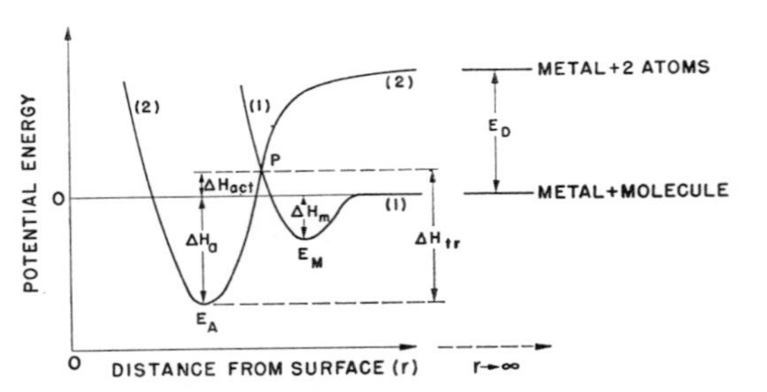

Fig. 5.1.a. Potential energy diagram for the adsorption of a diatomic molecule on a metal surface

* Since, as BRUNAUER [*101*] has shown, the isothermal heat of adsorption can be obtained from the equation for isosteric adsorption, one often identifies ΔH_{isoth} with ΔH_{isost} (see for example the review article by A. G. NASINI, and G. SAINI [*519*]). On the other hand, HÜCKEL defines ΔH_{isost} differently. In particular, for an ideal gas he writes

$$\Delta H_{\text{isost}} = \left(\frac{\partial E_g}{\partial T}\right)_N .$$

** This true activation energy ΔH_{tr}, which is made up of the heat of adsorption ΔH_a and the apparent activation energy ΔH_{Act}, is often referred to as the desorption heat ΔH_{des} in discussions of desorption processes (see for example DE BOER [*82*]). From this definition it follows that the heat of desorption is greater

The rate of the reaction is determined by the magnitude of the apparent activation energy ΔH_{Act}. A large amount of experimental material shows that the smaller the contribution of heteropolar binding forces, the greater the apparent activation energy ΔH_{Act}. Thus, no activation energy has been observed for the adsorption of alkali atoms on metal surfaces (see for example under Fig. 8.2.a and the discussion of EYRING et al. [*309, 310*]). But such systems as H_2 and N_2 on wolfram surfaces also require no activation energy for the adsorption process, as has been shown, for example by SCHUIT and DE BOER [*642*] and by BECKER and HARTMAN [*60*].

5.2. Heats of Adsorption for Physical Adsorption

Some of the more recent experimental data on heats of physical adsorption will now be discussed. Most of these data were obtained by use of modern techniques (field desorption methods and adsorption spectrometers) so that surface conditions are well defined (clean, single-crystal surfaces under very high vacuum). The values of physical heats of adsorption are reviewed and discussed in the summaries of BEEBE [*63*], BRUNAUER [*101*], DE BOER [*83*], TRAPNELL [*728*], CULVER and TOMPKINS [*131*], and YOUNG and CROWELL [*788a*]. The older results must be used with caution, however, since they are liable to large uncertainties because of poorly defined surface conditions. For example, the surface may have been gas-covered because of inadequate vacuum techniques, it may have been an unknown combination of different crystal planes, or the like.

Among the modern methods for investigating desorption processes and determining desorption heats (which for the case of nonactivated adsorption are equal to the adsorption heats) is the method of field desorption. This has been applied with success in recent years.

MÜLLER [*513*] showed that when a high electric field ε ($\varepsilon \approx 4-5\,\mathrm{V/Å}$) is applied to an emitter (such as a wolfram needle) the desorption of the particles adsorbed on the emitter can be observed. MÜLLER's expression for the rate constant K for ionic desorption is

$$K = \nu \exp\left(-\Delta H_i(\varepsilon)/kT\right), \tag{5.2–1}$$

where $\nu \approx 10^{13}\,\mathrm{sec}^{-1}$ (the order of magnitude of a vibration frequency), and $\Delta H_i(\varepsilon)$ is the heat of ionic desorption given by

$$\Delta H_i(\varepsilon) = \Delta H_i(\varepsilon = 0) - 3.8\,(ne)^{3/2}\varepsilon^{1/2}, \tag{5.2–2}$$

than the heat of adsorption whenever there is an activation energy. In Fig. 5.1.a. the spacing between curve *1* and curve *2* at an infinite distance from the surface is equal to the dissociation energy E_D of the molecule. The distance from the energy minimum E_M up to the zero of potential energy is the physical heat of adsorption ΔH_m for molecular adsorption.

with

$$\Delta H_i(\varepsilon = 0) = \Delta H_a(\varepsilon = 0) + e(I - \varphi). \qquad (5.2\text{–}3)$$

Here $3.8\,(ne)^{3/2}\varepsilon^{1/2}$ is the SCHOTTKY term with ne the charge of the desorbed ion and ε the electric field intensity applied at the emitter; $\Delta H_a(\varepsilon = 0)$ is the heat of desorption for a neutral atom (or molecule) and, in the case of nonactivated adsorption, is equal to the heat of adsorption; eI is the ionization energy of the atom; and $e\varphi$ is the work function of the metal surface. The second expression had been formulated earlier by SCHOTTKY [*640*].

One way of obtaining the heats of adsorption is to plot log K against $1/T$ and determine ΔH_i from the slope of the curve. On the other hand, one can plot $\varepsilon^{1/2}$ as ordinate against T as abscissa and determine ΔH_i from the intercept of the curve on the axis of ordinates, as was done for example by MÜLLER [*513*] in investigating the desorption of barium from wolfram.

GOMER [*247*] has carefully investigated these considerations of MÜLLER and has shown that Eq. (5.2–1) holds only for systems for which the difference $(I - \varphi)\,e$ has small positive values, as for example for certain electronegative heteropolar compounds. For the case that $I \gg \varphi$, as for example in physical adsorption of noble gases on metal surfaces, Eq. (5.2–1) must be replaced by

$$K = \nu(1 - e^{-t/\tau}) \exp(-\Delta H_i(\varepsilon)/kT), \qquad (5.2\text{–}4)$$

where $P(t) = 1 - e^{-t/\tau}$ represents the probability of ionization in the time interval t, and τ is the mean time which an electron of the adatom requires to tunnel through the potential barrier between the adatom and the metal surface. The heats of adsorption ΔH of various noble gases on wolfram and tantalum surfaces, obtained by the field desorption method by EHRLICH and HUDDA [*172, 174*] and by GOMER [*246, 248*], are listed in Table 5.2.A. Table 5.2.B gives the heats of adsorption of argon on various metal surfaces, but most of these were determined by other methods and cannot be considered as reliable as those in Table 5.2.A. The experimental values of the heats of adsorption in Table 5.2.B show only a very slight dependence on the nature of the surface. This tends to confirm the LENNARD-JONES assumption (Chap. 4) which takes no account of the nature of the surface.

Table 5.2.A. *Heat of adsorption ΔH (kcal/mole) of A, Kr, and Xe on W and Ta surfaces*

	A	Kr	Xe
W	1.9 [a] ≈1.9 [b]	5.9 [a] <4.5 [b]	10.0 [a] 8–9 [b]
Ta			5.3 [a]

[a] Data from GOMER [*246, 248*].
[b] Data from EHRLICH et al. [*172, 174*].

Table 5.2.B. *Heat of adsorption ΔH (kcal/mole) of argon on surfaces of W, Cu, Ag, Pt, Fe, and Ni*

ΔH (kcal/mole)							
W	Cu	Ag	Pt	Fe	Fe Not reduced	Ni	Ni Not reduced
1.9[a,b]	~3.1[c]	3.5[d]	3.3[d]	3.2[d]	3.3[d]	3.0[e]	4.7

[a] Data from GOMER [*246, 248*]. [b] Data from EHRLICH et al. [*172, 174*].
[c] RHODIN [*590*]. [d] ARMBRUSTER and AUSTIN [*31*].
[e] ZETTLEMOYER, YUNG-FANG YU, and CHESSIK [*803*].

A comparison of the experimentally determined heats of adsorption for argon and nitrogen on platinum surfaces with calculated values based on the equations of LENNARD-JONES [*431*], BARDEEN [*48*], and MARGENAU and POLLARD [*457*] is given in Table 5.2.C.

Table 5.2.C. *Comparison of experimental and theoretical values of the heats of adsorption ΔH (cal/mole) of A and N_2 on a platinum surface*

	ΔH theoretical			ΔH, exptl.
	MARGENAU and POLLARD [*457*]	BARDEEN [*48*]	LENNARD-JONES [*431, 432*]	
A	2050	1320	3000	3200[a]
N_2	2100	1800	3350	3380[a]

[a] ARMBRUSTER and AUSTIN [*31*].

The values have been calculated from the potentials computed by BRUNAUER [*101*]. The LENNARD-JONES values are seen to come closest to the experimental results, while the values of BARDEEN are smaller by about a factor of two. This is to be expected from a comparison of Eqs. (4.1–5) and (4.1–6), as has already been pointed out in Chap. 4. The LENNARD-JONES values for the adsorption of krypton on various metal surfaces are also closer to the experimental values obtained by PIEROTTI and HALSEY [*569*] than are the values computed from the formulas of BARDEEN and of MARGENAU and POLLARD.

However, PIEROTTI and HALSEY pointed out that their experimental values actually accord even better with the equations of BARRER [*50*] and KIRKWOOD [*385*], although these equations were derived for non-metallic surfaces*. This likewise indicates that the nature of the solid

* According to these equations, the interaction energy is given by

$$E = \frac{\pi N}{6} \frac{1}{R^3} 6 m_e c^2 \alpha_1 \alpha_2 \left(\frac{\alpha_1}{\chi_1} + \frac{\alpha_2}{\chi_2}\right)^{-1},$$

where α_1 and α_2 and χ_1 and χ_2 are the polarizabilities and diamagnetic susceptibilities for the adsorbates and adsorber, m_e is the electron mass, and c is the velocity of light. For further details, see R. M. BARRER [50] and J. G. KIRKWOOD [*385*].

surface has no decisive influence on the value of the heat of adsorption in the case of physical adsorption. On the other hand, a slight influence of the surface is certainly indicated by the few modern data (Table 5.2.B). In conclusion, it may be said that none of the various theoretical treatments give a truly satisfactory reproduction of the few admissible experimental data.

5.3. Heats of Adsorption for Weak Chemisorption

If hydrogen molecules strike a metal surface they can, as shown in Fig. 5.1.a, be physically adsorbed as molecules [in which case they follow the potential curve (*1*)], or they may be weakly chemisorbed as hydrogen atoms [in which case they follow potential curve (*2*)]. As shown in Fig. 5.1.a, the minimum E_A of the potential energy for curve (*2*) is below that for curve (*1*). From this it follows that the values of the heats of adsorption are smaller for physical adsorption than for weak chemisorption. A comparison of the values in Tables 5.2.A and 5.3.A confirms this. Although the heats of adsorption have values up to about 10 kcal/mole in the case of physical adsorption, for weak chemisorption the values may be as high as 50 kcal/mole or even more.

Obviously, the magnitude of the heat of adsorption is by no means a sure indicator of the type of binding that is predominant in a given case. This is to be determined rather from a combination of experimental results for example, from the magnitude and direction of the surface dipole (from contact potential measurements), from changes in the paramagnetism of the surface system, or from changes in the electrical resistance of thin metal films on which gases are adsorbed (SUHRMANN et al. [*694, 695*]).

Unfortunately our present knowledge of the potential curve between adatom and surface is not accurate enough to lead to any exact theory for calculating effective binding energies. However, attempts have been made to set up semi-empirical relations for determining such binding energies and thus determining heats of adsorption.

Thus, ELEY [*180, 181*] has shown that a modification of PAULING's [*547*] proposed expression.

$$D(A, B) = \tfrac{1}{2}\,[D(A, A) + D(B, B)] + 23.6\,(x_A - x_B)^2\,, \qquad (5.3\text{–}1)$$

which gives the binding energy $D(A, B)$ of two covalently bound atoms A and B, is also applicable to the covalent binding of an adatom on a metal surface. In Eq. (5.3–1), x_A and x_B are the electronegativities of atoms A and B, defined below.

Thus, as shown by ELEY, the equation for the case of adsorption of hydrogen on wolfram surfaces is

$$D(W, H) = \tfrac{1}{2}\,[D(W, W) + D(H, H)] + \Delta S\,. \qquad (5.3\text{–}2)$$

The first term, the interaction energy $D(W, W)$ of two wolfram atoms, can be estimated according to ELEY from the sublimation energy E_s. This estimate takes account of the interaction of nearest neighbors by making the crude approximation that there are 12 such for all metals (although, strictly speaking, this is true only for the face-centered cubic lattice). Then, $D(W, W) = (2/12) E_s \approx 33.8$ kcal/mole. EHRLICH [*173*] has recently criticized this procedure for determining the interaction energy $D(W, W)$ and has proposed the use of the surface energy in place of the sublimation energy. However, the difference between the procedures of ELEY and EHRLICH should not be very great. It must be noticed that the sublimation energy involves the nearest-neighbor interaction throughout a 4π solid angle, while the surface energy involves only 2π. (For this reason the sublimation energy is approximately twice the surface energy.) On the other hand, ELEY's consistent practice of multiplying the sublimation energy by $^1/_6$ takes this 4π nearest-neighbor interaction into account in first approximation and thus reduces the value of $D(W, W)$ correspondingly.

The electronegativity term $\Delta S = (x_W - x_H)$ is determined in first approximation by the dipole moment μ of the compound, i.e., $x_W - x_H = \mu(\theta = 0)$, where θ is the coverage degree. ELEY makes the further approximation that $\mu(\theta = 1) = \mu(\theta = 0)$. STEVENSON [*684*] proposed a different method which uses the electron work functions for determining the electronegativity term.

As shown by ELEY, the binding energy of hydrogen on wolfram is

$$D(W, H)_{\text{calc}} = \tfrac{1}{2}(33.8 + 103.2) + 4.9 = 73.4 \text{ kcal/mole}, \quad (5.3\text{–}3)$$

$$D(W, H)_{\text{exp}} = 74.1 \text{ kcal/mole}. \quad (5.3\text{–}4)$$

The agreement is satisfactory. The heat of adsorption ΔH_a for $\theta = 0$ (the difference between the minimum energy E_A and the zero of potential in Fig. 5.1.a) for diatomic molecules on metal surfaces is then given by

$$\Delta H_a = 2D(A, B) - D(B, B) \quad (5.3\text{–}5)$$

or, for hydrogen on wolfram, by

$$\Delta H_a = 2D(W, H) - D(H, H) = 146.8 - 103.2 = 43.6 \text{ kcal/mole}. \quad (5.3\text{–}6)$$

This value agrees well with the experimental value of 45 kcal/mole.

The experimental and theoretical values of the differential heats of adsorption for various metal hydrides are compared in Table 5.3.A. The theoretical values were calculated for $\theta = 0$. Further examples of the successful application of ELEY's equation can be found in the tables of TRAPNELL [*728*].

Table 5.3.A. *Experimental and theoretical values of the differential heats of adsorption for various metal hydrides (kcal/mole)*

	W	Ta	Cr	Fe	Ni	Rh	Cu	Pt
ΔH_a (exp)	45 [a]	45 [c] 39 [d]	45 [c]	32 [c]	29 [c]	28	35 [e] 9 [f]	30 [b]
ΔH_a (theor)	43.6 [h]	32 [h]	15.7 [h]	17 [h]	17.2 [h]	23 [h]	13.6 [h]	36.9 [g]

[a] WAHBA and KEMBALL [*748*].
[b] TAYLOR, KISTIAKOWSKI, and PERRY [*704*].
[c] BEECK [*65, 68*].
[d] Calculated from values given in Table 2 in the article by DE BOER [*83*].
[e] KWAN [*412*].
[f] WARD [*752*]. (The value is quoted for Cu powder!)
[g] TRAPNELL [*728*].
[h] ELEY [*180*].

An attempt has also been made to use the "charge transfer, no bond" theory in its modern form as developed by MULLIKEN [*517*] for computing heats of adsorption when there is covalent binding. According to MULLIKEN, the ground state of the adsorption complex MX is given by a linear combination of the wave function $\psi(M^0X^0)$ for the no-bond state and that for the structure M^-X^+ (for the case in which the metal is an electron acceptor), i.e., by

$$\psi(MX) = a\psi(M^0X^0) + b\psi(M^-X^+)\,. \tag{5.3–7}$$

After maximizing the coefficients a and b and normalizing Eq. (5.3–7) and using the additional assumption that the overlap integral $\int \psi(M^0X^0)\,\psi(M^-X^+)\,d\tau$ is equal to zero, one finds with MATSEN et al. [*462*] that

$$\Delta H = \tfrac{1}{2}\left\{-\left[E(M^-X^+) - E(M^0X^0)\right] + \sqrt{\left[E(M^-X^+) - E(M^0X^0)\right]^2 + 4\beta^2}\right\}. \tag{5.3–8}$$

If the metal is an electron acceptor, the term $[E(M^-X^+) - E(M^0X^0)]$ can be represented by

$$E(M^-X^+) - E(M^0X^0) = e(I - \varphi) - e^2/4r\,, \tag{5.3–9}$$

where eI is the ionization energy of the gas, $e\varphi$ is the work function of the metal surface, and $e^2/4r$ takes account of the image forces. The quantity β in Eq. (5.3–8) is an interaction integral for the resonance mixture of the wave functions of the ground state. As shown by MIGNOLET [*479*], it can be related to the surface potential.

BRODD [*95*], who has recently used Eq. (5.3–8) to compute the heats of adsorption of various metal hydrides, has found good agreement with the experimental data. GUNDRY and TOMPKINS [*267*], however, criticized these results and pointed out errors and arbitrary procedures in BRODD's calculation [such as the incorrect manner of introducing the dissociation energy of hydrogen into Eq. (5.3–8), the arbitrary choice of the distance of nearest approach r of the hydrogen adatom to the metal surface, and the like]. These authors show that after the necessary corrections are made to BRODD's calculations there is no sort of agreement between experimental and theoretical values, and the applicability of Eq. (5.3–8) to covalent bindings is brought into question.

A final brief comment should be made on the effect of surface heterogeneities on the heats of adsorption. We have already pointed out in section 1.1.3 that induced heterogeneities influence the value of heats of adsorption when the binding is predominantly covalent. Thus, TRAPNELL found that ΔH, the heat of chemisorption of hydrogen on wolfram, decreased by about 43 kcal/mole as the degree of coverage changed from $\theta = 0$ to $\theta = 1$. SCHUITT and DE BOER [*642*], as well as BEECK [*68*] found that at 0° C the heat of chemisorption of hydrogen on nickel decreased by about 14.4 kcal/mole as θ increased from 0 to 1. MAXTED and HASSID [*463a*] and KWAN [*412*] found that the heat of adsorption of hydrogen on platinum was independent of the coverage degree.

Various attempts have been made to interpret the decrease of the heat of chemisorption with increasing degree of coverage θ. Thus, TEMKIN [*713a*] attempted to compute these changes of the heat chemisorption on the basis of a model of a two-dimensional surface electron gas which follows the same statistical distribution laws and exclusion principle as the three-dimensional free electron gas. He found the expression

$$\Delta H = \frac{h^2 \sigma_0}{4 \pi m_e} \theta , \tag{5.3–10}$$

where h is PLANCK's constant, m_e is the electron mass, and σ_0 is the number of atoms adsorbed in a monatomic layer. However, as pointed out by DE BOER, the heat of chemisorption does not change as rapidly as indicated by this equation. Certainly TEMKIN's theory, which takes no account of the nature of the metal surface except indirectly in the value of the quantity σ_0, is too simple.

On the other hand, DE BOER showed that in many cases the changes of work function of a metal with increasing coverage degree are responsible for the changes in chemisorption. In addition, there may also be effects of surface diffusion processes (although BEECK [*65*, *68*] has shown that for H_2 on Fe at −180° C the diffusion is negligible and ΔH is independent of θ) as well as diffusion processes of hydrogen in the metal (as discussed by BOUDART [*87*]).

5.4. Heats of Adsorption for Strong Chemisorption

The occurrence of strong chemisorption can be observed, for example, in the collision of alkali atoms with surfaces of the refractory metals, where heteropolar binding is predominant. As we have already pointed out in Section 4.3.1, for clean surfaces the heat of adsorption for strong chemisorption (which, in the case of a monatomic gas, is equal to the binding energy of the adion, on the surface) consists of two parts. The first part comes from the transfer of an electron from the adatom to the metal surface, or the inverse process. The second contribution is the work done by the image force when the ionized adatom is brought to the equilibrium distance r_0 from the metal surface. On the basis of such classical electrostatic arguments DE BOER [*82, 84*] has expressed the heat of adsorption ΔH_a for zero coverage by the relation

$$\Delta H_a = e(\varphi - I) + e^2/4r. \tag{5.4–1}$$

The meaning of the two terms is understandable from what has been said previously in section 4.3.1. The first term $e(\varphi - I)$ is the difference between the energy levels A^* and I^* at an infinite distance from the surface (Fig. 4.3.1.b). It takes account of the heat liberated in the process of electron transfer from the adsorbate atom to the surface. The second term, which takes into account the image force, is the difference of the energy levels I^* (adion at infinite distance from the surface) and I (minimum potential of the "ionic curve") in Fig. 4.3.1.b. Note that in the case of adsorption of sodium on wolfram the heat of desorption is equal to the heat of adsorption, since the adatom in the desorption process follows the potential curve ISA^* (Fig. 4.3.1.b) and is desorbed as an atom. In this case the adsorption and desorption processes are reversible. In the case of the adsorption and desorption of Cs on wolfram, the situation is different. As Fig. 8.2.a shows, here the ionic curve is below the atomic curve. Even though the heat of adsorption is again given by $\Delta H_a = E_i + e(\varphi - I)$, the heat of desorption is given by ΔH (des) $= E_i$. This means that the heat of desorption is less than the heat of adsorption.

HIGUCHI, REE, and EYRING [*309, 310*] have used quantum mechanical considerations to obtain an expression for pure heteropolar binding. Eq. (5.4–1), which DE BOER derived on the basis of classical electrostatic arguments, was obtained by them as an approximation. HIGUCHI et al. showed that it is not generally permissible to set the equilibrium distance r_0 equal to the radius r_i of the adion, as DE BOER had done in evaluating Eq. (5.4–1). One obtains more reliable values for the equilibrium distance r_0 from a determination of the contact potential V_0 for which

$$V_0 = 4\pi\sigma_m e r_0, \tag{5.4–2}$$

where σ_m is the number of adsorption sites per unit area, and e is the charge on the electron. For example, for sodium and potassium on wolfram, $r_0 \approx 2r_i$. This explains the fact that for Na on W the heat of adsorption computed by DE BOER from Eq. (5.4–1) is equal to 63 kcal/mole and is twice as large as the value (28.6 kcal/mole) computed by HIGUCHI et al. The later value is in good agreement with the experimental value of 32 kcal/mole (BOSWORTH [*86*]).

HIGUCHI et al. have used Eq. (5.4–1) with the value of r_0 from Eq. (5.4–2) to compute the heats of adsorption for some alkali metals and alkaline earths on wolfram surfaces for the case of $\theta = 0$. Table 5.4.A

Table 5.4.A. *Calculated (Higuchi et al. [309, 310]) and experimental values of heats of adsorption for strong chemisorption*

System	$e\overline{\varphi}$ (eV)	eI (eV)	$e^2/4r_0$ (eV)	r_0 Å	r_i Å	ΔH_a (calc) (eV)	ΔH_a (exptl) (eV)
Na-W	4.52	5.12	1.84	1.96	0.97	1.24	1.39 [a] (2.73) [b]
K-W	4.52	4.32	1.83	1.97	1.33	2.03	(2.55) [b]
Cs-W	4.52	3.87	2.19	1.64	1.67	2.83	2.94 [c]
Sr-W	4.52	5.68	4.00	0.90	1.32	2.84	3.70 [d]
Ba-W	4.52	5.21	2.55	1.41	1.53	1.86	3.50 [d]

[a] BOSWORTH [*86*].
[b] STARODUBTSEV [*678*].
[c] TAYLOR and LANGMUIR [*705*].
[d] MOORE and ALLISON [*493*].

compares these computed values with the experimental results obtained by various authors. The values in brackets are the heats of desorption determined by STARODUBTSEV [*678*]. The deviation of the values of the heats of adsorption ΔH_a from the heats of desorption ΔH (des) for the system Na-W (ref. [*678*]) is surprising since the potential curve (Fig. 4.3.1.b) leads one to expect that $\Delta H\,(\text{des}) = \Delta H_a$.

There is no indication of a misuse of units (for example, kcal/mole in place of kcal/g-atom), although the deviation corresponds closely to this factor of two. Remarkably enough, Eq. (5.4–2) appears to be unsuitable for use in computing the heats of adsorption of alkaline earths on wolfram. HIGUCHI et al. have attempted to show that for this system there are contributions from both covalent and heteropolar binding and that the heats of adsorption can be better calculated by a modified ELEY equation. Thus, they found for Ba-W that $\Delta H_a\,(\text{calc}) = 3.80$ eV; and for Sr-W, $\Delta H_a(\text{calc}) = 5.5$ eV. But the deviations between the experimental and computed values are very large for the system Sr-W. Values of the heats of adsorption in strong chemisorption can be found for other systems from the summaries of TRAPNELL [*728*], HIGUCHI et al. [*310*], and EHRLICH [*174*].

As already mentioned in Chap. 4, the heats of adsorption in strong chemisorption depend strongly on the coverage degree θ. Various attempts to explain this θ dependence theoretically were also noted. It turned out that neither the TAYLOR-LANGMUIR equation [Eq. (4.3.3–1)] nor the two relations given by DE BOER [*82, 83*] ($\Delta H_a \propto a\theta$ or $\Delta H_a \propto a\theta^{3/2}$) reproduce the θ dependence correctly over the whole range $0 \leqslant \theta \leqslant 1$. HIGUCHI et al., however, derived a relation [Eq. (5.4–12)] which agrees well with the experimental values for such Cs-W systems for the range $0.05 \leqslant \theta \leqslant 1$.

As an approximation, one can regard the dipole layer which is formed on the surface by adsorption of $\sigma_m \theta$ ions as a parallel-plate condenser with a plate separation equal to r_0, the equilibrium distance of the adion from the metal surface, and with a medium having a dielectric constant ε. From electrostatics, we know that the total field energy E_{total} in a volume element $d\tau$ between the plates is given by

$$E_{\text{total}} = \frac{1}{8\pi} \int \varepsilon \mathfrak{D} \, d\tau \,, \tag{5.4–3}$$

where ε is the electric field strength and $\mathfrak{D}$ is the dielectric displacement. This leads at once to the total energy per unit area, so that the energy density E_j is found to be

$$E_j = \frac{1}{8\pi} \frac{\mathfrak{D}^2}{\varepsilon} r_0 \,. \tag{5.4–4}$$

The energy density E_{j0} produced by a single adsorbed ion when there is a degree of coverage θ is then given by

$$E_{j0} = \frac{\partial E_j}{\partial(\sigma_m \theta)} = \frac{r_0}{8\pi\sigma_m} \left[\frac{2\mathfrak{D}}{\varepsilon} \frac{\partial \mathfrak{D}}{\partial \theta} - \frac{\mathfrak{D}^2}{\varepsilon^2} \frac{\partial \varepsilon}{\partial \theta} \right]. \tag{5.4–5}$$

If $\mathfrak{D}$ and ε in this equation are replaced by the expressions

$$\mathfrak{D} = c\mu\theta \tag{5.4–6}$$

$$\varepsilon = 1 + \alpha c \theta \,, \tag{5.4–7}$$

where $c = 4\pi\sigma_m/r_0$, σ_m is the number of adsorption sites per unit area, and α is the polarizability of the adion, then according to HIGUCHI et al. the result is

$$E_{j0} = \frac{c\mu^2\theta}{1 + c\alpha\theta} - \frac{c\mu^2\theta \cdot c\alpha\theta}{2(1 + c\alpha\theta)^2} \,, \tag{5.4–8}$$

where μ is the electric dipole moment of the adion-metal surface system. In addition to this energy contribution, they also take account of the attractive forces which arise from the interaction that may occur between an adion and the image forces of the nearest-neighbor adions. These attractive forces are, however, reduced by the repulsive forces between the adions. The authors take account of these repulsive forces approximately in that they summarily give each adion an effective charge which is only a quarter of its elementary charge. For the contribution to the energy from these additional attractive forces, they get the expression

$$E_{\text{attr}} = \tfrac{1}{4}\,\sigma_1\,\theta\,\frac{e^2}{\varepsilon r}\,dS \cdot \cos\psi\,. \tag{5.4–9}$$

Here σ_1 is the number of adsorption sites per unit area, r is the distance between the representative ion and the surface element dS (Fig. 5.4.a), and $dS = 2\pi r^2 \tan\psi\, d\psi$ is a ring in the xy plane, where ψ is the angle

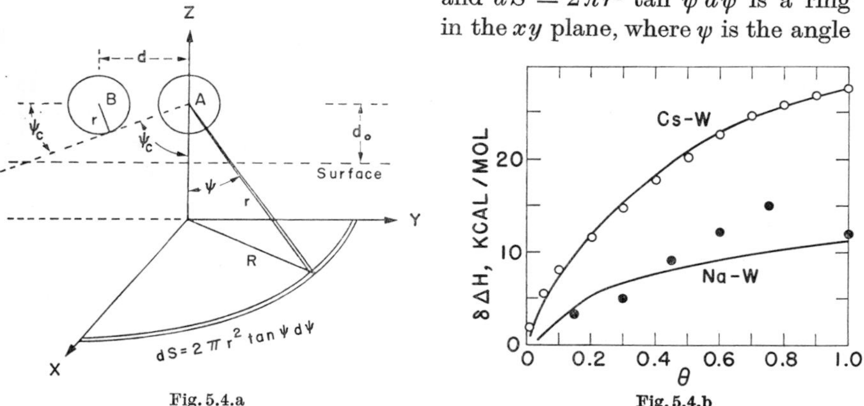

Fig. 5.4.a Fig. 5.4.b

Fig. 5.4.a. Schematic diagram to illustrate the interaction of an adion A and its neighboring adsorbed ions, such as B, with a metal surface (Higuchi et al. [*309*])

Fig. 5.4.b. The decrease of the heat of desorption $\delta\Delta H$ as a function of the surface coverage θ. The solid lines indicate the values calculated by Higuchi et al. [*310*]. The experimental points for *Cs* on wolfram and Na on wolfram are obtained from data given in reference [*705*] and by Langmuir [J. Amer. chem. Soc. **54**, 2798 (1932)]

between the z axis and the radius vector $\vec{r}$. Substitution of the expression for dS into Eq. (5.4–9) and integration of the latter yield

$$E_{\text{attr}} = \frac{c\mu^2\theta}{1 + c\alpha\theta} \cdot \tfrac{1}{4} \ln\left(\frac{d}{r_i\sqrt{\theta}}\right). \tag{5.4–10}$$

Here r_i is the radius of the adion and d the distance between adions when $\theta = 1$.

The heat of adsorption per unit area for a degree of coverage θ is then

$$\Delta H_a(\theta) = \Delta H_a(\theta = 0) + \delta\Delta H = \Delta H_a + E_{j0} + E_{\text{attr}} \tag{5.4–11}$$

$$\Delta H_a(\theta) = \Delta H_a(\theta = 0) + \frac{c\mu^2\theta}{1 + c\alpha\theta}\left[1 - \frac{c\alpha\theta}{2(1 + c\alpha\theta)} - \tfrac{1}{4}\ln\left(\frac{d}{r_i\sqrt{\theta}}\right)\right]. \tag{5.4–12}$$

HIGUCHI et al. have computed values of $\delta \Delta H$ for the systems Na-W and Cs-W on the basis of Eqs. (5.4–8) and (5.4–10). Figure 5.4.b, in which values of $\delta \Delta H$ are plotted against θ, gives the results of their computations. For the system Cs-W the agreement with the experimental values of TAYLOR and LANGMUIR [*705*] is to be regarded as good over the whole range of θ, when one considers that Eqs. (5.4–8) and (5.4.–10) have no free parameters (except for the effective charge).

6. Inelastic Collisions of Atoms and Molecules with Metal Surfaces: The Accommodation Coefficient

6.1. Definition and General Remarks

When a gas atom or molecule whose energy corresponds to the temperature T_0 strikes the surface of a solid at temperature $T_1 \neq T_0$, an exchange of energy can occur as a result of the interaction of the surface atoms with the incoming atoms (adatoms) or molecules. After remaining on the surface in accordance with their mean adsorption lifetime, the adatoms desorb from the surface with an average energy kT_2 (Fig. 6.1.a). As a measure of this energy exchange, KNUDSEN [*393, 397*] has defined a "thermal accommodation coefficient" α. If the energy exchange between the incident particle with mean energy E_0 and the surface with molecular energy E_1 is incomplete, so that the particles leave the surface with an energy E_2 which lies between E_0 and E_1, the accommodation coefficient* is defined as

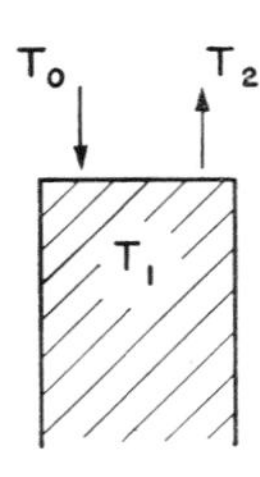

Fig. 6.1.a. Schematic diagram of the energy exchange between gas particles which have an average energy kT_0 and strike a metal surface at a temperature T_1 and leave it again with an average energy kT_2

$$\alpha = \lim_{E_1 \to E_0} \frac{E_2 - E_0}{E_1 - E_0} . \qquad (6.1\text{–}1)$$

In the most usual form of the equation defining the accommodation coefficient, the temperature is introduced in place of the energy so that

$$\alpha = \lim_{T_1 \to T_0} \frac{T_2 - T_0}{T_1 - T_0} . \qquad (6.1\text{–}2)$$

* Note that the general case of accommodation of a molecule with a surface involves a total accommodation coefficient $\bar{\alpha}$ which is made up of the three partial accommodation coefficients for exchange of translational (α_{tr}), rotational (α_{rot}), and vibrational (α_{vib}) energy, i.e.,

$$\bar{\alpha} = \frac{\alpha_{\text{tr}} C_{\text{tr}} + \alpha_{\text{rot}} C_{\text{rot}} + \alpha_{\text{vib}} C_{\text{vib}}}{C_{\text{tr}} + C_{\text{rot}} + C_{\text{vib}}} , \qquad (6.1\text{–}1\text{a})$$

where C_{tr}, C_{rot}, C_{vib} are the respective contributions to the molar specific heat.

It should be mentioned that BLODGETT and LANGMUIR [78] emphasize that this definition is strictly accurate only if the temperature T_2 of the particles leaving the surface is well defined, that is, if the particles have an undistorted MAXWELL distribution of velocities. The distribution will be distorted, however, except in the case that $T_0 \approx T_1$.

The accommodation coefficient also depends on the temperature, and in fact, as indicated in the above definition of the AC*, it depends on the magnitude of the temperature difference ΔT between the gas and the surface, as well as the particular temperature region in which this ΔT lies. The difference ΔT should be as small as possible.

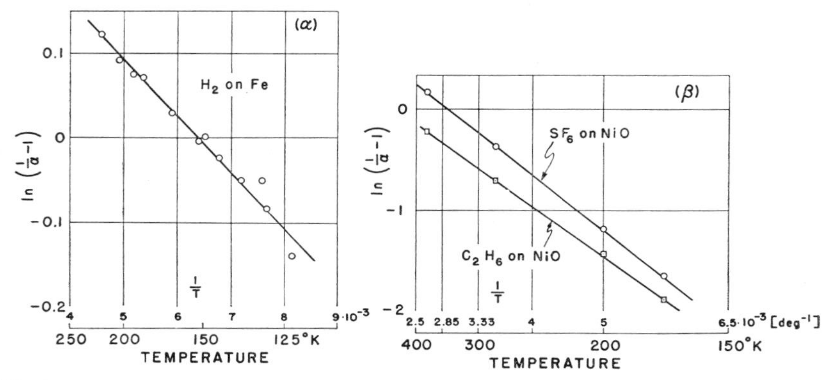

Fig. 6.1.b. Temperature dependence of the accommodation coefficient α in accordance with Eq. (6.1–3) (curve α, ROWLEY and EVANS [618]; curves β, HUNSMANN [330])

EUCKEN and BERTRAM [195] have expressed the temperature dependence of the accommodation coefficient α by the relation

$$\ln\left(\frac{1}{\alpha} - 1\right) = \frac{Q - E_a}{RT} + \text{constant}, \qquad (6.1\text{–}3)$$

where the activation energy E_a is obtained from the FRENKEL [227] equation $\tau = \tau_0 \exp(E_a/RT)$ which expresses the temperature dependence of the mean adsorption time τ (Chapter 8), and Q is given by the analogous equation $\beta = \beta_0 \exp(Q/RT)$ for the temperature dependence of the relaxation time** β. This relation (6.1–3) reproduces the measurements very well in many cases as illustrated (Fig. 6.1.b) by the experimental data of ROWLEY and EVANS [618], and HUNSMANN [330]. The relation appears to apply equally well to clean and gas-covered surfaces. AMDUR and GUILDNER [15] have recently reported measurements of the ac-

* In the following we shall abbreviate the accommodation coefficient as AC.

** The relaxation time β is the time required for the temperature difference between adsorbed gas molecules and wall to fall to one eth of its initial value.

commodation coefficients for systems consisting of noble gases, hydrogen, oxygen, and deuterium on gas-covered wolfram, nickel, or platinum surfaces at temperatures in the range from 9° to 100° C. They represented their results by the different, and indeed linear, relation

$$\alpha_t = \alpha_{25°}\,[1 + \delta\,(t - 25°)], \tag{6.1–4}$$

where the temperature coefficient δ for the systems mentioned above lies in the range between $+2.2 \times 10^{-4}$ and -9.5×10^{-4} degree^{-1}. This equation reproduces their experimental results within approximately $\pm 1\%$. If one plots, however, the AC-values of several other authors (listed in Tables 6.3.1.1.A, 6.3.1.1.B and 6.3.2.1.A) against temperature, the temperature dependence of the AC in most cases is found to be non-linear. It is to be noted in this connection, however, that unfortunately many experimenters have not borne in mind that the AC depends not only on the actual *temperatures* of the wire and the gas, but also on the magnitude of the selected temperature *difference* between the wire and the gas*. In his investigations, Amdur chose $\Delta T \approx 2°$, which is more in line with the smallness of ΔT required by the definition (6.1–2) than are the differences up to several hundred degrees used by other authors.

Some of the theories of the AC, which as will be seen later (Section 6.4) are mostly based on crude models and are limited in their application to the interaction of noble gas atoms on metal surfaces, give a much more complicated temperature dependence of the AC than do the above empirical formulas. For example, Jackson and Mott [*352*]** suggest the formula

$$\alpha = \frac{m}{M}\,\frac{\pi}{\nu_m{}^3}\int_0^{\nu_m}\frac{(h\nu)^2\nu^2 - H_1^{(1)}\,(i\,h\nu/2\,kT)}{(kT)^2\sinh\,(h\nu/2\,kT)}\,d\nu\,, \tag{6.1–5}$$

where m and M are the masses of the gas atom and surface atom, respectively, ν is one of the vibration frequencies of the atoms in this solid surface, and $\nu_m = k\Theta/h$ is the maximum vibration frequency at the Debye temperature. The Hankel functions $H_1^{(1)}\,(ix)$, with $x = h\nu/2\,kT$, are tabulated in such references as Jahnke and Emde [*356*].

On the other hand, Brown [*98*] used the formula of Devonshire [*144*] who represented the potential V on the surface by the function given by Morse, namely,

$$V = D\left[\exp\frac{-2\,(z - Z - z_0)}{\varrho} - 2\exp\frac{-(z - Z - z_0)}{\varrho}\right], \tag{6.1–6}$$

* Thus, in many cases, the gas temperature has been kept constant and only the wire temperature has been raised.

** Jackson and Mott make the crude simplification that the atoms are hard spheres so that the interaction potential between the adatom and the surface is $V = 0$ for $r > r_0$ and $V = \infty$ for $r < r_0$, where r_0 is the distance of nearest approach for the atoms.

where Z is the amplitude of the lattice vibrations, z is the coordinate of the gas atom in the direction normal to the surface, z_0 is the equilibrium position of the surface atom, and ϱ is a constant of the order of 10^{-8} cm. He thus included both attractive and repulsive forces. If V is expanded in powers of the amplitude Z and only the constant and the linear term are retained, the result is

$$V = D\left[\exp\frac{-2(z-z_0)}{\varrho} - 2\exp\frac{-(z-z_0)}{\varrho}\right] + \frac{2DZ}{\varrho}\left[\exp\frac{-2(z-z_0)}{\varrho} - \exp\frac{-(z-z_0)}{\varrho}\right]. \qquad (6.1\text{–}7)$$

Since the amplitude Z appears only to the first power, the probability that the lattice emits a quantum is given by a dipole formula. The zeroth-order wave functions are simply those that correspond to a diatomic molecule in the process of dissociation. If a DEBYE distribution of the eigenvibrations of the crystal is assumed, then a complicated computation of the transition probability shows that the AC is given by the formula

$$\alpha = \frac{24\pi^4 m^2}{\varrho^2 h^2 \nu_m{}^3 M}\int_0^{\nu_m}\left(\frac{h\nu}{kT}\right)^3 \frac{\nu^2 d\nu}{\exp(h\nu/kT)-1} \times \int_0^{\infty}\left[\frac{\sinh 2\pi\mu\,\sinh 2\pi\mu'(A_\mu + A_{\mu'})^2}{(\cosh 2\pi\mu - \cosh 2\pi\mu')^2 A_\mu A_{\mu'}}\right]\exp\frac{-E}{kT}\,dE\,, \qquad (6.1\text{–}8)$$

where

$$A_\mu = |\Gamma(-d+\tfrac{1}{2}+i\mu)|^2 \quad \text{and} \quad A_{\mu'} = |\Gamma(-d+\tfrac{1}{2}+i\mu')|^2, \qquad (6.1\text{–}9)$$

and

$$d = \frac{2\pi(2mD)^{1/2}}{\varrho h}, \quad \mu = \frac{2\pi(2mE)^{1/2}}{\varrho h}, \quad \text{and} \quad \mu' = \frac{2\pi(2m[E+h\nu])^{1/2}}{\varrho h}. \qquad (6.1\text{–}10)$$

Here the parameter D is given by the minimum of the potential energy curve $V(z)$ at $z = z_0$ and is approximately equal to the heat of adsorption for a low degree of gas coverage on the surface. If the zero-point vibrational energy of an adsorbed gas atom is subtracted from D, the result is the heat of adsorption at absolute zero. The total energy of a gas atom is E, ϱ is a constant, and the other symbols are defined as in formula (6.1–6). The gamma function $\Gamma(\chi)$ with $\chi = -d + \frac{1}{2} + i\mu$ can also be obtained from the function tables of JAHNKE and EMDE.

The quantities D and ϱ were adjusted to the experimental data. The degree to which the experimental results are fitted by formula (6.1–8) is shown in Fig. 6.1.c. Curves (1) and (2) in β and γ show the theoretical values calculated by BROWN for two choices of D and ϱ, and the values calculated by JACKSON and MOTT are shown for comparison as curve (1) of part α. However, the experimental points in α are much better represented by curve (2), obtained by dividing the calculated values of JACKSON and MOTT by three. The final explanation of the temperature dependence of the AC must, however, await the development of a quantitative theory based on more realistic assumptions concerning the mechanism of the interaction between adatom and surface.

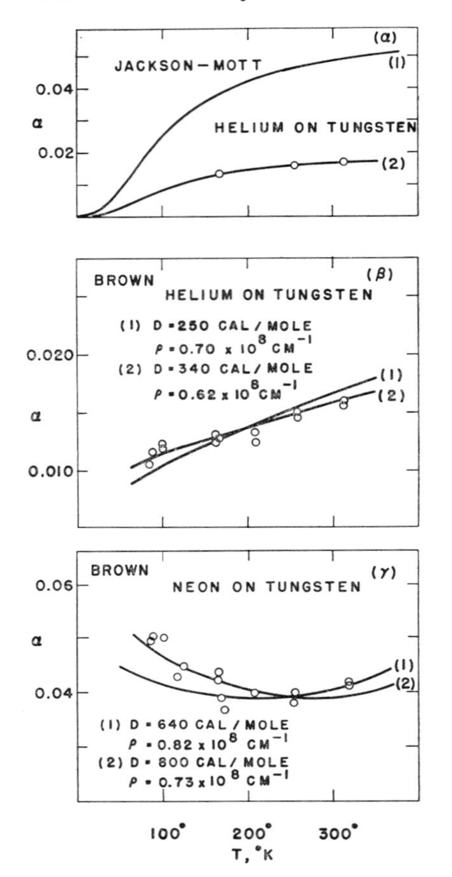

Fig. 6.1.c. Comparison of experimental and theoretical values of the temperature dependence of the accommodation coefficient α. In graph (α), curve (1) represents the values calculated by JACKSON and MOTT for helium on tungsten. Curve (2) of the same diagram has been obtained by dividing the calculated values of JACKSON and MOTT by 3. The circles on curve 2 represent experimental points. The curves *1* and 2 in diagrams β and γ have been calculated by BROWN [*98*] for the accommodation of helium and neon, respectively, on tungsten. These curves were obtained by substituting two sets of parameters D and ϱ into Eq. (6.1–8). The circle represent experimental values obtained by SILVERNAIL [*654a*]

6.2. Methods for Measuring Accommodation Coefficients

6.2.1. Measurements in the Low-Pressure Range

6.2.1.1. Molecular-Beam Method

McFEE and MARCUS [*466*] recently used the following method in their determinations of the thermal accommodation coefficients of potassium on various surfaces, such as mechanically polished copper, gold, quartz, and wolfram, and cleavage surfaces of magnesium oxide and lithium fluoride. In their experimental arrangement (shown schematically in Fig. 6.2.1.1.a) these authors introduced a beam of potassium atoms from an oven whose temperature varied between 550° K and 750° K into their experimental apparatus and allowed it to strike the surface under investigation at an angle of 45°. A vacuum of $\approx 10^{-6}$ mm Hg was maintained. The incident beam had approximately a Maxwellian velocity distribution, and the authors made no attempt to monochromatize it. They measured the velocity distribution of the atomic beam which was reflected* or desorbed from this surface at an angle of 45°. Since, as implied in the definition of the AC, the reflected beam has an effective temperature lying between that of the incoming beam and that of the surface, MARCUS and MCFEE in first approximation assumed that the velocity distribution of the reflected beam can be represented by a linear superposition of two Maxwellian distributions, one of which corresponds to the temperature of the incoming beam and the other to that of the reflecting surface. The actual result of incomplete accommodation, however, would presumably be a reflected beam with a non-Maxwellian velocity distribution which approaches a Maxwellian distribution as the temperature differences between the incident gas atoms and the reflecting surface approaches zero.

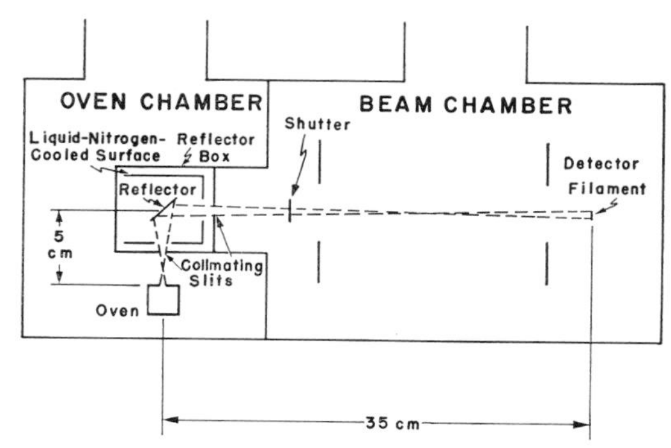

Fig. 6.2.1.1. a. Schematic diagram (top view) of beam apparatus used by MCFEE and MARCUS [466] for measurements of the reflected beam

Although the experimental results obtained with this method will be considered in more detail in section 6.3, let us as an example discuss the results of the measurements with a potassium beam striking a LiF surface. The results of these investigations, which were carried out for reflector temperatures up to 915° K, show an interesting deviation of the reflected beam from a Maxwellian velocity distribution as shown in Fig. 6.2.1.1.b. To make sure that this non-Maxwellian distribution is actually due to incomplete accommodation (and not, for example, to "local heating" of the crystal surface) the temperature of the incoming beam was varied (by use of a double-oven system). If the accommodation is incomplete, this temperature change should alter the velocity distribution of the reflected beam also. For a constant surface temperature, MCFEE and MARCUS observed that increasing the temperature of the

* In the following, the word "reflection" will be used when the atoms or molecules leave the surface before reaching the surface temperature. The "desorbed particles" will be those that were adsorbed for a long enough time to take on the surface temperature.

incident beam shifted the distribution in the reflected beam toward higher velocities (Fig. 6.2.1.1.b). Negative results in their control measurements on quartz reflector surfaces eliminated the possibility of local heating of the crystal surface by the incident beam; the velocity distribution was not shifted by continued exposure to the beam.

The authors assigned "temperatures" to the non-Maxwellian velocity distribution of the reflected beam in such a way that the computed MAXWELL distributions agreed with the measured distribution on the high-velocity side. Then for various temperatures of the incident beam,

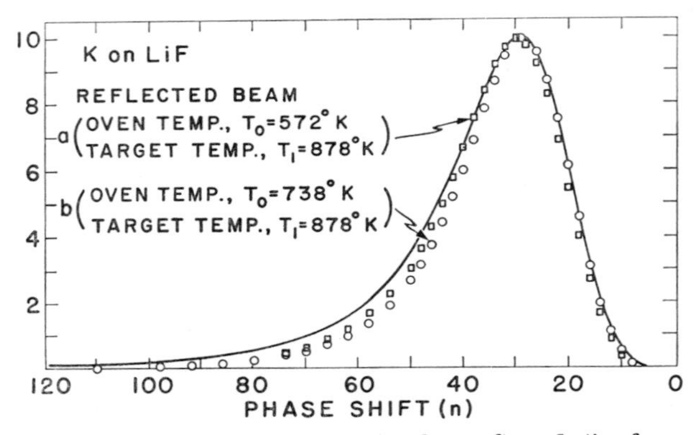

Fig. 6.2.1.1.b. The velocity distribution of a potassium beam after reflection from a LiF crystal surface. The two curves refer to two different temperatures T_0 (572° K and 738° K) of the gas impinging on the surface while the crystal surface was held at a constant temperature T_1 = 878° K (McFEE and MARCUS [466])

they experimentally determined the corresponding shifts of the velocity distribution of the reflected beam. The value of the thermal accommodation coefficient computed from this was approximately 0.7 ± 0.1 in the region of reflector temperatures around 900° K. Despite the scatter of the AC values, Fig. 6.2.1.1.b shows that the accommodation is higher at lower reflector temperatures. This is indicated by the fact that the velocity distribution of the reflected beam becomes more nearly Maxwellian at lower reflector temperatures. Unfortunately, the arrangement used by the authors did not permit measurement of the angular distribution of the reflected K atoms, although such experiments (especially if specular reflection occurs) would have better established the assumption of incomplete accommodation.

The authors point out that the shape of the non-Maxwellian velocity distribution of the reflected beam cannot be obtained by a simple linear superposition of two Maxwellian distributions corresponding respectively to the temperatures of the incident beam and the reflector. The experimentally observed non-Maxwellian velocity distribution shows a smaller half-width than would be obtained from such a linear superposition.

6.2.1.2. Heat-Conductivity Method

The AC has also been determined by measurements of the thermal conductivity in gases at such low pressures that the gas molecules go directly from a warm wall (a hot wire) to a cold wall without collision in the intervening interval. In this case, heat is transferred by the gas-wire and gas-wall accommodation processes and not by the collision of the gas atoms with one another. This method gives results characteristic of gas-covered surfaces.

As a rule the measurements are made by placing a small metal wire at a temperature T_1 on the axis of a cylindrical glass vessel at a temperature $T_0 < T_1$. The gas pressure is so low that the mean free path of the gas atom is large compared with the diameter of the vessel. The accommodation coefficient is determined from the heat loss of the wire. For a monatomic gas, for example, this is done in the following way.

If n gas atoms of mass m and temperature T_0 strike unit surface area of a wire in unit time, then each gas atom transfers energy $2kT_0$ per collision* and the rate of energy transfer to the wire in the stationary state is

$$Q_0 = 2nkT_0 = \frac{P}{\sqrt{2\pi mkT_0}} \cdot 2kT_0 \,. \qquad (6.2.1.2\text{–}1)$$

The gas atoms that leave the wire after only partial accommodations (i.e., those that do not come to thermal equilibrium with the wire at temperature T_1) have a mean energy corresponding to a "temperature" T_2, where $T_0 < T_2 < T_1$. If it is assumed that the number of molecules leaving the wire is equal to the number arriving at unit area per second and that the molecules have an undistorted Maxwellian velocity distribution, as is approximately the case for small values of $T_1 - T_0$ and for a monatomic gas, then the energy Q_2 which the atoms carry away from the wire per unit area per second is

$$Q_2 = \frac{P}{\sqrt{2\pi mkT_0}} \cdot 2kT_2 \,. \qquad (6.2.1.2\text{–}2)$$

Thus since T_1 is only slightly larger than T_0, the net energy loss from the wire per unit area per second is

$$\begin{aligned} Q = Q_2 - Q_0 &= \frac{P}{\sqrt{2\pi mkT_0}} \cdot 2k(T_2 - T_0) \\ &= \frac{P}{\sqrt{2\pi mkT_0}} \cdot 2k\alpha(T_1 - T_0) \,. \qquad (6.2.1.2\text{–}3) \end{aligned}$$

---

* Since relatively more fast gas atoms impinge on the wire, the average energy transferred by n atoms is larger than $\frac{3}{2}nkT_0$ by the amount $\frac{1}{2}nkT_0$ (as shown, for example, by MASSEY and BURHOP [*461*], p. 612).

Except for the accommodation coefficient α, all the quantities in this expression can be measured*. The rate of heat loss Q can be determined from the electrical energy supplied to the wire and the dimensions of the wire. Heat losses due to radiation and heat conduction must be taken into account (as discussed, for example, by AMDUR, JONES and PEARLMAN [*13*] or THOMAS and OLMER [*714, 715*]). Of far greater importance is the condition of the surface with respect to its physical, chemical, and induced heterogeneity.

Thus, for example, for the case of a rough metal surface which a gas atom strikes n times, ROBERTS [*592*] has shown that the measured accommodation coefficient α_{exp} is related to the AC of an ideal smooth surface by the expression

$$\alpha_{\text{exp}} = 1 - (1 - \alpha)^n . \qquad (6.2.1.2\text{–}4)$$

ROBERTS [*592, 593, 594*] also pointed out that previous measurements of the AC had almost always been made on surfaces covered with layers of adsorbed material, often of unknown composition.

ROBERTS therefore degassed the measuring cell and also the metal surface (for example, wolfram wire which could be heated to 2000° K—the "flash-filament" method). Moreover, a highly purified noble gas was used so that neither the impinging atoms of the gas nor foreign atoms present in trace amounts would form chemisorbed layers on the metal surface during the course of the experiment at the temperatures used.

Table 6.2.1.2.A. *Accommodation coefficient for noble gases on gascovered tungsten surfaces*; WACHMAN [*747*]

	Stepwise	Parallel	Evacuation
He on hydrogen-covered W	0.0420	0.0402	0.0401
Ne on hydrogen-covered W	0.113	0.116	0.108
He on deuterium-covered W	0.0453	0.0450	0.0463
Ne on deuterium-covered W	0.120	0.118	0.112

The "flash-filament" method is, however, applicable only for certain metals, since the surfaces of most metals are covered with oxide which cannot be degassed below the melting point (as discussed in more detail in Chapter 3). EGGLETON and TOMPKINS [*170*] therefore attempted to reduce such metal oxides in a hydrogen atmosphere. However, their experiment done with an iron wire showed that a hydrogen film was formed on the surface.

The method of determining the thermal accommodation by measurement of the thermal conductivity at low pressures was also used by

* For a more detailed description of the measuring technique, see for example FREDLUNG [*226*] and SCHÄFER [*631*].

WACHMAN [*747*], who investigated the thermal accommodation of noble gas atoms on surfaces covered with different gases. He determined the AC's in three independent ways: the "stepwise" method, the "parallel" method, and the "evacuation" method.

In the stepwise method, the noble gas under investigation was first introduced into the measuring tube to a pressure of from 2×10^{-2} up to 8×10^{-2} mm Hg, and then the wire was heated at 2000° C for 15—20 min. A 15-min delay allowed the wire to cool to the desired temperature for the experiment. The first AC measurement was then made with the surface of the wire prepared in this way. Immediately thereafter, the tube was filled to a pressure of 2×10^{-6}—2×10^{-5} mm Hg with the gas to be adsorbed on the surface. The thermal AC was then measured for this case in which the surface was covered mainly with a known gas. (Fig. 6.3.1.1.b in section 6.3.1.1 shows an example of measurements by the stepwise method.)

In the parallel method the wire is first baked out in vacuum. Then, instead of the gas whose accommodation coefficient is to be studied, the adsorption gas is admitted to a pressure of about 0.002–0.1 mm Hg (so that, when the impinging gas is later introduced, the two gases would contribute about equally to the heat transport). The AC of the adsorption gas with the metal surface is then determined. Finally the impinging gas is introduced and the AC of the gas mixture is measured. In determining the AC of the incoming gas, it is assumed that the two gases participate independently of one another (i.e., "in parallel") in the heat transfer.

The evacuation method is ordinarily used after applying the stepwise or parallel method. Here the gas mixture that is present in the experimental tube is pumped out and the heating current through the wire is turned off. Then the impinging gas is let in and the AC determined. In the case of stable adsorption layers on the surface, this serves as a check on the stepwise method or the parallel method.

6.2.1.3. Recoil Methods

Two other methods that are applicable in the low-pressure region deserve mention. WEHNER [*768*] recently attempted to draw some inferences concerning the accommodation of mercury ions on various metals (Ti, V, Cr, Mn, Zr, Cb, and Mo). In his experiments, the metals to be studied were used as the negative electrode in a mercury discharge. From the measured force exerted on this electrode by the mercury-ion bombardment, he then sought to infer the accommodation coefficient of mercury ions on these metal surfaces. Unfortunately, the resultant force on the electrode also includes contributions from other effects such as sputtering, radiometer forces, and negative-ion emission. Because of these effects, it is difficult to make precise statements about the AC. This is expecially true, as WEHNER shows, for oxide-covered surfaces. If the results for these are interpreted solely in terms of accommodation effects, the accommodation on oxide-coated surfaces is found to be less than on clean surfaces. But this contradicts numerous other experimental results (see Table 6.3.1.1.C).

Another method of determining the AC was used by KNUDSEN [*395, 396*]. He mounted radiometer vanes on the arms of a torsion balance and measured the force exerted by molecules of the residual gas as they rebounded from the blackened sides of the vanes (mean free path large compared with the dimensions of the vessel). This force, which intrinsically depends only on the translational energy of the impinging molecules, is equal to the difference between the recoils from molecules departing from the two sides, and is thus proportional to the difference between the accommodation coefficients on the blackened side and the bright side. In this way and also by the thermal-conduction method, KNUDSEN determined the accommodation coefficients for helium on platinum surfaces.

6.2.2. Measurements at High Pressures: the Temperature Jump

Attempts have also been made to study accommodation coefficients at pressures at which the mean free path is smaller than the dimensions of the apparatus. In this case, a discontinuity in the temperature near the gas-surface interface which results from the "temperature jump" in the neighborhood of the wall is used to measure the AC. The occurrence of this temperature jump can be explained briefly as follows.

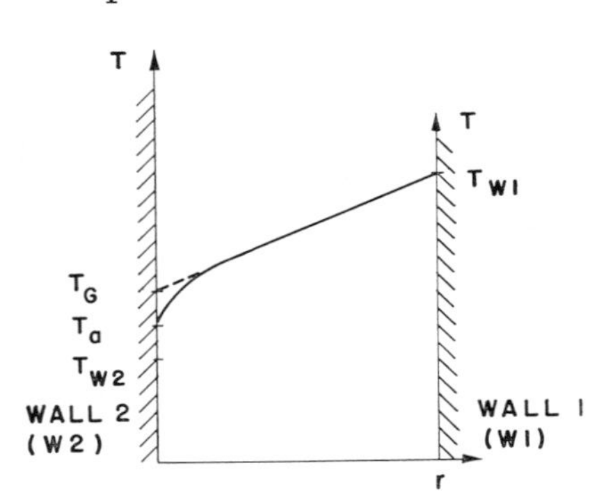

Fig. 6.2.2.a. Heat conduction by a gas between two parallel walls W_1 and W_2 separated by a distance r and which are held at temperatures T_{W_1} and T_{W_2}, respectively. T_G is the temperature the gas would have at the interface if the temperature gradient along the normal to the wall surface, $\partial T/\partial r$, continued without change up to the wall itself. T_a is the actual temperature of the gas at the interface

Consider a gas conducting heat between two parallel walls W_1 and W_2, that are separated from one another by a distance r and have different temperatures T_{W1} and T_{W2}. Assume that the mean free path is much smaller than the separation of the walls. The law of heat conduction would seem to require the temperature in the gas to increase linearly with distance from the colder plate. However, the observed temperature gradient deviates from linearity in the neighborhood of the wall (Fig. 6.2.2.a). The reason for this is the following. First consider the collisions of molecules with the colder wall and assume that initially the desorbed molecules have come into thermodynamic equilibrium with the surface and have an average energy corresponding to the surface temperature T_{W2}. Before collision with a wall these molecules have the temperature T_λ corresponding to a distance $\frac{2}{3}\lambda$ from the wall, the average distance from the wall to the position of the last collision in the gas. Since heat is transported to the colder wall, the molecules must carry away less energy from the wall than they bring there, that is, $T_{W2} < T_\lambda$. In a stationary distribution, the thermal flux $\frac{1}{4}\, n\bar{v}mc_v \times (T_\lambda - T_{W2})$ brought to the walls must be equal to $g\, \partial T/\partial r$, where r is the distance normal to the surface and g is the thermal conductivity, i.e.,

$$\tfrac{1}{4}\, n\bar{v}mc_v(T_\lambda - T_{W2}) = g\frac{\partial T}{\partial r}. \qquad (6.2.2\text{–}1)$$

It has become customary to introduce an auxiliary mathematical quantity, the so-called "temperature jump", in place of the steep temperature gradient in the neighborhood of the surface, which appears explicitly in Eq. (6.2.2–1). In other words, it is assumed that the conduction of heat from one wall to the other follows the heat-conduction law with the same value of g, and that at the surface itself there is a discontinuous change in temperature. The "temperature jump" or temperature discontinuity at the wall is defined as the difference between the actual surface temperature T_{W1} or T_{W2} and the temperature T_G obtained by applying the heat-conduction law to the interior of the gas and extrapolating to the wall surface. Thus the temperature jump is defined by the equation

$$T_G - T_{W2} = C_r \frac{\partial T}{\partial r}. \qquad (6.2.2\text{–}2)$$

The temperature jump coefficient C_r is proportional to the mean free path and also depends on the accommodation. If the AC is very small, the jump can take on values which correspond to more than 20 times the mean free path λ. If the ratio of the thermal conductivity at the pressure under investigation to that at very high pressure (where the thermal conductivity is independent of pressure in first approximation) is plotted against the reciprocal pressure $1/p$, the graph is a straight line which can be calibrated and used to determine the AC. This method has been used by Soddy and Berry [*665, 666*], Archer [*20, 21*], Hercus and Sutherland [*306*], Weber [*756*], Taylor and Johnston [*706*], Thomas and Golicke [*717*], Ubisch [*731, 732*], and Welander [*777*]. However, as already pointed out by Eucken [*197*] and Schäfer [*631*], this method has many disadvantages. First of all, operation in the high-pressure region makes it practically impossible to know the degree and the type of gas coverage on the metal surface. Moreover, the formation of double molecules at high pressure leads to contributions to the pressure dependence of the heat conduction; and these must be taken into account in evaluating the results. The point is that the double molecules dissociate more readily at the heated wire and again recombine on their way to the outer wall, so that they take more heat away from the wire than they do in the absence of this effect, and thus give a higher thermal conductivity. (For further details see GAZULLA and PEREZ [*230*], and SCHÄFER and GAZULLA [*629*].)

EUCKEN has also pointed out that in evaluating the final results the formulas are only approximate since the mean free path that appears here is simply an auxiliary quantity in the kinetic theory of gases and must be supplemented by the exact distribution function.

6.3. Experimental Results on Accommodation Coefficients

6.3.1. Monatomic Gases

6.3.1.1. Experimental Data

Data for the accommodation of monatomic gases on metal surfaces have been collected almost exclusively for noble gas atoms. From Table 6.3.1.1.A one finds that, except for helium, accommodation coefficients decrease with increasing temperature. According to the equation

$$\log\left(\frac{1}{\alpha} - 1\right) = \frac{Q - E_a}{RT} + \text{const.}\,, \qquad (6.3.1.1\text{–}1)$$

this means that the activation energy E_a is greater than Q (cf. section 6.1). It should be noted that the AC's of light noble gases have very small values, even though the accommodation is almost 100% for the translational energy (the only form of energy available to these atoms). This is partially explained by the fact that a certain percentage of the atoms are "elastically" reflected and thus do not participate in the exchange of energy.

From Tables 6.3.1.1.B and 6.3.1.1.C one also sees that in many cases the AC is increased if the surface is already covered by an adsorbed gas layer. The explanation probably lies in a "pseudo adsorption" mechanism in the collisions between the gas atoms and the wall. Instead of being reflected elastically from an "infinitely heavy" particle in the metal wall, the (usually) "light" gas molecules rebound from similar molecules adsorbed on the surface. Since in the later case the colliding particles have the same (or comparable) mass, the energy transfer is large—as expected from the classical laws of impact.

The values of the accommodation coefficient often increase during an experimental run, apparently in proportion to the formation of foreign atom layers on the surface. This is illustrated in Fig. 6.3.1.1.a (EGGLETON and TOMPKINS [*170*]), which shows the increase in the accommodation coefficient $\alpha(\theta)$ of Ne on an iron surface as the adsorption of oxygen increases.

The extent to which the rise in the AC depends on the kind of foreign layer covering the surface and on the degree of covering θ has not been investigated in most of the previous work. Only recently have more systematic studies of the dependence of the AC on the nature of the covering layer been undertaken by WACHMANN [*747*] and by EGGLETON and TOMPKINS [*170*]. WACHMANN studied the thermal AC of helium and neon on wolfram surfaces which were first freed of adsorption layers by the "flash filament method" and then covered with adsorption layers of known gases.

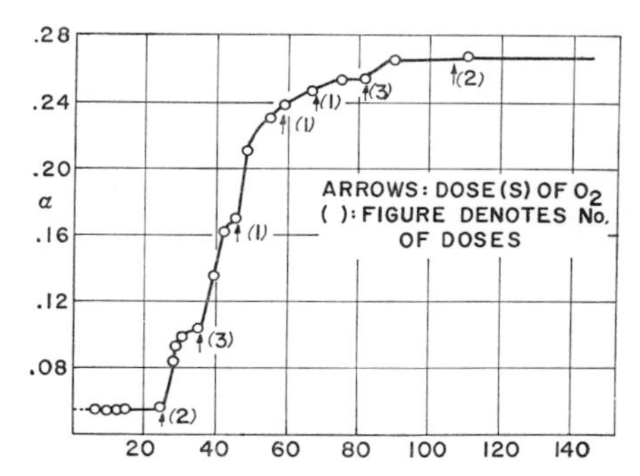

Fig. 6.3.1.1.a. Changes of the accommodation coefficient α of neon on an iron surface on admission of oxygen in small doses (the actual amount of gas in a dose not reported) (EGGLETON and TOMPKINS [*170*])

Table 6.3.1.1.B gives an idea of the dependence of the AC on the kind of foreign atom layers covering tungsten surfaces. In the case of the AC of He on oxygen-covered tungsten surfaces, an attempt was made to attain various degrees of covering with oxygen: (a) an atomic layer, (b) a composite film (first oxygen film), and (c) an upper molecular film (second oxygen film), as ROBERTS and MORRISON [*506*] had already sought to do. WACHMANN's results indicate three (or possibly four) different types of oxygen films on wolfram surfaces, which correspond to AC values (He on oxide-covered tungsten) of 0.185, 0.107, 0.065, and possibly 0.085. The AC values of 0.185 and 0.107 in this set for He seem to correspond to the AC values obtained for neon by ROBERTS and MORRISON, who ascribe them to the "upper molecular film" or the "composite oxygen film". Although the so-called "upper molecular film" remained stable up to an outgassing temperature of 1000° K in WACHMANN's experiment, for ROBERTS this "film" seemed to evaporate even at 355° K. The AC value of 0.085 for He seems to agree with the AC value for neon on oxide-covered tungsten which ROBERTS and MORRISON ascribe to the "atomic film". MORRISON and ROBERTS found that this film remained stable up to baking temperatures of approximately 1750° K. A similar behavior, was found for the oxygen film corresponding to the AC value of 0.065, but not for the film corresponding to the AC value of 0.085.

Table 6.3.1.1.A. *Accommodation coefficient α for monatomic gases on only slightly gas-covered surfaces*

1 Surface	2 Condition of surface	3 Surface temperature T_g (°K)	4 Incident gas	5 Gas temperature T_g (°K)	6 $\Delta T = T_s - T_g$ (°C)	7 Accommodation coefficient α	8 Author
W	Initially degassed but probably slightly gas covered during the run	79	He	(see ΔT)	10 to 30	0.025	ROBERTS [*594*]
W		103.1	He	83.1	20	0.015_1	THOMAS, SCHOFIELD [*718*]
W		195	He	(see ΔT)	10 to 30	0.046	ROBERTS [*594*]
W		156.1	He	138.1	18	0.016_2	THOMAS, SCHOFIELD [*718*]
W		211.1	He	193.1	18	0.016_2	THOMAS, SCHOFIELD [*718*]
W		259.1	He	243.1		0.017_2	THOMAS, SCHOFIELD [*718*]
W		295	He	(see ΔT)	10 to 30	0.057	ROBERTS [*594*]
W		Not specified	He	Room temp.	Not specified	0.069	MICHELS, WHYTE [*475*]
W		303	He	Not specified		0.054	SPIVAK, ZACHARJIN [*792*]
W		320	He	303	17	0.016_7	THOMAS, SCHOFIELD [*718*]
W		320	He	83.1—303.1		0.016_9	THOMAS, SCHOFIELD [*718*]
W		329.8	He	Not specified		0.020_0	WACHMAN [*747*]
W		1870	He	Room temp.		0.035_9	DEPOORTER [*143a, 143b*]

W	Initially degassed but probably slightly gas covered during the run	79	Ne	(see ΔT)	10 to 30	0.08	ROBERTS [*594*]
W		195	Ne	(see ΔT)	10 to 30	0.08	ROBERTS [*594*]
W		295	Ne	(see ΔT)	10 to 30	0.07	ROBERTS [*594*]
W		300	Ne	(see ΔT)	20	0.057	MORRISON, ROBERTS [*506*]
W		303	Ne	Not specified	Not specified	0.06	SPIVAK, ZAITZEV [*671*]
W		303	Ne	Not specified	Not specified	0.08	ROBERTS [*594*]
W		Room temp.	Ne	Not specified	Not specified	0.057	VAN CLEAVE [*115*]
W		329.8	Ne	Not specified	Not specified	0.055_1	WACHMAN [*747*]
W		1710	Ne	Near toom temp.		0.045	DEPOORTER [*143a*, *143b*]
W		1880	Ne	Near room temp.		0.054	DEPOORTER [*143a*, *143b*]
W		1986	Ne	Near room temp.		0.052	DEPOORTER [*143a*, *143b*]
W		2120	Ne	Near room temp.		0.047	DEPOORTER [*143a*, *143b*]
W		2350	Ne	Near room temp.		0.029	DEPOORTER [*143a*]
W		Room temp.	Ar		~10	0.82	MICHELS [*474*]
W		1860	A	Near room temp.		0.091	DEPOORTER [*143a*, *143b*]
W		2120	A	Near room temp.		0.085	DEPOORTER [*143a*, *143b*]

Table 6.3.1.1.A. (continued)

1 Surface	2 Condition of surface	3 Surface temperature T_s (° K)	4 Incident gas	5 Gas temperature T_g (° K)	6 $\Delta T = T_s - T_g$ (° C)	7 Accommodation coefficient α	8 Author
W	Initially degassed but probably slightly gas covered during the run	2240	A	Near room temp.		0.073	DEPOORTER [*143a, 143b*]
W		2335	A	Near room temp.		0.066	DEPOORTER [*143a, 143b*]
Pt		77.4	He	(see ΔT)	0	0.090	ROLF [*606*]
Pt		193.2	He	Not specified	Not specified	0.943	ROLF [*606*]
Pt		225	He	Not specified		0.038	BROWN [*98*]
Pt		273.2	He	(see ΔT)	0	0.071	ROLF [*606*]
Pt		~372.1	He	289.1	83	0.05	MANN, NEWELL [*456*]
Pt		373	He	Not specified	Not specified	0.072	ROLF [*606*]
Pt		312	A	305	7	0.869_5	THOMAS, BROWN [*716*]
Pt		405	A	305		0.648_6	THOMAS, BROWN [*716*]
Pt		303.1	Hg	305	α extrapolated to $\Delta T = 0$	1.0	THOMAS, OLMER [*714*]
Pt		405	A	305	100	0.648_7	THOMAS, BROWN [*716*]
Pt		312	Kr	(see ΔT)	7	0.90_2	THOMAS, BROWN [*716*]

Pt	Initially degassed but probably slightly gas covered during the run	405	Kr	(see ΔT)	100	0.69_1	THOMAS, BROWN [716]
Ni		90	He	(see ΔT)	10	0.048	RAINES [580]
Ni		195	He	(see ΔT)	10	0.060	RAINES [580]
Ni		273	He	(see ΔT)	10	0.071	RAINES [580]
Ni		369	He	(see ΔT)	10	0.077	RAINES [580]
Fe		303.1—333.1	Ne	(see ΔT)	30	0.053	EGGLETON, TOMPKINS [170]
Be		335.2	He	305.2	30	0.145	BROWN [98]
Be		335.2	Ne	305.2	30	0.315	BROWN [98]
Al		418	He	(see ΔT)	$\approx$40 to 180	0.073	FAUST [207]
Al		418	Ne	(see ΔT)	40 to 180	0.159	FAUST [207]
Al		483	He	(see ΔT)	40 to 180	0.074	FAUST [207]
Al		483	Ne	(see ΔT)	40 to 180	0.163	FAUST [207]

Table 6.3.1.1.B. *Accommodation coefficient α for monatomic gases on surfaces covered with known gases*

1 Surface	2 Condition of surface	3 Surface temperature T_s (°K)	4 Incident gas	5 Gas temperature T_g (°K)	6 $\Delta T = T_s - T_g$ (°C)	7 Accommodation coefficient α	8 Author
W	H-covered	329.7	He			0.040_7	Wachman [*747*]
W	H-covered	329.7	Ne			0.112	Wachman [*747*]
W	H-covered	295	Ne			0.17	Roberts [*595*]
W	D-covered	329.7	He	304.7		0.045_8	Wachman [*747*]
W	O-covered	303.1	He			0.36	Spivak, Zacharjin [*792*]
W	O-covered	303.1	Ne			0.27	Roberts [*596*]
W	D-covered	329.7	Ne	304.7		0.117	Wachman [*747*]
W	O-covered	329.7	He	304.7		0.107	Wachman [*747*]
W	O-covered	329.7	Ne	304.7		0.26_4	Wachman [*747*]
W	O-covered	329.7	He	304.7		0.18_5	Wachman [*747*]
W	O-covered	329.7	Ne	304.7		0.40_6	Wachman [*747*]
W	O-covered	unknown	Ne			0.20	van Cleave [*115*]
W	O-covered	unknown	Ne			0.30	Spivak [*671*]
W	O-covered	300	Ne		20	0.22 to 0.36 (depending on coverage degree)	Morrison, Roberts [*506*]
W	N-covered	313.1	Ne			0.385	Spivak, Zacharjin [*792*]
W	N-covered	329.7	He	304.7		0.039_6	Wachman [*747*]
W	N-covered	329.7	Ne	304.7		0.117	Wachman [*747*]
W	CO_2-covered	329.7	He	31.7	25 to 28	0.040_4	Wachman [*747*]
W	CH_4-covered	329.7	He	31.7	25 to 28	0.03_4	Wachman [*747*]
W	C_2H_6-covered	329.7	He	31.7	25 to 28	0.025_8	Wachman [*747*]
W	C_2H_4-covered		He	31.7	25 to 28	0.098_6	Wachman [*747*]
Pt	H_2-covered	373.1	He		89	0.245	Mann, Newell [*456*]
Pt	D_2-covered	373.1	He			0.242	Mann, Newell [*456*]
Fe	H_2-covered	303.1—333.1	Ne			0.099	Eggleton, Tompkins [*170*]
Fe	O_2-covered	303.1—333.1	Ne			0.268	Eggleton, Tompkins [*170*]
Fe	N_2-covered	303.1—333.1	Ne			0.438	Eggleton, Tompkins [*170*]

The variation of the helium AC for wolfram surfaces is shown in Fig. 6.3.1.1.b. It varies from a value of 0.019 for initially degassed surfaces up to a value of about 0.042 for the hydrogen covering obtained after approximately 1.8×10^{-2} cm²-ML* of hydrogen was admitted at each of the times indicated by a_2, b_2, c_2, d_2 in the figure. In a study of the AC values for Ne and He on originally degassed (but probably slightly gas covered during the run) and strongly gas-covered wolfram surfaces, WACHMANN found only slight differences in the ratio of the AC of Ne to

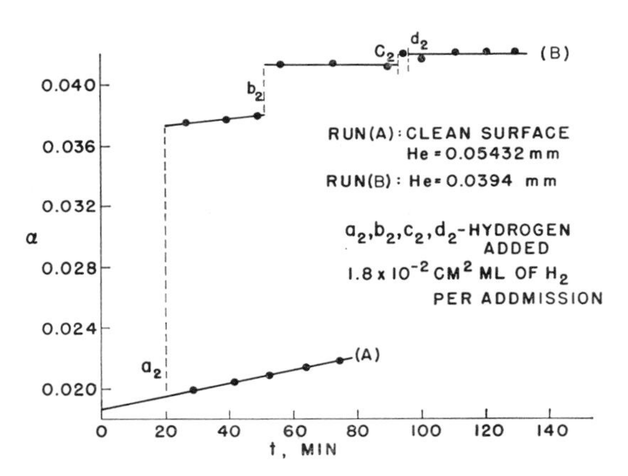

Fig. 6.3.1.1.b. Variation of the accommodation coefficient α of helium with time. Curve A is for a clean wolfram surface. Curve B is for the case in which hydrogen is admitted in doses of 1.8×10^{-2} cm ML, where ML denotes monolayers. (After WACHMAN [747]

that of He between the degassed wolfram surface and the surfaces covered with various gases. The values lie between 2.14 (wolfram surface covered with C_2H_4) and 2.76 (degassed, but slightly gas-covered wolfram surfaces). It should be mentioned that BROWN [98] also showed that the ratio of the AC of Ne to that of He varied only within quite narrow limits (from about 2.16 to 2.5) for such different degassed surfaces as Be, Al, and W. This contradicts the theoretical expectations of BAULE [55] and MICHELS [476] as will be discussed in more detail in the next section.

EGGLETON and TOMPKINS studied the time dependence of the AC for neon on an iron surface which initially was clean but then became gas-covered as a result of the admission of accurately measured amounts of highly purified gases such as H_2, O_2, or N_2.

Fig. 6.3.1.1.c represents the observed adsorption of H_2 on Fe surfaces on which the added quantity of H_2 was initially the low value corresponding to 3×10^{-4} mm Hg. From the fact that the AC reached a saturation

* The number of gas molecules introduced is divided by the number of molecules necessary to produce a monomolecular adsorption layer on a cm² of surface (cm²-ML).

Table 6.3.1.1.C. *Accommodation coefficient α for monatomic gases on gas-covered surfaces*

1 Surface	2 Condition of surface	3 Surface temperature T_s (° K)	4 Incident gas	5 Gas temperature T_g (° K)	6 $\Delta T = T_s - T_g$ (° C)	7 Accommodation coefficient α	8 Author
W	Surface covered with gases of unknown composition	373	He	293		0.454	von Ubisch [*731*]
W		295	He			0.19	Roberts [*593*]
W		298	He		≈ 2	0.470	Amdur, Guildner [*15*]
W		303	He			0.53	Michels [*474*]
W		298	Ne			0.868	Amdur, Guildner [*15*]
W		373	Ne	293		0.80	von Ubisch [*731*]
W		Room temp.	Ar		10	1.00	Michels [*474*]
W		298	A			0.967	Amdur, Guildner [*15*]
W		373	A	293		0.96	von Ubisch [*731*]
W		373	Kr	293		0.96	von Ubisch [*731*]
Pt		~ 77	He	≈ 77		0.43	Rolf [*606*]
Pt		~193	He	≈193		0.071	Rolf [*606*]
Pt		~273	He	≈273		0.170	Rolf [*606*]

Pt	Surface covered with gases of unknown composition	290	He			0.36	Chavken [113]
Pt	Surface covered with gases of unknown composition	293.1	He			0.338	Knudsen [394]
Pt	Surface covered with gases of unknown composition	293.1	He		20	0.05	Dickens [145]
Pt	Surface covered with gases of unknown composition	314.3	He	303.1	11.2	0.17_6 0.17_8	Thomas [717]
Pt	Surface covered with gases of unknown composition	324.5	He	202.1	21.4	0.17_3 0.17_6	Thomas [717]
Pt	Surface covered with gases of unknown composition	353.5	He	303.1	50.4	0.17_3 0.17_4	Thomas [717]
Pt	Surface covered with gases of unknown composition	~373	He			0.170	Rolf [606]
Pt	Surface covered with gases of unknown composition	373.1	He		84	0.35	Mann [455]
Pt	Surface covered with gases of unknown composition	373	He	293		0.402	Chavken [113]
Pt	Surface covered with gases of unknown composition	298	He			0.368	Amdur, Guildner [15]
Pt	Surface covered with gases of unknown composition	293.1	Ne			0.653	Knudsen [394]
Pt	Surface covered with gases of unknown composition	314.3	Ne	303.1	11.2	0.446 0.33	Thomas, Golicke [717]
Pt	Surface covered with gases of unknown composition	324.5	Ne	303.1	21.4	0.43 0.326	Thomas, Golicke [717]
Pt	Surface covered with gases of unknown composition	298	Ne			0.816	Amdur, Guldner [15]
Pt	Surface covered with gases of unknown composition	329.1	Ne			0.55	Thomas, Brown [716]
Pt	Surface covered with gases of unknown composition	353.5	Ne	303.1	50.4	0.422 0.321	Thomas, Golicke [717]

Table 6.3.1.1.C. (continued)

1 Surface	2 Condition of surface	3 Surface temperature T_s (°K)	4 Incident gas	5 Gas temperature T_g (°K)	6 $\Delta T = T_s - T_g$ (°C)	7 Accommodation coefficient α	8 Author
Pt	Surface covered with gases of unknown composition	373	Ne	293		0.78	VON UBISCH [731]
Pt		290	Ne			0.67	CHAVKEN [113]
Pt		290	Ar			0.81	CHAVKEN [113]
Pt		293.1	A		20	0.89	DICKENS [145]
Pt		298	A			0.925	AMDUR, GUILDNER [15]
Pt		373	A	293		1.00	VON UBISCH [731]
Pt		290	Kr			0.84	CHAVKEN [113]
Pt		373	Kr	293		0.97	VON UBISCH [731]
Pt		290	Xe			0.86	CHAVKEN [113]
Ni		90	He			0.413	RAINES [580]
Ni		195	He			0.423	RAINES [580]
Ni		273	He			0.360	RAINES [580]
Ni		298	He			0.385	AMDUR, GUILDNER [15]

Ni	Surface covered with gases of unknown composition	369	He			0.343	RAINES [580]
Ni		298	Ne			0.824	AMDUR, GUILDNER [15]
Ni		298	A			0.935	AMDUR, GUILDNER [15]
Fe		303.1—333.1	Ne			0.26	EGGLETON, TOMPKINS [170]

value and from the amount of H_2 necessary for this, EGGLETON and TOMPKINS concluded that a complete monatomic H_2 layer is built up on the iron surface.

EGGLETON, TOMPKINS, and WANFORD [*171*] have very carefully repeated the measurements of ROBERTS on the AC of Ne on wolfram, and have shown that BREMNER [*93, 94*] was not justified in his criticism of the very small values that ROBERTS found for the AC on degassed wolfram surfaces. Finally, we discuss in the following the results which MARCUS and MCFEE [*466*] have obtained from measurements of the velocity distribution of potassium beams reflected from copper, wolfram, and magnesium oxide surfaces covered with various gases.

Fig. 6.3.1.1.c. Time variation of the accommodation coefficient α for neon on an iron surface on admission of 3×10^{-4} mm Hg of hydrogen. (The hydrogen was admitted at the time indicated by the arrow.) (EGGLETON and TOMPKINS [*170*])

a) K Beam on Gas-Covered Copper Surfaces

The measurements were carried out with the reflector surface at various temperatures (500 to 700° K) accurately determined with a thermocouple.

The measured velocity distribution of the reflected beam was almost completely Maxwellian and corresponded to the measured surface temperature to within 2% (cf. Fig. 6.3.1.1.d). From this the authors concluded that the incoming potassium atom accommodates almost completely with the gas-covered surface, a result that is in complete agreement with other experimental results for gas-covered surfaces (cf. Table 6.3.1.1.C).

b) K Beam on Polycrystalline Gas-Covered Wolfram Surfaces

During this measurement the temperature of the incoming potassium beam (approximately 530° K) was held much below that of the reflecting wolfram surface (measurements up to 2135° K) so that even slight deviations from thermal accommodation would have resulted in considerable deviation from a Maxwellian velocity distribution in the reflected beam. As one sees from Fig. 6.3.1.1.d, the experimentally observed velocity distribution is entirely Maxwellian and its temperature corresponds to the temperature of the surface. Thus, even for this relatively high surface temperature, the thermal accommodation of the incoming K beam with the wolfram surface is still complete.

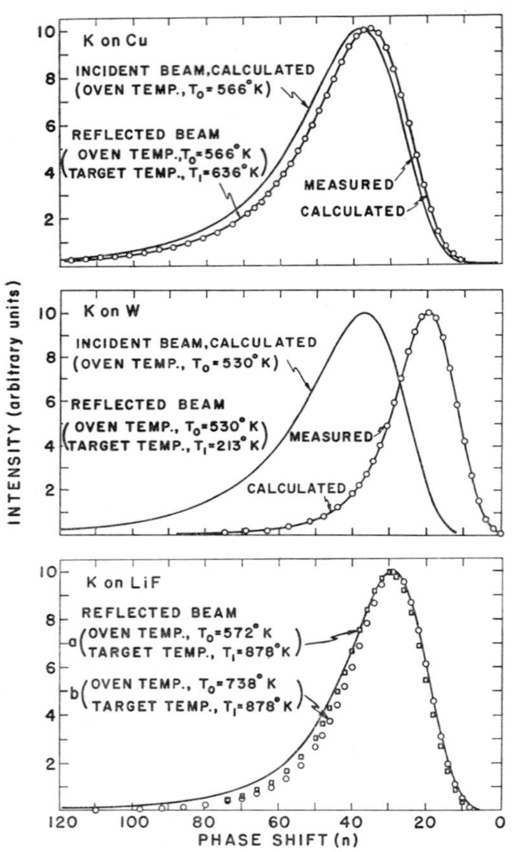

Fig. 6.3.1.1.d. Curves for the velocity distribution of a potassium beam after reflection from Cu, W, and LiF surfaces for the different indicated temperatures of the target surface and incident beam. The solid lines represent Maxwellian distributions calculated for a particular gas temperature and the circles and squares indicate measured points (McFee and Marcus [466])

c) K Beam on a Magnesium Oxide Crystal Surface

In these studies the surface temperature of the reflector was held at values up to 950° K where, because of the poor thermal conductivity, the surface temperature could no longer be measured precisely enough with a thermocouple. The temperatures found from the velocity distribution of the reflected beam were always considerably higher than the temperatures determined with the thermocouple. The authors found an almost perfectly Maxwellian distribution in the reflected beam and, despite the uncertainty in the determination of the surface temperature, regarded this as evidence that the thermal accommodation is almost complete. The results obtained by MARCUS and McFEE, namely that the accommodation of potassium atoms with the three different gas-covered surfaces is practically complete, is in agreement with the results of other authors who have used other methods of measurement.

From all of the results presented here one sees how sensitively the exchange of energy between a gas and a metal surface depends on the condition of the surface.

6.3.1.2. Theoretical Considerations

Various attempts have been made to compute the AC theoretically from the properties of the gas and the surface. BAULE [*55*] has based a derivation on purely classical considerations concerning the elastic collision of an incident gas atom of mass m and the surface atom of mass M. From the simple laws of impact, on the basic assumption that a single surface atom participates in only one collision, he expressed the AC by the relation

$$\alpha = 1 - \frac{m^2 + M^2}{(m + M)^2} \,. \qquad (6.3.1.2\text{–}1)$$

However, this relation represents the experimental data rather poorly (for example, for Ne on clean W, the AC is $\alpha_{\text{calc}} = 0.94$, while at 295° K the measured value is $\alpha_{\text{exp}} = 0.07$) and does not give the temperature dependence of the AC at all. For further criticism of this formula, see for example EGGLETON et al. [*170*] and SPIVAK et al. [*671, 672*].

MICHELS [*476*] developed a semiclassical theory on very simplified assumptions (particularly a one-dimensional model of the metal surface) which, like the considerations of COMPTON [*121*] for the AC of positive ions, led to the relation

$$\alpha = \frac{4\,m\,M}{(m + M)^2} \exp\left(-\,\theta'/T\right) , \qquad (6.3.1.2\text{–}2)$$

where θ' is the characteristic COMPTON temperature*. This formula also failed to properly represent the experimental results. For example, for Ne on W it predicts an AC that is 4—5 times the value for He on W, although the experimental ratio is only about 2.7. From Fig. 6.3.1.2.a one sees for example that the formula predicts considerably larger values of the ratio of the AC for Ne to that for He on the same surface (Be, Al,

* The specific heat of a solid is given by $c_v = 3Nk\,[(\theta'/T) + 1] \exp(-\,\theta'/T)$, where N and k are the AVOGADRO number and BOLTZMANN constant and θ' is the above-mentioned COMPTON temperature (for example, $\theta' \approx 148°$ K for wolfram COMPTON [*120*]).

or W) than are found experimentally. The relation given above varies only slightly with the mass of the metal atom.

LANDAU [*417*] started from the assumption that the incident atoms have such a large mass that, within particular temperature ranges, the energy change in the collision is small compared with the total energy of the atom. Then, from considerations based on the correspondence principle, he arrived at an expression for α. As corrected by DEVONSHIRE [*144*] this relation is

$$\alpha = \frac{3}{8\, m^{1/2} M} \left(\frac{2\, h^2\, T}{\pi\, a^2\, k\, \Theta^2} \right)^{3/2}. \quad (6.3.1.2\text{–}3)$$

In this equation Θ is the characteristic DEBYE temperature and a is the "characteristic length" in the assumed form $V = A \exp(-r/a)$ for the potential energy of the atom in the force field of the surface. The assumptions made, however, do not permit application of this equation to light atoms. (For a further discussion, see DEVONSHIRE [*144*].)

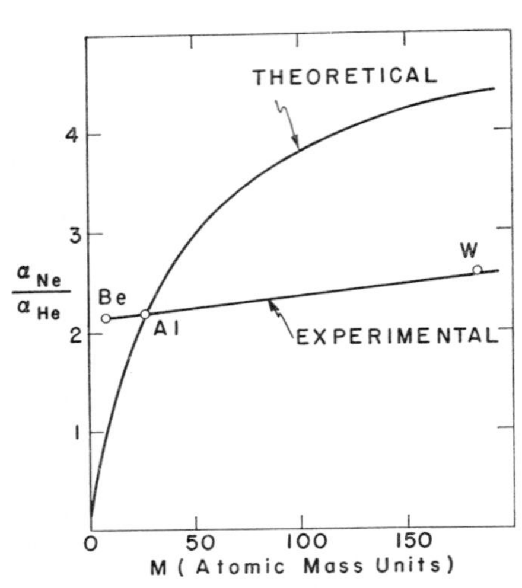

Fig. 6.3.1.2.a. The ratio of the accommodation coefficients of neon and helium on the same solid surface at the same temperature, plotted against the mass of the target atom expressed in atomic mass units. The theoretical curve was calculated on the basis of Eq. (6.3.1.2–2) and the experimental data were obtained for Be by BROWN [*98*], for Al by FAUST [*208*] and for W by SILVERNAIL [*654a*]

The first quantum mechanical calculations of the AC of monatomic gases were carried out by ZENER [*799*, *801*], and by JACKSON et al. [*351*, *352*]. They made the simplifying assumption that only motions perpendicular to the surface are to be considered and that the interaction potential between the gas atom and the metal surface is entirely repulsive. The potential chosen was

$$V = A \exp\left[\frac{-(Z-z)}{\varrho}\right], \quad (6.3.1.2\text{–}4)$$

where z is the coordinate of the gas atom in the direction perpendicular to the surface, Z is that of the atom in the solid and ϱ is a constant of the order of 10^{-8} cm. The transition probability, for the case in which the collision of a gas atom having energy E causes a surface atom in vibrational state n to make the transition to state n' so that the rebounding molecule receives the energy $E^* = E + (n - n')h$, was calculated by the method used in the corresponding problem of molecular vibrational motion resulting from atomic collisions in the gas phase (cf. MASSEY and BURHOP [*461*]). According to JACKSON [*352*], the most frequent transitions of this type are those in which only a single quantum is exchanged between the atom and some one of the crystal vibrations. BROWN [98] pointed out that the experimental values of SILVERNAIL [654a] for the AC of He on wolfram agree with the theoretical values of JACKSON and MOTT if these theoretical values are divided by 3 (Fig. 6.1.c).

As already discussed in more detail in section 6.1, DEVONSHIRE [144] made use of very simplified assumptions to derive Eq. (6.1–8) for the accommodation coefficient. The formula is so complicated that the relative contributions of the different terms are very difficult to discern. Several cases for various values of the interaction parameter have therefore been computed and are presented graphically in Fig. 6.3.1.2.b. The calculated and experimental values obtained by BROWN [98] for the function $\alpha(T)$ for He and Ne on W have been shown already in Fig. 6.1.c. One should keep in mind that DEVONSHIRE derived his expression for $\Delta T = T_{\text{gas}} - T_{\text{surface}} = 0$, whereas the experimental results were obtained for $\Delta T > 0$.

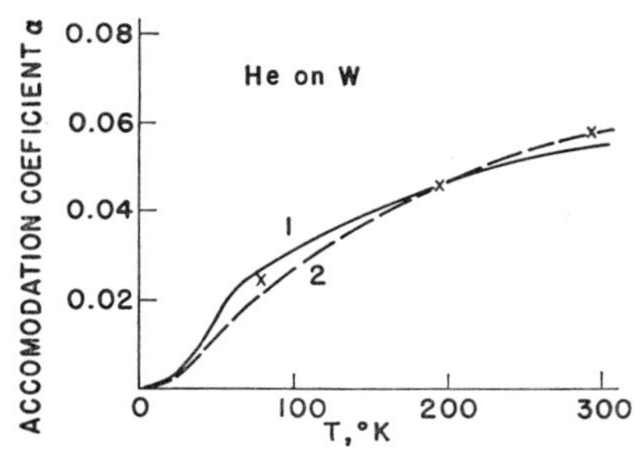

Fig. 6.3.1.2.b. Variation of the accommodation coefficient α with temperature for helium on wolfram. The crosses indicate the values measured by ROBERTS. The solid and the dotted curves represent the values calculated by DEVONSHIRE [144] with the following values of the interaction constants: Curve *1* (solid line): $D = 60$ cal/mole, $P = 1.16 \times 10^8$ cm^{-1}, roughness factor = 1.1; Curve *2* (dotted line): $D = 0$, $P = 2.00 \times 10^8$ cm^{-1}, roughness factor 1.06

STRACHAN [685] attempted to include a superposition of higher order multipoles in place of the dipole interaction (exchange of a single quantum) treated so far. He used the complete MORSE function without expanding it in powers of the vibration amplitude. However, his results show that, within the domain of applicability of the first-order perturbation theory used, the multiple-quantum transitions are extremely improbable.

Since, as shown in detail above, these quantum mechanically derived formulas for the AC are based upon extremely simplified assumptions*, their application is very limited. In conclusion it should be pointed out that, as seen from the formula (6.3.1.2–3) proposed by DEVONSHIRE, the AC of monatomic gases should be inversely proportional to the mass of the surface atom. This means that the adsorption of light atoms on the

* In this connection, an assumption made in most treatments should be pointed out. It is that the duration of a collision between the incident atom and the solid surface is of the same order of magnitude as for the collision of two atoms in the gas phase. However, as will be seen in Section 6.4, the adsorbed particles can have quite a long lifetime on the surface before they are desorbed.

surface increases the accommodation—in agreement with the experimental result (Tables 6.3.1.1.B and 6.3.1.1.C).

6.3.2. Diatomic Gases

6.3.2.1. Experimental Data

Tables 6.3.2.1.A and 6.3.2.1.B give the average thermal AC of diatomic gases for metal surfaces which probably are slightly covered with gas and certainly more heavily gas covered, respectively. Especially for the earlier work, one must assume that the values given by the authors for supposedly clean surfaces actually refer to surfaces with a slight gas covering. Table 6.3.2.1.B contains the average thermal AC values for metallic surfaces with a known kind of gas covering. By comparison of the two tables one finds that, as for monatomic gases, the AC values for surfaces with a stronger covering of foreign gas are higher than those for only a slight surface coverage.

6.3.2.2. Theoretical Considerations

In the case of diatomic molecules an exchange of rotational and vibrational energy, as well as the exchange of translational energy, can take place between the incoming molecule and the surface. The average thermal AC then is made up of the three partial AC's (α_{trans} for translation, α_{rot} for rotation, and α_{vib} for vibration) combined according to the relation

$$\bar{\alpha} = \frac{2\,R\,\alpha_{\text{trans}} + C_{\text{rot}}\,\alpha_{\text{rot}} + C_{\text{vib}}\,\alpha_{\text{vib}}}{C_{\text{trans}} + C_{\text{rot}} + C_{\text{vib}} + \frac{1}{2}\,R}\,, \qquad (6.3.2.2\text{–}1)$$

where R is the ideal gas constant, and C_{trans}, C_{rot}, and C_{vib} are the corresponding contributions to the molar specific heat. It is of interest to determine the partial AC's separately from experiment. Numerous measurements are available for molecules which, at the low temperature used, have only rotational energy in addition to translational energy.

Thus, for example, it is easy to carry out experiments on ortho- and para-hydrogen and to explain the resulting AC's on the basis of different degrees of accommodation of the rotational and translational energy, as was done by Bonhoeffer and Rowley [*617*] on degassed platinum. They found that the AC values at a given temperature were higher for para-hydrogen than for ortho-hydrogen. They concluded from this that for o-H_2 the rotation is only very slightly accommodated.

Knudsen investigated the accommodation of hydrogen on platinum surfaces by a completely different method, the radiometer method. With a platinum strip which first was bright on both sides and then bright on one side and blackened on the other, he determined the thermal accommodation and (by means of the deflection of a torsion balance) the

Table 6.3.2.1.A. *Accommodation coefficients for diatomic gases on gas-covered surfaces*

1	2	3	4	5	6	7	8
Surface	Condition of surface	Surface temperature T_s (°K)	Incident gas	Gas temperature T_g (°K)	$\Delta T = T_s - T_g$ (°C)	Accommodation coefficient α	Author
W	Probably slightly gas covered	200	H_2	91		0.180	BLODGETT, LANGMUIR [*78*]
		300		96		0.151	
		400		102		0.143	
		500		109		0.153	
W	Gas covered	298	H_2			0.357	AMDUR, GUILDNER [*15*]
W	Gas covered	298	D_2			0.454	AMDUR, GUILDNER [*15*]
W	Gas covered		N_2			0.868	AMDUR, GUILDNER [*15*]
W	Gas covered		O_2			0.905	AMDUR, GUILDNER [*15*]
W	Gas covered	373.1	H_2	294		0.357	UBISCH [*731*]
W	Gas covered	373.1	D_2	294		0.447	UBISCH [*731*]
W	Gas covered	373.1	N_2	294		0.864	UBISCH [*731*]
W	Gas covered	1870	N_2	Near room temp.		0.087	DEPOORTER [*143a, 143b*]
W	Gas covered	2120	N_2	Near room temp.		0.083	DEPOORTER [*143a, 143b*]
W	Gas covered	2240	N_2	Near room temp.		0.081	DEPOORTER [*143a, 143b*]
W	Gas covered	2335	N_2	Near room temp.		0.074	DEPOORTER [*143a, 143b*]
W	Gas covered	2511	N_2	Near room temp.		0.067	DEPOORTER [*143a, 143b*]
W	Gas covered	373.1	O_2	294		0.887	UBISCH [*731*]
Pt	Probably slightly gas covered	303.1	H_2		α, extrapolated to $\Delta T = 0$	0.220	THOMAS, OLMER [*715*]
Pt	Probably slightly gas covered	373.1	H_2			0.10	ROLF [*606*]

Table 6.3.2.1.A. (continued)

1 Surface	2 Condition of surface	3 Surface temperature T_s (°K)	4 Incident gas	5 Gas temperature T_g (°K)	6 $\Delta T = T_s - T_g$ (°C)	7 Accommodation coefficient α	8 Author
Pt	Probably slightly gas covered	109—400	H_2	88		0.36—0.22	ROWLEY, BONHOEFFER [617]
Pt		393.1	D_2			0.16	ROLF [606]
Pt		109	P-H_2		20	0.37	ROWLEY, BONHOEFFER [617]
Pt		129	P-H_3			0.35	ROWLEY, BONHOEFFER [617]
Pt		210	P-H_2			0.27	ROWLEY, BONHOEFFER [617]
Pt		373.1	O-H_2	289.1		0.011	MANN, NEWELL [456]
Pt		373.1	O_2	289.1		0.42	MANN, NEWELL [456]
Pt	Gas covered	373.1	H_2	289.1	84	0.20	MANN, NEWELL [456]
Pt		291	H_2		20	0.36	DICKENS [145]
		291	O_2		20	0.90	
Pt		297	N_2		20	0.89	DICKENS [145]
Pt		297	CO		20	0.92	DICKENS [145]
Pt		297	Air		20	0.90	DICKENS [145]
Pt		283	H_2.			0.287	AMDUR, GUILDNER [15]
Pt			D_2			0.377	AMDUR, GUILDNER [15]
Pt			N_2			0.816	AMDUR, GUILDNER [15]
Pt			O_2			0.853	AMDUR, GUILDNER [15]
Pt		290	H_2			0.28	CHAVKEN [113]
Pt		290	D_2			0.33	CHAVKEN [113]
Pt		290	O_2			0.77	CHAVKEN [113]
Pt		290	N_2			0.77	CHAVKEN [113]
Pt		290	CO			0.78	CHAVKEN [113]

Pt	Gas covered	373	H_2	294		0.332	UBISCH [731]
Pt			D_2	294		0.402	UBISCH [731]
Pt			N_2	294		0.843	UBISCH [731]
Pt			O_2	294		0.864	UBISCH [731]
Pt		283—295	H_2		20	0.291_6	ARCHER [21]
Pt		373	H_2	292	81	0.358	KNUDSEN [393]
Pt		283—300	D_2		20	0.376	ARCHER [21]
Ni		298	H_2			0.294	AMDUR, GUILDNER [15]
Ni		298	D_2			0.387	AMDUR, GUILDNER [15]
Ni			N_2			0.824	AMDUR, GUILDNER [15]
Ni			O_2			0.862	AMDUR, GUILDNER [15]

mechanical force experienced as a result of the recoils from the strip blackened on one side. The mechanical recoil effect is proportional to the difference between the partial accommodations of the translation energy on the blackened and bright sides of the strip. This difference can be determined absolutely from the geometric constants of the experimental arrangement, the recoil force per unit area being given by $\frac{1}{4} p \alpha (T_1 - T_0)/T_0$. Supplementary experiments on thermal accommodation by use of the thermal-conduction method then yield the difference of the total accommodations. KNUDSEN found that α_{trans} (black) $- \alpha_{\text{trans}}$ (bright) $= \alpha_{\text{total}}$ (black) $- \alpha_{\text{total}}$ (bright). From this he concluded that $\alpha_{\text{trans}} \approx \alpha_{\text{rot}}$ for hydrogen on clean platinum.

Another method for determining partial thermal-accommodation coefficients was proposed by SCHÄFER and RIGGERT [*633*]. This method, which is based on a proposal of EUCKEN and KRONE [*196*] for simultaneous determination of accommodation and molar specific heat, is also based on the heat transfer at low pressure. A narrow strip of metal is stretched parallel to and about 1 mm away from a thin wire of the same material inside a glass vessel. The average thermal AC [the sum of the translational, rotational, and vibrational contributions as in Eq. (6.3.2.2–1)] is then determined by using the previously described wire method. The strip is then heated to a temperature higher than that of the wire so that the wire receives additional heat, the amount being a function of the accommodations on the strip and the wire. It is obvious that here also the necessary corrections must be made for radiation losses and the like. SCHÄFER and RIGGERT restricted themselves to the case

Table 6.3.2.1.B. *Accommodation coefficient α for diatomic molecules on surfaces covered with known gases*

1 Surface	2 Condition of surface	3 Surface temperature T_s (°K)	4 Incident gas	5 Gas temperature T_g (°K)	6 $\Delta T = T_s - T_g$ (°C)	7 Accommodation coefficient α	8 Author
W	H-covered	329.8	H_2		25 to 28	0.16_5	WACHMAN [*747*]
W	D-covered	329.8	D_2		25 to 28	0.23_2	WACHMAN [*747*]
W	O-covered	329.8	H_2		25 to 28	0.21_0	WACHMAN [*747*]
W	O-covered	329.8	H_2		25 to 28	0.24	WACHMAN [*747*]
W	N-covered	329.8	N_2		25 to 28	0.62_4	WACHMAN [*747*]
W		150	H_2	92		0.42	BLODGETT, LANGMUIR [*78*]
W		300	H_2	100		0.19	BLODGETT, LANGMUIR [*78*]
W		500	H_2	114		0.19	BLODGETT, LANGMUIR [*78*]
W		700	H_2	127		0.20	BLODGETT, LANGMUIR [*78*]
W		900	H_2	143		0.23	BLODGETT, LANGMUIR [*78*]
Pt		297.1	O_2			0.90	DICKENS [*145*]
Pt		373.1	O_2	≈ 19	≈81	$0.84 - 0.83_5$	KNUDSEN [*393*]
Fe	O-covered	303.1—333.1	O_2			0.69 —0.72	EGGLETON [*170*]
Fe		303.1—333.1	N_2			$0.65_4 - 0.68_4$	EGGLETON [*170*]
Fe		303.1—333.1	H_2			$0.09_0 - 0.09_3$	EGGLETON [*170*]
Ni		(113)	N_2			0.097	EUCKEN, BERTRAM [*195*]
Ni		173	N_2	170		0.95	EUCKEN, BERTRAM [*195*]
Ni		199	N_2	170		0.95	EUCKEN, BERTRAM [*195*]
Ni		279	N_2	170		0.92	EUCKEN, BERTRAM [*195*]

in which the contribution of the vibrational energy of the molecule to the total molar heat can be neglected. In this case, the two values of the accommodation coefficient ($\bar{\alpha}_w$ from the wire method and $\bar{\alpha}_{rw}$ from the ribbon-wire method) can be used to determine the partial AC's for translation and rotation. Table 6.3.2.1.C gives such partial AC values as determined by Schäfer and Riggert for N_2 and O_2 on gas-covered gold surfaces.

Table 6.3.2.1.C. *Partial accommodation coefficients for N_2 and O_2 molecules on gas-covered gold surfaces [633, 635]*

	N_2				O_2			
Temp. (° C)	$\bar{\alpha}_w$	$\bar{\alpha}_{rw}$	α_{trans}	α_{rot}	$\bar{\alpha}_w$	$\bar{\alpha}_{rw}$	α_{trans}	α_{rot}
20	0.851	0.854	0.89	0.78	0.856	0.860	0.90	0.77
44	0.824	0.835	0.89	0.68	0.839	0.846	0.89	0.77
66	0.800	0.818	0.88	0.63	0.821	0.834	0.89	0.67
91	0.776	0.800	0.87	0.58	0.795	0.816	0.88	0.60
116	0.755	0.782	0.86	0.54	0.762	0.798	0.87	0.53

6.3.3. Polyatomic Gases

6.3.3.1. Experimental Data

As a basis for the discussion of the results on the accommodation of polyatomic molecules on metal surfaces, the values of the total accommodation coefficients (not separated into partial AC's) for different gases and surfaces are presented in Table 6.3.3.1.A. The accommodation coefficients all have high values since they are for gas-covered surfaces. Unfortunately, no results on gas-free or only slightly gas-covered metal surfaces seem to be available.

With the strip-wire method described in Section **6.3.2**, Schäfer et al. also determined the partial AC's for translational, rotational, and also (under supplementary assumptions) the vibrational energy.

6.3.3.2. Theoretical Considerations

Molecules that have also vibrational energy present a difficulty because one must determine three partial AC's.

This is not possible without some arbitrariness. Schäfer et al. [633] discuss the plausibility of additional assumptions such as

$$\alpha_{rot} = \alpha_{trans}, \quad \text{or} \quad \alpha_{rot} = K\alpha_{trans}, \quad \text{or} \quad \alpha_{trans} = 1 \,.$$

The plausibility of the first relation is supported by results on sound dispersion, which show that the energy exchange between translation and vibration is slow, whereas rotational energy quickly comes to equilibrium with translational energy. It naturally is questionable to what

Table 6.3.3.1.A. *Accommodation coefficient α for polyatomic molecules on gas-covered surfaces*

1 Surface	2 Condition of surface	3 Surface temperature T_s (°K)	4 Incident gas	5 Gas temperature T_g (°K)	6 $\Delta T = T_s - T_g$ (°C)	7 Accommodation coefficient α	8 Author
W	CO_2-covered	329.8	CO_2			0.990	Wachman [747]
W	Gas covered	1860	CO_2			0.136	DePoorter [143a, 143b]
W		2120	CO_2			0.126	DePoorter [143a, 143b]
Pt		373.1	CO_2		80	0.868	Knudsen [393]
Pt		48	CO_2		8	0.45	Archer [20]
Pt		243	CS_2			0.97	Grau [254, 255]
Pt		273	CS_2			0.95	Grau [254, 255]
Pt		290	CS_2			0.93	Grau [254, 255]
Pt		304	CS_2			0.92	Grau [245, 255]
Pt		313	CS_2			0.89	Grau [254, 255]
Pt		359	CS_2			0.84	Grau [254, 255]
Pt		195	COS			(0.96)	Strohmeier [689]
Pt		250	COS			(0.78)	Strohmeier [689]
Pt		273	COS			(0.66)	Strohmeier [689]
Pt		299	COS			0.50	Strohmeier [689]
Pt		293.1	N_2O			0.850	Schäfer, Gerstacker [636]
Pt		338.1	N_2O			0.745	Schäfer, Gerstacker [636]
Pt		383.1	N_2O			0.665	Schäfer, Gerstacker [636]
Pt		314.3	CO_2			0.781	Thomas, Golicke [717]
Pt		324.4	CO_2	303.1	21.4	0.786	Thomas, Golicke [717]
Pt		353.5	CO_2		50.4	0.771	Thomas, Golicke [717]
Pt		293.1	N_2O			0.845	Schäfer, Gerstacker [636]
Pt		338.1	N_2O			0.805	Schäfer, Gerstacker [636]
Pt		383.1	N_2O			0.770	Schäfer, Gerstacker [636]
Pt		303.1	CO_2		30	0.76	Thomas, Olmer [715]
Pt		373.1	H_2S			0.92	Ubisch [731, 732]
Pt		373.1	SO_2	294		0.96	Ubisch [731, 732]
Pt		373.1	CH			0.807	Ubisch [731, 732]

Pt	Gas covered	373.1	C_2H_2			0.837	Ubisch [731, 732]
Pt		373.1	C_2H_4			0.90_6	Ubisch [731, 732]
Pt		373.1	C_2H_6			0.95	Ubisch [731, 732]
Pt		373.1	C_3H_2			1.00	Ubisch [731, 732]
Pt		273.1	CO_2			0.815	Schäfer, Klingenberg [634]
Pt		290	CO_2	2		0.69	Chavken [113]
Pt		293.1	CO_2			0.794	Schäfer, Klingenberg [634]
Pt		313.1	CO_2			0.773	Schäfer, Klingenberg [634]
Pt		333.1	CO_2			0.752	Schäfer, Klingenberg [634]
Pt		353.1	CO_2			0.73_0	Schäfer, Klingenberg [634]
Pt		373.1	CO_2			0.70_9	Schäfer, Klingenberg [634]
Pt		293	C_2H_6			0.97	Schäfer [632]
Pt		341	C_2H_6			0.85	Schäfer [632]
Pt		370	C_2H_6			0.77	Schäfer [632]
73% Pt, 27% Cu		293	C_2H_6			0.94	Schäfer [632]
73% Pt, 27% Cu		341	C_2H_6			0.87	Schäfer [632]
73% Pt, 27% Cu		370	C_2H_6			0.81	Schäfer [632]
59% Pt, 41% Cu		293	C_2H_6			0.89	Schäfer [632]
59% Pt, 41% Cu		341	C_2H_6			0.83	Schäfer [632]
59% Pt, 41% Cu		370	C_2H_6			0.80	Schäfer [632]
33% Pt, 67% Cu		293	C_2H_6			0.91	Schäfer [632]
33% Pt, 67% Cu		341	C_2H_6			0.85	Schäfer [632]
33% Pt, 67% Cu		370	C_2H_6			0.83	Schäfer [632]
Cu		293	C_2H_6			0.95	Schäfer [632]
Cu		341	C_2H_6			0.86	Schäfer [632]

Table 6.3.3.1.A. (continued)

1 Surface	2 Condition of surface	3 Surface temperature T_s (°K)	4 Incident gas	5 Gas temperature T_g (°K)	6 $\Delta T = T_s - T_g$ (°C)	7 Accommodation coefficient α	8 Author
Cu	Gas covered	370	C_2H_6			0.80	Schäfer [632]
Ag		293.1	N_2O			0.85_5	Schäfer, Gerstacker [636]
Ag		338.1	N_2O			0.82_5	Schäfer, Gerstacker [636]
Ag		383.1	N_2O			0.79_5	Schäfer, Gerstacker [636]
Ni	Oxide-coated	173	C_2H_6			0.99_1	Eucken, Bertram [195]
Ni		194.5	C_2H_6			0.97_8	Hunsmann [330]
Ni		199	C_2H_6			0.98_1	Eucken, Bertram [195]
Ni		221	C_2H_6			0.95_3	Hunsmann [330]
Ni		279	C_2H_6			0.94_5	Eucken, Bertram [195]
Ni		281.3	C_2H_6			0.85	Hunsmann [330]
Ni		380.2	C_2H_6			0.656	Hunsmann [330]
Ni		152	SF_6			0.98_8	Eucken, Bertram [195]
Ni		173	SF_6			0.96_1	Eucken, Bertram [195]
Ni		194.5	SF_6			0.960	Hunsmann [330]
Ni		199	SF_6			0.83_5	Eucken, Bertram [195]
Ni		276.5	SF_6			0.740	Hunsmann [330]
Ni		279	SF_6			0.99_4	Eucken, Bertram [195]
Ni		378	SF_6			0.367	Hunsmann [330]
Ni		152	CO_2			0.99_1	Eucken, Bertram [195]
Ni		173	CO_2			0.98_7	Eucken, Bertram [195]
Ni		199	CO_2			0.97	Eucken, Bertram [195]
Ni		279	CO_2	170		0.93_3	Eucken, Bertram [195]
Ni		197.6	N_2O			0.922	Hunsmann [330]
Ni		218.9	N_2O			0.874	Hunsmann [330]
Ni		273.1	N_2O			0.726	Hunsmann [330]
Ni		291.3	N_2O			0.667	Hunsmann [330]
Ni		378	N_2O			0.458	Hunsmann [330]

extent these relations can be carried over from the collision of two gas molecules to the collision of a gas molecule with a solid surface. In any case one could probably consider the adsorption on the surface to be comparable to the formation of a double molecule from a smaller and a larger molecule in a gas. On the basis of quantum mechanical considerations, ZENER ([*799*] has shown that in the collision of two gas molecules the rotational energies are readily exchanged while vibrational energies are exchanged only with difficulty.

The second proposed relation, $\alpha_{rot} = K \alpha_{trans}$, is based on the assumption that for the N_2 molecule the vibrational energy can be neglected so that the constant $K(N_2)$ can be computed. Once $K(N_2)$ is known, the quotients $K(\chi) = \alpha_{rot}(\chi)/\alpha_{trans}(\chi)$ can also be determined for other gases.

The third suggested condition, $\alpha_{trans} = 1$, corresponds to an assumption made by EUCKEN and BERTRAM [*195*] in their investigations.

The partial AC's for CO_2, computed by SCHÄFER and KLINGENBERG [*634*], are presented in Table 6.3.3.2.A. so that the values calculated under the three suggested assumptions can be compared.

There is no unified theory for the energy exchange of adsorbed polyatomic molecules on a surface.

Table 6.3.3.2.A. *Partial accommodation coefficients for CO_2 on platinum* [*634*]

Surface temperature	0° C	20° C	40° C	60° C	80° C	100° C
α_w	0.815	0.794	0.773	0.752	0.700	0.709
α_{rw}	0.875	0.869	0.864	0.858	0.853	0.848
Assumption a						
$\alpha_{trans} = \alpha_{rot}$	0.931	0.930	0.929	0.927	0.924	0.919
α_{vib}	0.393	0.357	0.325	0.296	0.268	0.243
Assumption b						
K	0.966	0.957	0.953	0.949	0.945	0.939
α_{trans}	0.934	0.941	0.944	0.943	0.944	0.937
α_{rot}	0.905	0.902	0.900	0.897	0.895	0.888
α_{vib}	0.399	0.364	0.328	0.303	0.272	0.267
Assumption c						
α_{trans}	1.00	1.00	1.00	1.00	1.00	1.00
α_{rot}	0.747	0.747	0.744	0.744	0.737	0.720
α_{vib}	0.452	0.398	0.356	0.376	0.297	0.272

6.4. Determination of Relaxation Times for the Process of Energy Exchange between the Normal Vibrational States of the System Comprising Adsorbed Molecule and Metal Surface

In this connection, attention should be called to certain attempts to determine the relaxation times for the energy-exchange process, i.e., the time required for the temperature difference between the adsorbed gas molecule and the wall to fall to one eth of its value. According to EUCKEN et al. [*195*], the average thermal AC, the mean residence time τ of the adsorbed particle on the surface, and the relaxation time β are related by the expression*

$$\overline{\alpha} = \frac{\tau}{\tau + \beta} . \tag{6.4–1}$$

Since the quantities α and τ can be experimentally determined (by the methods of measurement described in Chaps. 6 and 8) β can then be computed from them. As examples, Table 6.4.A. gives values of β determined in this fashion by HUNSMANN [*330*].

These relaxation times for the process of energy exchange between admolecule and surface may be compared with those between molecule and molecule (as computed, for example, from sound-dispersion measurements). The values of β in the latter case must still be multiplied by the ratio of the actual duration of a collision to the time between collisions; but when this is done one finds, interestingly enough, that the relaxation times are of the same order of magnitude in both cases.

The relaxation times for systems such as CO_2, COS, or CS_2 on gas-covered platinum have been investigated in a similar way by SCHÄFER et al. [*630, 689*]. They sought, under simplifying assumptions, to determine even partial relaxation times (e.g., for the vibrational energy). SCHÄFER [*631*] has also tried to compute the probabilities for energy trans-

---

* If t is the time interval during which a molecule remains adsorbed on a surface, then during this time the quantity of heat $q_t = q_\infty (1 - e^{-t/\beta})$ will be transferred to the surface. Here q_∞ is the maximum amount of heat which can be transferred; that is, the amount transferred in a time long enough for the molecule to completely attain the temperature of the surface. The constant β is the relaxation time for the process. The quantity of heat given off per molecule in the time interval dt is then $dq_t = q_\infty (1/\beta) e^{-t/\beta} dt$. At time $t = 0$, it is assumed that N_0 molecules simultaneously strike the surface and then evaporate once more in accordance with the equation $N_t = N_0 e^{-t/\tau}$, where τ is the mean residence time on the surface. Then the heat transferred during the time interval dt by the molecules present at time t is given by

$$d Q_t = N_t d q_t = q_\infty N_0 \frac{1}{\beta} \exp\left[-t\left(\frac{1}{\beta} + \frac{1}{\tau}\right)\right] dt . \tag{6.4–1a}$$

Integration with respect to t gives the total quantity of heat transferred,

$$Q_t = Q_\infty \frac{\tau}{\tau + \beta} . \tag{6.4.–1b}$$

Substitution of this into the definition of the AC leads to the EUCKEN equation, namely,

$$\overline{\alpha} = \frac{Q_t}{Q_\infty} = \frac{\tau}{\tau + \beta} . \tag{6.4.–1c}$$

It should be pointed out that this derivation of Eq. (6.4.–1c) assumes that the desorption process from the surface is a first-order reaction.

Table 6.4.A. *Values of the relaxation time β for various gases adsorbed on oxidized nickel surfaces* (HUNSMANN [*330*])

	T (° K)	τ (sec)	α	β (sec)
SF_6 on oxidized nickel surface	194.5	4.87×10^{-7}	0.960	2.2×10^{-8}
	235.8	8.95×10^{-8}	0.880	1.2×10^{-8}
	276.5	2.33×10^{-9}	0.740	8.2×10^{-9}
	378	2.2×10^{-9}	0.367	3.3×10^{-9}
N_2O on oxidized nickel surface	197.6	4.0×10^{-7}	0.922	3.4×10^{-7}
	218.9	9.1×10^{-7}	0.874	1.3×10^{-7}
	273.1	3.8×10^{-8}	0.726	1.4×10^{-8}
	291.3	1.6×10^{-8}	0.667	7.5×10^{-9}
	278	1.1×10^{-9}	0.458	1.2×10^{-9}
C_2H_2 on oxidized nickel surface	194.5	2.7×10^{-7}	0.978	6.0×10^{-9}
	221	7.4×10^{-8}	0.953	3.7×10^{-9}
	281.3	1.2×10^{-8}	0.854	2.0×10^{-9}
	350.2	1.8×10^{-9}	0.656	1.0×10^{-9}

fer between normal vibrations of an admolecule and the vibration of the surface atoms. His considerations contain quite drastic simplifying assumptions*, both with regard to the model assumed (only pure electrical interaction between the admolecule and the surface, namely, an interaction between an electric dipole and a point charge) and in the approximations in the subsequent quantum mechanical perturbation calculation. A corresponding uncertainty attends the attempt to compare the theoretical values of the relaxation times thus obtained (which apply to energy exchange between an admolecule and the atom of a clean metal surface) with the experimentally determined relaxation times (which are obtained from measurements on gas-covered surfaces, not with the atoms of a "clean metal surface" as is assumed in the computation).

7. Elastic Collisions of Atoms and Molecules with Metal Surfaces

The preceding chapters have discussed the cases in which an atom or molecule making an inelastic collision with a surface either has a partial exchange of energy with the surface (thermal accommodation coefficient α greater than 0 but smaller than 1) or has a complete energy exchange in an absorption process (thermal accommodation coefficient equal to unity). The following is a brief discussion of the case of elastic collision in

* One simplifying assumption is, for example, that the frequencies of the normal vibrations of the adsorbed molecules are the same as those of these molecules in the gaseous state.

which the atom or molecule impinging on the surface immediately leaves without energy exchange ($\alpha = 0$) including the case of specular reflection in which it leaves so that the incident and reflected beams make equal angles with the normal to the surface.

As in optics, the condition for specular reflection is that at the glancing angle θ the projection of the "average" height h of surface irregularities be small compared to the wavelength; that is, the, requirement

$$h \sin \theta < \lambda \, . \tag{7–1}$$

Since the DE BROGLIE wavelength for hydrogen molecules or helium atoms at room temperature is of the order of 10^{-8} cm, it would be desirable to keep the average height of the surface irregularities no larger than this order of magnitude to obtain easily attainable glancing angles. The cleavage surfaces of ionic crystals provide such smooth surfaces, the irregularities being of the order of 10^{-8} cm and coming primarily from the thermal vibration of the lattice ions about their equilibrium positions. This is the reason why most investigations of elastic reflection of atomic or molecular beams from solid surfaces have been carried out with such cleavage surfaces of ionic crystals.

Thus ESTERMANN and STERN [*193*] found specular reflection of helium atoms from a LiF cleavage surface at room temperature for a glancing angle of less than 30°. The present article confines itself to the investigation of metallic surfaces. For this reason the numerous investigations of cleavage surfaces on ionic crystals will not be presented in any more detail. Instead, the reader is referred to the comprehensive presentations by ESTERMANN [*194*], RAMSEY [*581*], and MASSEY and BURHOP [*461*].

Very few investigations of the phenomenon on metal surfaces are available at present. A prime reason for this is the fact that it is difficult to obtain sufficiently flat polished metal surfaces. Thus the average height h of the surface irregularities for good mechanically polished surfaces (Finish 2, ASA Standards) is of the order of 5×10^{-6} cm. When a hydrogen molecular beam is reflected from such a surface at room temperature, condition (7–1) is satisfied only for glancing angles $\theta < 0.001°$. In their investigation of the reflection of beams of hydrogen molecules and beams of helium atoms from polished steel surfaces or speculum metal surfaces (copper-tin alloys), KNAUER and STERN [*388*] actually were able to show the occurrence of specular reflection for glancing angles of the order of 0.001°. Their values for the reflecting power (ratio of maximum intensity of the reflected beam to that of the incoming beam) for hydrogen molecules impinging on a speculum metal surface at room temperature for various glancing angles are given in Table 7.A. These authors were unable to observe any specular reflection from a brass surface.

Since this first fundamental work of KNAUER and STERN, several decades have passed, but recently HURLBUT et al. [*331, 332, 333*] have taken up such investigations anew. HURLBUT [*331*] has investigated the reflection of nitrogen molecular beams from polished steel and aluminium surfaces at room temperature and at 100°C for a range of glancing angles from 15.58° to 90°. The results show no elastic reflection pattern but

Table 7.A. *Values of the reflecting power for the case of hydrogen molecules impinging on a speculum metal surface* (KNAUER and STERN [*388*])

Glancing angle (degrees)	1×10^{-3}	1.5×10^{-3}	2×10^{-3}	2.25×10^{-3}
Reflecting power (%)	5	3	1.5	0.15

rather a diffuse distribution of the re-emitted particles. This result is by no means surprising since the measurements were made under conditions such that the DE BROGLIE wavelength for nitrogen was of the order of 10^{-9} cm and the average height h of irregularities was probably not better than 10^{-5} cm, so that the chosen glancing angle of 15.58° is approximately five orders of magnitude too large to fulfill condition (7–1). Moreover, the second condition for elastic reflection, that the mean adsorption lifetime of the incident atom or molecule on the surface must be small, is probably not fulfilled for the nitrogen-steel or nitrogen-aluminium system. It was already pointed out in Chaps. 3, 4, and 6 that the case of nitrogen on a metal surface (the only one considered so far) involves an adsorption process with weak chemisorption and that the value of the measured accommodation coefficient ($\alpha \approx 0.8$) indicates a practically complete energy exchange between the nitrogen and the metal surface. Thus the diffuse distribution of particles coming from the surface arises in part from the diffuse scattering due to surface roughness and in part from the cosine distribution of the desorption process.

HURLBUT and BECK [*333*], in a recent investigation of the reflection of nitrogen molecules from liquid gallium and indium surfaces, found a slight deviation from diffuse scattering in the case of a gallium surface. But since the nature of the surface is inadequately known for these measurements it is very unlikely that condition (7–1) is satisfied even in this case for the glancing angles used (87.5°—49.9°). This becomes even more questionable since the authors also report a deviation from diffuse scattering in the case of an even more irregular surface of liquid gallium which has been covered with hydrogen. Moreover, one would expect that adsorption processes (mean adsorption lifetime considerably greater than 10^{-13} sec) will also give rise to an inelastic collision process in the nitrogen-gallium system discussed above.

In conclusion, it may be said that there is surprisingly little material available at present concerning elastic reflection of atoms and molecules from metal surfaces and that a much more vigorous effort in this direction is highly desirable.

8. Emission of Positive Ions Formed at Metal Surfaces (Surface Ionization)

8.1. Theoretical Considerations

When atoms or molecules from a molecular beam or from a vapor impinge on an incandescent metal surface, they may evaporate partly as neutral atoms and partly as positive or negative ions after a mean residence time τ_a or τ_i respectively. LANGMUIR and KINGDON [*420*] first observed the formation of positive cesium ions on an incandescent wolfram surface. Subsequently, numerous studies of positive surface ionization have been conducted, since this effect opens interesting possibilities for ion production as well as for the detection of molecular and atomic beams. The developments in recent years have followed two main directions. On the one hand there have been attempts to extend the classes of atoms and metal surfaces for which surface ionization occurs and the SAHA-LANGMUIR equation is valid. On the other hand, and to a much smaller extent, there has been an attempt to study the mechanism of surface ionization itself, as for example by direct determinations of desorption probabilities for ions and atoms, and for the charge-transfer probabilities.

The following is a survey of surface ionization phenomena with special attention to the newer results. Detailed presentations, in particular of the older data, can be found in review articles and monographs such as those of ZANDBERG and IONOV [*797*], DOBREZOW [*153*], ARIFOW and SCHUPPE [*22*], and REIMANN [*584*].

8.1.1. Equation for Emission from a Homogeneous Surface in the Absence of External Electrical Fields

Two quantities of interest in the study of surface ionization are the degree of ionization

$$\alpha = N_+/N_0, \tag{8.1.1–1}$$

and the ionization coefficient

$$\beta = N_+/N. \tag{8.1.1–2}$$

In these equations N_+ is the number of positive ions leaving the hot metal surface per unit area per second, N_0 is the number of neutral atoms emitted from the same surface element in this same time, and N is the number of atoms impinging on the surface per unit area per second. In the stationary state these must be related by

$$N = N_0 + N_+ . \tag{8.1.1–3}$$

Thus the relation between the ionization coefficient β and the degree of ionization α is

$$\beta = \frac{N_+}{N} = \frac{N_+}{N_0 + N_+} = \frac{\alpha}{1 + \alpha} . \tag{8.1.1–4}$$

From this it follows that $\beta \approx \alpha$ in cases for which $\alpha \ll 1$ (as for Ag or Cu on W), and that $\beta \approx 1$ for $\alpha \gg 1$ (as for Cs on W). The degree of ionization α depends on various parameters, e.g., the surface temperature T, the work function $e\varphi$ of the metal surface, the nature of the surface with respect to heterogeneities (physical, chemical, or induced heterogeneities as discussed in Sections 1.1.1, 1.1.2, 1.1.3), the coverage degree θ of the surface, the ionization energy of the impinging atoms and their electronic states, and finally the strength of the externally applied electric field ε. LANGMUIR and KINGDON [*420*] have used SAHA's result, for the simplified case of a homogeneous metal surface in the absence of an external electric field to express the temperature dependence of the ionization degree α by the relation

$$\alpha(\varepsilon = 0) = \frac{g_+}{g_0} \exp\left[\frac{e(\varphi - I)}{kT}\right] . \tag{8.1.1–5}$$

In this equation (g_+/g_0) is the ratio of the statistical weights of the ionic and atomic states. For the case of alkali atoms, $(g_+/g_0) = \frac{1}{2}$ because the atom has two states (parallel or antiparallel spin of the valence electron) while the ion has only one state. The other abbreviations correspond to those in earlier sections.

The authors derived Eq. (8.1.1–5) under the simplifying assumption of thermodynamic equilibrium between the atoms, ions, and electrons striking the surface and then being desorbed. Such an equilibrium can exist if the particles rebounding from the surface remain on it long enough to come to thermodynamic equilibrium. Results concerning accommodation coefficients (see chapter 6) imply that this requirement is satisfied for many systems, e.g., for alkali atoms on wolfram and copper surfaces. IONOV [*341*], by measuring the energy distribution of positive ions emitted from a surface (e.g., for K on W), has confirmed the correctness of this assumption. On the other hand, for potassium and potassium halides on platinum surfaces, DATZ and TAYLOR [*137*] observed that fewer potassium ions are formed by surface ionization than would be expected from the SAHA-LANGMUIR theory. In their interpretation, the authors claimed that a sizable fraction of the impinging atoms did not come into thermal equilibrium with the surface. However, the results of the more direct investigations of the inelastic scattering of alkali atoms on Pt surfaces (see chapter 6) showed that this fraction is much smaller than the authors assumed. Disturbances of the thermal equilibrium may result from the application of an external electric field, as will be discussed more thoroughly in Section 8.3.2.5. Corrections have been made in an attempt to take account of such deviations from the assumptions underlying Eq. (8.1.1.–5) Such a "corrected" equation for the ionization degree α in zero external electric field ($\varepsilon = 0$) is

$$\alpha(\varepsilon = 0) = \frac{\dot{N}_+}{\dot{N}_0} = \frac{P_+ N_+}{P_0 N_0} = \frac{G_+}{G_0} \left[\frac{1 - r_+}{1 - r_0}\right] \exp \frac{e a}{k} \exp \left[\frac{-e(I - \psi)}{kT}\right]. \tag{8.1.1–6}$$

Here $\dot{N}_+$ and $\dot{N}_0$ are the number of desorbing ions or atoms per unit area and per unit time, P_+ and P_0 are the corresponding desorption probabilities, and G_+ and G_0 are the statistical weights for the ions or atoms for the case in which the latter can also occur in an excited state whose excitation energy is comparable to kT. According to MOROSOW [*505*], the applicable relations are

$$G_+ = g_+ + \sum_s g_+^{(s)} \exp \frac{-\Delta E_+^{(s)}}{kT}, \tag{8.1.1–7}$$

$$G_0 = g_0 + \sum_s g_0^{(s)} \exp \frac{-\Delta E_0^{(s)}}{kT}. \tag{8.1.1–8}$$

Here $g_+^{(s)}$ and $g_0^{(s)}$ are the statistical weights of the sth excited ionic or atomic states, and $\Delta E_+^{(s)}$ and $\Delta E_0^{(s)}$ are the excitation energies of these states relative to the ground state of the ion or atom. One sees that when $\Delta E_{+,0}^{(s)} \gg kT$, the summed terms in Eqs. (8.1.1–7) and (8.1.1–8) can be neglected and the ratio G_+/G_0 reduces to the ratio g_+/g_0 of Eq. (8.1.1–5).

In Eq. (8.1.1–6), r_+ and r_0 are the reflection coefficients of the surface for ions and atoms, respectively, (see for example COPLEY and PHIPPS [*124, 125*]). The quantities a/k and ψ in Eq. (8.1.1–6), which take account of the temperature dependence of the work function $e\varphi$, are defined by $\varphi(T) = \varphi(T_0) + a(T - T_0)$ and $\psi = \varphi(T_0) - aT_0$. For homogeneous surfaces this temperature dependence of the work function can be neglected in many cases.

It should also be noted that DOBREZOW [*148*] has derived Eq. (8.1.1–6) from statistical arguments and has sought (DOBREZOW [*154, 155*]) to answer the objections that AVAKYANTS [*35, 36*] raised against this derivation.

8.1.2. Equation for Emission from a Homogeneous Surface in an External Electrical Field

If the positive surface ionization occurs in an external electric field $\mathcal{E}$ which accelerates the ion, the heat of evaporation E_i of the ions will be reduced by the SCHOTTKY term $e\sqrt{e\mathcal{E}}$. The heat of vaporization can at most be reduced by the amount of the image force $e^2/4r_0$ when the external field $\mathcal{E}$ compensates the image field at the equilibrium distance r_0. The correspondingly corrected SAHA-LANGMUIR equation is

$$\alpha = \alpha(\mathcal{E} = 0) \exp\left[\frac{e\sqrt{e\mathcal{E}}}{kT}\right]. \qquad (8.1.2\text{–}1)$$

This relation was confirmed experimentally by DOBREZOW [*150*].

On application of extremely high electric fields $\mathcal{E}$, of the order of 10^8 V/cm, the field ion desorption discovered by MÜLLER [*508*] becomes important. Then Eq. (8.1.2–1) must be supplemented by two additional

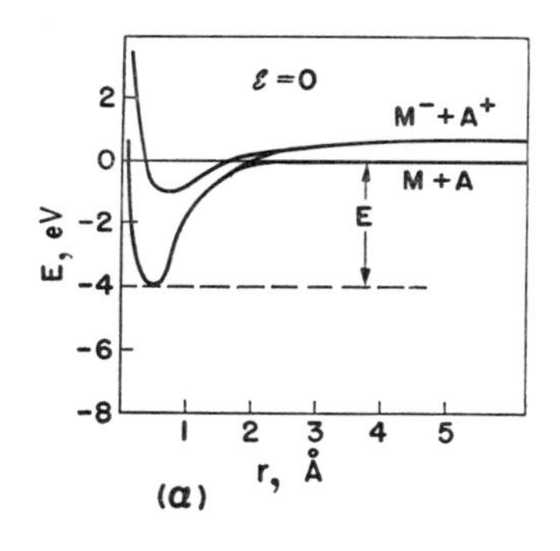

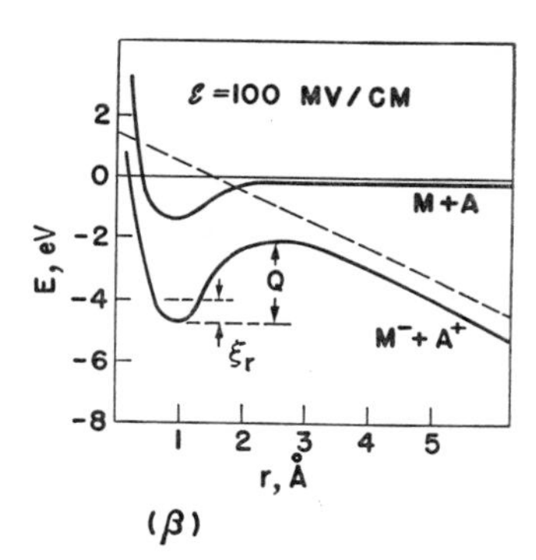

Fig. 8.1.2.a. α u. β. Potential energy diagram for an electropositive metal ($eI > e\Phi$) adsorbed on a metal surface. Diagram (α) is characteristic of the adsorption process in the absence of an external electrical field. Diagram (β) is characteristic of the process in an external electrical field $\mathcal{E} = 100$ MV/c

terms. The first term $\frac{1}{2}\mathcal{E}^2(\alpha_0 - \alpha_i)$ takes account of the difference in the polarization energies of the adatom and the ion, where α_0 and α_i are the corresponding polarizabilities. The second term $\mathcal{E}r_0$ takes account of the "field-bond energy" at the equilibrium separation r_0 (Fig. 8.1.2.a). The SAHA-LANGMUIR equation corrected for high electric fields is then

$$\alpha = \alpha(\mathcal{E} = 0) \exp\left[\frac{e\sqrt{e\mathcal{E}} + \frac{1}{2}\mathcal{E}^2(\alpha_0 - \alpha_i) + e\mathcal{E}r_0}{kT}\right]. \qquad (8.1.2\text{–}2)$$

At field intensities below 10^6 V/cm, the two additional terms can be neglected in comparison with the SCHOTTKY term.

8.1.3. Equation for Emission from Heterogeneous Surfaces in the Absence of External Electrical Fields

Equations (8.1.1–6) and (8.1.2–2) for the degree of ionization α, which were derived in the preceding sections, are strictly valid only for the ideal case of a clean homogeneous surface. However, as will be seen in the sequel, most investigations of surface ionization have been carried out on polycrystalline metal surfaces—frequently with an unknown covering of foreign gas. The influence of physical, chemical, and induced heterogeneity on the work function, and consequently on the degree of ionization α, must therefore be taken into account.

The work function $e\overline{\varphi}$ of a polycrystalline surface is the weighted mean of the values for the different single-crystal surfaces $e\varphi_\nu$, according to the formula $e\overline{\varphi} = e\sum_\nu f_\nu \varphi_\nu$. Here f_ν gives the fraction of the νth single-crystal surface in the total emitting metal surface and $e\varphi_\nu$ is the work function of the corresponding area. The extent to which the φ_ν differ from one another, for example for the case of wolfram, can be found from the results of SMITH [*660*].

On the assumption that the emission of a "patch" f_ν is independent of that of its neighbors, it follows that the total emission $(i_{tot})_\nu$ from the patch can be written in terms of the total emission current I_{tot} as

$$(i_{tot})_\nu = f_\nu I_{tot} = (i_+)_\nu + (i_0)_\nu . \qquad (8.1.3\text{–}1)$$

Then the expression for the ionization coefficient β_ν of such a patch is

$$\beta_\nu = \frac{(i_+)_\nu}{(i_{tot})_\nu} = \frac{(i_+)_\nu}{(i_+)_\nu + (i_0)_\nu} . \qquad (8.1.3\text{–}2)$$

Moreover, since

$$\beta = \frac{\alpha}{1+\alpha} = \frac{1}{1+(1/\alpha)} \quad \text{and} \quad \frac{1}{\alpha} = \frac{1}{\beta} - 1 , \qquad (8.1.3\text{–}3)$$

application of Eq. (8.1.1–6) leads to

$$\beta_\nu = \left[1 + \frac{G_0}{G_+}\,\frac{1-r_0}{1-r_+}\exp\frac{e[I-\varphi_\nu(T)]}{kT}\right]^{-1} . \qquad (8.1.3\text{–}4)$$

The ionization coefficient β of the composite surface is

$$\beta = \sum_\nu f_\nu \beta_\nu . \qquad (8.1.3\text{–}5)$$

From Eqs. (8.1.3–3)—(8.1.3–5), ZEMEL [*798*] finds that the reciprocal of the degree of ionization is given by

$$\frac{1}{\alpha} = \frac{1}{\sum_\nu f_\nu \beta_\nu} - 1 , \qquad (8.1.3\text{–}6)$$

$$\frac{1}{\alpha} = \sum_\nu \frac{1}{f_\nu}\left[1 + \frac{G_0}{G_+}\,\frac{1-r_0}{1-r_+}\exp\frac{e[I-\varphi_\nu(T)]}{kT}\right] - 1 . \qquad (8.1.3\text{–}7)$$

In considering ZEMEL's comparison of values computed from Eq. (8.1.3–7) with the experimental data obtained by PHIPPS et al. [*566, 568*] for systems such as potassium and sodium halides on polycrystalline wolfram, it must be noted that this author (without knowing the actual surface structure of wolfram) treated the "fractional patch areas" f_ν as well as the work functions of the patches as free parameters (and arbitrarily used some of the φ_ν values given by HERRING and NICHOLS [*307*] in obtaining a best fit to the experimental data. His Eq. (8.1.3–7) also does not consider possible changes in work function as a result of foreign gas covering of the surface.

ROMANOV and STARODUBTSEV [*608*], in their investigations of surface ionization of sodium and lithium on polycrystalline wolfram surfaces, have also interpreted their results in terms of the occurrence of patches with different work functions, and have assumed that the effect of foreign gas cover is negligible. This, however, is not justified. For example, for Cs absorbed on wolfram the work function is reduced by approximately 2.4 eV for a coverage degree $\theta \approx 0.40$ (Table VI in the article by EBERHAGEN [*169*]). The usual residual gases (nitrogen, hydrogen, and carbon monoxide increase the work function of wolfram. For the (113) plane, for instance, the changes produced by the different gases are:

$$W_{(113)}\text{-N}: \quad \Delta(e\varphi) \approx -0.35 \text{ eV}$$
$$W_{(113)}\text{-H}: \quad \Delta(e\varphi) \approx -0.45 \text{ eV}$$
$$W_{(113)}\text{-CO}: \quad \Delta(e\varphi) \approx -0.86 \text{ eV}\,.$$

For oxygen on polycrystalline wolfram, according to ref. [*244*] the increase could be as high as

$$\text{W—O}: \ \Delta(e\varphi) \approx -1.9 \text{ eV}\,.$$

Neither the best vacua (ca. 8×10^{-8} mm Hg) attained in the experiments of ROMANOV and STARODUBTSEV [*608*] nor the high surface temperatures (2200° C) used exclude a partial covering of the surface with oxygen (chapter 3).

Moreover, Eq. (8.1.3–7) fails to reproduce the rapid changes in the degree of ionization below a definite threshold temperature T_s. For Cs on W, DOBREZOW [*153*] gives $T_s \approx 1200°$ K; and for Cu on W, WEIERSHAUSEN [*774*] gives $T_s \approx 1300°$ K. These changes can be attributed in part to the increasing development of adsorption layers on the surface with decreasing surface temperature.

Unfortunately, there is still no generally valid relation which describes the changes in work function over the whole range of coverings $0 \leqslant \theta \leqslant 1$ for metal surfaces which act as electron acceptors or donors in adsorption systems. Although the authors mentioned above explain their results by use of the surface patch effect alone and disregard the surface covering, WEIERSHAUSEN [*774*] has recently taken the opposite tack in interpreting his results, which showed that the temperature dependence of the degree of ionization α deviated from the SAHA-LANGMUIR formula (8.1.1–5) in the case of Ag or Cu ionized on polycrystalline wolfram below 2500° K. He attributed this deviation to the adsorption of oxygen layers on the wolfram surface, and disregarded its polycrystalline structure.

No quantitative correction of Eq. (8.1.3–7) for this covering effect has yet been given. However, some qualitative statements can be made.

In the stationary state, let N represent the equal numbers of particles striking the surface and evaporating from it per unit area per unit time. Then the relation between N and the number n of particles adsorbed on the surface for the case $\varepsilon = 0$ is

$$n = N[K_i \exp(-E_i/kT) + K_a \exp(-E_a/kT)], \qquad (8.1.3\text{–}8)$$

where K_i and K_a are coefficients which depend only slightly on temperature, E_i and E_a are the isothermal heats of vaporization of ion and atom, respectively, and $E_i - E_a = e(\varphi - I)$. Equation (8.1.3–8) shows that the number n of particles adsorbed on the surface increases as T decreases.

For cases in which $(I - \varphi) < 0$, as for Rb or Cs atoms on a wolfram surface, the increase of n (or θ, the ratio of n to the number of available adsorption sites) resulting from a decrease in T leads to a reduction in φ. According to Eq. (8.1.3–8), this results in a reduction in the ion current. On the other hand, the direct effect of reducing T in Eq. (8.1.3–8) is to *increase* the ion current. Therefore, as the temperature decreases, the ion current would be expected to go through a maximum and then fall off rapidly since n (and therefore θ) depends exponentially on T. This behavior is confirmed experimentally (see Fig. 1 in the article by DATZ and TAYLOR [*136*], and Figs. 109 and 110 of the article by DOBREZOW [153]).

For cases in which $(I - \varphi) > 0$, as for Li, Na, or Cu atoms on wolfram surfaces, the increase in n with decreasing temperature T again leads to a reduction in φ. This causes a steeper drop in the ion current than would be expected from Eq. (8.1.3–8). However, when the wolfram surface is slightly oxidized ($\theta \ll 1$) so that the work function is increased, the oxygen atoms compete with the beam particles for adsorption positions on the surface at lower surface temperatures. However, the available results seem to show that these additional adsorbed beam particles have a negligible effect on the change in work function of polycrystalline surfaces above some definite degree of oxygen coverage. In this case the ion current may be expected to pass through a maximum as the surface temperature decreases. This is confirmed by the experimental results of WERNING [*778*] for the system Ba on wolfram oxide and of WEIERSHAUSEN [*774*] for Ag on wolfram oxide.

8.1.4. Equation for Emission from Heterogeneous Surfaces in an External Electrical Field

For surface ionization on a patch surface (polycrystalline surface) the degree of ionization α is greatest on those patches whose work function φ is largest. Between regions with different work functions, however, there is an electric field (patch field) which results from differences in contact potential (Fig. 8.1.4.a). The direction of this field is such that it accelerates the ions which are emitted from the regions of low work function $\varphi_{\min}$ and, conversely, retards the ion emission from surface regions with larger work function $\varphi_{\max}$. If an external electric field ε is applied to the surface in order to accelerate the ions away from the surface, this compensates the "patch field" and increases the ion current (anomalous SCHOTTKY effect). For the case of very high electric fields ($\varepsilon > 10^6$ V/cm) one must take account of polarization effects, compensation of image force, etc. For a patch surface in a high external electrical field, one finds from Eqs. (8.1.3–4) and (8.1.3–7) that the ionization coefficient β can be represented by

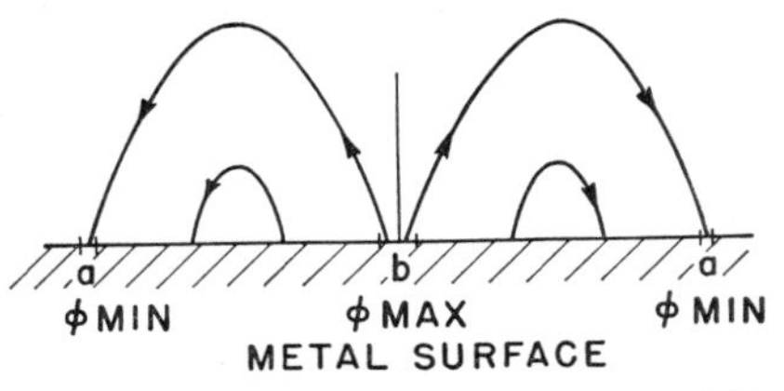

Fig. 8.1.4.a. Schematic diagram of patch fields caused by the existence of surface patches with different work function (e.g., $\varphi_{\min}$, $\varphi_{\max}$) on a polycrystalline metal surface

$$\beta = \sum_{\nu} \frac{f_\nu}{1 + A \exp\left[(e/kT)\,(I - \varphi_\nu - \psi)\right]}, \tag{8.1.4–1}$$

where

$$A = \frac{G_+}{G_0}\,\frac{(1 - r_+)}{(1 - r_0)},$$

and

$$\psi = \sqrt{e\,\varepsilon} + \varepsilon r_0 + \frac{\varepsilon^2}{2e}\,(\alpha_0 - \alpha_i).$$

From this it follows that the positive-ion current i_+ is

$$i_+ = \sum_{\nu} \frac{e f_\nu N_0 A}{A + \exp\left[(e/kT)\,(I - \varphi_\nu - \psi)\right]}. \tag{8.1.4–2}$$

In Eqs. (8.1.4–1) and (8.1.4–2), following ZANDBERG and IONOV [797], it is assumed that the work function for each patch increases by the same amount ψ with increasing ε. These authors divide the surface patches into two groups. The first group satisfies the inequality $e(I - \varphi_\nu - \psi) \gg kT$ and is designated by a subscript μ; the others, which do not satisfy this inequality, are designated by the subscript λ.

Then Eq. (8.1.4–2) can also be written as

$$i_+ = eAN\left[\sum_\mu f_\mu \exp\left(\frac{e(\varphi_\mu + \psi - I)}{kT}\right) + \sum_\lambda \frac{f_\lambda}{A + \exp[e(I - \varphi_\lambda - \psi)/kT]}\right]. \tag{8.1.4–3}$$

This equation represents the dependence of surface ionization on temperature and external electric field, as will be discussed in more detail when it is compared with the experimental results in Section 8.3.2.5.

A few cases in which Eq. (8.1.4–3) can be simplified will now be discussed. For case 1 in which $e[I - (\varphi_\mu + \psi)] \ll kT$, the second term in Eq. (8.1.4–3) can be dropped so that

$$i_+ \approx e\bar{A}\exp\left[\frac{e(\overline{\varphi} + \psi - I)}{kT}\right], \tag{8.1.4–4}$$

where

$$\bar{A} = AN\sum_\mu f_\mu \exp\left[\frac{e(\varphi_\mu - \overline{\varphi})}{kT}\right]. \tag{8.1.4–5}$$

Here $e\overline{\varphi} = e\sum_\mu f_\mu \varphi_\mu$ is an "effective" work function.

It is evident from Eq. (8.1.4–4) that the graph of $\ln i_+$ vs $(1/T)$ is no longer a straight line as it was for the SAHA-LANGMUIR equation. Moreover, the ion current $i_+ = F(T)$ for case 1 can be represented by

$$i_+ = i_+(\varepsilon = 0)\exp\left(\frac{e\psi}{kT}\right), \tag{8.1.4–6}$$

where

$$i_+(\varepsilon = 0) = e\bar{A}\exp\left[\frac{e(\overline{\varphi} - I)}{kT}\right]. \tag{8.1.4–7}$$

This means that the graph of $T \ln i_+$ vs ψ, for different values of T, gives a family of parallel lines.

For case 2 in which $e\{I - [(\varphi_\mu)_{\min} + \psi]\} \leqslant kT$, only the second term in Eq. (8.1.4–3) is significant so that

$$i_+ = eAN\sum_\lambda \frac{f_\lambda}{A + \exp[e(I - \varphi_\lambda - \psi)/kT]}. \tag{8.1.4–8}$$

From this follows the approximation

$$i_+ \approx \frac{eN\bar{f}}{1 + (1/A)\exp[e(I - \overline{\varphi} - \psi)/kT]}. \tag{8.1.4–9}$$

Hence the effective degree of ionization $\overline{\alpha}$ is

$$\overline{\alpha} = \bar{A}\exp\left[\frac{e(\varphi + \psi - I)}{kT}\right], \tag{8.1.4–10}$$

where

$$\bar{A} = AN \sum_{\lambda} f_\lambda \exp\left[\frac{e(\varphi_\lambda - \bar{\varphi})}{kT}\right]. \qquad (8.1.4\text{–}11)$$

From Eq. (8.1.4–9) it follows that the ion current i_+ decreases only slightly with increasing temperature. Its maximum range of variation is $eNf > i_+ > eNfA/(1 + A)$ for $0 \leqslant T \leqslant \infty$.

Finally, for the case in which $e(\varphi_\lambda + \psi - I) \gg kT$, the ion current i_+ can be represented approximately by $i_+ \approx eNf$. This means that every molecule that strikes the surface becomes ionized so that N can be determined.

8.2. Relation of the Saha-Langmuir Equation to the Frenkel Equation and to Charge-Transfer Probabilities

Consider next the possible connection between the temperature dependence of desorption probabilities for ions and atoms on metal surfaces, their probabilities of charge transfer, and the temperature dependence of the degree of ionization α. Attempts in this direction have been made by many authors (LANGMUIR and TAYLOR [*705*], GURNEY [*271*], STARODUBTSEV [*678*], BECKER [*61*], DOBREZOW [*153*], and HUGHES [*328*]) and have led to a variety of results. (See, for example, STARODUBTSEV [*678*], HUGHES [*328*], and DOBREZOW [*153*]).

The first topics to be discussed are the desorption and charge-transfer probabilities for ions and atoms on metal surfaces. The desorption of alkalis from metal surfaces will be taken as a specific example.

Alkali atoms adsorbed on a metal surface exchange electrons with the metal, and one can attempt to estimate the probabilities of the adatoms being desorbed as ions or neutrals. But the only atoms and ions that can be evaporated are those that are given an energy greater than or equal to the heat of evaporation (heat of evaporation E_a for neutrals or E_i for the ions [Fig. 4.3.1.b]). The probability of converting an adatom into an ion or neutral atom at a metal surface will first be considered in more detail.

From the picture of adsorption of electropositive atoms on metal surfaces, one sees that as a free atom approaches a metal surface there is a perturbation of the discrete energy levels of its external electrons. As a result of this perturbation, the allowed states are spread out into an energy band whose maximum $eI(r_c)$ is shifted away from the discrete energy level E for the valence electron of the free atom at an infinite distance from the surface. This splitting is greater for lower values of the ionization voltage I of the free atom and for smaller values of the distance of closest approach $r_{\min}$. Above a certain critical distance r_c, electron exchange is no longer possible because the potential barrier between the adatom and the metal surface becomes broader and higher.

For distances $r < r_c$, the electron exchange is so intense that atomic and ionic states of the adatom cannot be distinguished. Therefore the charge of the adatom for $r > r_c$ either is 0 if the electron is in a state with energy $E = E_c$ such that the adatom is neutralized, or the charge of the adatom is $+ ne$ (for electropositive adatoms) when the state $E = E_c$ is not occupied by an electron. From FERMI statistics, the probability that the level $E = E_c$ is occupied by an electron is

$$W_0 = W(E_c) = \{1 + \exp[(E_c - \eta)/kT]\}^{-1}, \tag{8.2–1}$$

where η is the electrochemical potential. The probability that this level is not occupied is

$$W_+ = 1 - W(E_c) = \{1 + \exp[-(E_c - \eta))/kT\}^{-1}. \tag{8.2–2}$$

The charge-transfer probability of a neutral atom is then proportional to the ratio of the probabilities defined in Eqs. (8.2–2) and (8.2–1), namely,

$$\frac{W_+}{W_0} = A\,\frac{1 + \exp[(E_c - \eta)/kT]}{1 + \exp[-(E_c - \eta)/kT]} = A\exp\left[\frac{E_c - \eta}{kT}\right]. \tag{8.2–3}$$

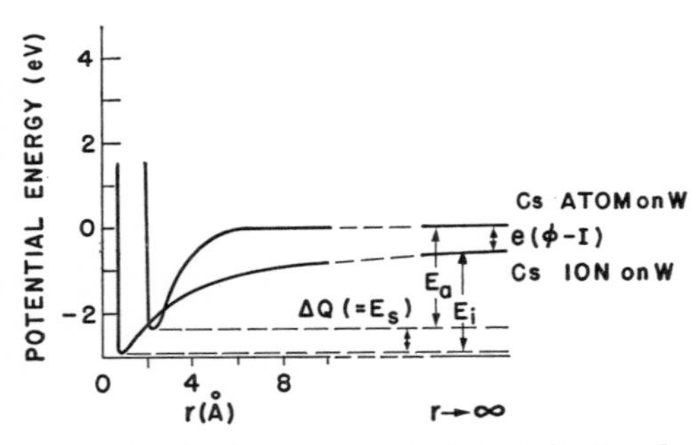

Fig. 8.2.a. Potential energy diagram for the adsorption of cesium atoms and ions on a wolfram surface

The coefficient $A = g_+/g_0$ is the ratio of the statistical weights of the ionic and atomic states, as defined in Eq. (8.1.1–5). The difference $E_c - \eta$ in the exponent is equal to $e(\varphi - I_{rc}) = \Delta Q$, where I_{rc} is the ionization potential for the adatom at a distance r_c. From the potential diagram in Fig. 8.2.a, one finds that $\Delta Q = E_s = e(\varphi - I_{rc}) = E_i - E_a - e(I - \varphi)$, a relation which DOBREZOW [*153*] obtained from consideration of an energy cycle. The quantities E_i and E_a are the heats of vaporization of ion and atom, which have already been used in Eq. (8.3.1–8). Thus Eq. (8.2–3) may also be written as

$$\frac{W_+}{W_0} = A\exp\left[\frac{E_i - E_a - e(I - \varphi)}{kT}\right]. \tag{8.2–4}$$

However, to find the ratio of the number of ions to the number of atoms evaporated from the surface (ionization degree α) the probability of charge transfer must still be multiplied by the probability of evaporation of ions and atoms. FRENKEL [*227*] showed that, in a MAXWELL distribution for a surface temperature T, the probability that the kinetic energy of a given atom or ion will exceed the evaporation energy E_i or E_a is

$$W(E_i, E_0) = \omega_0(E_i, E_a) \exp(E_{i,a}/kT) = 1/\tau_{i,a}. \qquad (8.2\text{–}5)$$

The probability of evaporation $W(E_i, E_a)$ for ions or atoms can be determined directly from experiment since, in the case of nonactivated adsorption, the mean residence time $\tau_{i,a}$ of the adsorbed particles on the surface is inversely proportional to this quantity. These quantities can be measured by the method described in Section 8.3.1.5. The $\omega_0(E_i, E_a)$ corresponds to a frequency of the order of a normal vibration of the adatom (or ion) in the adsorbed state. (For example, HUGHES [*328*] found for Rb^+: $\omega_0(E_i) = 2.85 \times 10^{+12}$ sec^{-1}; for Rb^0: $\omega_0(E_A) = 4 \times 10^{13}$ sec^{-1}).

The degree of ionization α is obtained by multiplying the probability for charge transfer, Eq. (8.2–4), by the probability that the particle will have enough energy to evaporate, Eq. (8.2–5). The result is

$$\alpha(\varepsilon = 0) = \frac{N_i}{N_a} = \frac{W_+}{W_0} \frac{W(E_i)}{W(E_a)} = A \exp\left[\frac{e(\varphi - I)}{kT}\right]. \qquad (8.2\text{–}6)$$

Equation (8.1.4–17) is the desired relation between the SAHA-LANGMUIR equation, the FRENKEL equation, and the ratio of the charge-transfer probabilities. This relation makes possible a more detailed study of the mechanism of surface ionization and, in particular, the charge-transfer probabilities than has previously been possible. In contrast to the numerous data for ionization coefficients and degree of ionization, there are so far no satisfactory experimental data available on the mechanism of surface ionization. For this reason a pulsed-molecular-beam mass spectrometer has been described (KAMINSKY [*361, 365*]) which will make possible the mass spectrometric determination of desorption probabilities for ions and atoms as well as the determination of ionization coefficients, so that the charge-transfer probabilities can be determined. Some of the more noteworthy features of this mass spectrometer will be discussed in Section 8.3.1.5.

As a final item in this connection, one should mention a relation found by STARODUBTSEV [*678*] between the degree of ionization or the ionization coefficient and desorption probabilities. He sets the degree of ionization α equal to $W(E_1)/W(E_a)$, the ratio of the desorption probabilities, and disregards the charge-transfer probability W_+/W_0. He obtains

$$\alpha = \frac{\dot{N}_+}{\dot{N}_0} = \frac{W(E_i)}{W(E_a)} = A \exp\left(\frac{E_a - E_i}{kT}\right). \qquad (8.2\text{–}7)$$

To bring this equation to the familiar form of the SAHA-LANGMUIR equation, STARODUBTSEV uses the well-known SCHOTTKY equation

$$E_a - E_i = (\varphi - I)e \qquad (8.2\text{–}8)$$

to replace the difference $E_a - E_i$ and obtains

$$\alpha = \frac{\dot{N}_+}{\dot{N}_0} = A \exp \frac{E_a - E_i}{kT} = A \exp \frac{e(\varphi - I)}{kT} . \tag{8.2–9}$$

The use of the SCHOTTKY equation (8.2–8) in Eq. (8.2–9) is, however, invalid since the former relates the heats of evaporation E_a of the atoms and E_i of the ions of the same element (for example Pt atoms and Pt ions from heated platinum) to the work function $e\varphi$ and the ionization energy eI of the same element, but does not give the relation for gases of one element adsorbed on the metal surface of another element. Thus, for example, the value of the heat of evaporation for Cs atoms on Cs is only $E_a = 0.7$ eV, whereas the value for evaporation of Cs atoms from a wolfram surface is $E_{a,W} = 2.83$ eV. From the potential diagram shown already in Fig. 8.2.a, one sees that the evaporation of adsorbed gases from a metal surface is governed, not by Eq. (8.1.4–19) but rather by

$$E_a - E_i = e(\varphi - I) - \Delta Q = e(\varphi - I) - E_s . \tag{8.2–10}$$

BECKER finds that for Cs on wolfram $\Delta Q \approx 0.5$ eV and is thus of the same order of magnitude as $e(\varphi - I)$. Here eI is the ionization energy of the adsorbed gas (Cs in our example), and $e\varphi$ is the work function of the metal. STARODUBTSEV's value of E_i for the evaporation of Na ions from wolfram must be corrected because he computes it from the experimental value $E_a = 2.73$ eV by (incorrectly) applying Eq. (8.2–8) to obtain

$$E_i = E_a + e(I - \varphi) = 3.3 \text{ eV} . \tag{8.2–11}$$

The same correction must be made in the corresponding value listed in Table 1 of DATZ and TAYLOR [*138*]. The correct expression is (see Fig. 4.3.1.6)

$$E_i = E_a + e(I - \varphi) - \Delta Q = 3.3 \text{ eV} - \Delta Q . \tag{8.2–12}$$

Unfortunately no reliable value for ΔQ is yet available for Na on W.

8.3. Experimental Investigation of Positive Surface Ionization (PSI)

8.3.1. Experimental Methods

8.3.1.1. "Bulb" Method without Mass Spectrometric Analysis

The so-called "bulb" method was used for instance by LANGMUIR [*424*], IVES [*345*], MEYER [*471*], BECKER [*61*], and KILIAN [*377*] in their first classic investigations of the surface ionization of cesium or potassium on wolfram or wolfram oxide surfaces. The apparatus, which was enclosed in a vacuum chamber, consisted of a cylindrical collector concentric with the heated wire which served as the ionizing surface. Usually the collector was divided transversely into three sections, only the central section being used for measuring the emission current while the others served as guard rings. A small amount of alkali vapor was admitted into the evacuated bulb after being purified by vacuum distillation. The vapor pressure of the alkali metal, and thus the rate at which the alkali atoms arrived at the ionizing surfaces, could be set by regulating the gas temperature by means of a thermostat surrounding the vessel. Note that the accuracy of the results on ionization coefficients depends on the correctness of the information about the rate of arrival (and thus about the gas temperature). The positive-ion current was determined by applying a suitable voltage to the collector. However, at high temperatures of the central wire it then appeared that photoelectrons as well as secondary electrons were emitted from the collector and these produced an undesired contribution to the recorded collector current. MORGULIS [*495*] applied an axial magnetic field to hinder the bombardment of photo- and secondary electrons on the ionizing wire, while ZANDBERG et al. [*796*] recently made use of a highly transparent grid to reduce the effects of secondary electrons.

Since the collector records only the total current, one cannot determine the contributions from ions arising from impurities in the gas or the metal surface itself. In this respect the mass spectrometric procedures offer an advantage.

8.3.1.2. Single-Filament Ion Source with Mass Spectrometer: Plain Emitter—Porous Emitter

In this method the substance to be ionized is usually deposited in the form of a solution (chloride, nitrate, or the like) on the ionization filament (e.g., wolfram) in a thin layer and dried. Subsequent heating of the filament causes this layer to evaporate. Above a certain surface temperature, it may evaporate partly as ions. A mass spectrometer is then used to distinguish the different types of evaporated ions, and because of the high sensitivity of such an instrument (especially when it uses an electron-multiplier as detector) very small amounts of substance can be detected.

For example, in the determination of the absolute isotopic composition of uranium STEVENS and HARKNESS [*683a*] used a single-filament surface-ionization technique in the following way. The sample was deposited in the form of a solution (uranyl nitrate dissolved in a 0.05 to 2 N nitric acid solution up to a concentration of about 5 mg of uranium per ml solution) onto a small tantalum ribbon (0.001 inch thick and 0.030 inch wide) and evaporated to dryness. The ribbon was placed in the ion source of a mass spectrometer and heated. With a filament prepared in this way, the authors were able to obtain an UO_2^+ ion beam current of 5×10^{-11} A from a 50 μg sample or uranium (using a 60° magnetic-sector-field mass spectrometer with a mean radius of curvature of 12 inches with a source slit 0.006 inch wide and an 8-keV accelerating voltage). The authors point out that they also used the multiple-filament technique (Section 8.3.1.3) and usually obtained the metal ions U^+ rather than the oxide ions UO_2^+ as in the single-filament technique. For more details the reader is referred to references [*683a*] and [*683b*].

The single-filament surface-ionization technique has been used by numerous authors (see also the review articles by Ewald and Hintenberger [*200*] and by Inghram and Hayden [*337*]) and has proved to be quite useful in routine analyses. However, the Saha-Langmuir equation cannot be properly tested by this method because of the complicated ionization and evaporation processes (as when the melt spreads over the ionization surface; if the melt forms a layer several atoms thick, the evaporating particles may interact more strongly with each other than with the carrier surface). This has been discussed by Dobrezow [*153*] and Kaminsky [*360*].

The use of powder anodes for thermal ionization also deserves brief mention in this connection. (For a more detailed discussion, see Ewald and Hintenberger [*200*].) These anodes, which were developed by Kunsman [*409, 410, 411*], Koch [*398*], Walcher [*751*], Schmidt [*638*], and Blewett and Jones [*77*], give a very high yield because of the large surface area. Thus Koch [*398*] loaded a previously vacuum-purified wolfram powder with a few volume percent of an alkali chloride and "activated this mixture by slow heating to 800° C. As a result of dissociation of the alkali chloride and evaporation of the chloride, an alkali metal layer is formed on the surface of the mixed powder; and this, on further heating becomes a source of ions of the alkali metal involved. In this way, ion currents up to 10 μA could be obtained for a period of 10—20 hrs.

Another technique to improve the ionization efficiency and to stabilize the ion emission from a single-filament source has been developed by Goris [*250a*]. He prepared a mixture of tantalum metal powder and tantalum pentoxide (approximately one part metal powder to three parts pentoxide, finely ground together) and applied it as an aqueous slurry to a tantalum filament. After the filament was heated (to about 1870° C) at a pressure of 10^{-6} mm Hg or less for a few seconds, a deposit of tantalum metal from the oxide joins the metal particles together and to the filament and gives a highly porous metal layer on the tantalum filament. Using such a filament in a micro analysis for uranium, he was able to maintain stable and intense UO_2^+ ion emission and to extend the lower detection limit by approximately a factor of 50 in comparison with a plain tantalum filament.

Finally, it should be mentioned that the emission of positive ions from porous emitter surfaces has recently become the subject of great interest in connection with the development of electrostatic propulsion units. Optimum operation of such units requires a maximum ionization efficiency with the emitter at a minimum operating temperature in order to insure maximum use of the fuel (i.e., alkali vapors) and a high power efficiency (see such references as E. STUHLINGER [*692a*]). Since high current densities in the ion beam when fuel material flows steadily to the ionizing surface seems to be provided by porous surfaces, both experimental studies (such as those of O. K. HUSMANN et al. [*333a, 333b, 333c, 556a*], STAVISSKII et al. [*426a, 681a*], and SNOKE et al [*664a*]) and theoretical studies (such as those of D. ZUCCARO et al. [*806a*], T. W. REYNOLDS et al. [*588b*], G. M. NAZARIN et al. [*519a*]) have been started. The experimental results available so far do not allow a general description of the highly complex phenomenon of surface ionization on porous surfaces. The process is governed by numerous important parameters such as the ionization potential of the gas, the work function of the emitter surface, the emitter temperature, the external electrical field (drawing-out field for ions), and the rate of vapor flow through the porous emitter. The last of these depends in turn on such parameters as grain size, grain form, thickness of the emitter plate, consolidation of the porous material, and the vapor pressure.

HUSMANN studied the surface ionization of alkali vapors (especially Cs) on porous emitters such as W, Mo, Ta, and Re. His emitter plates had thicknesses of from 0.020 to 0.104 inch (HUSMANN [*333c*]) and were placed in turn in an ion-optical system suggested by PIERCE [*568a*] to suppress beam spreading due to space charge and to extract ions at high current densities. With one particular porous wolfram plate (with an average pore diameter of 3.20 μ, an average spacing between pores of 15 μ, an average number of pores per surface area of 3×10^5 per cm^2, and a helium transmission coefficient* for a 0.020 inch thickness of porous plate of 1.3×10^4), he was able to obtain a Cs-ion current density of 18 mA/cm^2 in continuous operation over several hours. A comparison of porous emitters such as wolfram, rhenium, molybdenum, and tantalum up to ion current densities of 10 mA/cm^2 indicated that wolfram offered the maximum ionization at minimum emitter temperatures and the highest power efficiency. His experimental results also indicate that the number of pores per unit area is relatively more important for a high ionization efficiency than the choice of the pore diameter, but reduction of pore size is limited by the clogging of the pores. He states that a reasonable choice seems to be 3×10^6 pores per cm^2 (with an

---

* The helium transmission coefficient defines a ratio of the number of helium atoms passing through the porous plate to the number of helium atoms impinging on the inlet side of the porous plate, determined over an emitter temperature range from 300° to 1600° K (HUSMANN [*333a*]).

average pore diameter between 1.8 μ and 3.2 μ) since for such a porous wolfram emitter he found that the ionization efficiency für Cs was more than 97% at a current density of 10 mA/cm^2. In order to achieve both high ionization and high conductance of the alkali vapor, he suggested that the porous emitter should be prepared from powders whose grains were spherical and of quite uniform size. If treated properly, such powders would also offer the advantages of dense packing and small tendency to fuse into an impervious aggregate.

8.3.1.3. *Multiple-Filament Source with Mass Spectrometer*

In order to distinguish between the various evaporation and ionization processes which occur with a single-filament source, INGHRAM and CHUPKA [*336*] proposed a very simple and practical solution—a triple-filament source. It consists of an evaporation filament, an ionization filament, and a filament to ensure symmetric potentials. This method of separating the evaporation region and ionization region is also the principle of the beam method which was used even earlier and has greater accuracy for the determination of the number of atoms striking the surface used for ionization (Section 8.3.1.4). The ion source recently developed by HINTENBERGER et al. [*314, 746*] can be regarded as a transition type between the beam source and the triple-filament ion source. As seen from Fig. 8.3.1.3.a, the atoms and molecules to be ionized evaporate from a crucible (large evaporator surface with a small effusion hole) where the vapor pressure is not so well defined, as, for example, the beam density is when a KNUDSEN cell (with laminar flow) is used in the beam method. The use of a crucible makes possible a longer duration of evaporation than can be obtained with the evaporation filament of a triple-filament source. The ion source of HINTENBERGER et al. has the further advantage that the surface available for ionization (an annular cavity between heated coaxial cylinders R_1 and R_2) is greater than that of the ionization-filament source. Ion formation is also favored by the multiple collisions which can occur in this source (important, for example, in the ionization of KCl on Pt). However, because of the distribution of the "drawing out" potential for the ions between cylinders R_1 and R_2, only a certain fraction of the ions so formed will have a chance to leave the exit slit in R_1. This partly nullifies the initial advantage of more capious ion

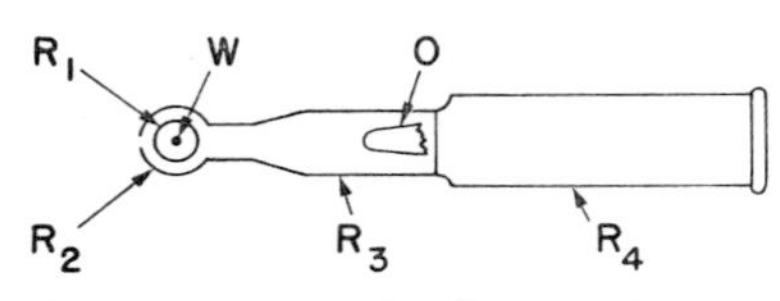

Fig. 8.3.1.3.a. Schematic diagram of an ion source reported by HINTENBERGER and LANG [*314*]. The hot coaxial cylinders R_1 and R_2 serve as the actual ionization surfaces. Wire W is the heating element of cylinder R_1. The oven O which evaporates the material to be ionized is fitted into the tube R_4. The tube R_3 connects the evaporation and the ionization region. (The temperature distribution is $T_{R_1} > T_{R_3} > T_{R_4}$)

emission. With their ion source, HINTENBERGER et al. were able to detect extremely small amounts of alkali metal (e.g., amounts of Cs of the order of 5×10^{-13} g).

The triple-filament source has been successfully applied to the study of the temperature dependence of the SAHA-LANGMUIR equation, as in the recent work of WEIERSHAUSEN [*774*] on the surface ionization of copper and silver on wolfram.

8.3.1.4. Beam Methods

COPLEY and PHIPPS [*124, 125*] and DOBREZOW [*147*] were the first to use a well-defined molecular or atomic beam produced from an effusion oven, collimated, and allowed to strike an ionizing surface. Because they were able to prevent the collimated beam from striking the collector surface, they could disregard the effect of a secondary electron current. The experimental results which they obtained for surface temperatures up to 3000° K were in good agreement with the SAHA-LANGMUIR equation. The beam method was then used by numerous investigators for studies of surface ionization, some of the more recent examples being the work of DATZ and TAYLOR [*136, 137*], WERNING [*778*] and ZANDBERG [*795*]. In this connection, reference should also be made to the numerous studies in which surface ionization is used solely as a means of detection in studies made with molecular beams, such as those in high frequency spectroscopy. KOPFERMANN [*405*] gives an excellent survey of these.

8.3.1.5. Pulsed-Beam Methods

Very few experimental investigations of the mechanism of surface ionization have been carried out so far. There exist at present only very few and not too consistent results concerning the probabilities of desorption of ions and atoms from metal surfaces. STARODUBTSEV [*678*], KNAUER [*389*], BULL and MARSHALL [*104*], HUGHES and LEVINSTEIN [*327*], SCHEER and FINE [*636a, 636b*] have used a pulsed molecular-beam apparatus in an attempt to determine the mean residence time of ions and atoms on metal surfaces. The beam of particles incident on the ionizing filament is interrupted by a vibrator or a rotating slotted disk. Then the mean residence time (the reciprocal of the probability of evaporation) can be determined from the exponential decay of the current of ions desorbing from the ionizing filament. The fluctuations in the experimental values of these mean residence times obtained so far are so large, even for a given adsorption system (such as the system K on W studied by STARODUBTSEV [*678*], KNAUER [*389*], and BULL and MARSHALL [*104*]), that no reliable statements can be made about probabilities of desorption (see table 8.3.1.5.A). The reason for these fluctuations is that

Table 8.3.1.5.A. *Mean adsorption lifetime τ_i for different alkali metals adsorbed on a wolfram surface*

T (°K)	Na on W	K on W			Rb on W	Cs on W	
	STARODUBTSEV [*678*] τ_i (sec)	STARODUBTSEV [*678*] τ_i (sec)	BULL and MARSHALL [*104*] τ_i (sec)	KNAUER [*389*] τ_i (sec)	HUGHES [*327*] τ_i (sec)	KNAUER [*389*] τ_i (sec)	SCHEER and FINE [*636a*] τ_i (sec)
1100					10^{-3}		2.6×10^{-3}
1250		1.16×10^{-3}			5×10^{-5}		4.3×10^{-4}
1360						10^{-3}	
1400	1.3×10^{-3}						
1490			3×10^{-5}				
1600						10^{-5}	
1690				10^{-3}			
2200				10^{-5}			
Q_{obs} (kcal/mole)	75	57		66	45	82	46.1

many relevant parameters have not been taken into account. Among these are the covering of the surface by foreign gases (vacuum usually *ca.* 10^{-6} mm Hg), polycrystalline surfaces with unknown structure (physical and chemical heterogeneities), the SKOUPY effect [*655*], the anomalous SCHOTTKY effect, the temperature drop from center to ends of the heated ionizing filament, and others (as discussed in more detail by KAMINSKY [*361, 365*]. Only HUGHES and LEVINSTEIN and to some extent SCHEER and FINE took adequate care to have well-defined surface conditions. Since none of these authors used a mass spectrometer as detector, they could make no statements concerning the composition (monomers and polymers) and purity of the incident beam, concerning isotope effects of the desorption probabilities, or concerning the reaction products that arise from interaction of the beam particles with the surface.

In order to make measurements of this type with the necessary accuracy, the pulsed-molecular-beam mass spectrometer shown schematically in Fig. 8.3.1.5.a has been built (KAMINSKY [*361, 365*]). A beam of molecules or atoms emerging from a double-oven system passes through a system of collimating and ion-deflecting plates and is interrupted mechanically by a rotating slotted disk P_1. The pulses (each consisting of n_1^0 molecules) impinge on a heated single-crystal wire f_1 placed in an ultra-high vacuum ($p < 10^{-9}$ mm Hg). Some of the atoms adsorbed during a particular pulse then desorb as ions and give rise to a decreasing ion current at the collector. This signal is amplified and put on an oscilloscope which is triggered by a photocell. A certain number n_2^+ of desorbed ions are detected at the collector of the mass spectrometer. A certain number n_2^0 of the particles desorb from f_1 as neutral atoms and impinge on a second heated filament f_2, which is at a higher temperature than f_1, and are ionized there. For cases in which no surface ionization occurs at f_2, an electron-impact source is provided. The detector used is a secondary-electron multiplier (SEM). The device SE in Fig. 8.3.1.5.a serves to measure the electron emission from the wire. To study the isotope effect on desorption probabilities in a different experimental setup (KAMINSKY [*361*]), a monoisotopic ion beam (produced by passage

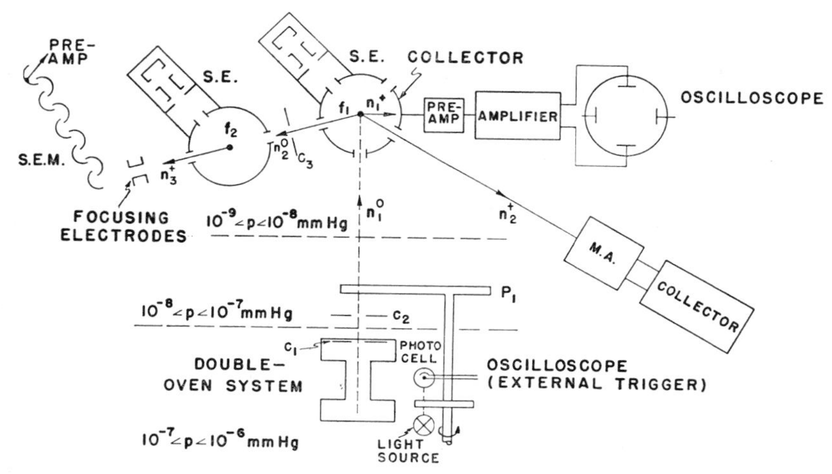

Fig. 8.3.1.5.a. Schematic diagram of the pulsed-molecular-beam mass spectrometer according to KAMINSKY [*361, 365*]

through the mass spectrometer) is allowed to strike the same wire arrangement, and the collector current resulting from ions desorbed from the filament f_1 is measured. For further details concerning the machine (velocity selector, etc.), see KAMINSKY [*361*] and [*365*].

8.3.2. Experimental Results

8.3.2.1. Positive Surface Ionization for Alkali Metal Atoms in Weak Electric Fields

The positive surface ionization of alkali metals (Li, Na, K, Rb, and Cs) on various surfaces (such as polycrystalline wolfram, molybdenum, tantalum, platinum, rhenium, nickel, and thoriated or platinized wolfram) has been studied for many decades. Table 8.3.2.1.A is an attempt to survey the numerous investigations of these systems by various authors. In addition to the type of gas which was ionized and the nature of the ionizing surface the table also, so far as possible, gives indications of

Table 8.3.2.1.A. *Ionization of Alkali Atoms on Metal Surfaces*

Atoms Li, Na, K, Rb, Cs	Polycrystalline Surfaces unless otherwise Indicated	Condition of Surface (Vacuum, mm Hg)	Temperature Range (° K)	Ion Current as a Function of Temperature or Electric Field	Remarks	Reference
Na, Rb	W, Single Crystal	5×10^{-9}	1200—2000	$f(T), f(\mathcal{E})$		F. L. Hughes, H. Levinstein [327]
Li	W	$2-3\times10^{-7}$	1800—2500	$f(T), f(\mathcal{E})$	$\mathcal{E}$=1.3MV/cm	E. Ya. Zandberg [795]
K	W	1×10^{-7}	500—2500	$f(T)$		E. Ya. Zandberg [794]
Li, Na, K	W	8×10^{-8}	1000—2200	$f(T), f(\mathcal{E})$		A. M. Romanov [609]
Na	W	$10^{-6}-10^{-7}$	~1200—2200	$f(T)$		A. M. Romanov, S. V. Starodubtsev [610]
K	W	$10^{-6}-10^{-7}$	1000—2000	$f(T), f(\mathcal{E})$	$\mathcal{E}$=10MV/cm	N. L. Ionov [342]
Li, Na, K, Rb,Cs	W, Pt, W-Pt	2×10^{-7}	1100—2100 1100—2800	$f(T)$		S. Datz, E. H. Taylor [136]
Rb	W, Pt-W	Not Stated	Not Stated			R. H. Tomlinson, A. K. das Gupta [724]
K	Pt	Not Stated	Not Stated		Note	C. F. Robinson [597]
Li	W	$3-5\times10^{-7}$	1200—2200	$f(T)$		A. M. Romanov, S. V. Starodubtsev [607]
Li, K	Ta, W, Sn					M. A. Eremeev [187]
Na	Thoriat. W			$f(T), f(\mathcal{E})$		L. D. Konozenko [404]
Na	W		800—1250	$f(T), f(\mathcal{E})$		R. C. L. Bosworth [86]
Li	Pt (gauze)		1270—1400	$f(T)$		T. P. Blewett, E. J. Jones [77]
Li, Na	Pt	$1-2\times10^{-6}$	Not Stated			M. B. Sampson, W. Bleakney [627]
K	W		1350—2170	$f(\mathcal{E}), f(T)$		T. E. Phipps, M. J. Copley [563, 564]
Na	W	$10^{-6}-10^{-3}$	1400—2400	$f(T), f(\mathcal{E})$		N. B. Morgulis [495]
Na	W, thoriat. W, Mo			$f(T)$		L. N. Dobretzow [147]
Na, Cs	Re, W		1000—2900	$f(T)$		H. Altertum, K. Krebs, R. Rompe [12]
Cs	Re, W	$\sim10^{-8}$	1070—1194(W) 951—1081(Re)	$f(T)$		M. D. Scheer, J. Fine [636a, 636b]
Cs	W		1100—1200	$f(T)$		I. Langmuir [422]
K	Mo		1000—1850	$f(T)$		R. C. Evans [199]
K, Cs	Ni		950—1350	$f(\mathcal{E}), f(T)$		P. B. Moon [491]
K	W, Mo, Ta		1000—2000	$f(T)$		E. Meyer [471]
K, Rb	W		~1200	$f(\mathcal{E}), f(T)$		T. J. Kilian [377]

vacuum conditions, the temperature range investigated (often a very uncertain indication since the data have to be taken from unclear figures), and the particular purpose of the investigation (e.g., the dependence of ion current on surface temperature or electric field).

It is surprising how often investigations of a given ionizing system have been repeated by different authors. The most popular ionization surface was polycrystalline wolfram, though many of the authors unfortunately did not consider the "patch character" of the surface [see Eqs. (8.1.4–3) and (8.1.4–4)] in discussing their results. The influence of foreign gas coverage of the surface is likewise disregarded. This is especially true of the earlier work (say up to COPLEY's work in 1953), in which the bulb method was used.

As background for a discussion of some of the details of positive surface ionization of alkali metals on metal surfaces, Table 8.3.2.1.B

Table 8.3.2.1.B. *Values of the degree of ionization α for alkali metal atoms on polycrystalline wolfram surfaces, computed from Eq. (5) with $e\varphi = 4.52$ eV*

Temperature (°K)	Li $I = 5.40$ eV	Na $I = 5.12$ eV	K $I = 4.32$ eV	Rb $I = 4.10$ eV	Cs $I = 3.88$ eV
1000	1.8×10^{-5}	5.0×10^{-4}	6.3	103.9	790.0
1500	5.5×10^{-4}	5.0×10^{-3}	2.2	35.8	72.0
2000	3.0×10^{-3}	1.6×10^{-2}	1.6	11.4	19.9
2500	8.4×10^{-3}	3.2×10^{-2}	1.3	7.0	9.8

presents the values computed from Eq. (8.1.1–5) for such atoms on wolfram surfaces at four different temperatures. For this computation, the value used for the work function was $e\varphi = 4.52$ eV, an experimental figure found by many authors as a mean value for polycrystalline surfaces. Table 8.3.2.1.B shows the rapid fall-off of the degree of ionization in the series Cs, Rb, K, Na, Li; and the small value of α for Li on wolfram makes it clear why the corresponding experimental values could be found only recently with improved systems for ion detection.

The ionization of lithium atoms on polycrystalline wolfram surfaces has been investigated recently by DATZ and TAYLOR [*136*], ROMANOV [*609*], and ZANDBERG and IONOV [*796*]. For this case, $eI < e\varphi$ and $\alpha \ll 1$, so that the ion current i follows the SAHA-LANGMUIR equation with the simplifications for N and A already used in Eqs. (8.1.1–4) and (8.1.1–5) namely,

$$i = eNA \exp\left[-\frac{e(I-\varphi)}{kT}\right]. \qquad (8.3.2.1\text{–}1)$$

From this it follows that the logarithm of the ion current i falls off linearly with $1/kT$, and that the difference $e(I - \varphi)$ is the negative slope of the line. The above-mentioned authors (references [*136*], [*609*], and [*796*]) have verified this relation experimentally for the high-temperature region 2100—2500° K. At lower surface temperatures, ROMANOV [*609*] and ZANDBERG and IONOV [*796*] found the deviations from linearity shown in Fig. 8.3.2.1.a. ROMANOV made the interesting observation that the curve of log i vs $1/kT$ at low temperatures (ca. 1400° K) passes through a slight maximum. As will be seen, this behavior is also observed for other systems with $(I - \varphi) < 0$, e.g., for Cu, Ag, and Ba on W (references [*774*] and [*778*]). ROMANOV and ZANDBERG and IONOV attribute these deviations from linearity entirely to adsorption processes (and the resultant changes in the work function $e\varphi_{\max}$ of the "patches") without more thoroughly investigating the occurrence of the maximum. In Section 8.1.3, in agreement with an interpretation by ZANDBERG, the occurrence of such a maximum was explained qualitatively as a result of oxygen (which raises $e\varphi_{\max}$) and alkalis (which lower $e\varphi_{\max}$) competing for the adsorption sites.

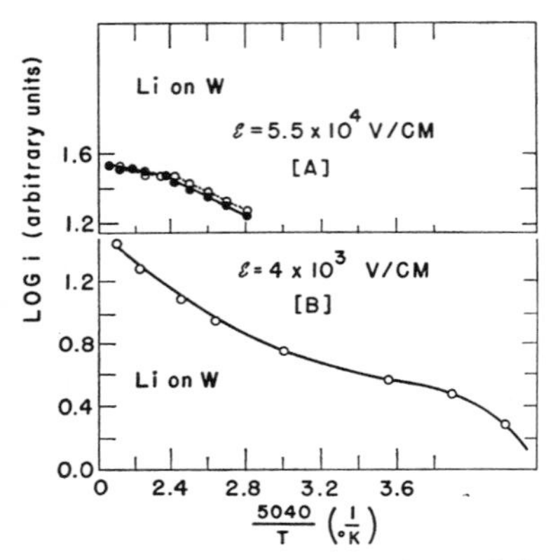

Fig. 8.3.2.1.a. The dependence of the logarithm of the lithium ion current on the temperature of a wolfram surface. Curve *A*, data reported by ZANDBERG and IONOV [*796*] for $\mathcal{E} = 5.5 \times 10^4$ V/cm; curve *B*, data reported by ROMANOV [*609*] for $\mathcal{E} = 4 \times 10^3$ V/cm

ROMANOV [*609*] determined the difference $e(I - \varphi)$ in the high-temperature region to be 0.6—0.7 eV from the slope of the curve $\log i = f(1/T)$. This value is somewhat below the value of 0.88 eV which is obtained if the value $e\varphi = 4.52$ eV is assumed as an "average value" for the work function of the polycrystalline wolfram surface. But since the patch character of the W surface actually used by ROMANOV is unknown, the assumption of the value 4.52 eV involves an uncertainty which is carried over to the value found for $e(I - \varphi)$.

In interpreting their results, both ROMANOV [*609*] and ZANDBERG and IONOV [*796*] make some theoretical arguments concerning the influence of the patch character of the surface on ion current. These are, in principle, identical with the arguments given by ZEMEL. Unfortunately in applying such arguments to the interpretation of the measurements, they failed to consider that the average value of the work function of a polycrystalline surface as obtained from measurements of thermionic emission (EM) is not identical with the value from positive surface-ionization (PSI) measurements, since in the first case the emission from patches with $e\varphi_{\min}$ dominates, whereas in the second case it is the patches with $e\varphi_{\max}$ that dominate. The average values of the work functions for the same polycrystalline surface are therefore given by

$$e\overline{\varphi}_{\mathrm{EM}} = e \sum_{\mu} f_{\mu} \varphi_{\mu,\,\mathrm{min}} + e \sum_{\lambda} f_{\lambda} \varphi_{\lambda} \,, \qquad (8.3.2.1\text{–}2)$$

$$e\overline{\varphi}_{\mathrm{PSI}} = e \sum_{\nu} f_{\nu} \varphi_{\nu,\,\mathrm{max}} + e \sum_{k} f_{k} \varphi_{k} \,, \qquad (8.3.2.1\text{–}3)$$

from which it is evident that $e\varphi_{\mathrm{PSI}} \neq e\varphi_{\mathrm{EM}}$.

In the ionization of lithium on pure platinum and on 92% Pt-8% W surfaces, DATZ and TAYLOR find that their experimentally determined ionization coefficients have much smaller values than are to be expected from Eq. (8.1.1–6) and that the slope of the "SAHA-LANGMUIR line" does not agree with the theoretically expected value. They tried to explain this by stating that a certain fraction of the particles striking the surface are reflected without coming into thermal equilibrium with it. They assume the same type of reflection for other alkali metal atoms, the fraction reflected depending on the kind of atom. For example, at $T = 1850°$ K they determined the reflection coefficients to decrease in the order K > Li > Rb > Cs. However, the temperature dependence (Fig. 8.3.2.1.b) of the reflection coefficient r (a 100% increase in r between 1800 and 2000° K for Li on platinum) given by these authors is improbably strong for a reflection process. This leads one to suspect that possibly effects other than reflection are responsible for the low value of the ionization coefficient. Thus KIM HENG PONG and SOKOLSKAYA [378] have attributed the low value which they found for the ionization coefficient of Na on Pt to the development of highly chemisorbed Na layers on the platinum, which, because of their semiconductor character, hinder the ionization process. A more likely explanation, however, is that the adsorbed Na layers lower the work function of the metal and thus decrease the degree of ionization.

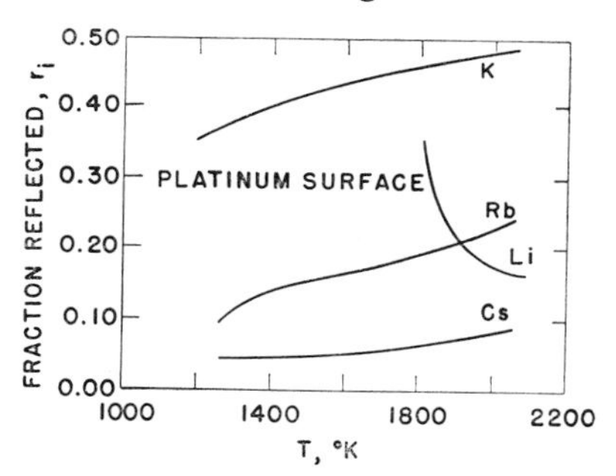

Fig. 8.3.2.1.b. "Reflection" of alkali metals on a Pt surface (DATZ and TAYLOR [136])

The resolution of such questions requires further investigation of surface ionization on platinum surfaces and, in particular, direct measurements of reflection (as have been carried out, for example, by MARCUS and MC FEE [466]).

The ionization of sodium on wolfram, is a case for which $eI > e\varphi$ so that $\alpha \ll 1$ and the logarithm of the ion current should vary linearly with $1/T$. In the high-temperature range (2600—2100° K) the curve of $\log i$ vs $1/T$ actually is linear, as has been confirmed experimentally by

ROMANOV and STARODUBTSEV [*608*], ROMANOV [*609*], and DATZ and TAYLOR [*136*]. However, ROMANOV and STARODUBTSEV [*608*] found that under their vacuum conditions (case A: 10^{-6} mm Hg; case B: 10^{-7} mm Hg, see Fig. 8.3.2.1.c) the curve of $\log i$ vs $1/T$ passes through a maximum with decreasing temperature. As already mentioned, such a behavior has also been observed for other systems, such as lithium on wolfram, and appears actually to be typical for investigations in the region of partial pressures above 10^{-7} mm Hg for oxidizing gases. The interpretation given above for the appearance of such a maximum for Li on wolfram also holds for Na on wolfram since $eI > e\varphi$ in both cases. From the slope of the linear part of the $\log i$ vs $1/T$ curve, DATZ and TAYLOR [*136*] determined the difference $e(I - \varphi)$ for Na on wolfram to be 0.44 eV. This result implies a value of 4.70 eV for the work function, which deviates only, slightly from the "mean value" for electron emission from polycrystalline W surfaces, and which is understandable as being the result of the presence of physical and induced heterogeneities on the surface. DOBREZOW [*150*], in his investigations of the surface ionization of Na on thoriated wolfram, was able to show definitely that the ionization on such a patch surface occurs predominantly from the patches with the largest work function.

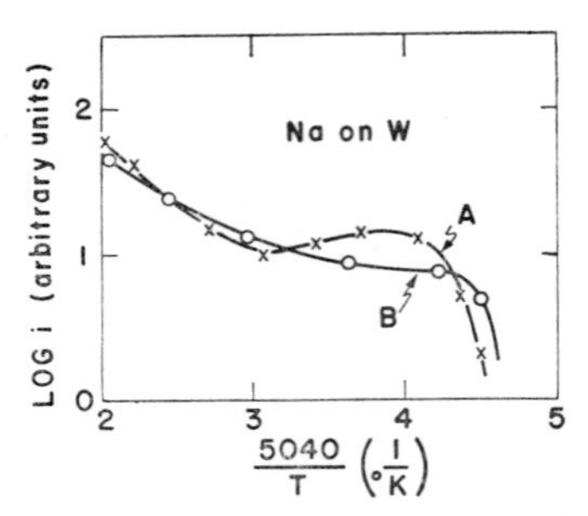

Fig. 8.3.2.1.c. The dependence of the logarithm of the sodium ion current on the temperature of a wolfram surface. Curve *A*, total pressure $p = 10^{-6}$ mm Hg; curve *B*, total pressure $p = 10^{-7}$ mm Hg. (After ROMANOV and STARODUBTSEV [*608*])

HUGHES, LEVINSTEIN, and KAPLAN [*326*] have recently studied the surface ionization of sodium on a wolfram surface with practically uniform work function [superstructure of (112) surfaces with $\varphi_{(112)} = 5.20$ eV and microstructure of (110) faces with $\varphi_{(110)} = 5.27$ eV]. From the slope of the "SAHA-LANGMUIR line" they determined the effective work function to be $e\varphi = 5.25 \pm 0.05$ eV, which is in good agreement with the corresponding values determined from thermionic electron emission.

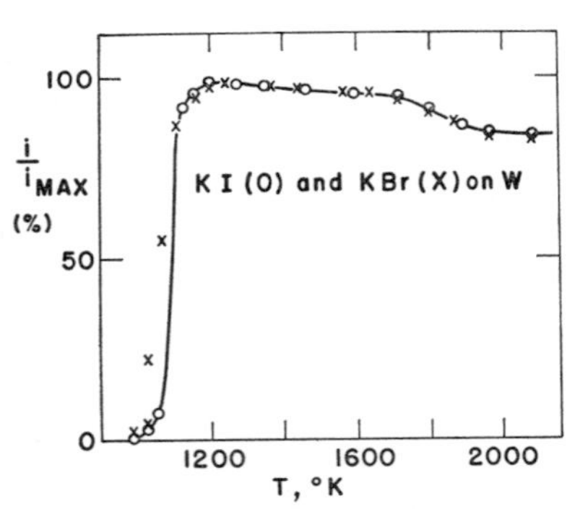

Fig. 8.3.2.1.d. Variation of the ratio of the ion current to the maximum ion current with the surface temperature for the ionization of potassium on wolfram (IONOV [*341*])

On the other hand, the ionization of potassium atoms on wolfram is the converse of the examples of Li and Na given above in that $eI < e\varphi$

so that $e(\varphi - I) \approx kT$. According to the SAHA-LANGMUIR equation this means that the ion current decreases with increasing temperature above a certain high value of the temperature ($T > 1150°$ K). Such a behavior has actually been found by various authors (IONOV [*341*], DATZ and TAYLOR, and DOBREZOW [*153*]). Fig. 8.3.2.1.d represents the temperature behavior of the ion current as determined by IONOV, which for temperatures above 1150° K gives the behavior expected from the SAHA-LANGMUIR equation. The same author also studied the threshold temperature T_s (for the onset of ion emission) as a function of the voltage applied between the ionizing filament and the accelerating grid. He found that T_s decreased from ca. 900° K (at 0.6 kV) to ca. 300° K (at 14.8 kV); but he could not, unfortunately, give any accurate estimates of the effective intensity of the electric field at the ionizing filament. Hence it is difficult to decide the extent to which the field-ion desorption mechanism plays an essential role in his experiments.

In the ionization of potassium atoms on Pt or on 92% Pt-8% W surfaces, DATZ and TAYLOR [*136*] found that the measured ion currents were much smaller than was expected from the SAHA-LANGMUIR equation. As in the case of lithium, they interpreted this result as being the consequence of a strong reflection of potassium atoms from the Pt as well as from the 92% Pt-8% W surface (Fig. 8.3.2.1.b).

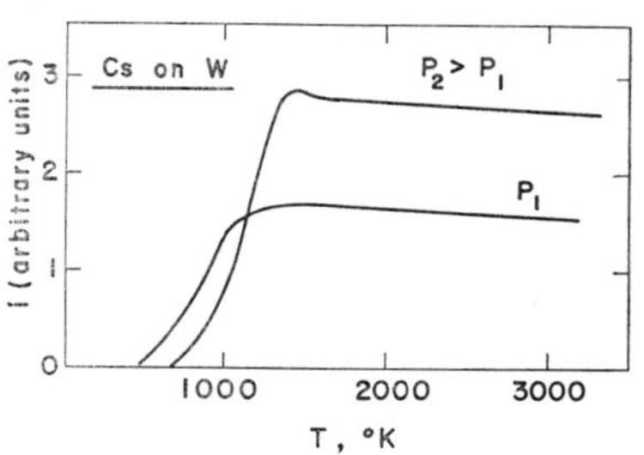

Fig. 8.3.2.1.e. The dependence of the ion current i on the surface temperature T for the ionization of Cs on W for two different cesium partial pressures p_1 and p_2 (DOBREZOW [*153*])

In the ionization of rubidium atoms on polycrystalline wolfram surfaces ($eI < e\varphi$), DATZ and TAYLOR found (in good agreement with the SAHA-LANGMUIR equation) that the ionization coefficient decreased with increasing temperature above a surface temperature of about 1200° K, where the slope of the graph of β vs $1/T$ is in relatively good agreement with the values expected from Eqs. (8.1.1–4) and (8.1.1–6), namely, $eI = 4.20$ eV, $e\varphi \approx 4.70$ eV. For the case of ionization of rubidium atoms on Pt or 92% Pt-8% W surfaces, on the contrary, they find a much lower ion current than was to be expected, a behavior which in fact they also observe for other alkali atoms on such surfaces.

Among the earliest and most thoroughly investigated processes is that of cesium atoms on polycrystalline wolfram surfaces (Table 8.3.2.1.A). Since in this case $e(I - \varphi) \ll kT$, it is to be expected from the SAHA-LANGMUIR equation that the ionization coefficient should be 100% throughout the whole temperature range of the investigation. The emitted ion current for a constant number of cesium atoms impinging on the heated wolfram surface is therefore approximately independent of temperature. Actually, DOBREZOW [*153*] and DATZ and TAYLOR [*136*] have claimed just this behavior over the broad temperature range $1000^\circ\,\mathrm{K} < T < 3000^\circ\,\mathrm{K}$ (reference [*153*]) and at 2200° K (reference [*136*]). At lower temperatures, the ion current drops off sharply because of work function changes initiated by adsorption processes (Fig. 8.3.2.1.e).

8.3.2.2. Positive Surface Ionization of Some Alkali-Salt Molecules on Metal Surfaces in Weak External Fields

A considerable amount of experimental information about the ionization of alkali-salt molecules on metal surfaces is available as may be seen from Table 8.3.2.2.A. In most cases, polycrystalline wolfram has been used as the ionizing surface. In the ionization of potassium metal atoms and potassium salt molecules on polycrystalline wolfram surfaces, many authors (IONOV [*342*], DATZ and TAYLOR [*137*], STARODUBTSEV [*677*], and ZIMM and MAYER [*805*]) have observed that the temperature dependence of the current of potassium ions is largely independent of whether the potassium is in the form of metal atoms or potassium salt molecules (compare Fig. 8.3.2.1.c and Fig. 8.3.2.2.a). Figs. 8.3.2.2.a and 8.3.2.2.b represent results of this type which were obtained by IONOV [*342*] (wolfram surface) and by DATZ and TAYLOR [*137*] (oxidized wolfram surface). A comparison of the curves for the same salt (KBr, KI) shows certain similarities. One also sees that the ion current becomes measurable at about the same surface temperature ($T_s \approx 1150^\circ$ K) for both the potassium atoms (Fig. 8.3.2.1.c) and the potassium salt molecules (except for the case of KF). With increasing surface temperature, the ion current then rises steeply and runs through a maximum around 1200—1300° K. The ion current decreases only slightly at higher temperatures.

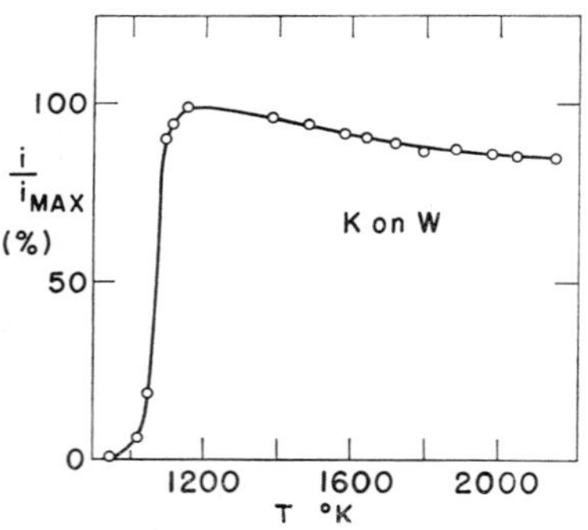

Fig. 8.3.2.2.a. Variations of the ratio of the ion current i to the maximum ion current i_{max} with the surface temperature for the ionization of KI and KBr molecules (indicated by ○ and ×, respectively) on wolfram (IONOV [*342*])

Evidence has been found that potassium salt molecules incident on a heated wolfram surface are dissociated and only then are desorbed, and that in the meantime they come into thermal equilibrium with the surface. This has been shown by numerous experimental studies (for example, by mass spectrometric analysis of the types of ions emitted or

Table 8.3.2.2.A. *Ionization of Alkali Salt Molecules on Metal Surfaces*

Molecule					Polycrystalline Surfaces	Temperature Range (°K)	Condition of Surface (Vacuum, mmHg)	Ion Current as a Function of Temperature or Electric Field	References
Lithium Halide	Sodium Halide	Potassium Halide	Rubidium Halide	Cesium Halide					
LiCl	NaCl				W	1800—2500	$2—3\times10^{-7}$	$f(\varepsilon), f(T)$ $\left(\varepsilon = 1\,\frac{MV}{cm}\right)$	E. YA ZANDBERG [795]
		KCl			W, Ta				N. I. IONOV, V. I. KARATAEV [343a]
		KCl		CsCl	W	800—2500	10^{-7}	$f(\varepsilon), f(T)$ $\left(\varepsilon = 2\,\frac{MV}{cm}\right)$	E. YA. ZANDBERG [794]
		KCl, KBr, KI	RbCl		W, Pt, W-Pt	1100—2200	2×10^{-7}	$f(T)$	S. DATZ, TAYLOR [136]
	NaCl	KCl, KBr, KI	RbCl	CsCl	W	1300—2650		$f(T)$	N. I. IONOV [341]
	NaCl	KCl			W	Not Stated			R. J. HAYDEN [295]
	NaCl	KCl	RbCl	CsCl	W	400—2400		$f(T)$	S. V. STARODUBTSEV [676,677]
	NaCl, NaBr	KCl, KBr, KI	RbCl, RbBr	CsCl	W	—		$f(T)$	W. M. DUKELSKI, IONOV [165]
	NaCl, NaBr, NaI				W, W-O	1000—2600	Partial pressure of oxidizing gases, less than 10^{-10} mm Hg	$f(T)$	T. E. PHIPPS, JOHNSON [568]
		KCl, KBr, KI			W, W-O	1000—2400		$f(T)$	T. E. PHIPPS, M. J. COPLEY, J. O. HENDRICKS [304]
		KI			W, W-O	1700—2200		$f(T)$	T. E. PHIPPS, M. J. COPLEY [566]
Alkali halide (not specified)					W				W. H. RODEBUSCH, W. F. HENRY [598, 599]

by determining the velocity distribution of ions emitted from the filament [*341*, *343a*]).

The sharp drop in ion current below the "threshold temperature" is explained partially by the increasing adsorption of alkali atoms on the surface (resulting in a large drop in work function). The small difference between the ionization curves for potassium atoms and potassium salt molecules in the temperature range 1300°—1600° K may result from the fact that at these temperatures the halogen atoms still form a double

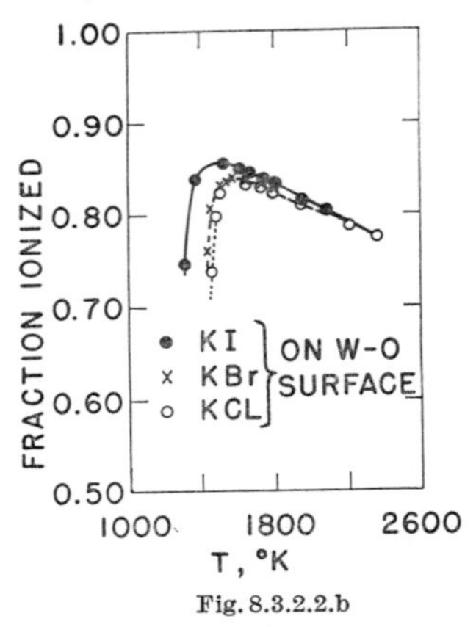

Fig. 8.3.2.2.b

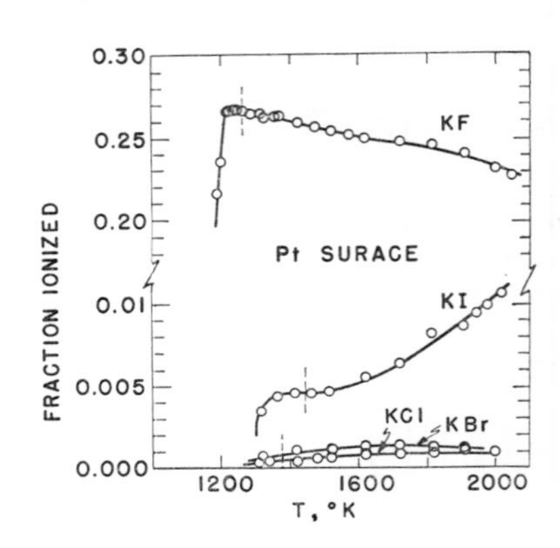

Fig. 8.3.2.2.c

Fig. 8.3.2.2.b. The dependence of the fraction of potassium ions ("fraction ionized" is here equivalent to the ionization coefficient β) formed at an oxidized wolfram surface on the surface temperature T for the case that different alkali halide molecules strike the surface (Datz and Taylor [*137*])

Fig. 8.3.2.2.c. The dependence of the fraction of potassium atoms converted into ions at a platinum surface on the surface temperature T for the case that different alkali halide molecules strike the surface (Datz and Taylor [*137*]). The "fraction ionized" is here equivalent to the ionization coefficient β

layer on the wolfram surface so that the work function is raised slightly. Since such a difference is not seen at higher temperatures (1800 to 2000° K), disintegration of this double layer must be assumed.

In their investigations of the ionization process and the mean residence time τ of rubidium ions and atoms on wolfram single crystals, Hughes and Levinstein [*327*] found that, as in the case of the ionization of K on W, the value of τ is independent of whether the impinging beam consists of rubidium metal atoms or rubidium salt molecules.

Thus although for wolfram surfaces the ion emission current is practically independent of whether the incident alkali beam is atomic or molecular this, interestingly enough, is no longer the case for platinum surfaces, as has been shown by Datz and Taylor [*137*]. As seen from Fig. 8.3.2.2.c, the temperature dependence of the ionization coefficient varies strongly with the kind of potassium salt incident on the surface. Moreover, the values of the ionization coefficient are far below those expected from the SAHA-LANGMUIR equation, Eqs. (8.1.1–4) and (8.1.1–6). As in the case of ionization of alkali atoms on platinum surfaces, here also the authors interpret these results as perhaps being due to reflection.

Such indications are still vague, and further experimental information, especially direct determinations of the velocity distribution of the emitted particles, is greatly to be desired.

8.3.2.3. Positive Surface Ionization of Alkaline Earth Elements and Compounds in Weak External Fields

The surface ionization of alkaline earth elements and compounds has received attention in recent years, as seen from Table 8.3.2.3.A. In particular, as was the case for the ionization of the alkalis, wolfram and wolfram oxide have been the surfaces most often used. A critical examination of the experimental results frequently reveals serious discrepancies between the results of different authors. As a typical example of this, the results on ionization of barium on W and W-O surfaces will be mentioned.

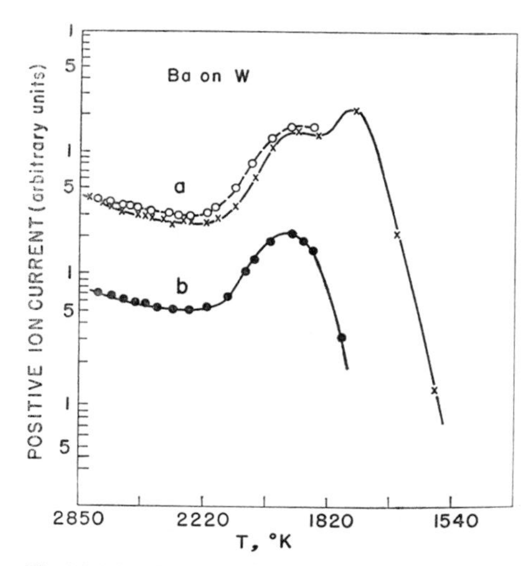

Fig. 8.3.2.3.a. Temperature characteristic of the ionization of Ba on W. The measured values shown in curve *b* were obtained after 4 hrs of baking the wolfram surface at 2200° K. The values in curve *a* were taken one week after the run for curve *b* in two independent runs. (WERNING [*778*])

The work of GUTHRIE [*276*] is among the first on the ionization of Ba on polycrystalline wolfram surfaces. He found that in the high-temperature region the ion current is independent of surface temperature; i.e., in contrast with the theoretical prediction $e(I - \varphi) = 0.69$ eV, the slope of the "SAHA-LANGMUIR line" gave the value 0. However, the sensitivity of his ion-detecting system did not permit any exact measurements.

MOROSOW [*505*] also investigated the system Ba on polycrystalline wolfram, and found that in the high-temperature region the slope of the SAHA-LANGMUIR line is not equal to 0 but rather 0.39 eV. This value is still far below the theoretically expected value of 0.69 eV.

Although MOROSOW made corrections to take account of the possible occurrence of excited states of barium ions and atoms (DARWIN-FOWLER

Table 8.3.2.3.A. *Ionization of Alkaline Earth Metals on Metal Surfaces*

Atom	Polycrystalline Surfaces unless otherwise Indicated	Condition of Surface (Vacuum, mm Hg)	Temperature Range (°K)	Ion Current as a Function of Temperature	References
Ba	Re	$\sim 1^{-8}$	2099—2309	$f(T)$	M. D. Scheer, J. Fine [636b]
Sr, Ca	W, Single Crystal	$3-8\times10^{-9}$	1500—2800	$f(T)$	F. L. Reynolds [587]
Mg, Ca, Sr, Ba	W, W-O	10^{-6} oxygen partial pressure	~1500—2250	$f(T)$	Yu. K. Szhenov [700]
Sr, Ba	W, Ta, Re	$\sim 10^{-7}$	~1500—2500	$f(T)$	J. R. Werning [778]
Mg, Ca, Sr	W-O	—	—		V. K. Gorshkov [251]
Mg, Ca, Sr	W-O	$10^{-5}-10^{-6}$ oxygen partial pressure	~2000—2600	$f(T)$	Yu. K. Szhenov [699]
Sr	W, Pt	Not Stated	Not Stated		R. H. Tomlinson, A. K. Das Gupta [724]
BeO	W	—	—		N. Sasaki, T. Yoasa [625]
SrO, BaO	W, Mo, Pt, Ta, Ni	3×10^{-7}	~1250—1700	$f(T)$	L. T. Aldrich [4]
Mg	W, W-O	—	—	$f(T)$	S. V. Starodubtsev [680]
$Ca(NO_3)_2$, $Sr(NO_3)_2$, $Ba(NO_3)_2$	W	Not Stated	Not Stated		R. J. Hayden [295]
Mg, Ca	W-O, CaO- and MgO-coated W	Not Stated	Upper Limit ~3000	$f(T)$	I. N. Dobretzow, S. V. Starodubtsev, J. I. Timokhina [149]
Ba	W	—	—		G. A. Morosow [505]
Sr, Ba	SrO-, BaO-coated W	$1-2\times10^{-6}$	Not Stated		J. P. Blewett, M. B. Sampson [75]
Ba	W		~—2500	$f(T)$	J. P. Blewett, E. J. Jones [77]
Sr, Ba	Pt	$1-2\times10^{-6}$	Not Stated		M. B. Sampson, Walker Bleakney [624]
Ba	W		~1700—2600	$f(T)$	N. N. Guthrie [276]
Ba	BaO-, CaO-, SrO-coated	1×10^{-8}	~1000—1700	$f(T)$	J. A. Becker [58]
Ca	W-O	Not Stated	1,400	$f(T)$	K. H. Kingdon [382]

theory), these were not sufficient to explain the large discrepancy between the experimental and theoretical values. In the low-temperature region, this author observed a considerable increase in ion yield which he attributed to increasing oxygen adsorption on the surface. This behavior in the region of lower temperatures ($T < 2100°$ K) was qualitatively confirmed by WERNING [*778*]. WERNING observed in his mass spectrometric investigations that the curve of log i vs $1/T$ went through a maximum with decreasing temperatures (Fig. 8.3.2.3.a). Similar temperature dependences were also observed for other systems (such as Li, Na on W and as shown in Fig. 8.3.2.3.b for Ag and Cu on W), as pointed out in Section 8.3.2.1. Thus such a behavior seems to be typical of gas-covered polycrystalline surfaces. A qualitative interpretation of this temperature dependence for the case of $eI \gg e\varphi$, as in these systems, has already been given in Section 8.1.3.

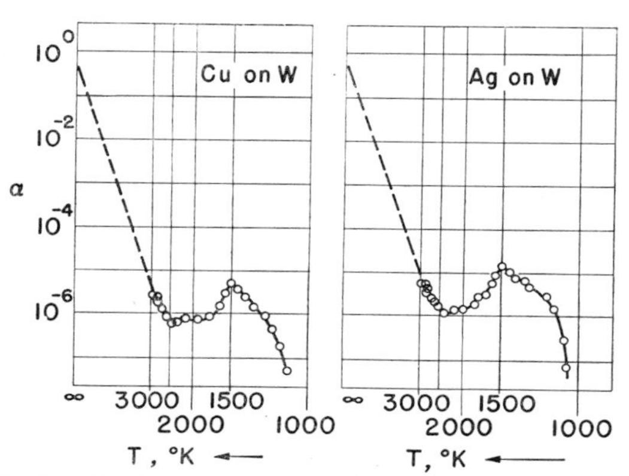

Fig. 8.3.2.3.b. The temperature characteristic of the ionization degree α for the ionization of Cu and Ag on W (WEIERSHAUSEN [*774*])

WERNING also found indications of surface reactions of barium with the fluorine which is present in the wolfram. In contrast to the results of GUTHRIE [*276*] and of MOROSOW [*505*], his experimental results in the region of higher temperatures were in good agreement with the theory. From the slope of the SAHA-LANGMUIR line (with $\overline{\varphi} = 4.58$ eV), he determined the value $e(I - \varphi) = 0.60$ eV. The results obtained by SZHENOV [*700*] confirm this good agreement of the experimental values with the theory. However, in contrast to the results of WERNING, he finds that the curve of log i vs $1/T$ (curve A of Fig. 8.3.2.3.c) is linear over the entire temperature range from about 1600 to 2200° K. Since SZHENOV does not clearly state the surface conditions under which he made his measurements, it is difficult to discuss the contradictory results.

In the surface ionization of barium and strontium on tantalum surfaces and of strontium on rhenium surfaces, WERNING found that the experimental data in the high temperature region ($T > 2500°$ K) are correctly represented by the SAHA-LANGMUIR theory. From the slope of the SAHA-LANGMUIR line he found that the work function of tantalum has a value in the range 4.14—4.26 eV, and that of rhenium is

5.17 ± 0.2 eV (with $eI_{Sr} = 5.69$ eV). These values are in good agreement with those from measurements of thermionic emission (see, for example, EBERHAGEN [*169*] and SIMS [*654*]).

DOBREZOW, STARODUBTSEV and TIMOCHINA [*149*] studied the surface ionization of magnesium and calcium on surfaces covered with oxygen and surfaces covered with layers of CaO or MgO. The observed currents were orders of magnitude greater than expected from the theory. In a repetition of their measurements with an improved mass spectrometric method [*680*], they found that in the high-temperature region ($T > 2000°$ K) the slope of the curve of $\log i$ vs $1/T$ (where i is the ion current) followed Eq. (8.1.1–5) approximately, but the coefficient A was several orders of magnitude larger than expected. The attempt of MORGULIS [*501*] to interpret this result on the basis of the semiconductor properties of the MgO and CaO layers was discarded by DOBREZOW [*153*]. In the opinion of the latter, such semiconductor layers would only hinder ion emission, since electron transitions are made more difficult by the presence of forbidden zones in the semiconductor. Here again more exact experimental material is needed to clarify these questions.

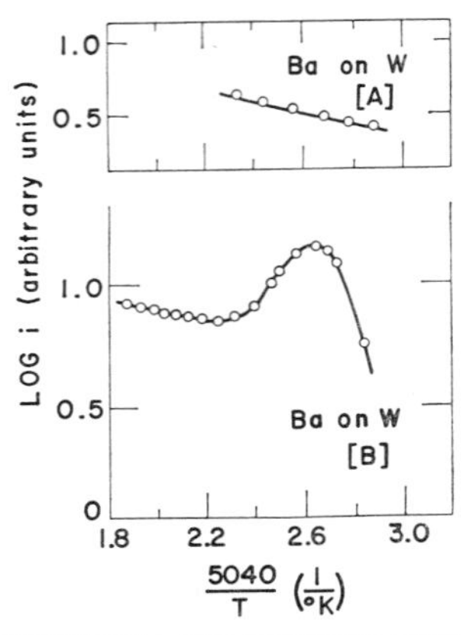

Fig. 8.3.2.3.c. The dependence of the logarithm of the barium ion current ($\log i$) on the temperature T of a wolfram surface. Curve A according to measurements of SZHENOV [*700*]); curve B according to measurements of WERNING [*778*]

8.3.2.4. Positive Surface Ionization of Other Elements and Compounds in Weak External Fields

Since investigations of positive surface ionization of elements and compounds other than the ones previously given are very numerous, they cannot be discussed individually in any detail. But, as an aid in finding the results of such investigations, the ionization systems are summarized in Table 8.3.2.4.A.

In the discussion which follows, only the ionization of uranium on polycrystalline wolfram surfaces will be considered in detail as an example. RAUH [*582*] used the surface-ionization method, but without a mass spectrometer, to determine the ionization energy of uranium. He observed $eI = 4.7$ eV, a somewhat larger value than the 4.0 eV found by KEISS, HUMPHRIES and LAUN [*376*].

More recently WERNING [*778*] has made improved mass spectrometric studies of surface ionization of uranium on polycrystalline wolf-

Table 8.3.2.4.A. *Ionization of Various Elements and Compounds on Metal Surfaces*

Type of Atom or Molecule	Polycrystalline Surfaces	Condition of Surface (Vacuum, mm Hg)	Temperature Range (°K)	Ion Current as a Function of Temperature	References
Pr, Nd	W	Not Stated	2300—2800	$f(T)$	N. I. Ionov, M. A. Mittsev [343]
UCl_4, UF_4	W	$\sim 10^{-6}$	2130—2800	$f(T)$	I. N. Bakulina, N. I. Ionov [45]
Nd, U	W, Ta, Re	$\sim 10^{-7}$	2000—2500	$f(T)$	J. R. Werning [778]
Cu, Ag	W-O	5×10^{-6}—2×10^{-5} Oxygen partial pressure	1100—2800	$f(T)$	W. Weiershausen [774]
Pu, U	W (crucible)	Not Stated	—3000		A. L. Harkness [291]
Ce, La, Pr, Nd, Sm		6×10^{-7}—1×10^{-6}	2420—2620	$f(T)$	R. N. Ivanov, G. M. Kukawaze [344]
Na_2S	W		1800—2300	$f(T)$	I. N. Bakulina, N. I. Ionov [44]
KCN, KCNS	W	Not Stated	2210—2440	$f(T)$	I. N. Bakulina, N. I. Ionov [42]
La, LaO	La_2O_3-coated W	10^{-7}	Not Stated		J. H. Reynolds [587]
Al	W	10^{-5}	up to 1780		Hin Lew [435]
$Zr(NO_3)_2$, $Ce(NO_3)_2$, $La(NO_3)_2$, $Pr(NO_3)_2$, $Sm(NO_3)_2$, $Eu(NO_3)_2$	W	Not Stated	Not Stated		R. I. Hayden [295]
$La_2(O)_3$, $Pr_2(O)_3$, $Nd_2(O)_3$, Zn, Bi	CaO- and MgO-coated W				L. N. Dobretzow, S. V. Starodubtzev, J. I. Timokhina [149]
Ce, La, Pr, Nd, Y, Cb	*	Not Stated	Not Stated		L. G. Lewis, M. W. Garrison, D. King, R. J. Hayden [437]
Ga_2O_3, In_2O_3, SiO_2, Ce, Al_2O_3, U, Mn, Y	W	Not Stated	900—1800	$f(T)$	J. P. Blewett, E. J. Jones [77]
InO		Not Stated	Not Stated		J. P. Blewett, M. B. Sampson [75]
Ga, In	Pt	1—2×10^{-6}			M. B. Sampson, Walker Bleakney [624]
In, Tl	W-O	Not Stated	1000		L. Brata, C. F. Powell [577]
Cu, Bi	W-O	Not Stated	1400		K. H. Kingdon [382]

* Platinized wolfram covered with the oxide of the elements involved.

ram surfaces. In the region of high surface temperatures (2800—2600° K) he was able to determine that $e(I - \varphi) = 1.67 \pm 0.02$ eV from the linear part of the curve of log i vs $1/T$. If a value of 4.58 eV is assumed for the average work function of polycrystalline wolfram, this gives a value of 6.25 ± 0.02 eV for the ionization energy of a uranium atom. This value deviates widely from the two values given above. But recently BAKULINA and IONOV [45] have again found a high value for the ionization energy of uranium. These authors carried out their mass spectrometric studies of surface ionization of UF_4 and UCl_4 on polycrystalline wolfram surfaces by the following difference method. They bombarded an incandescent wolfram surface alternately with beams of molecules of known ionization energy (LiCl, LiF) and with beams of UCl_4 and UF_4 molecules (Fig. 8.3.2.4.a). Then from the SAHA-LANGMUIR equation, the ratio of the positive ion current of lithium to that of uranium is

$$\frac{i_{\mathrm{Li}}}{i_{\mathrm{U}}} = \frac{A_{\mathrm{Li}} N_{\mathrm{Li}}}{A_{\mathrm{U}} N_{\mathrm{U}}} \exp\left[\frac{e}{kT} \, (I_{\mathrm{U}} - I_{\mathrm{Li}})\right], \qquad (8.3.2.4\text{–}1)$$

$$\log \frac{i_{\mathrm{Li}}}{i_{\mathrm{U}}} = \log \frac{N_{\mathrm{Li}} A_{\mathrm{Li}}}{N_{\mathrm{U}} A_{\mathrm{U}}} + \frac{5040}{T} \, (I_{\mathrm{U}} - I_{\mathrm{Li}}), \qquad (8.3.2.4\text{–}2)$$

where T is in °K and eI in eV. The positive-ion currents were investigated over the temperature range from around 2800 to 2130° K (see Fig. 8.3.2.4.a). The curve of log $(i_T/i_{2650°\mathrm{K}})$ vs $1/T$ for Li^+ and U^+ is seen to be linear, as expected from theory. From the plot of log $(i_{\mathrm{Li}}/i_{\mathrm{U}})$ vs $1/T$, the authors then determined that $(I_{\mathrm{U}} - I_{\mathrm{Li}}) = 0.68 \pm 0.08$ eV. From this, using 5.40 eV for the ionization energy of lithium, they found the ionization energy of uranium to be $I_{\mathrm{U}} = 5.40 + 0.68 = 6.08 \pm 0.08$ eV.

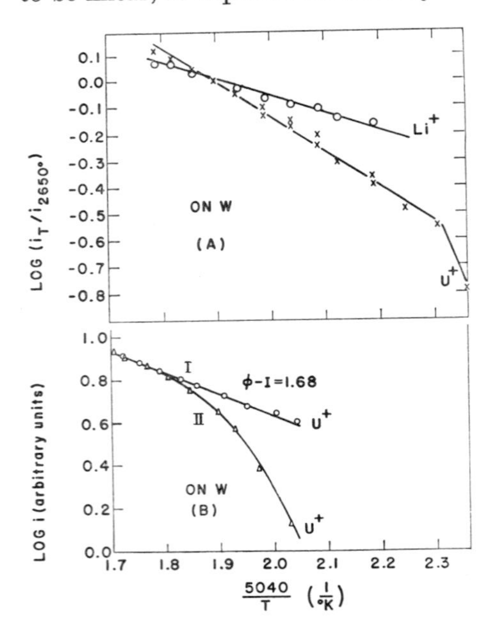

Fig. 8.3.2.4.a. Temperature characteristic of the ionization of uranium and lithium on wolfram surfaces. (Note the difference in the units plotted in the ordinates.) In plot A, the logarithm of the ratio of the ion current i_T at surface temperature T to that of 2650° K (BAKULINA et al. [45]) is plotted against $1/T$. In graph B (WERNING [778]), the logarithm of the ion current is plotted as a function of $1/T$. Curve *I* of graph B represents the initial run at the source pressure of $\approx 1.0 \times 10^{-7}$ mm Hg. The data represented in curve *II* were taken after heating and outgassing the source region

8.3.2.5. *Positive Surface Ionization in Strong External Electric Fields—Field Desorption*

The dependence of positive-ion emission on the strength of the applied electric field was already observed in the first investigations of PSI, but the results were often incorrectly interpreted. Thus many authors who used polycrystalline surfaces did not take into account the presence of surface fields ε_s which arise as a result of potential differences between "patches" on the surface (see Fig. 8.1.4.a) and which can reach high values. (For example, $\varepsilon_s \approx 10^4$ V/cm when the mean separation of such patches is $\bar{r}_0 \approx 10^3$ Å, $\overline{\Delta\varphi} \approx 0.1$ V.) In this case the anomalous SCHOTTKY effect on polycrystalline surfaces can mask the results of measurements which are carried out up to fields of 10^4 V/cm and simulate a different dependence of the ion current on the external electrical field ε than would be expected from the normal SCHOTTKY effect [Eq. (8.1.2–1)].

Thus in the ionization of sodium on polycrystalline wolfram in electric fields up to 1.5×10^4 V/cm, MORGULIS [*495*] observed that the logarithm of the ion current varied linearly with the applied field, instead of with the square root of the field. He erroneously interpreted his results by claiming that with increasing field strength the valence electron level (of the adatom) at the critical distance x_{cr} from the metal surface is displaced by an amount $e\varepsilon x_{cr}$. Such an effect is actually observed only at much higher field strengths (around 10^8 V/cm). (See, for example, ZANDBERG [*794*].) DOBREZOW [*148*] pointed out that the critical distance obtained from the results of MORGULIS would be $x_{cr} = 10^{-6}$ cm, which is much too high. In agreement with the results of MORGULIS but in contradiction to Eq. (8.1.2–1), a linear relationship between the logarithm of the ion current and the applied field was found by PHIPPS and COPLEY [*565*] for the ionization of potassium on polycrystalline wolfram at fields up to 1.5×10^4 V/cm, and by ROMANOV and STARODUBTSEV [*608, 610*] for the ionization of lithium and sodium on polycrystalline wolfram at fields up to 5×10^4 V/cm (Fig. 8.3.2.5.a).

DOBREZOW [*148*] and KONOZENKO [*404*], however, suggested that their results for the ionization of potassium and sodium on wolfram for $\varepsilon \approx 10^5$ V/cm and of sodium on thoriated wolfram for $\varepsilon \approx 4 \times 10^4$ V/cm could best be reproduced by the relation $i \propto \sqrt{e\varepsilon}$. Actually their results are not well reproduced by such a relation either.

The reasons for such discrepancies between the experimentally determined field dependence of log i in the above-mentioned experiments and the dependence expected from Eq. (8.1.2–1) probably are that (a) the applied field intensities still do not compensate the surface fields, as already indicated, and (b) in several experiments stray fields were not eliminated.

Only by applying much higher fields (of the order of magnitude of 10^6 V/cm), as for example in the experiments of ZANDBERG [794], and simultaneously using high-transmission grids (98% transmission) to suppress stray currents, is it possible to verify experimentally the theoretically predicted linear relation between log i and $\sqrt{e\mathcal{E}}$ (normal SCHOTTKY effect) for this range of field strengths. As an example of this, the field dependence found by ZANDBERG for the lithium ion current emitted from a polycrystalline wolfram surface is reproduced in Fig. 8.3.2.5.b. The observed linearity of the log i vs $\sqrt{e\mathcal{E}}$ curve up to field strengths of 2 MV/cm shows that these high field strengths still are unable to compensate the image forces (i.e., $\mathcal{E}$ still is not high enough that $e\sqrt{e\mathcal{E}} = e^2/4r_0$. For this reason ZANDBERG estimated an upper limit for the equilibrium distance r_0, which for $\mathcal{E} \geqslant 2\times10^6$ V/cm was

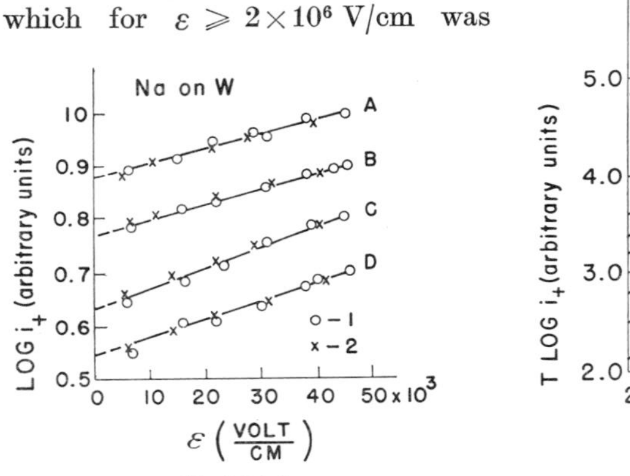

Fig. 8.3.2.5.a

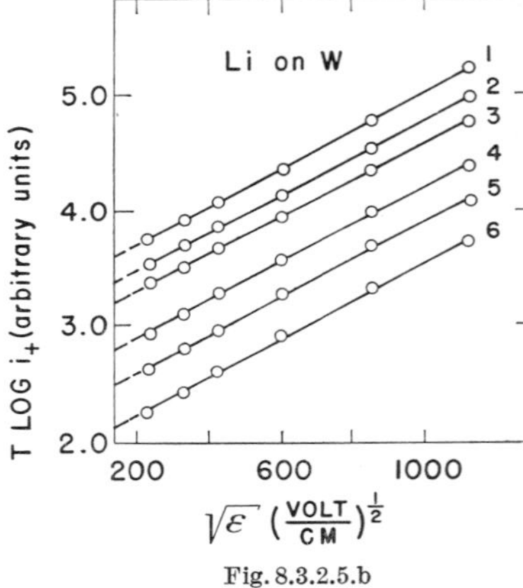

Fig. 8.3.2.5.b

Fig. 8.3.2.5.a. The dependence of the logarithm of the ion current (log i) on the external electrical field $\mathcal{E}$ for the ionization of sodium on wolfram for different surface temperatures and for different intensities of the incident neutral beam. Curve A, $T = 1870°$ K; curve B, $T = 1500°$ K; curve C, $T = 1420°$ K; curve D, $T = 1300°$ K. Points o indicate neutral-beam intensities $j = 1.5\times10^{14}$ atoms/cm² sec and points x indicate $j = 3.2\times10^{15}$ atoms/cm² sec (ROMANOV and STARODUBTSEV [608]))

Fig. 8.3.2.5.b. Plot of T log i_+ vs $\sqrt{\mathcal{E}}$ (where i_+ is the incident current, $\mathcal{E}$ is the external electric field), for different surface temperatures T (ZANDBERG and IONOV [796]). Curve *1*, $T = 2460°$ K; curve *2* $T = 2340°$ K; curve *3*, $t = 2240°$ K; curve *4*, $T = 2060°$ K; curve *5*, $T = 1940°$ K; curve *6*, $T = 1800°$ K

$r_0 = 1/(4\sqrt{\mathcal{E}/e}) \leqslant 6.7\times10^{-8}$ cm. IONOV's results [342] on the field dependence of the threshold temperature for the onset of ion emission likewise indicate that the equilibrium distance r_0 is of the order of magnitude of an atomic radius.

In his measurements, ZANDBERG investigated both the electric fields and the surface temperature as parameters, the number of atoms incident on the ionizing surface being held constant. The result is a family of straight lines, as illustrated in Fig. 8.3.2.5.c, in which log i is plotted as a function of $1/T$ with ε as parameter for the case of ionization of NaCl on polycrystalline wolfram. The negative slope of these lines decreases with increasing field strength ε. This result indicates that the first sum in Eq. (8.1.4–4) determines the temperature dependence of the ion current for NaCl on polycrystalline wolfram. For the ionization of NaCl, the mean values of the work function of polycrystalline wolfram (as determined from the slope of the lines) lie in the range 4.86—4.70 eV when the field strength is in the range of $3.7 \times 10^5 - 1.3 \times 10^6$ V/cm. Below field strengths of 3.7×10^5 V/cm, the value $e\overline{\varphi}_W = 3.86$ eV remains unchanged.

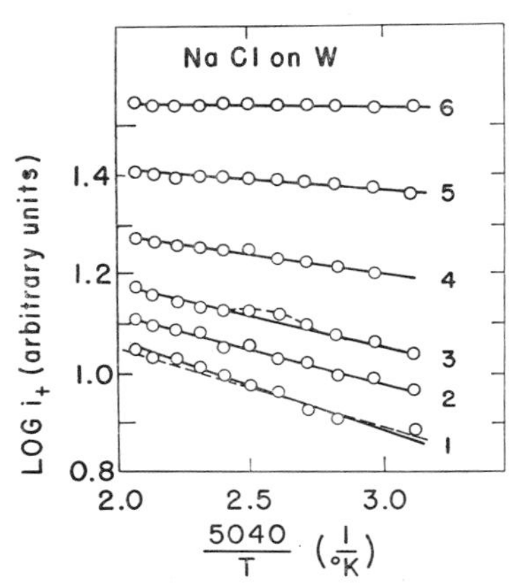

Fig. 8.3.2.5.c. The dependence of the logarithm of the ion current on the surface temperature T for the ionization of NaCl on wolfram for different electric fields (ZANDBERG [*795*]). Curve *1*, $\mathcal{E} = 5.5 \times 10^4$ Volt/cm (solid line; main run; dashed line, repeated measurement at end of run); curve *2*, $\mathcal{E} = 1.1 \times 10^5$ Volt/cm; curve *3*, $\mathcal{E} = 1.83 \times 10^5$ Volt/cm; curve *4*, $\mathcal{E} = 3.66 \times 10^5$ Volt/cm; curve *5*, $\mathcal{E} = 7.32 \times 10^5$ Volt/cm; curve *6*, $\mathcal{E} = 1.28 \times 10^6$ Volt/cm

MÜLLER [*508*] first investigated the well-known field emission desorption of electropositive ions, such as Ba^+ and Ba^{++}, from wolfram surfaces at still higher electric fields ($\approx 10^7 - 3 \times 10^8$ V/cm). In this region of field strengths, not only the SCHOTTKY term but also the linear and quadratic terms in ε become more important in Eq. (8.1.2–2). In their interpretation of the results on field desorption of barium, on wolfram, MÜLLER [*513, 514*] and GOMER [*247*] considered the contribution of the polarization term (the term proportional to ε^2) in the exponent of Eq. (8.1.2–2) to be negligible. They were able to give a good interpretation of the results by appropriately modifying Eq. (8.1.2–2).

These studies have been followed by numerous others, including, for example, the field desorption of electronegative ions, and, in certain cases, mass spectrometric analyses of the desorbed species of molecular ions. For further details, the reader is referred to a recent publication by MÜLLER [*514*], which gives an excellent survey of such studies.

9. Formation and Emission of Negative Ions at Metal Surfaces

9.1. Theoretical Considerations

9.1.1. Emission Equation for a Homogeneous Surface

For the emission of negative ions from incandescent metal surfaces, thermodynamic and statistical arguments can be used to derive emission equations analogous to the SAHA-LANGMUIR equation (8.1.1–5). If an atom adsorbed on the metal surface has an electron affinity S, negative-ion formation occurs in the inverse of the process of positive-ion formation. Instead of giving off a valence electron to the metal, in this case an electron from the metal is taken up by the adatom. The charge-transfer probability, the probability that the adatom passes through the critical distance r_c as an ion instead of as a neutral atom, is then given in analogy to Eq. (8.2–3) by

$$\left(\frac{W_-}{W_0}\right)_{r_c} = A \exp\left(\frac{E_c - \eta_0}{kT}\right), \qquad (9.1.1\text{–}1)$$

with $E_c - \eta_0 = e[S(r_c) - \varphi]$ and $eS(r_c) = eS + E_{i-} - E_a$. In analogy to Eq. (8.2–5), the evaporation probabilities $W(E_{i-})$ for negative ions and $W(E_a)$ for neutral atoms are given by

$$W(E_{i-}, E_a) = \omega_0(E_{i-}, E_a) \exp\left(\frac{E_{i-,a}}{kT}\right) = \frac{1}{\tau_{i-,a}}, \qquad (9.1.1\text{–}2)$$

where the abbreviations are the same as those in Eq. (8.2–5).

Multiplying the probabilities given in Eqs. (9.1.1–1) and (9.1.1–2) yields the degree of ionization α for negative-ion emission at zero external electric field, namely,

$$\alpha(\varepsilon = 0) = \left(\frac{W_-}{W_0}\right)_{r_c} \left[\frac{W(E_{i-})}{W(E_a)}\right] = A_- \exp\frac{e(S - \varphi)}{kT}. \qquad (9.1.1\text{–}3)$$

For the case in which high electric fields are applied, field-dependent terms must be included in the exponent of Eq. (9.1.1–3) as they were in Eq. (8.1.2–2). The result is

$$\alpha(\varepsilon) = \alpha(\varepsilon = 0) \exp\left\{\frac{e}{kT}\left[\sqrt{e\varepsilon} + \varepsilon r_0 + \frac{\varepsilon^2}{2e}(\alpha_0 - \alpha_i)\right]\right\}. \qquad (9.1.1\text{–}4)$$

An experimental test of this equation is difficult because the field-emission currents of electrons at field strengths of 10^6—10^8 V/cm are several orders of magnitude greater than the currents of negative ions. On the other hand, the effect of the SCHOTTKY term $(e\sqrt{e\varepsilon})$, which already is an effect in the region of lower field strengths (10^4—10^6 V/cm), is more easily observed experimentally.

Then for homogeneous surfaces, the negative-ion current i_- can be represented by

$$i_- = \frac{eNfA_-}{A_- + \exp[(e/kT)(\varphi - S - \sqrt{e\mathcal{E}})]} . \qquad (9.1.1\text{–}5)$$

For the case in which $e(\varphi - S - e\mathcal{E}) \gg kT$, this equation simplifies to

$$i_- \approx eNfA_- \exp\left(\frac{e(S + \sqrt{e\mathcal{E}} - \varphi)}{kT}\right) . \qquad (9.1.1\text{–}6)$$

An experimental test of this equation was attempted by DUKELSKII and IONOV [*167*].

9.1.2. Emission Equation for Heterogeneous Surfaces

In contrast to the case of positive ions, for which the surface regions giving the predominant contributions are those with the highest work function φ, the emission of negative ions from heated polycrystalline surfaces ("patch" surfaces) is similar to the thermionic emission of electrons. By considering the distribution of different work-function regions on the surface $(f_\lambda \varphi_\lambda)$, ZEMEL [*798*] derived an ionization coefficient analogous to Eq. (8.1.4–8). For the negative-ion current he found

$$i_- = \sum_\lambda \frac{ef_\lambda NA}{A_- + \exp[(e/kT)(\varphi_\lambda - S - \sqrt{e\mathcal{E}})]} . \qquad (9.1.2\text{–}1)$$

For the case in which $e(\varphi_{\lambda\,\min} - S - \sqrt{e\mathcal{E}} \gg kT$, this equation simplifies to

$$i_- = eNA_- \exp\left[\frac{e(S + \sqrt{e\mathcal{E}})}{kT}\right] \sum_\lambda f_\lambda \exp\left(\frac{-e\varphi_\lambda}{kT}\right) . \qquad (9.1.2\text{–}2)$$

This equation may also be written as

$$i_- = e\bar{A}_- \exp \frac{e(S - \sqrt{e\mathcal{E}} - \bar{\varphi}_-)}{kT} , \qquad (9.1.2\text{–}3)$$

$$\bar{A}_- = NA_- \sum_\lambda f_\lambda \exp \frac{e(\bar{\varphi}_- - \varphi_\lambda)}{kT} . \qquad (9.1.2\text{–}4)$$

Here $e\bar{\varphi}_-$ is the "average" work function of the polycrystalline surface. The work function $e\bar{\varphi}_e$ determined from thermoelectron emission will be equal to $e\bar{\varphi}_-$ if the thermionic constants B and $\bar{A}_-$ are equal. The effect of foreign gas coverage on negative-ion emission has not yet been included quantitatively.

9.2. Experimental Methods

The experimental determination of negative-ion currents is made difficult by several influences. Since relatively high temperatures of the metal surfaces are needed for ion emission, thermionic electrons are also emitted and make a contribution to the negative-ion current. Furthermore, the negative ions formed when these emitted electrons combine with gas atoms obscure the effects of surface ionization. Finally, when high electric fields are applied, the field emission of electrons must be taken into account.

Various methods have been used to separate the effects of thermoelectron emission from ion emission. SCHUPPE et al. [*355, 722*] attempted to keep the surface temperatures so low that the contribution of the thermoelectron current was small. But the increasing amount of foreign gas on the surface at the lower temperatures and the associated change in work function falsified the results. To suppress the electron current, SUTTON and MAYER [*698*] applied a magnetic field parallel to the cathode. The electron trajectories were bent by this field and the electrons, instead of impinging on the anode which is coaxial with the cathode, struck a radially mounted grid. The paths of the heavier negative ions are hardly influenced and the ions reach the anode. IONOV [*339, 341*] also used a magnetic deflecting field for separating electron and negative-ion currents. The most suitable method for investigating negative-ion emission is the mass spectrometric one (as applied, for example, by BAKULINA and IONOV [*43*], IONOV [*341*], and BAILEY [*41*], which not only permits the required separation of electron and ion currents, but also makes possible an analysis of the types of ions emitted from the surface. Thus, for example, IONOV [*341*] used a single-focusing mass spectrometer with 180° magnetic deflection to determine the temperature dependence of the negative-ion currents of F^-, Cl^-, Br^-, and I^- produced by surface ionization of NaCl, RbCl, KBr, and WCl_6 on a polycrystalline surface.

The electron affinity S of electronegative atoms can be determined by use of a procedure which was introduced by DUKELSKII and IONOV [*167*]. If N molecules MX impinge on a hot metal surface, N_+ ions of the type M_+ and N_- ions of the type X_- may be desorbed. Then on the assumptions that N has equal values in the two cases and that the surface is homogeneous, the ratio of the negative-ion to the positive-ion current is

$$\frac{i_-}{i_+} = \left[1 + \frac{1}{A} \exp \frac{e(I - \overline{\varphi} - \sqrt{e\,\mathcal{E}})}{kT}\right] \bar{A}_- \exp \frac{e(S + \sqrt{e\,\mathcal{E}} - \overline{\varphi})}{kT} . \qquad (9.2\text{–}1)$$

The corresponding expression for a heterogeneous surface is

$$\frac{i_-}{i_+} = \left[1 + \frac{1}{\bar{A}} \exp \frac{e(I - \overline{\varphi} - \sqrt{e\,\mathcal{E}})}{kT}\right] \frac{\bar{A}_-}{\bar{j}} \exp \frac{e(S + \sqrt{e\,\mathcal{E}} - \overline{\varphi}_-)}{kT} . \qquad (9.2\text{–}2)$$

The experimental determination of this ratio for different temperatures makes possible the evaluation of S. The values of S determined by DUKELSKII and IONOV in this fashion for various halogens are listed in Table 9.3B (next section).

Another frequently used method for determining S (MAYER et al. [*157, 484, 698, 805*], IONOV [*340*]) is to determine the temperature dependence of the ratio of the negative-ion current i_- to the thermoelectron current i_e.

For the temperature dependence of the electron current, the familiar RICHARDSON-DUSHMAN equation gives

$$i_e = B f T^2 \exp \frac{e(\sqrt{\mathcal{E} e} - \varphi)}{kT}. \tag{9.2–3}$$

From this and Eq. (9.1.1–5), the ratio of the electron current to the ion current for a homogeneous surface is found to be

$$\frac{i_e}{i_- T^2} = \frac{B}{e N A_-} \exp\left(\frac{-eS}{kT}\right). \tag{9.2–4}$$

This ratio is thus independent of φ and ε. Similarly, for a heterogeneous surface the ratio is

$$\frac{i_e}{i_- T^2} = \frac{\bar{C}}{e N A_-} \exp\left(\frac{-eS}{kT}\right), \tag{9.2–5}$$

with

$$\bar{C} = \frac{\sum\limits_{\lambda} B_\lambda f_\lambda \exp(e\varphi_\lambda/kT)}{\sum\limits_{\lambda} f_\lambda \exp(e\varphi/kT)}. \tag{9.2–6}$$

Although $\bar{C}$ is somewhat temperature dependent, over small temperature ranges the two equations (9.2–4) and (9.2–5) reproduce the temperature dependence of i_e/i_- equally well. The values of S determined by such measurements are also included in Table 9.3B (next section). Unfortunately the values of S obtained in this way are not too precise because of the spread in the experimental values.

Finally we may mention a method used by several authors [*41, 44*] for determining the difference between the electronegativities of two elements. The ratio of the ion currents of two ionic species X_1^-, X_2^-, under the additional condition $e(\varphi_{\lambda\,\min} - S - \sqrt{e\varepsilon}) \gg kT$ (the restriction already applied to obtain Eq. (9.1.1–6)], is found from Eq. (9.1.2–3) to be

$$\frac{i_{1-}}{i_{2-}} = \frac{N_1}{N_2}\,\frac{A_{1-}}{A_{2-}} \exp \frac{e(S_1 - S_2)}{kT}. \tag{9.2–7}$$

Thus if molecular beams of species MX_1 and MX_2 are allowed to impinge independently on the ionizing metal surface, and the temperature dependence of the ion currents i_{1-} and i_{2-} are then determined mass

Table 9.3.A. *Emission of Negative Ions from Metal Surfaces*

Type of Atom or Molecule					Poly-crystalline Surfaces	Condition of Surfaces (Vacuum, mm Hg)	Temperature Range (°K)	Ion Current as a Function of Temperature	References
F	Cl	Br	I	Mis-cellaneous					
F	Cl	Br	I		W	$\sim 10^{-6}$	2200—2500	$f(T)$	T. L. BAILEY [*41*]
				S	W	Not Stated	1800—2300	$f(T)$	I. N. BAKULINA, N. I. IONOV [*44*]
F	Cl	Br	I		W	Not Stated	1700—2300		I. N. BAKULINA, N. I. IONOV [*43*]
				S, CN	W	Not Stated	2210—2440		I. N. BAKULINA, N. I. IONOV [*42*]
F	Cl	Br	I		Thoriat. W	$\sim 10^{-6}$	1200—1500		J. W. TRISCHKA, D. T. F. MARPLE, A. WHITE [*730*]
	Cl				W	Not Stated	Not Stated		L. D. DAVIS, B. T. FELD, C. W. ZABEL, J. R. ZACHARIAS [*140*]
F					W	$4-12\times10^{-3}$	1990—2400	$f(T)$	M. METLAY, G. E. KIMBALL [*469*]
F	Cl	Br	I		W		2300—2600	$f(T)$	N. I. IONOV [*341*]
			I		W	10^{-6}	1800—2330		N. I. IONOV [*340*]
				O, O_2	W	$10^{-6}-10^{-1}$	2000—2200		D. T. VIER, J. E. MAYER [*741*]
		Br			W	$3.9\times10^{-4}-2\times10^{-2}$	1610—1966		P. M. DOTY, J. E. MAYER [*157*]
	Cl				W	$5\times10^{-4}-1\times10^{-2}$	1800—2300		K. J. MCCALLUM*, J. E. MAYER [*465*]
				CO	Ni-Cr	Not Stated	Not Stated		R. H. SLOAN, R. PRESS [*657*]
		Br			W	—	2310—2463		G. GLOCKLER, M. CALVIN [*240*]
			I		W	$8\times10^{-3}-3\times10^{-2}$	2020—2400		G. GLOCKLER, M. CALVIN [*239*]
			I		W	10^{-6}	2000		P. P. SUTTON, J. E. MAYER [*698*]

* MCCALLUM, K. J., and J. E. MAYER: J. chem. Phys. **11**, 56 (1943).

spectroscopially, the difference $S_1 - S_2$ can be determined from the graph of i_{1-}/i_{2-} vs $1/kT$. The values determined in this way by Bakulina and Ionov [43, 44] are listed in Table 9.3C (next section).

9.3. Experimental Results

Table 9.3.A lists the systems for which the formation of negative ions by surface ionization has been investigated by various authors. It can be seen that a preponderance of the investigations have been with halogen atoms on polycrystalline wolfram surfaces.

Table 9.3.B gives the values of the electronegativity S determined by the various authors for halogen atoms and also for oxygen and sulphur.

Table 9.3.B. *Values of the Electron Affinity S of Several Elements (eV)*

F^-	Cl^-	Br^-	I^-	O^-	S^-	Authors
		3.82	3.24			Glockler and Calvin [239, 240]
3.62 4.11	3.71	3.64	3.31			Dukelski and Ionov [165]
	4.00 3.72	3.5	3.14	3.1 2.3		Mayer et al. [157, 698, 741]
	3.70	3.34	3.12			Ionov [341]
					2.37	Bakulina and Ionov [43a]
3.58	3.75	3.48	3.18			Bailey [41]

Note that Glockner and Calvin [239, 240] as well as Mayer et al. [157, 484, 698, 805] used Eqs. (9.2–4) and (9.2–5) for the determination of S while Dukelski and Ionov [165], Ionov [341], and Bailey [41] determined it by such equations as (9.2–1). From the data of Bailey, one finds the interesting result that the electron affinity of chlorine is greater than that of fluorine. This conclusion agrees qualitatively with the result of Bakulina and Ionov [43], who used Eq. (9.2–7) to determine the difference of electronegativities and found $S_{Cl} - S_F = 0.21 \pm 0.03$ eV. Such differences in the values of S are listed in Table 9.3.C. The values in Table 2.3.C are in qualitative agreement with the results of Bailey. An extension of such investigations to other elements by means of the mass spectrometric method would be very desirable.

Table 9.3.C. *Experimentally Determined Values of Differences of Electronegativity (eV) according to Bakulina and Ionov* [43]

$S_{Cl} - S_{Br}$	0.21 ± 0.06
$S_{Cl} - S_I$	0.48 ± 0.03
$S_{Cl} - S_F$	0.21 ± 0.03
$S_{Br} - S_I$	0.27 ± 0.02
$S_{Br} - S_F$	0.03 ± 0.02
$S_F - S_I$	0.24 ± 0.04

10. Sputtering of Metal Surfaces by Ion Bombardment

10.1. Introduction

The emission of target particles under the impact of neutral or charged particles on metal targets has been the subject of investigation for more than a century. Thus it was realized several decades ago that the sputtering process played an important role in gas discharges. As seen from the excellent survey of the field, up to 1955, given by WEHNER [*760*], most of the older data are more qualitative in character because of inadequate control of experimental conditions (as explained in section 10.2). However since the appearance of WEHNER's summary*, numerous studies under more controlled conditions have been conducted. These have led to significant new quantitative results which revise many of the conclusions drawn from qualitative older data. The astonishingly rapid increase of interest in sputtering phenomena over the last few years may be explained by the increased recognition of the important role the sputtering process plays in such fields of activity as thermonuclear fusion plasmas (e.g., sputtered wall particles may cause undesirable cooling effects in the plasma), the corrosion of surfaces of satellites and ion-propulsion electrodes, ion getter pumps, the controlled deposition of thin films of almost any material, and surface cleaning.

It is important to distinguish between so-called "physical" and "chemical" (or reactive) sputtering. Chemical sputtering arises whenever a reactive ion and the target material form volatile compounds. In this case, the kinetic energy of the incident ion is of less importance. Physical sputtering on the other hand is caused by a collision process between incident ion and lattice atom. These processes depend on the energy region of the incident ion. In the latter case it is necessary to distinguish between "back" and "forward" sputtering. In "back" sputtering the target particles leave from the surface that is struck by the incident ion beam, while in "forward" sputtering they leave from the opposite surface (e.g., THOMPSON [*720*]). The following summary deals mostly with investigations of "back" sputtering.

10.2. Experimental Methods

10.2.1. Methods for Producing the Incident Ion

10.2.1.1. Glow Discharge

Sputtering was first observed (GROVE [*259*]) at the cathode of a glow discharge, as a consequence of which this field of study was given the name "cathode sputtering." Numerous qualitative experimental results (e.g., those by BIERLEIN et al. [*73*]) and others reviewed in the

* Added in proof: An extensive review article on sputtering phenomena by R. BEHRISCH [Ergebn. exakt. Naturwiss. **35**, 295 (1964)] has just appeared.

articles by KOEDAM [402], WEHNER [760], MASSEY and BURHOP [461], and GUENTHERSCHULZE [266]) have been obtained with the glow discharge technique, which operates in the pressure range from about 100 μ up to a few cm Hg. However, quantitative information concerning the influence of the numerous parameters which determine the sputtering process, such as the type and charge of the incident ion, the ion energy, the angle of incidence of ions on the target surface, etc., cannot be obtained by glow-discharge methods. The reasons for this are the following.

The pressure in the glow discharge is so high that the mean free paths (λ_i for the ions and λ_a for the sputtered target atoms) are small compared with the dimensions d of the container (λ_a, $\lambda_i < d$). Here multiple collisions are possible between the gas particles in the discharge. This can result in the formation of multiply-charged ions, the building up of molecule ions, and finally ionization of the sputtered target atoms as the result of charge exchange. The sputtered target atoms can also be ionized by the secondary electrons which are ejected from the target surface by the ion bombardment. A further consequence of the condition λ_a, $\lambda_i < d$ is that a fraction of the sputtered material (which, according to HIPPEL [316], can amount to 90% of the sputtered atoms) diffuses back to the cathode. This makes the determination of the sputtering yield practically impossible. How the type and charge of the bombarding ion affects the sputtering process can therefore not be determined quantitatively by the glow-discharge method. The ions produced in a glow discharge collide with the surface with a wide range of energies since they are produced in various parts of the cathode fall and have various masses (atoms and molecule ions) and charges. The energy dependence of the "sputtering ratio" can therefore not be investigated quantitatively by this method. As a result of multiple collisions in the discharge, the ions impinge on the cathode at different angles of incidence. Consequently the quantitative determination of the angle dependence is also made impossible. The angular distribution of the sputtered particles is also unobtainable by the glow-discharge method.

In summary, it may be said that this method is unsuitable for quantitative studies of the sputtering process.

10.2.1.2. Low-Pressure Gas Discharge in a Magnetic Field

In order to reduce some of the previously mentioned undesirable effects, such as multiple collisions of gas particles in the discharge and back-diffusion of sputtered target atoms to the target, PENNING and MOUBIS [553] sought to operate the gas discharge at lower pressures (e.g., $p \approx 0.014$ mm Hg). In this way they succeeded in increasing the mean free path λ_m of the neutral gas particles as well as that of the ions so that they were approximately the same as, or somewhat greater than, the dimensions of the container ($\lambda_a, \lambda_i \geqslant d$). On the other hand, this also increases the mean free path of the electrons which maintain the discharge, and consequently produces an undesirable reduction in the ionization of the gas. To overcome this difficulty, PENNING and MOUBIS applied a magnetic field parallel to the direction of the discharge so that the ionization process would be less affected by secondary electrons ejected from the target. By this method these authors were able to carry out sputtering investigations at high ion-current densities ($\approx 10\,\mathrm{mA/cm^2}$). The pressure used was sufficiently small to have no effect on the sputtering ratio. This method has been used by other authors in recent years. ARMSTRONG, MADSEN, and SYKES [*32*] have studied the sputtering of uranium, thorium, and their alloys by argon ions (4—4.5 keV) at a pressure of 40 μ Hg and with a magnetic field of 100 Oe in the discharge tube. NELSON [*520*] investigated the sputtering of Ag, Au, Cu, Pd, and Rh by argon ions ($\approx$ 2 keV) at pressures of the order of 1 μ. GILLAM [*236*] studied the sputtering of Cu_3Au alloys and silver-palladium alloys in an argon discharge (ca. 5 keV ion energy) at pressures of a few microns, with an axial magnetic field of $\approx$ 250 Oe. OGILVIE et al. [*531*] have studied the sputtering of polycrystalline as well as single-crystal silver by ions of He^+, A^+, Xe^+, O^+, and He^+, in the pressure range from a few tenths of a micron up to a few microns in an axial magnetic field of 250 Oe.

However, this method also fails to yield an unobjectionable determination of the sputtering process. Thus, the ions impinging on the target still have a wide range of energies (although already smaller than in method 10.2.1.1) and they may also occur either as singly- or multiply-charged ions. Furthermore in this method it is difficult to precisely determine the angle of incidence of the ions on the target surface.

Finally this method does not permit one to investigate the sputtering process at lower ion energies (of the order of 100 eV) although this is of extreme interest for the study of threshold values for sputtering.

10.2.1.3. Supplemented Low-Pressure Arc

In order to extend investigations of sputtering into the region of lower ion energies and lower pressures than are possible with the glow-discharge method, FETZ [*210*] and later, with some modifications, WEHNER amd his school [*763*] used a plasma method. A low-pressure ($\approx 0.1-1.0\,\mu$ Hg) plasma is produced by a mercury arc discharge between an anode and a mercury-pool cathode. The discharge is stabilized and maintained by means of an auxiliary discharge (auxiliary anode). A grid placed between the cathode and anode enables the experimenter to vary the plasma density in the anode region where the target is located. The target, introduced as an additional electrode in the plasma, is at a negative potential relative to the anode. A LANGMUIR sheath (dipole layer) is developed around the target. The sheath thickness can be determined from the LANGMUIR-CHILDS space-charge equation. The kinetic energy of the ions is controlled by the voltage applied between target and anode. However, in order to determine the actual energy of the ions bombarding the target, it is necessary to determine the potential difference between the anode and the outer edge of the LANGMUIR sheath and also between the LANGMUIR sheath and the target. This determination, which is especially important in the region of low ion energies, requires probe measurements.

The low-pressure arc technique is superior to the glow-discharge method in determining many of the parameters that are important for sputtering. Thus, for example, the angle of incidence of the bombarding ions at the target is better determined; for target thicknesses that are large compared with the thickness of the LANGMUIR sheath, the ions impinge on the target along the normal. Furthermore, since a lower voltage can now maintain the arc discharge, the number of multiply-charged ions is reduced. The disadvantage of the low-pressure arc technique and the previous methods is that they do not permit measurement of the secondary electrons ejected from the target by ion bombardment. Thus, the measured current at the target is too high and the experimental sputtering ratios are too low. Furthermore, in order to avoid building up a mercury film on the target surface, one must heat the target to a high temperature (400—800° C). This makes it difficult to measure the temperature dependence of the sputtering ratio. The conclusion appears to be that this method also is not suited for general quantitative investigations of sputtering, although quantitative results may be obtained in particular cases by experimental tricks (e.g., WEHNER [*769*]).

10.2.1.4. Ion Beams (Not Mass Analyzed)

In recent years, most investigations of sputtering have been carried out with ion beams rather than in a gas discharge. For a long time it was generally thought that ion beams could not be made intense enough to make good sputtering measurements within reasonable times of irradiation. However, the recent development of high-intensity ion sources (e.g., Duoplasmatrons) on the one hand and the use of highly sensitive detectors for the sputtered particles on the other (e.g., as described in Sections 10.2.2.2. to 10.2.2.6), have made the beam method into a practical tool. As a result, it is now possible to study sputtering over a wide energy range with only a narrow spread in energy of the ion beam. Also, it is easier to study the exact dependence of sputtering on the angle of incidence of the bombarding ions. In addition, passing such ion beams through a mass analyzer permits quantitative investigation of the dependence of sputtering on the mass of the incident ions (see next section). YURASOWA, PLESHIVTSEV, and ORFANOV [790] and also PLESHIVTSEV [572] have procuded hydrogen and argon ions with a modified Duoplasmatron source and focused them onto the target by means of an electrostatic lens system. They could obtain ion currents up to 30 mA at accelerating voltages of 50 kV. For measurements, however, they used currents of only 3—10 mA (up to 20 mA in the case of PLESHIVTSEV).

KOEDAM [402] produced the noble gas ions (A^+, Ne^+, Kr^+) for his studies by means of a modified PENNING discharge, whereby he could reach ion currents of approximately 0.5—1.0 mA even for ion energies of 200 eV. VEKSLER [739] produced Cs ions by surface ionization and obtained current densities of the order of 1—6 $\mu A/cm^2$ at the target. PEROVIC [555] used a modified plasmatron source to produce beams of argon ions and achieved current densities of 0.2—1 mA/cm^2 at the target.

10.2.1.5. Mass-Analyzed Ion Beams

Since about 1958, electromagnetic mass separators have frequently been used for sputtering investigations in the kilovolt range (e.g. references [9, 10], [274], [363], [529], [602], [787]). For this purpose the collector of such an analyzer is replaced by the target to be studied. The ion beams that sputter the target are then practically monoenergetic (e.g., in the experiments of YONTS et al. [787] the relative energy spread is $\Delta E/E \approx 4 \times 10^{-4}$). Such monoenergetic and monoisotopic beams make possible a more accurate study of such parameters of the sputtering process as the charge, mass, and energy of the ions. Furthermore, such ion beams can be made practically parallel in the neighborhood of the target so that (as, for example, in the experiments of ROL et al. [603]) the angle of incidence of the ions on the target can be specified within $\pm$ 1.5°. In studies with a mass analyzer, the residual pressure p in the target chamber could always be kept sufficiently low ($p < 10^{-5}$ mm Hg) that the back diffusion of sputtered particles to the target was suppressed. Because of the high ion-current densities (0.1 — 1.0 mA/cm^2) impinging on the target in most cases, especially in using the β calutron at Oak Ridge (YONTS et al. [787]) or the separators at Amsterdam (KISTEMAKER et al. [602, 603]), Gothenburg, Sweden (ALMEN et al. [9]), and Moscow (GUSEVA et al. [275], MOLCHANOV et al. [489]), the development of adsorption layers on the target surface was suppressed and the sputtering rates were not particularly affected by the absolute magnitude of the ion current. (This is discussed in more detail in Section 10.3.1.2.5.) KAMINSKY [361, 362, 363] has studied sputtering in the high-energy region (0.1—4 MeV) by use of mass-selected ion beams from a COCKCROFT-WALTON generator and a VAN DE GRAAFF accelerator in conjunction with a magnetic analyzer. The current densities obtainable were in the region of 300 to 4,000 $\mu A/cm^2$.

10.2.2. Methods of Determining the Sputtering Yield

10.2.2.1. Weight Change of Target and/or Collector

A common way to determine the number of sputtered target atoms is to measure the weight loss Δm as the result of sputtering of the target (e.g., PLESHIVTSEV [*572*], BADER [*40*], YONTS [*786*], PITKIN [*571*], YONTS et al. [*787*], MOLCHANOV et al. [*489*], GUSEVA [*274*], GUSEV et al. [*275*], ROL et al. [*603*], PITKIN et al. [*570*], ALMEN et al. [*9, 10*]). If the incident-ion current I at the target and the bombardment time t are also known, then the sputtering ratio S (atoms/ion) can be determined from the relation.

$$S = 26.6 \frac{\Delta m}{A I t}, \qquad (10.2.2.1\text{–}1)$$

where Δm is in μg, I in μA, and t in hr; and A is the mass number of the target atom.

In determining Δm one should realize that there are several effects which can influence its value. Thus on the one hand the ions impinging on the target may be captured in the target lattice and increase the mass of the target by the amount Δm_1. This effect is especially important at the higher ion energies. ALMEN and BRUCE [*9*] showed, for example, that a titanium target bombarded with 45-keV Kr ions can collect incident particles up to a saturation value of 4.7 μg/cm^2. Since the weight measurements are usually carried out in air, adsorption layers may develop on the target surface, and these increase the weight by an additional amount Δm_2. Then Eq. (10.2.2.1–1) is replaced by

$$S = \frac{26.6\,\Delta M - \Delta m_1 - \Delta m_2}{A I t}, \qquad (10.2.2.1\text{–}2)$$

where ΔM is the experimentally determined total change in weight. In many cases the experiments are performed in such a way (e.g., large exposure times, t or large currents I) that $\Delta M \gg \Delta m_1, \Delta m_2$. This is for instance the case in the experiments of BADER, WITTEBORN, and SNOUSER [*40*] who obtained weight changes in the range of 500—2000 μg and the experiments of PLESHIVTSEV [*572*] where Δm was in the range 3 to 35 mg.

Another possible way of determining the number of sputtered target atoms is to measure the increase in weight Δm of the collector (COBIC and PEROVIC [*117, 555*]). Here, however, one must take precautions against two sources of error in Δm. On the one hand, because of the geometric properties of the system, not all the sputtered target atoms will reach the collector; and, on the other hand, the residual gas may adsorb as a surface layer on the collector. If the sputtering rate is very small (for example $S \approx 10^{-3}$ for 1-MeV protons on Cu) these two methods are not usable because of the long exposure time required.

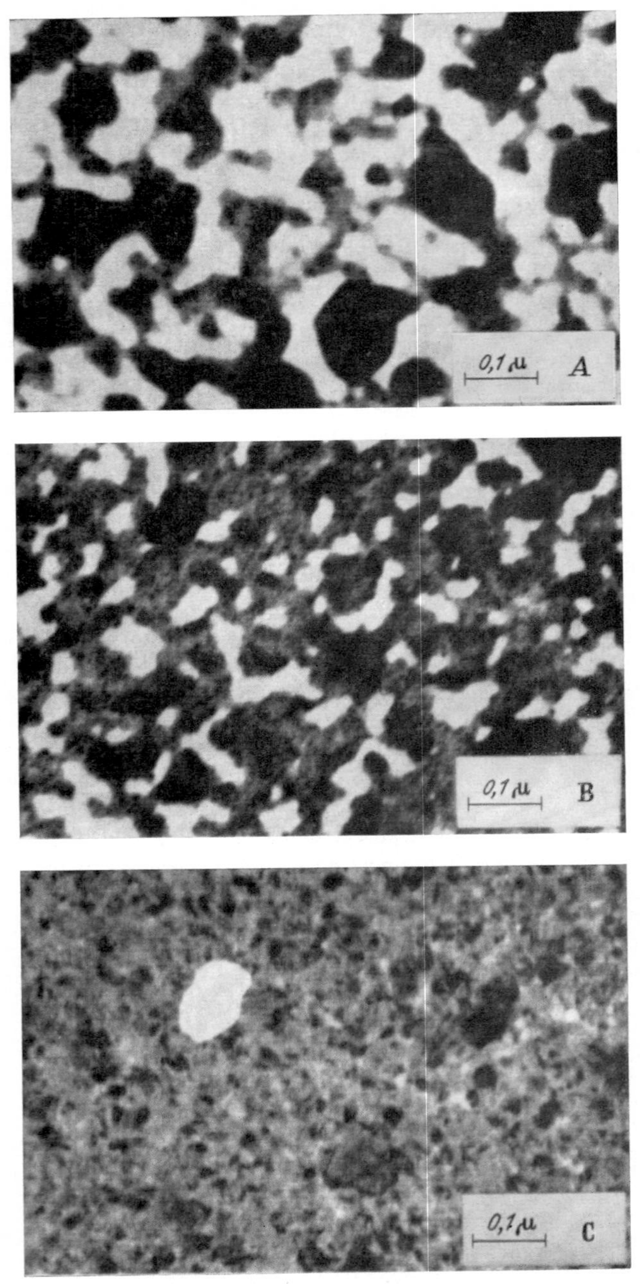

Fig. 10.2.2.2.a

10.2.2.2. Optical Transmission through Sputtered Layers

A more sensitive technique than that of Section 10.2.2.1 is to determine the number of sputtered particles by optical transmission. Here the sputtered particles are condensed on quartz, mica, or glass collectors and the thickness of the atomic layers developed can then be determined quite accurately by optical transmission (e.g., in the case of silver on glass at $\lambda = 6700$ Å, KOEDAM [*402*] determined the thickness to within 3 atom layers). However, it is first necessary to know the relation between the layer thickness of the particular adsorbed substance and the optical transmission. The monograph of HEAVENS [*303*] contains relations for layers of numerous substances. However, one must note that the data given in this monograph apply to films produced by evaporation. KOEDAM [*402*] and also ROTHMAN et al. [*616*] gave impressive evidence that the condensed films produced by evaporation and by sputtering have different structures. Thus, for example, silver films produced by evaporation have a much coarser grain structure than do those produced by sputtering (see Fig. 10.2.2.2.a). For silver and copper layers, KOEDAM [*399, 401*] found corresponding differences in transmission for the same number of condensed particles per cm² of collector surface. He also determined experimentally that on glass the condensation coefficient for these layers was unity, i.e., all the particles incident on the collector surface remain there. In addition to its relatively high sensitivity, the optical transmission method has the further advantage that it can very easily give measurements of the angular distribution of the sputtered particles (as, for example, in the work of ROL et al. [*605*], KOEDAM [*402*], and WEHNER et al. [*770*]). The use of the transmission method is illustrated in Fig. 10.3.2.1.1.a, which shows the sputtering pattern from the (100) plane of copper single crystal bombarded at normal incidence by 500-eV ions of various noble gases. This figure clearly shows the preferred directions of sputtering along the directions of the lines along which the atoms are most densely packed—the [101], [001] lines (see Section 10.3.2.1.1).

10.2.2.3. Radioactive-Tracer Method

If the target contains a radioactive isotope, the number of sputtered target atoms which are deposited on the collector can be determined from the radioactivity of the layer. Here it is assumed that the isotopic ratios are sufficiently well known and that all the isotopes of the particular target element are sputtered equally. O'BRIAIN et al. [*529*], who used

Fig. 10.2.2.2.a. Electron micrographs of thin silver layers (thickness 100 Å) formed by evaporation (in 2 sec) and by cathode sputtering (in 20 min) on a glass plate (KOEDAM [*402*]). Micrograph A: Silver layer formed by evaporation. Micrograph B: Silver layer formed by evaporation. The collector was bombarded with ions while the layer was being deposited. Micrograph C: Silver layer formed by sputtering

Ag^{110} as tracer in a silver target, could detect as little as 10^{-8} g of sputtered silver on the collector. The angular distribution of the sputtered material was also determined with good accuracy. The method is unfortunately limited to target materials for which radioactive isotopes of sufficiently long life can be obtained. The disadvantage of using such long-lived radioactive isotopes is the possible radioactive contamination of the apparatus.

10.2.2.4. Surface Ionization

BRADLEY [*88*] employed surface ionization to reveal the presence of sputtered sodium and potassium ions on a hot platinum electrode. The method is very sensitive, but applicable only to a restricted number of materials (cf. Chaps. 8 and 9). An accurate determination of the number of sputtered particles is restricted to systems with 100% surface ionization, and then only when all ions are collected.

10.2.2.5. Mass-Spectrometric Detection

A number of authors (HONIG [*322*], BRADLEY [*90*], VEKSLER [*739*], STANTON [*674*] and KAMINSKY [*363a*]) have used mass spectrometers to show the sputtering of target particles in the form of ions. HONIG, BRADLEY and KAMINSKY were also able to detect sputtered neutral particles after ionizing them by electron impact in the mass spectrometer. However, since the probability of ionization is very small because of the relatively high kinetic energy ($\approx$ 10 eV) of the neutral particles, such ion currents are low. Even though this method does not lend itself easily to a determination of the total number of sputtered particles, it gives interesting information on the sputtered atomic and molecular species and on the ratio of ions to neutral particles.

10.2.2.6. Optical Spectroscopy

Another very sensitive method of detecting sputtered target atoms is the spectrometric method of STUART et al. [*690, 691*]. In their experimental arrangement, the target is in a plasma. The sputtered target atoms can be excited in the plasma, where their emission spectrum can then be observed—superposed, of course, on the emission spectrum of the plasma. The intensity of the emission lines of the target material was determined by using a monochromator and photomultiplier. With this method the authors could, for numerous metals, determine the sputtering ratio S down to values of $S \approx 10^{-4}$ atom/ion.

10.3. Experimental Results

10.3.1. Sputtering of Neutral Target Atoms from Polycrystalline Targets

10.3.1.1. Threshold Values for Sputtering

Measurements of sputtering yields from primary ions with very low energies have been conducted by several authors ([*88, 329, 381, 419, 446, 467, 502, 504, 690, 736, 758, 762, 764*]) in order to investigate the possible existence of an onset energy for the sputtering process. Such investigations are associated, however, with great experimental difficulties. In the energy region of the threshold, the amount of sputtered material is so small that conventional methods (Section 10.2.2.1) do not yield a sufficiently accurate measurement of the sputtering yields. Since the sputtering ratio in the threshold region does not increase linearly with increasing energy E of the primary ions, but rather varies with E^2 (as shown, for example, by Wehner [*762*] and Langsberg [*419*]), a linear extrapolation of experimental data taken at medium energies of the primary ions will not give meaningful threshold values (Langsberg [*419*]). Other difficulties encountered in the region of low primary-ion energies and of small sputtering yields are the formation of impurity layers at the surface and the focusing of ions onto the target. In order to overcome some of these difficulties, more sensitive methods of detecting sputtered particles have been developed recently, such as a radioactive tracer (Morgulis and Tishenko [*502*]), surface ionization (Bradley [*88*]), optical spectroscopy (Stuart and Wehner [*690*]), or the measurement of the frequency change of a plated quartz crystal to determine its weight loss under ion bombardment (McKeown [*467*]). Fig. 10.3.1.1.a illustrates some of the $S(E)$ curves obtained for the threshold region with more sensitive detection methods. Experimental data obtained by such sensitive methods indicate that the sputtering thresholds (range 5—40 eV) have much smaller values than were reported in the past (Wehner [*760, 762, 764*]

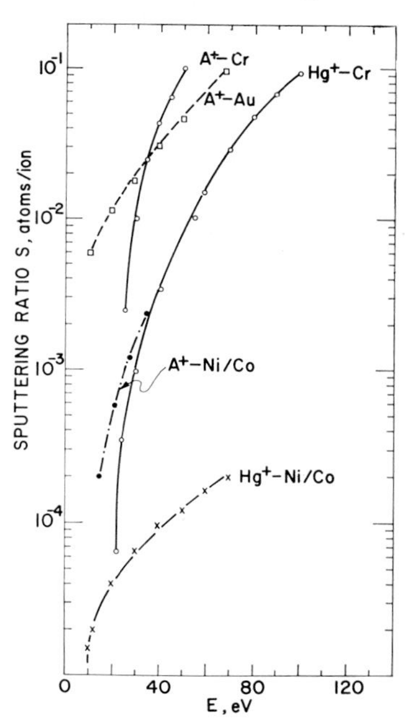

Fig. 10.3.1.1.a. Variation of the sputtering ratio S with incident-ion energy E in the low-ion-energy region for different ion-target systems: A^+-Cr, Hg^+-Cr, (Stuart and Wehner [*690*]); A^+-Au, (McKeown [*467*]); A^+ and Hg^+ on the alloy Ni + 1% Co^{60} (Morgulis and Tishenko [*502*])

and LANGSBERG [*419*]) (range 50—150 eV). As examples of the rather drastic corrections which have to be made (STUART and WEHNER [*692*]) for some targets, under Hg^+ bombardment the threshold value for Th changed from 120 to 30 eV, for V from 120 to 25 eV, for Ta from 120 to 30 eV, and for Zr from 120 to 30 eV. Table 10.3.1.1.A compares the threshold values determined by several authors (STUART and WEHNER [*690*],

Table 10.3.1.1.A. *Average sputtering thresholds for ions incident normally on polycrystalline targets*

Bombarding ion	Target	Crystal structure	Heat of sublimation (eV)	Exper. threshold values E_t (eV)		
				STUART et al. [*692*]	MORGULIS et al. [*504*]	MCKEOWN [*467*]
A^+	Ag	fcc	2.8	15	4	10
	Au	fcc	3.9	20		
	Ta	bcc	8.0	26	13	
	W	bcc	8.8	33	13	
	Co	hcp	4.4	25	6	
Xe^+	Ag	fcc	2.8	17		
	Au	fcc	3.9	18		
	Ta	bcc	8.0	30		
	W	bcc	8.8	30		
	Co	hcp	4.4	22		
Hg^+	Ag	fcc	2.8	(40)*		
	Au	fcc	3.9	(40)*		
	Ta	bcc	8.0	30		
	W	bcc	8.8	30		
	Co	hcp	4.4	37	7.2	

* G. K. WEHNER [*762*]

MORGULIS and TISHCHENKO [*502*], MCKEOWN [*467*]) who used the more sensitive detection methods for targets with different crystal structures and for different bombarding ions. For a more complete listing of average threshold values for polycrystalline surfaces, the reader is referred to the articles of STUART et al. [*692*] and HARRISON et al. [*293*]. Small values of the sputtering threshold are also reported by BRADLEY [*88*], who found $E_t = 5$ eV for Ne^+ on Na, $E_t \approx 10$ eV for A^+ on Na, and $E_t \approx 25$ eV for Xe^+ on Na. The values listed in Table 10.3.1.1.A should not, however, be considered as actual average threshold energies, but rather as upper limits of the sputtering onset, since they have been extrapolated from yield data at medium energies (around 20—80 eV). The large discrepancy between the values of STUART et al. and MORGULIS et al. for Hg^+ on Co might be due, as suggested by HARRISON et al. to a higher concentration of Hg^{++} ions in the experiment of MORGULIS.

Table 10.3.1.1.A shows the interesting result that the threshold values are almost independent of the ratio of masses of incident ion and target atom, although the lowest values for a given target should be expected when incident and target atoms are most nearly equal in mass. There seems to be a trend that materials with low heats of sublimation (such as Ag, Au) have lower average threshold values than those with high heats of sublimation (e.g., Ta, W).

Stuart et al. [*692*] pointed out that the threshold energies might be linked to the displacement energies in solids. Table 10.3.1.1.B gives

Table 10.3.1.1.B. *Calculated values for the maximum transferable energy, for incident argon ions with energies close to the treshold energy* E_t

From Stuart and Wehner [*692*]

Target	Cu	Ag	Au	Ta	Mo	W
E_{max}(eV)	24	20	17	22	21	24

some values of the maximum energy $E_{max} = [4\,mM/(m+M)^2]\,E$ that can be transferred to the lattice from an incident A^+ ion whose energy is close to the average threshold E_t for a polyrystalline surface. These values are close to the thresholds for the displacement of lattice atoms (Dienes and Vineyard [*146*]), which for most metals range between 20 and 25 eV. Harrison et al. [*293*] calculated thresholds for single-crystal surfaces by a semiquantitative theory, using a simple collision model and taking Silsbee's chaining mechanism (see Section 10.4.2) into account. With the aid of simplifying assumptions (such as averaging over all possible crystal orientations), these authors extended their calculations to polycrystalline surfaces. Their calculated values of E_t were consistently lower than the experimental values reported by Stuart et al. [*692*]. In a few cases, such calculated values of E_t were even smaller than the heats of sublimation E_S of the target material (e.g., for A^+ on Ag, $(E_t)_{theor} = 1.62$ eV and $E_S = 2.84$ eV; for A^+ on Ta, $(E_t)_{theor} = 5.12$ eV and $E_S = 8.02$ eV) and were of the order of magnitude of surface-atom bond energies. However the existing experimental data are not accurate enough to allow any conclusions to be drawn with respect to the concept of sputtering thresholds and the sputtering process itself. More investigations, especially of single-crystal surfaces, are urgently needed.

10.3.1.2. Sputtering Yields in the Hard-Sphere Collision Region

10.3.1.2.1. Dependence on the Energy of the Incident Ion

a) Energy Range 0.1–1.0 keV. While the threshold value for the sputtering process (see Table 10.3.1.1.A) is not very different for extremely different types of materials, there are nevertheless quite marked deviations in the rate at which the sputtering yields S of different materials rise with increase in primary ion energy. The function $S(E)$ for the sputtering of a particular ion may show a marked dependence on the ion energy E. This may actually happen within an energy interval in which one particular collision process (for example hard-sphere, weak-screaning, or Rutherford collision) predominates. In order to exhibit such regularities more clearly, the energy range characteristic of hard-sphere collisions will be discussed below in two different parts. Fig. 10.3.1.2.1a shows the relation between sputtering ratio S and ion energy E from approximately 50—250 eV as reported by KOEDAM [*400*] in his experiments with a polycrystalline silver target bombarded at normal incidence with He^+,

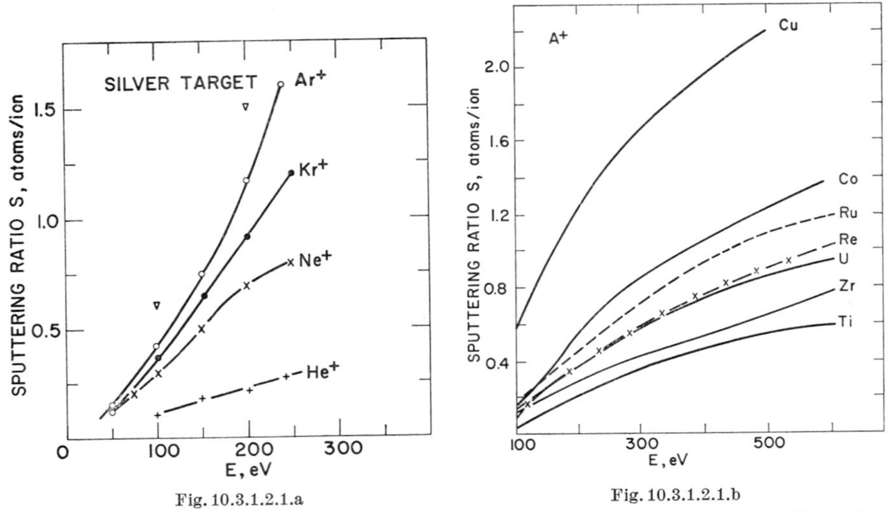

Fig. 10.3.1.2.1.a

Fig. 10.3.1.2.1.b

Fig. 10.3.1.2.1.a. Variation of the sputtering ratio S with incident-ion energy E for noble-gas ions incident on a silver sufrace (KOEDAM [*400*])

Fig. 10.3.1.2.1.b. Variation of the sputtering ratio S with incident-ion energy E for argon ions incident on various target materials (WEHNER [*769*])

Ne^+, A^+, and Kr^+ ions. The function $S(E)$ is approximately linear in the energy range from 150 to 250 eV for A^+ and Kr^+ ion bombardment and is roughly quadratic at lower energies ($\approx$ 50—150 eV). WEHNER and MEDICUS [*759*] observed a similar quadratic behavior of the function $S^*(E)$, where S^* is the sputtering ratio uncorrected for secondary-electron emission, in the sputtering of platinum by mercury ions in the low-energy region. For Ne^+ ions, the function $S(E)$ is approximately linear in the range from ca. 80 to 180 eV. Above this region the curve appears to bend over into a saturation region.

In the case of He^+ ion bombardment, the function $S(E)$ is linear over the energy range 100—250 eV. The absolute values of S for various noble gases show the expected behavior for He^+, Ne^+, and Kr^+ in that the value of S increases with increasing mass and atomic number of the incident ion. In the case of bombardment by A^+ and Kr^+, however, the yield curve for argon is above that for krypton despite the higher mass and charge numbers of krypton. A similar behavior is also observed by WEHNER [*769*] in his sputtering experiments on various targets with A^+ and Xe^+ ions (see Figs. 10.3.1.2.1.b and 10.3.1.2.1.c). A comparison of the sputtering-yield curves for A^+ and Xe^+ bombardement of the same targets shows that the yield curve for A^+ in the low-energy region is higher than for Xe^+. The curves cross at higher energies, the crossing

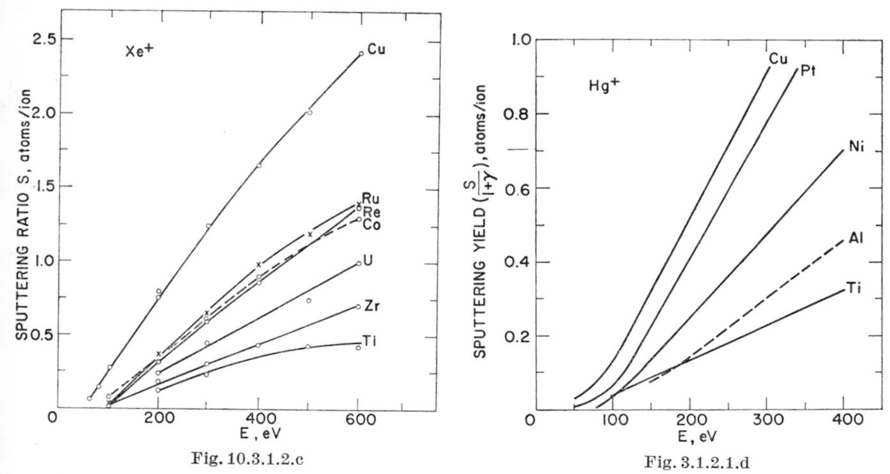

Fig. 10.3.1.2.c

Fig. 3.1.2.1.d

Fig. 10.3.1.2.1.c. Variation of the sputtering ratio S with incident-ion energy E for xenon ions incident on various target materials (WEHNER [*769*])

Fig. 10.3.1.2.1.d. Variation of the sputtering yield $S/(1+\gamma)$ with incident-ion energy E for mercury ions incident on various target materials (WEHNER [*763, 764*])

point for targets with lower atomic weights occurring at higher energies ($E_s \approx 600$ eV) than for those with higher atomic weight ($E_s \approx 150$ eV). This "cross-over" behavior is confirmed by the investigations of other authors, e.g., BRADLEY [*88*] for the case of A^+ and Xe^+ ions on Na and K targets and OECHSNER [*530*] for A^+ and Xe^+ ions on Mo targets. As an example of sputtering, Fig. 10.3.1.2.1.d shows the results of investigations with low-energy ions other than noble-gas ions, as reported by WEHNER [763, *764*]. The curves give the yield $S(E)/[1+\gamma(E)]$, uncorrected for the secondary electron yield γ, for various targets under bombardment with mercury ions. A comparison of these yield curves with those for A^+ and Xe^+ bombardment (see Fig. 10.3.1.2.1.b and 10.3.1.2.1.c) for the same target show that so long as the mass m of the incident ion is smaller than the mass M of the target atom ($m < M$) the Hg^+ yield curves lie lower than those for A^+ and Xe^+ ions.

These results indicate that the value of the sputtering ratio has a complicated energy dependence in the low-energy region. Various authors have predicted quite different energy dependences of S. For example, KEYWELL [*375*] predicted $S \propto E^{1/2}$, LANGSBERG [*419*] predicted $S \propto E^2$ at the lower energies and $S \propto E$ at the higher, and BRADLEY [*88*] predicted $S \propto \ln E$ for light ions and $S \propto E^{1/2}$ for heavy ions. However, as will be shown in more detail in section 10.4, none of the existing sputtering theories that should be applicable to this energy range can correctly reproduce all the experimental results. Even the more quantitative sputtering theories of ROL et al. [*603*] and of PEASE [*550*] cannot correctly reproduce

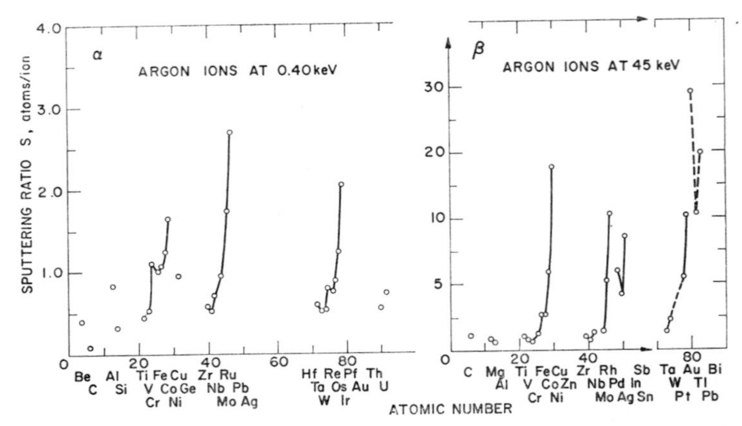

Fig. 10.3.1.2.1.e. The sputtering ratio S as a function of the atomic number of the target material. Diagram α is for the case of argon ions incident at 0.40 keV (WEHNER [*769*]), diagram β for the case of argon ions incident at 45 keV (ALMEN and BRUCE [*9*])

the results for this energy range. Thus, none of the predictions concerning the behavior of the function $S(E)$ in this energy range are correct for the general case. Nevertheless they may be applicable in special cases.

In a qualitative interpretation, WEHNER [*769*] pointed out that plotting the yield for argon ions at 0.40 keV as a function of the atomic number of the target material shows an interesting relation (see curve α of Fig. 10.3.1.2.1.e). The yields increase within a given period of the periodic system as the d shells are filled, and Cu, Ag, and Au have the highest sputtering yields. This holds for bombardment with A^+ and Ne^+ as well as for Hg^+ ions (see Figs. 10—12 in WEHNER's report [*769*]). For all three gases, the only exception is the chromium target for which the sputtering yield behaves more like that of Co. Curve β of Fig. 10.3.1.2.1.e illustrates corresponding results for high-energy argon ions (45 keV).

b) Energy Range 1—60 keV. Figs. 10.3.1.2.1.f and 10.3.1.2.1.g show the curves of $S(E)$ for the energy range from 1—60 keV for the sputtering of copper and silver targets by various noble gas ions. It is interesting to compare these curves with the corresponding curves for lower ion energies (Figs. 10.3.1.2.1.a and 10.3.1.2.1.b). For the case of A^+ ion bombardment of silver and copper targets at higher energies, the yields for the silver target are seen to be much above those for the copper target, whereas at lower energies (100—250 eV) quite the reverse is true. For energies above about 8 keV, the yields for both targets increase with increasing mass of the corresponding noble gas ion; whereas at lower

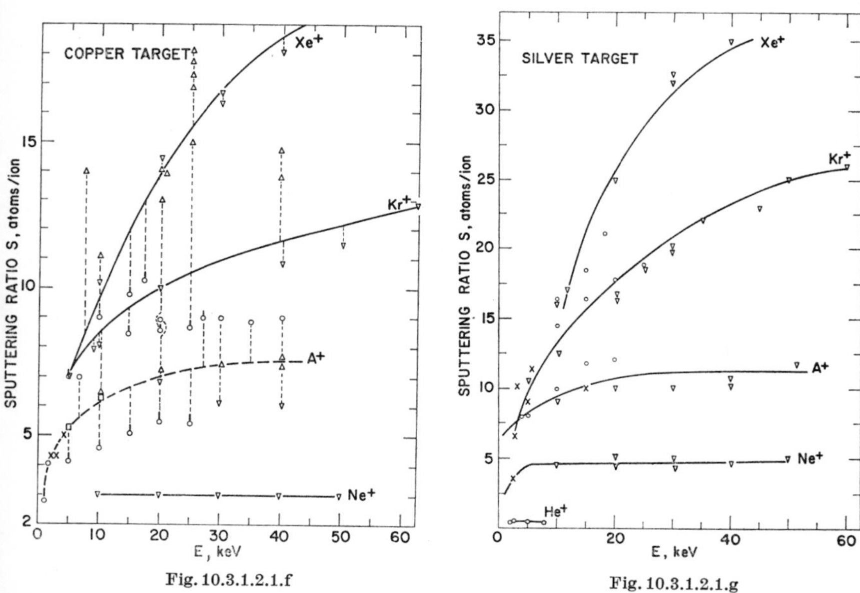

Fig. 10.3.1.2.1.f

Fig. 10.3.1.2.1.g

Fig. 10.3.1.2.1.f. Variation of the sputtering ratio S with incident-ion energy E for noble-gas ions incident on a copper target. Points ×, KEYWELL [375]; points ○, YONTS et al. [787]; points ♁ GUSEVA [274]; points △, PITKIN et al. [570]; points ▽, ALMEN et al. [9]; points □, ROL et al. [603]

Fig. 10.3.1.2.1.g. Variation of the sputtering ratio S with incident-ion energy E for noble-gas ions, incident on a silver target. Points ×, KEYWELL [375]; points ○, YONTS et al. [787]; points ▽, ALMEN et al. [9]

energies (100—300 eV), the yields for Kr^+ and Xe^+ ions drop even below those for the A^+ ion. Furthermore it appears that the $S(E)$ curves at high values bend over into a saturation region where the characteristic feature is that the lighter the incident ion, the lower the ion energy at which the saturation value is achieved.

For the case in which copper targets are sputtered with A^+ ions, a system which has been investigated by numerous authors, there are clearly marked discrepancies in the various experimental data. Whereas the experimental results of KEYWELL [375], ROL et al. [603, 604] and ALMEN et al. [9] do not differ very much from one another, those of

GUSEVA [274] show a marked deviation toward one side and those of YONTS et al. [787] and PITKIN et al. [570] deviate by 30—40% in the other direction. Such deviations in the experimental results could arise from various circumstances (such as the degree of development of surface layers, residual gas pressure, ion-current densities, or possibly such effects as capture of the incident ion in the lattice of the target with resulting distortion of the lattice structure, or increasing roughness of the surface under protracted bombardment, i.e., surface etching effects). Fig. 10.3.1.2.1.h shows $S(E)$ curves for A^+ bombardment of various

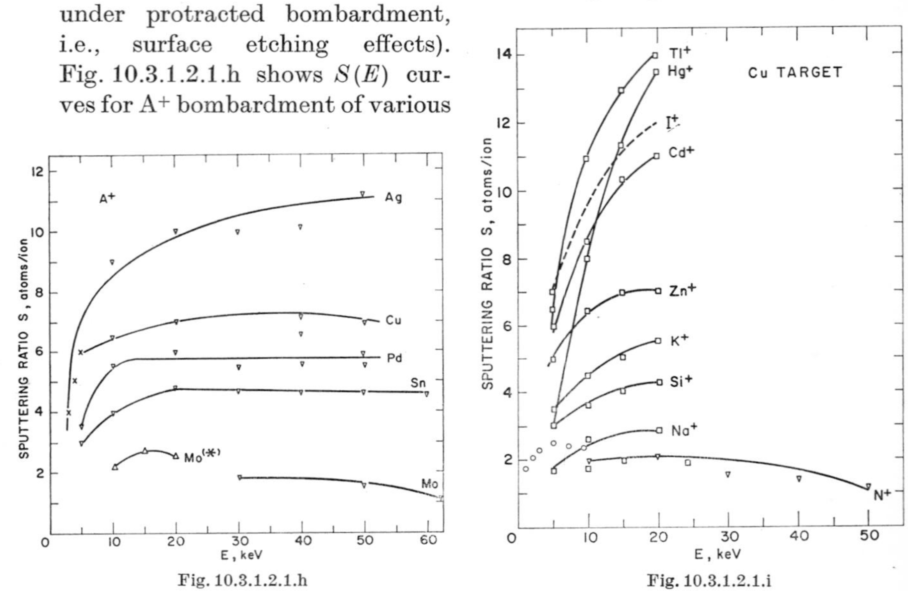

Fig. 10.3.1.2.1.h

Fig. 10.3.1.2.1.i

Fig. 10.3.1.2.1.h. Variation of the sputtering ratio S with incident-ion energy E for argon ions incident on various targets. Points ×, KEYWELL [375]; points △, PITKIN et al. [570] (values of S not corrected for the trapping of incident particles in the lattice); points ▽, ALMEN et al. [9]

Fig. 10.3.1.2.1.i. Variation of the sputtering ratio S with incident-ion energy E for a copper target bombarded by various kinds of ions. Points □, ROL et al. [603]; points ▽, ALMEN et al. [9]; points ○, BADER et al. [40] (for N^+ on Cu)

target materials which, except for copper, have only slightly different mass numbers (95—118) and atomic numbers (42—50). The sputtering ratio, however, varies considerably from one target material to another. If the sputtering ratio for 45—keV A^+ ions is plotted as a function of the atomic number of the target material (curve β of Fig. 10.3.1.2.1.e), the S values for targets within a given period of the periodic table are seen to increase as the d shell fills; and Cu, Ag, and Au have the highest values. This is just as in the low-energy case (curve α of Fig. 10.3.1.2.1.e). As an example of target sputtering with ions other than those of the noble gases, Fig. 10.3.1.2.1.i shows curves of the function $S(E)$ for the sputtering of a copper target with various metal ions as well as I^+, Si^+, and N^+ ions.

Here, just as for the case of bombardment with a noble gas, the lighter the bombarding ion the lower the energy at which the $S(E)$ curve goes to saturation. Above approximately 15 keV, the yields tend to increase with increasing mass and atomic number of the bombarding ions. Fig. 10.3.1.2.1.k shows $S(E)$ curves for the sputtering of various targets with nitrogen ions to emphasize the effect of the target material. The curves for $S(E)$ bend over into the saturation region at energies above 2 keV; and in the case of Cu, Ni, and Mo targets they actually show a slight downward trend for energies above 6 keV. Comparing the yields S for various targets shows that neither the crystal structure (Cu and Ni are fcc crystals) nor the mass M of the target atom nor its atomic number Z_T is a dominant influence on the sputtering process, since $S_{W,\,Mo} < S_{Fe,\,Ni} < S_{Cu}$, whereas $(M, Z_T)_{W,\,Mo} > (M, Z_T)_{Cu} > (M, Z_T)_{Fe,\,Ni}$. On the other hand, the variation of the yields with the atomic number of the target material is similar to that under bombardment by noble gas ions (curves α and β of Fig. 10.3.1.2.1.e) in that it can be correlated with the filling of the d shell within a period of the periodic system. As will be shown in more detail in Section 10.4.1, none of the theories applicable to the hard-sphere-collision region can correctly reproduce the experimental results over the whole energy region discussed here. However, the theory proposed by Rol et al. [*604*], when its free parameter is properly fitted to the experimental results, allows calculation of the sputtering yields for noble gas ions on various targets (Almen et al. [*9*]) within about 20% error in the energy range from 10—60 keV.

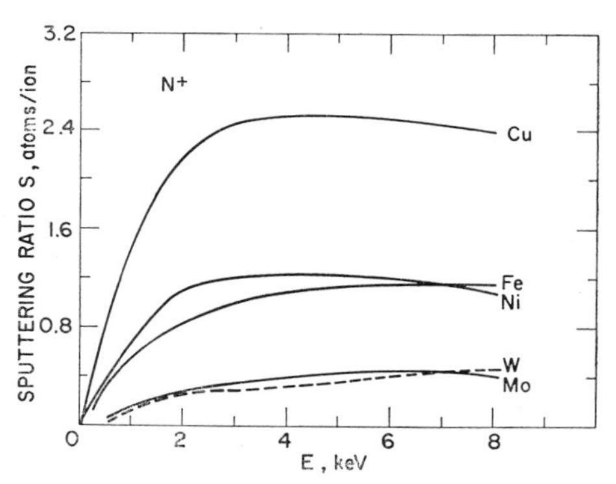

Fig. 10.3.1.2.1.k. Variation of the sputtering ratio S with incident-ion energy E for nitrogen ions incident on various targets (Bader et al. [*40*])

10.3.1.2.2. Dependence on Angle of Incidence

The first experimental evidence on the angular dependence of sputtering with low-energy ions was obtained by Fetz [*211*], but 15 years elapsed before systematic investigations of this effect were undertaken by many authors. In most cases these results have been quite qualitative because of inadequate knowledge of various pertinent factors. In particular, in nearly all experiments conducted so far the angle of incidence could not be determined accurately because of uncertainties about the parallelism and uniformity of the incident beam and about the macroscopic flatness and microscopic roughness of the target surface, the roughness being due either to the initial preparation or to protracted ion bombardment of the surface. A precise determination of the angle dependence is also impeded by

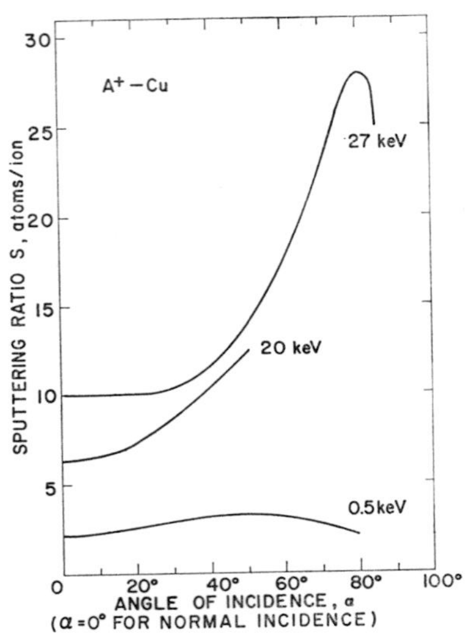

Fig. 10.3.1.2.2.a. Variation of the sputtering ratio S with the angle of incidence α for argon ions incident on copper with energies of 27 keV (MOLCHANOV [*489*]), 20 keV (ROL et al. [*604*]), and 0.5 keV (WEHNER [*767*])

(a) ion reflection (Section 11.3.4) which according to MOLCHANOV et al. [*489*] increases markedly at large angles of incidence (e.g., 22% of a beam of 27-keV A^+ ions incident on Cu at 84° is reflected); and (b) secondary electron emission (Chap. 14) as a result of ion bombardment, which also depends on the angle of incidence of the ions and makes an undesired contribution to the target current. Since these perturbing factors are only partially corrected for in the individual papers, the experimental

Table 10.3.1.2.2.A. *Sputtering ratio $S(\alpha)$ for a Cu target bombarded by various ions incident at different energies and angles* $Q = S(50°)/S(0°)$

Ion energy (keV)	Ne^+			A^+			Kr^+			Xe^+			N^+			Si^+			Tl^+		
	0°	50°	Q	0°	50°	Q	0°	50°	Q	0°	50°	Q	0°	50°	Q	0°	50°	Q	0°	50°	Q
5				5.6	8.7	1.5_5							1.7	3.0	1.7_6	3.0	5.5	1.8_3			
10				6.3	11.0	1.7_4							1.9	3.4	1.7_9	3.6	7.4	2.0_5			
15				6.4	11.6	1.8_1							2.1	3.4	1.6_2	4.0	7.8	1.9_5			
20				6.4	12.1	1.8_9				15.5	25.5	1.65	1.7	3.4	2.0_0	4.3	8.2	1.9_0	13.8	30.0	2.17
25				6.5	12.4	1.9_0															
45	3.1	5.5	1.7_7				11	23.5	2.13												

results should not be regarded as quantitatively significant, but they do serve to give useful qualitative conclusions.

Fig. 10.3.1.2.2.a shows the dependence of the sputtering coefficients S on the angle of incidence α for argon ions of various energies impinging on a copper target. One sees that the sputtering ratio S increases markedly with increasing incidence angle α. Thus, the value of S for 27-keV argon ions (MOLCHANOV et al. [489]) increases by more than 280% in the range $0° < \alpha < 70°$. This marked dependence of S on α decreases, however, for lower ion energies, as one sees from the curves for 27 keV (MOLCHANOV et al. [*489*], 20 keV (ROL et al. [*604*]), and 0.5 keV (WEHNER [*767*]), and from the data of Table 10.3.1.2.2.A. While MOLCHANOV et al. concluded

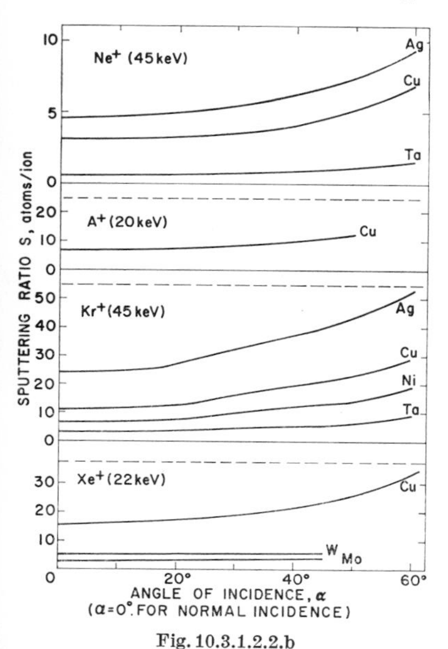

Fig. 10.3.1.2.2.b

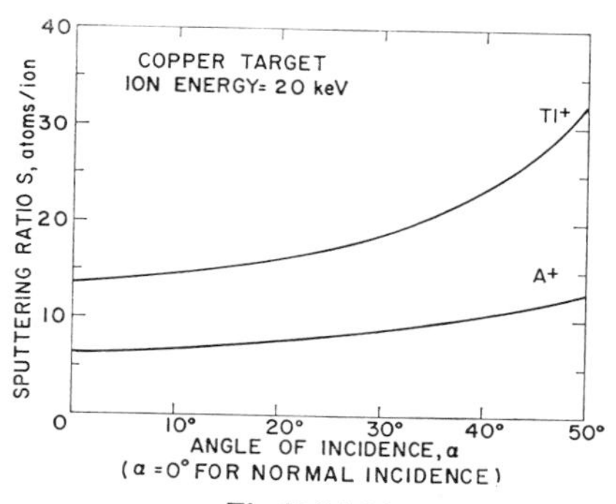

Fig. 10.3.1.2.2.c

Fig. 10.3.1.2.2.b. Variation of the sputtering ratio S with the angle of incidence α for various target materials bombarded by different noble-gas ions: Ne^+ and Kr^+ at 45 ke V(ALMEN et al. [*10*]), A^+ at 20 keV (ROL et al. [*602*]), Xe^+ at 22 keV (PITKIN [*571*])

Fig. 10.3.1.2.2.c. Variation of the sputtering ratio S with the angle of incidence α, for 20-keV Tl^+ and A^+ ions incident on copper (ROL et al. [*603*])

that their experimental determination of the variation of $S(\alpha)$ for angles α up to 70° was best fitted by the formula $S = S_0/\cos\alpha$, where S_0 is the sputtering ratio for normal incidence of the ion on the surface, ROL et al. [*603*] find that their results (the curve for 20-keV A^+ ions) for incidence angles up to 45° is best fitted by the formula $S = S_0 (2 - \cos\alpha)/\cos\alpha$. At the larger angles of incidence ($\alpha \lesssim 70°$) the function ($S(\alpha)$ goes through a maximum, as one can see from the curves for 27-keV and 0.5-keV argon ions. This behavior may be caused in part by the marked increase in the reflection of the primary ion beam at larger angles of incidence (as will be discussed further in Chap. 11). MOLCHANOV et al. observed that at $\alpha = 70°$ no argon ions (at 27 keV) are reflected from the copper surface, but 6% are reflected for $\alpha = 78°$, 17% for $\alpha = 82°$, and 22% for $\alpha = 84°$. Quantitative corrections of the $S(\alpha)$ values for A^+ on copper (MOLCHANOV et al. on the basis of the reflection coefficients could not, however, completely explain the observed behavior of of the function $S(\alpha)$ for $\alpha > 70°$. Fig. 10.3.1.2.2.b shows the function $S(\alpha)$ for the sputtering of various target materials by the noble gas ions Xe^+ at 22 keV (PITKIN [*571*]), Kr^+ at 45 keV (ALMEN and BRUCE [*10*]), A^+ at 20 keV (ROL et al. [*603*]), and Ne^+ at 45 keV (ALMEN and BRUCE [*10*]). Comparison of the results for the copper target shows that the functions $S(\alpha)$ begin to increase at smaller and smaller angles of incidence as the atomic weight of the bombarding ions is increased. On the other hand, at larger angles the ratio $S(\alpha)/S(0°)$ for bombardment

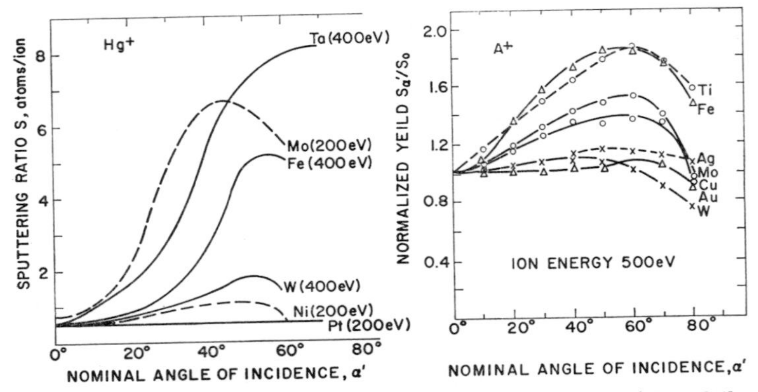

Fig. 10.3.1.2.2.d. Variation of the sputtering ratio S, or the normalized ratio $S_{\alpha'}/S_0$, with the nominal angle of incidence α' for Hg^+ and A^+ ion bombardment of various target materials (WEHNER [*767*, *768*]). (The use of a spherical target prevented an exact determination of α)

by noble gas ions appears to vary only slightly with increasing atomic weight of the ions. For the case in which a copper target is sputtered by 20-keV Tl^+ and A^+ ions (ROL et al. [*603*]), Fig. 10.3.1.2.2.c shows that the heavier Tl^+ has the stronger angular dependence and that, as can be seen from Table 10.3.1.2.2.A, the ratio $S(50°)/S(0°)$ is greater for Tl^+ ions than for A^+. ALMEN et al. [*10*] found that in the angular range from 0° to 60° the $S(\alpha)$ curves for Kr^+ and Ne^+ bombardment of various targets can be best represented by the relation $S_\alpha = S_0 \cos \alpha^{3/2}$. Only the curves for a silver target did not follow this relation.

Fig. 10.3.1.2.2.d gives WEHNER's curves of $S(\alpha')$ for the sputtering of various targets with low-energy (0.5 keV) Hg^+ and A^+ ions, where α' should be regarded only as the nominal angle of incidence since WEHNER's method of sputtering from spherical targets did not permit an exact determination of the angle of incidence. Although the experiments may also be affected by such things as partial surface covering and insufficient collimation of the ion beam, the results nevertheless are usable for a qualitative discussion of the effect of target material on the angular distribution.

Comparison of the curves for the Hg^+ and A^+ ion bombardment in Fig. 10.3.1.2.2.d shows first of all that the angular effect for bombardment with A^+ ions is smaller than that for Hg^+ ions. Furthermore one sees that in the angular range $10° < \alpha < 60°$, the angular effect for a target of a given lattice type (e.g., A_1 = fcc, A_2 = bcc, A_3 = hexagonal close packed) becomes larger with decreasing atomic weight. Thus, one finds, for example, for lattice type A_1 that $S(\alpha)_{Cu} > S(\alpha)_{Ag} > S(\alpha)_{Au}$, or for type A_3 that $S(\alpha)_{Ti} > S(\alpha)_{Zr} > S(\alpha)_{Hf}$. It is interesting to note that this relation is no longer found for bombardment with higher energy ions, e.g., 45-keV Kr^+ and Ne^+ ions on Ag and Cu targets. Here for example $S(\alpha)_{Ag} > S(\alpha)_{Cu}$. For low-energy ion bombardment, the target materials showing the smallest angular effect are those that give the largest sputtering yields (e.g., Cu, Ag, Au, Ni, Pt), while the substances that are only weakly sputtered (Mo, Fe, Ta) show a very strong angular effect. The complicated behavior of the angular effect has not yet been satisfactorily incorporated in a general sputtering theory (Section 10.4).

10.3.1.2.3. Dependence on Target Temperature

The first investigations (Blechschmidt [*74*]), General Electric [*586*] of the effect of target temperature on the sputtering ratio were carried out by the glow-discharge method (Section 10.2.1.1.), which as already noted does not give measurements under well-defined conditions. Therefore, the results of the above mentioned authors, which showed that S is practically independent of temperature, cannot be considered to be reliable. With the somewhat more reliable low-pressure method (Section 10.2.1.2.), Fetz [*211*] found that the sputtering ratio doubled when the temperature of a molybdenum target bombarded with low-energy mercury ions was increased from 400—1,000° C. Wehner [*762*] observed a small increase in the sputtering ratio when the temperature of a platinum target bombarded with low-energy ions was increased from 300—650° C. This increase of the function $S(T)$ with increasing T led Wehner to the conclusion that mercury films, which at lower temperatures hindered the sputtering of the target material, were desorbed from the target surface. This interpretation seems likely in view of other investigations (Wehner [*772*], Schiefer [*637*]) of the bombardment of various targets with noble gas ions, since here the values of S are practically independent of target temperature up to about 850° C. It is interesting that in the case of an iron target bombarded with A^+ and Kr^+ ions, a decrease of sputtering yield with target temperatures was observed only above 900° C. This decrease may be related to the phase change of iron from a body-centered to a face-centered lattice (at 906° C). Fig. 10.3.1.2.3.a reproduces the measurements of ALMEN et al. [9] for Ag, Pt, and Ni targets bombarded with high-energy (45 keV) Kr^+ ions. For silver targets at temperatures up to 600° C, S remains independent of temperature and then rises as the result of a superposed evaporation process. For Ni and Pt targets, on the other hand, S decreases slowly with increasing temperature in the range from 200—600° C. This has been related to the annealing of the metal at high temperatures. As a result of this process, damage to the target lattice by the incident ion is more or less repaired before a new ion strikes close to the point of incidence of the first.

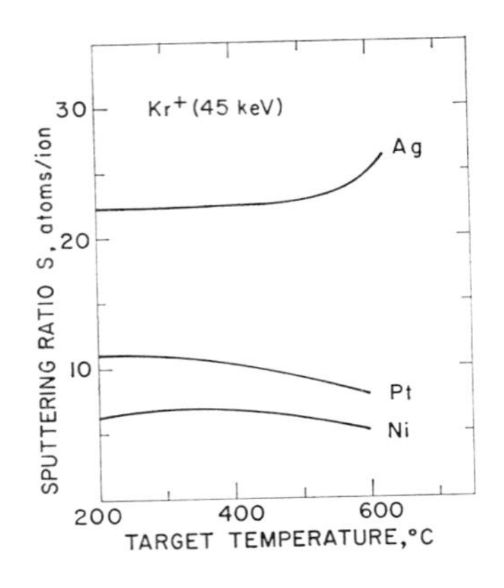

Fig. 10.3.1.2.3.a. Variation of the sputtering ratio S with the temperature of the target for Kr^+ ion bombardment of Ag, Pt, and Ni targets (ALMEN et al. [9])

10.3.1.2.4. Self-Sputtering

For a better understanding of the sputtering process, it is of interest to bombard a target of mass M with ions of the same element and then determine what is called the "self-sputtering" ratio S_s. Such investigations have been carried out by DOBREZOW and KARNAUKHOVA [152], GUSEVA [274], and ALMEN and BRUCE [10]. In determining the number of sputtered target atoms from the weight lost by the target, as was done by GUSEVA and by ALMEN and BRUCE, the self-sputtering ratio S_s is found from the relation

$$S_s = 26.6 \frac{\Delta m}{A I t} + f_i, \tag{10.3.1.2.4–1}$$

where Δm is the weight lost by the target (in μg), A the atomic weight of the target, I the ion current (in μA), and t the time of bombardment (in hours). The quantity f_i is the fraction of the incoming ions that is captured into the target lattice. According to ALMEN et al. [10], $f_i = 1$, for all "low-sputtering metals." For "high-sputtering metals" the exact knowledge of f_i ($0 \leqslant f_i \leqslant 1$) is unnecessary, since the first term in Eq. (10.3.1.2.4–1) is dominant. ALMEN et al. here assumed the value $f_i = 1$.

Fig. 10.3.1.2.4.a shows the curves of $S_s(E)$ for Cu and Ag as given by GUSEVA [274] (curves a) and ALMEN et al. [10] (curves b). For the energy range from 5—60 keV, the values of S_s for Cu and Ag lie far above the value 1, and the function of $S_s(E)$ increases with increasing ion energy. For Ag, the values of S_s in this range lie higher than for Cu. In this connection, it is of interest to compare the curves in Fig. 10.3.1.2.4.a with those in Fig. 10.3.1.2.1.g. In the case of Ag targets, one sees that the sputtering yields become larger with increasing atomic weight of the incident ions, the order being $S(A^+) < S(Kr^+) < S_s(Ag^+) < S(Xe^+)$. It is uncertain whether or not this correlation holds also for Cu targets [i.e., $S_s(Cu^+) < S(Kr^+)$] because different authors disagree rather seriously in their experimental curves for Kr^+ and Cu^+. GUSEVA's result that $S(Cu^+) > S(Xe^+)$ seems improbable when compared with the results of the authors.

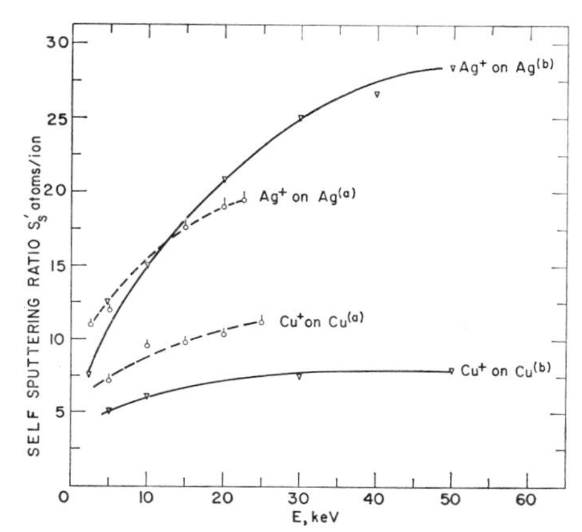

Fig. 10.3.1.2.4.a. Variation of the self-sputtering ratio S_s with the energy E of the incident ion. Curves (*a*), GUSEVA [*274*]; curves (*b*), ALMEN et al. [*10*]

ALMEN et al. assuming hard-sphere collisions between the incident ion and target ion, calculated the value of the function $S_s(E)$ for Ag and Cu on the basis of the sputtering theory developed by ROL et al. [*604*]. A more complete discussion of the results is postponed to Section 10.4. In this calculation, the authors fitted two constants to the experimental measurements. Their resulting equation for S_s in (atoms/ion) was

$$S_s = 4.2 \times 10^{-10} n_0 R^2 E^{\frac{1}{4}} \exp\left(-11.2\, E_B / 2\sqrt{M}\right), \quad (10.3.1.2.4\text{–}2)$$

where n_0 is the number of target atoms per unit volume (in m^{-3}), R is the collision diameter at ion energy E (in m), E is the incident ion energy (in eV), E_B is the binding energy of the target atoms (in eV), and M is the mass number of the element sputtered. The agreement between their experimental results (the measured points marked ∇ on curves *b* of Fig. 10.3.1.2.4.a) and those calculated from Eq. (10.3.1.2.4–2) (the solid lines in curves *b*) is quite good for both Ag and Cu over this energy range. The energy dependence of the function $S(E)$, which by Eq. (10.3.1.2.4–2) is given by $S(E) \propto ER^2(E)$, explains why the curve approaches a saturation value at higher energies since $R \propto \ln\{b/[ER(E)]\}$ for a screened Coulomb field (BOHR [*85*]). In Fig. 10.3.1.2.4.b, which represents the measurements of ALMEN et al., the selfsputtering ratio S_s for 45-keV ions is plotted as a function of the atomic number. It is evident that the values of S_s are periodic in the sense that S_s increases as the nth d shell and the $(n+1)$th s shell are filled. The maxima lie at Zn (filled 3 d and 4 s shells), Cd (filled 4 d and 5 s shells), and (by extrapolation) Hg (filled 5 d and 6 s shells). From the similar behavior for bombardment of targets by noble gas ions (Fig. 10.3.1.2.1.e), we conclude that self-sputtering is not in principle different from any other sputtering process.

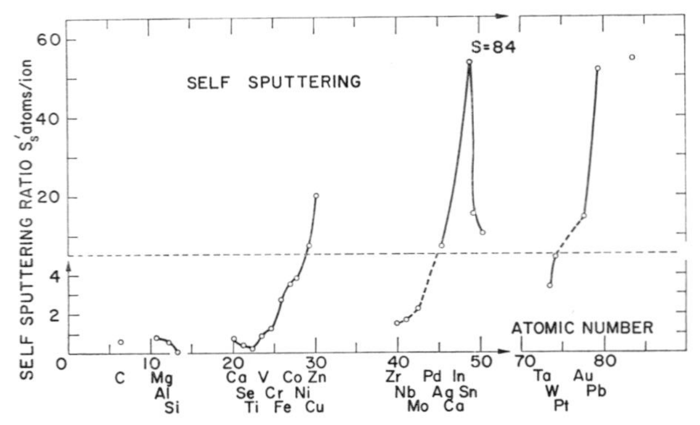

Fig. 10.3.1.2.4.b. The self-sputtering ratio S_s for 45-keV ions as a function of the atomic number of the sputtering system ($Z_{target} = Z_{ion}$) (ALMEN et al. [*10*])

10.3.1.2.5. Dependence on the Incident-Ion Current Density

ALMEN et al. [*10*], using a mass analyser, investigated the sputtering of Cu and Ag targets by 45-keV ions of the krypton isotopes Kr^{82+}, Kr^{84+}, and Kr^{86+}, as a function of ion-current density j. The ion-current densities are proportional to the isotopic abundances (Kr^{82}, 11.5%; Kr^{84}, 57.1%; Kr^{86}, 17,5%), since the focusing of the ion beam on the collector target is approximately the same for all the mass lines considered here. The authors found the same sputtering ratio for all the isotopes, despite their different current densities j. From their experimental results they concluded that the value of the sputtering ratio S is independent of ion-current density in the range $10 < j < 100\ \mu A/cm^2$. GRØNLUND and MOORE [*258*], who bombarded Ag targets with H_2^+ molecular ions ($2 < E < 12$ keV), found that the sputtering ratio S is practically independent of j in the range $30 < j < 220\ \mu A/cm^2$. Fig. 10.3.1.2.5.a shows the behavior of the function $S(j)$ as determined by ROL et al. [*603*] for 20-keV A^+ ions on a Cu target. One sees that, for current densities greater than $20\ \mu A/cm^2$, the slope dS/dj becomes equal to 0. Fig. 10.3.1.2.5.a also includes the values of $S(j)$ given by GUSEVA [*274*], who bombarded Cu targets with 20-keV Kr^+ and Si^+ ions at various ion-current densities. The slopes of the $S(j)$ curves are seen to become 0 for ion-current densities in the region $35 < j < 250\ \mu A/cm^2$ for the Kr^+-Cu system and in the region $170 < j < 600\ \mu A/cm^2$ for the Si^+-Cu system.

The marked change in the function $S(j)$ in the region of lower ion-current densities ($j < 10\ \mu A/cm^2$) may result from such influences as the following. The ion beam impinging on the target surface can serve to remove adsorbed layers of foreign gases (Chap. 3), whose presence on the surface hinders the sputtering of the target. The following estimate may illustrate this. For current density $j = 10\ \mu A/cm^2$ and a sputtering ratio ≈ 9 atoms/ion (for Kr^+ on Cu at 20 keV, GUSEVA approximately 6×10^{14} atoms/cm²-sec are sputtered from the target; but at the same time, at a pressure of $\approx 10^{-6}$ mm Hg in the target region, about 1.8×10^{15} atoms/cm²-sec impinge on the surface. If these residual gas atoms are chemisorbed on the surface, the sputter ng process may be seriously hindered by the building up of adsorbed gas layers. Only by increasing the current density (in this case up to about 30 μA/cm²) will the number of atoms leaving the surface exceed the number of residual gas atoms impinging on it. Alternatively, the same result can be achieved by reducing the residual gas pressure. In the latter case, the layer of impurity must be cleaned off the surface at least once, e.g., by ion bombardment or the flash-filament technique (Chap. 3). Another effect may also contribute to the change of the function $S(j)$ in the region of low ion-current densities. ALMEN et al. [*10*] found that for many substances the ions striking the target are captured into the target lattice until their concentration reaches some saturation value. As a result of their partial incorporation into the lattice, the properties of the lattice deviate from those in the undistorted condition and the sputtering ratio changes correspondingly. The results obtained in the low-energy region by various authors (PENNING et al. [*553*], FETZ [*211*], and WEHNER [*762*]) will not be considered more thoroughly here since they are not corrected for secondary-electron emission and cannot be regarded as reliable.

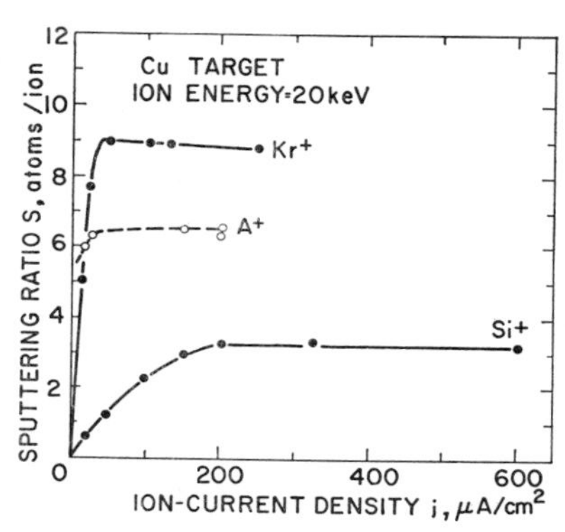

Fig. 10.3.1.2.5.a. The dependence of the sputtering ratio S on the incident-ion-current density j for different 20-keV ions striking a Cu target. Curve for A^+ on Cu, ROL et al. [*603*]; curves for Kr^+, Si^+ on Cu, GUSEVA [*274*]

10.3.1.2.6. Dependence on Residual Gas Pressure

The effects of adsorbed gas layers on the target were involved in the preceding section to explain the variation of the sputtering ratio S with the ion-current density j. These same effects can at least partly explain the dependence of S on the gas pressure p, which has been observed by numerous authors. The earlier investigations (e.g., GUENTHERSCHULZE et al. [*260*], MEYER et al. [*472*], PENNING et al. [*553*], SEELIGER [*647*],

GUENTHERSCHULZE [*264*], and WEHNER et al. [*759*]) of the pressure dependence of S will not be discussed further here; instead, the reader is referred to the dicsussion by WEHNER [*760*]. The results are definitely affected in a very complicated way by many effects resulting from the measuring techniques used (e.g., the low-pressure discharge discussed in Section 10.2.1.2. or the glow discharge discussed in Section 10.2.1.1). Among these effects of increasing pressure are the increasing back-diffusion of sputtered particles to the target, the increasing number of multiple collisions of ions in the cathode fall, and the effects of the LANGMUIR sheath (which may lead to different angles of incidence of the ions striking the target as well as to changes in their mean energy).

In the modern investigations (ALMEN et al. [*10*] and YONTS et al. [*787*]), the function $S(p)$ was studied by the ion-beam method, in some cases even combined with an isotope separator, so that more definite experimental conditions could be assured.

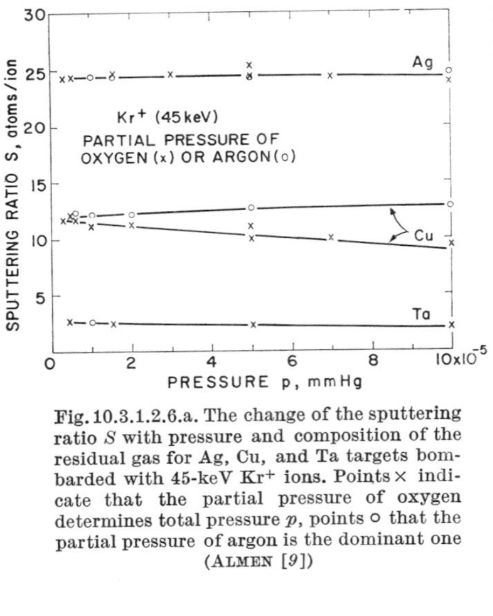

Fig. 10.3.1.2.6.a. The change of the sputtering ratio S with pressure and composition of the residual gas for Ag, Cu, and Ta targets bombarded with 45-keV Kr^+ ions. Points × indicate that the partial pressure of oxygen determines total pressure p, points o that the partial pressure of argon is the dominant one (ALMEN [*9*])

Fig. 10.3.1.2.6.a shows the results obtained by ALMEN et al. [*9*] for the sputtering ratio S with pressure and composition of the residual gas for Ag, Cu, and Ta targets bombarded with 45-keV Kr^+ ions. [In one case the gas pressure was determined from the partial pressure of oxygen (points marked x), in the other case from the partial pressure of argon (points marked o).] For the cases in which argon is the dominant constituent of the residual gas, one sees that the value of S is almost independent of pressure throughout the entire range investigated. On the other hand, for cases in which oxygen or nitrogen dominates in the residual gas, one observes a decrease of S with increasing residual gas pressure for the Cu and Ta targets, whereas the sputtering of the noble metal silver is practically constant.

In bombarding Cu targets with 30-keV A^+ ions, YONTS et al. [*787*] also observed that the sputtering ratio decreased from $S = 8.84 \pm 0.02$ atoms/ion in the pressure range $2 \times 10^{-5} < p < 4 \times 10^{-5}$ mm Hg to $S = 7.92 \pm 0.01$ atoms/ion in the range $4 \times 10^{-5} < p < 8 \times 10^{-5}$ mm Hg. The composition of the residual gas was not given in more detail by these authors. A similar decrease of S with increasing residual gas pressure was also reported by BADER et al. [*40*]. The latter authors, who bombarded targets of five different metals (Cu, Ni, Fe, Mo, W) with nitrogen ions (0—8 keV), found that S decreased about 10% when the partial pressure of nitrogen was reduced from 6×10^{-5} to 6×10^{-6} mm Hg, the total partial pressure of other gases being 4×10^{-7} mm Hg. A similar behavior was also found by WEHNER in his studies of sputtering by ions at lower energies. In Fig. 10.3.1.2.6.b the sputtering ratio S^* (uncorrected for secondary-electron emission) for Ni targets bombarded with 105-eV A^+ ions is plotted as a function of gas pressure. One notes a decrease of S^* with increasing pressure.

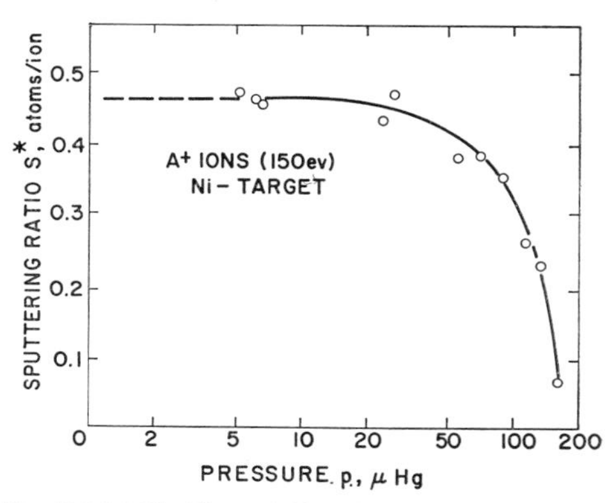

Fig. 10.3.1.2.6.b. The variation of the sputtering ratio S^* (uncorrected for secondary-electron emission) with the gas pressure, for Ni targets bombarded with 150-eV A^+ ions (WEHNER [*759, 760*])

These various observations (WEHNER [*759, 760*], YONTS et al. [*787*], and BADER et al. [*40*]) of a change of the function $S(p)$ with residual gas pressure may well be due mainly to the building up of adsorbed layers on the target surface, where they hinder the sputtering of the target.

10.3.1.2.7. Angular Distribution of Sputtered Particles

Investigations of the angular distributions of sputtered particles from polycrystalline metal targets at various ion energies and angles of incidence are of great interest for a rigorous study of the underlying mechanism of the sputtering process, namely, the mechanism of energy transfer between incident ion and target atom. Such results permit one, for example, to decide between various proposed theories such as the evaporation theory or the momentum-transfer theory of sputtering (discussed in more detail in Section 10.4). For example, if the evaporation theory is correct, one would expect a cosine distribution of the sputtered particles. A knowledge of the angular distribution is also necessary in determining the sputtering yield from the thickness of the deposit of sputtered particles on the collector. It is also of great importance in such applications of sputtering as the controlled production of very uniform thin films.

The first investigations of this kind were carried out by SEELIGER and SOMMERMEYER [*646*]], who bombarded a silver target with A^+ ions with energies in the neighborhood of 10 keV. The results showed that the angular distribution of the sputtered particles followed KNUDSEN's cosine law, independent of the angle of incidence of the ions on the target. For a long time this result was accepted as a proof of the validity of the evaporation theory of sputtering. The general validity of the conclusion of SEELIGER and SOMMERMEYER was first challenged by the investigations of WEHNER [*761*, *762*], whose studies of the sputtering of single-crystal target surfaces will be discussed in more detail in Section 10.3.2. WEHNER

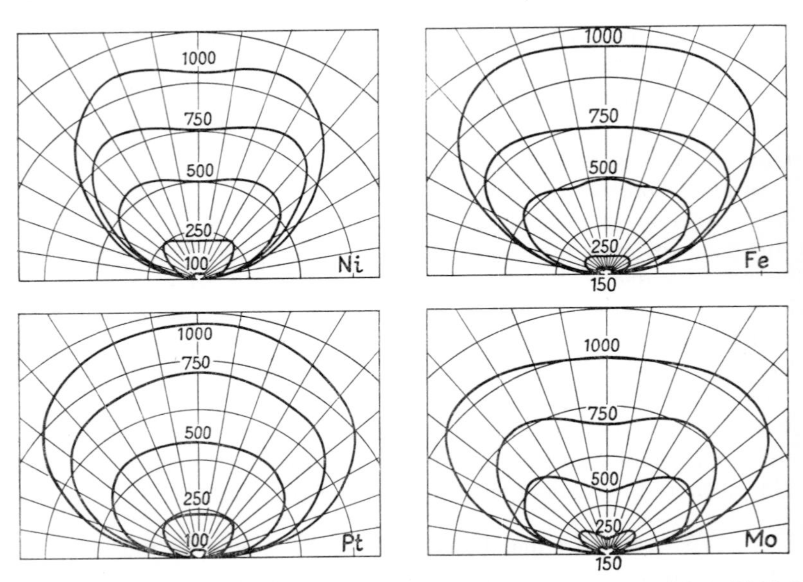

Fig. 10.3.1.2.7.a. Polar diagram of the angular distribution of material sputtered from Ni, Pt, Fe, and Mo targets by Hg^+ ions at normal incidence with energies of 150 eV (or 100 eV), 250 eV, 500 eV, 750 eV, and 1000 eV (WEHNER et al. [*770*])

found that the particles sputtered from a single crystal are emitted preferentially in certain directions. In order to check the extent to which the angular distributions deviate from the cosine law in polycrystalline materials, WEHNER et al. [*770*] bombarded Ni, Pt, Mo, and Fe targets with low-energy Hg^+ ions (100—1,000 eV). WEHNER's results for ion energies of 250, 500, and 1,000 eV are shown in Fig. 10.3.1.2.7.a in the form of polar diagrams. In this representation, the areas enclosed by the various polar curves are proportional to the sputtering yields. The courses of the individual polar curves in Fig. 10.3.1.2.7.a show that actually there are quite large deviations from the cosine distribution. Relative to the cosine distribution, more particles are scattered in directions parallel to the surface and fewer in the direction of the normal. (WEHNER termed this a "below-cosine" distribution.) This effect is more marked for the Mo and Fe targets (bcc lattice) than for the Ni and Pt targets (fcc lattice). From Fig. 10.3.1.2.7.a one draws the further interesting conclusion that as the ion energy increases the angular distribution approaches closer to a cosine distribution. Investigations at much higher ion energies (for example, those of Rol et al. [*603*] at 20 keV, Cobic et al. [*117*] at 17 keV, and Kaminsky [*363*] at 800 keV) indicate that with increasing ion energy the angular distribution goes over from the "below cosine" distribution to an "above cosine" distribution, i.e., relative to the cosine distribution more particles are then sputtered in the direction of the surface normal and fewer in directions parallel to the surface.

Fig. 10.3.1.2.7.b shows the angular distributions found by Rol et al. [*603*] for copper sputtered by 20-keV A^+ ions incident at an angle

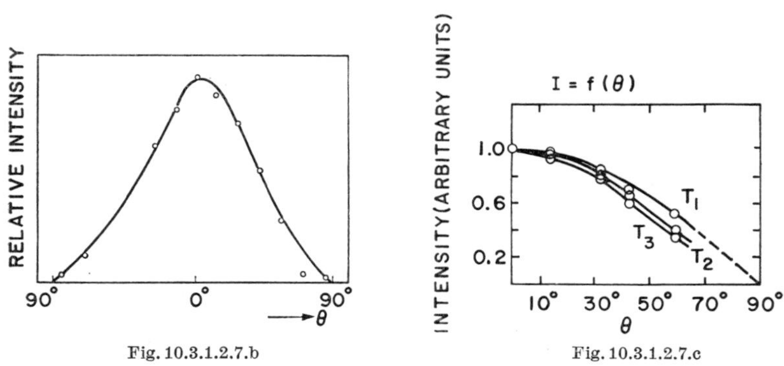

Fig. 10.3.1.2.7.b

Fig. 10.3.1.2.7.c

Fig. 10.3.1.2.7.b. The angular distribution of material sputtered from copper by bombardment with 20-keV Ar^+ ions. The angle of incidence α was 50° (Rol et al. [*603*])

Fig. 10.3.1.2.7.c. The angular distribution of material sputtered from a copper target bombarded with 17-keV A^+ ions, for different target temperatures $T_1 = 100°C$, $T_2 = 25°C$, $T_3 = -160°C$. The dotted line represents a cosine-law distribution. (After Cobic and Perovic [*117*])

of 50°. The distribution is closer to a Gaussian than to a cosine distribution. The shape of this distribution is independent of the angle of incidence of the ions relative to the normal of the target surface, a result which has been confirmed also by other authors (Rol et al. [*603*] and Cobic et al. [*117*]). By bombarding Ni and Mo targets at oblique incidence with lower energy Hg^+ ions (250 eV) Wehner [*770*] found that the angular distribution depended on the angle of incidence of the ions. The particles are preferentially sputtered in the direction in which the ion beam would be directed if it were specularly reflected from the target surface. Grønlund and Moore [*258*] observe a similar distribution in the sputtering of a silver target by 9-keV D^+ ions incident at an angle of 60°. In investigations of the sputtering of silver targets by noble gas ions of lower energies (Ne^+, A^+, and Kr^+ at 100—250 eV), Koedam [*402*] found (in contrast to the results of WEHNER with lower energy Hg^+ ions) that the angular distribution of the sputtered particles follows a cosine law. The same result was found by GRØNLUND et al. in the bombardment of Ag targets with 4-keV Ne^+ ions incident at an angle of 60°.

COBIC and PEROVIC [*117*] investigated the angular distribution of copper target atoms sputtered by 17-keV A^+ ions at target temperatures of 100° C, 25° C, and −160° C. Although at 100° C the experimental angular distribution follows a cosine law very well, the deviations from such a distribution increase with decreasing target temperature (Fig. 10.3.1.2.7.c). The authors also investigated the dependence of the angular distribution of sputtered copper target atoms on the duration of the irradiation with 17-keV Xe^+ ions, keeping the ion-current density fixed for target temperatures of 100° C and −160° C. They found that with increasing duration of the irradiation (0.36 mA-hr, 0.87 mA-hr, and 2.78 mA-hr) the angular distribution becomes closer to a cosine distribution and that this effect is practically independent of target temperature. Possibly this effect may arise from increasing surface roughness resulting from etching by continued ion bombardment, as will be discussed in more detail in Section 10.3.4.

Current sputtering theories (Section 10.4) are unable to explain the aforementioned phenomena in the angular distribution. However, the results show that an evaporation theory is not suitable for describing the sputtering process.

10.3.2.1.8. Mean Velocity of Sputtered Particles

Investigations to determine the average velocity of sputtered target atoms and, of even greater significance, their velocity distribution are of great interest for a better understanding of the mechanisms underlying the sputtering process. The results of such studies will reveal the extent to which sputtering depends on local evaporation processes ("hot spot" theory, Section 10.4.1), since sputtered particles should then be observed to have thermal velocities.

The earlier investigations, which lead to generally contradictory results, will not be discussed here but have been summarized by WEHNER [*760*]. Unfortunately, because of inadequate methods of measurement (e.g., the glow-discharge or low-pressure plasma methods which were criticized in Sections 10.2.1.1 and 10.2.1.2) and crude approximations in evaluating the data, some more recent measurements are not reliable in absolute value to better than an order of magnitude. The following discussion of such data will therefore be very brief. SPORN [*673*] observed in an oxygen glow discharge that luminous layers developed around the cathode which he had covered with various surface coatings (e.g., MgO, Na_2O, Li_2O, K, Na, and Cu). From the emitted spectral lines in these luminous layers, he concluded that at least part of the target atoms that leave the surface must be in excited states. From a knowledge of the lifetimes of the excited states of the sputtered target atoms and a determination of the mean free path of such excited target atoms by measuring the change in intensity of the luminous layer with ion energy, SPORN calculated an average velocity of the sputtered excited target atoms. He found that the mean velocity $\bar{V}$ of sputtered Li, Na, and Mg target atoms has values between 0.6×10^6 and 1.4×10^6 cm/sec (corresponding to 4.5—7 eV) and that these values are practically independent of the energy of the impinging oxygen ions in the voltage range investigated (400—4000 V).

WEHNER [*760*], using a different measuring technique, allowed the sputtered target atoms in a low-pressure plasma to impinge on a collector mounted on an arm of a sensitive torsion balance. From the resulting force on the collector he determined the mean velocity $\bar{V}$. For Ni target atoms, sputtered by Hg^+ ions with energies from 600 to 900 eV, the average velocity was approximately $\bar{V} = 6.5\times10^5$ cm/sec (corresponding to 12.7 eV); whereas for W target atoms it was 3.5×10^5 cm/sec (corresponding to 12 eV) and for Pt target atoms 4×10^5 cm/sec (corresponding to 16 eV).

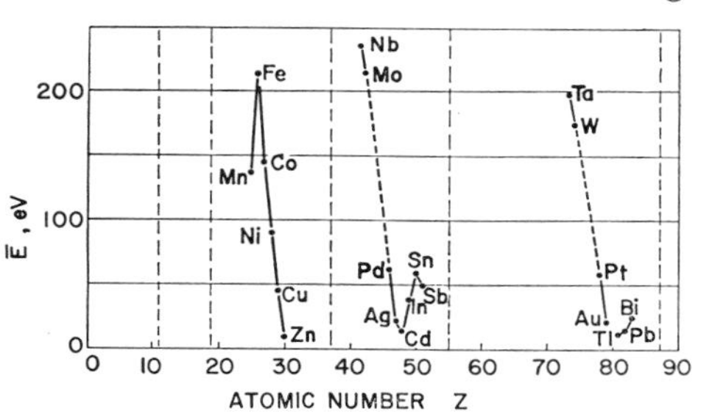

Fig. 10.3.1.2.8.a. Mean kinetic energy $\bar{E}$ of particles sputtered from metal targets of atomic number Z by 35-keV xenon ion bombardment under normal incidence (KOPITZKI and STIER [*405b*])

In other studies WEHNER [*768*] determined the forces which the sputtered target atoms exerted on a collector attached to a torsion balance. In a bombardment with 300-eV Hg^+ ions, he found $\bar{V} = 3.3\times10^5$ cm/sec (corresponding to 3.6 eV) for sputtered Cu target atoms, 2.9×10^5 cm/sec (corresponding to 4.7 eV) for Ag target atoms, and 2.2×10^5 cm/sec (corresponding to 5 eV) for Au target atoms.

More extensive studies of the average kinetic energy $\bar{E}$ of sputtered particles for numerous target materials bombarded by various noble gas ions with energies varying from 20 keV and for different angles of incidence have been conducted by KOPITZKI et al. [*405a—c*]. In Fig. 10.3.1.2.8.a, the average kinetic energy $\bar{E}$ of particles sputtered from various metal targets under 35-keV xenon-ion bombardment is plotted vs the atomic number Z. The value of $\bar{E}$ obtained are always considerably larger than the sublimation energy and a certain periodicity in the values of $\bar{E}$ corresponding to the periodic table of the elements can be observed. Thus $\bar{E}$ always has a minimum value for the elements of the second subgroups (i.e., for Zn, Cd, Tl). Comparison of Fig. 10.3.1.2.1.e with Fig. 10.3.1.2.8.a shows that the value of the mean kinetic energy is the largest for those target materials which have the smallest sputtering rates. Another interesting result is that for the investigated range of incident-ion energies (20—60 keV) the average energy of the particles leaving a gold target under bombardment by normally incident noble gas ions depends only very slightly on the incident-ion energy (Fig. 10.3.1.2.8.b). The value of $\bar{E}$ however depends more strongly on the angle of incidence showing a decrease with increasing angle of incidence (Fig. 10.3.1.2.8.c). The results seem to lead to the not easily understandable conclusion that the value of $\bar{E}$ increases with increasing penetration depth of the incident ion (the energy $\bar{E}$ is higher for lighter incident ions and for more nearly normal incidence).

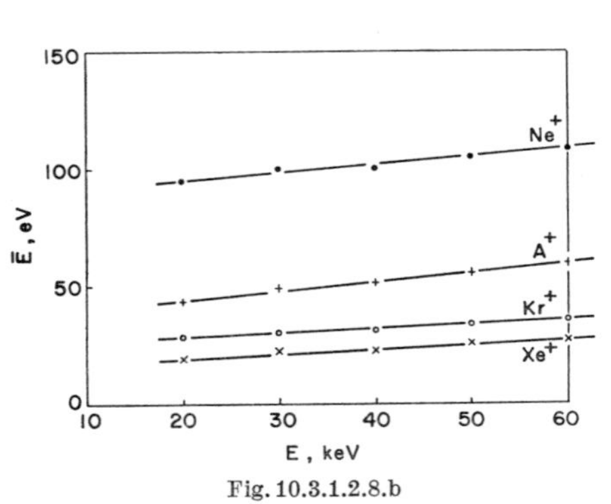

Fig. 10.3.1.2.8.b

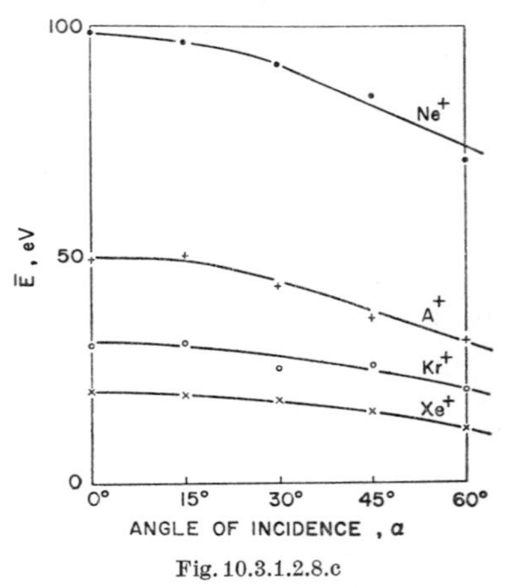

Fig. 10.3.1.2.8.c

Fig. 10.3.1.2.8.b. Mean kinetic energy $\bar{E}$ of particles sputtered from a gold target by bombardment with various normally incident noble gas ions with energy $\bar{E}$ (Kopitzki and Stier [*405b*])

Fig. 10.3.1.2.8.c. Mean kinetic energy $\bar{E}$ of particles sputtered from a gold target under bombardment by 35-keV ions of various noble gases in dependence on the angle of incidence α (Kopitzki and Stier [*405b*])

Kopitzki et al. [*405c*], who conducted similar experiments also with copper monocrystals, observed that the values of $\bar{E}$ were smaller for the more densely packed crystal directions (e.g. $\bar{E}_{[110]} < \bar{E}_{[100]} < \bar{E}_{[111]}$ for Xe-ions bombarding a Cu (100) plane over the energy range from approximately 25 keV to 60 keV).

Finally, Thompson and Nelson [*721a*] investigated the velocity distribution of particles ejected from a gold target under bombardment by noble gas ions (42.5-keV A^+ ions and 43-keV Xe^+ ions). A 70-μA ion beam was focused on a strip of polycrystalline gold 1 mm wide at an angle of incidence of 70°, the sputtered atoms were collimated into a beam by a 0.5-mm slit, and their velocity distribution was observed by means of a mechanical velocity analyzer. After transformation of the velocity distribution curve into an energy spectrum, the authors observed that for Xe-ion bombardment the spectrum showed a peak at 0.15 eV ($\pm$0.03 eV) and that 12% of the total number of particles ejected had energies below 0.5 eV. For Argon-ion bombardment, the spectrum showed a peak at 0.20 eV ($\pm$ 0.05 eV) and only 4% for the total number of sputtered particles had energies below 0.5 eV. The authors suggested that the ejection of these small portions of particles with near thermal energies was due to evaporation from thermal spikes.

As mentioned at the beginning of this section, only the order of magnitude of the values for the mean kinetic energy reported may be considered reliable. Since the average velocities obtained in these studies are much greater than thermal, the evaporation theory must be considered incapable of describing the sputtering process adequately.

10.3.1.3. Sputtering Yields in the Weak-Screening Region

10.3.1.3.1. Dependence on the Incident-Ion Energy

If the kinetic energy of the ions bombarding the target becomes so large that the collision process involves a partial penetration of the electron cloud of the lattice atoms, the value of the characteristic radius for the screened electron cloud becomes of importance in the collision process.

This characteristic radius is given by $a = a_0 (Z_1^{2/3} + Z_2^{2/3})^{-1/2}$, where a_0 is the Bohr radius of the hydrogen atom and Z_1 and Z_2 are the atomic numbers of the two colliding atoms. The minimum energy that can be transferred to a target atom in such a collision is given by $E_{\min} = E_B E_d / E$, where E_B is the energy limit above which the collision process follows the Rutherford scattering law [Eq. (10.4.2–5) in Section 10.4.2], E_d is the threshold energy for the displacement of lattice atoms, and E is the kinetic energy of the incident ions. Such screened Coulomb collisions occur in the energy range $E_A < E < E_B$, where E_A [Eq. (10.4.2—9) in Section 10.4.2] is the energy limit below which the collisions become approximately of the hardsphere type. Table 10.4.2.A in Section 10.4.2 contains values for the limiting energies E_A and E_B, calculated for various ions bombarding Cu and Ag targets.

Fig. 10.3.1.3.1.a gives the values determined by Grønlund and Moore [*258*] for the sputtering ratio $S(E)$ for Ag targets bombarded by atomic and molecular hydrogen and deuterium ions. In the energy range $E_A < E < E_B$, the value of the sputtering ratio S is seen to depend only slightly on ion energy. This also holds for other systems, e.g., for Cu targets bombarded with helium or deuterium ions. Fig. 10.3.1.3.1.b shows the results obtained by Yonts et al. [*787*] and Gusev et al. [*275*] for the sputtering of Cu targets with deuterium ions. The agreement between the values found by these authors can be regarded as satisfactory.

Fig. 10.3.1.3.1.b also includes the curve representing the function $S(E)$ for helium bombardment of Cu targets as reported by YONTS et al.. On the other hand, the results given by PLESHIVTSEV [572] for the sputtering of Cu targets by hydrogen ions are not shown since they lie completely outside the range of the results of the other authors (GRØNLUND

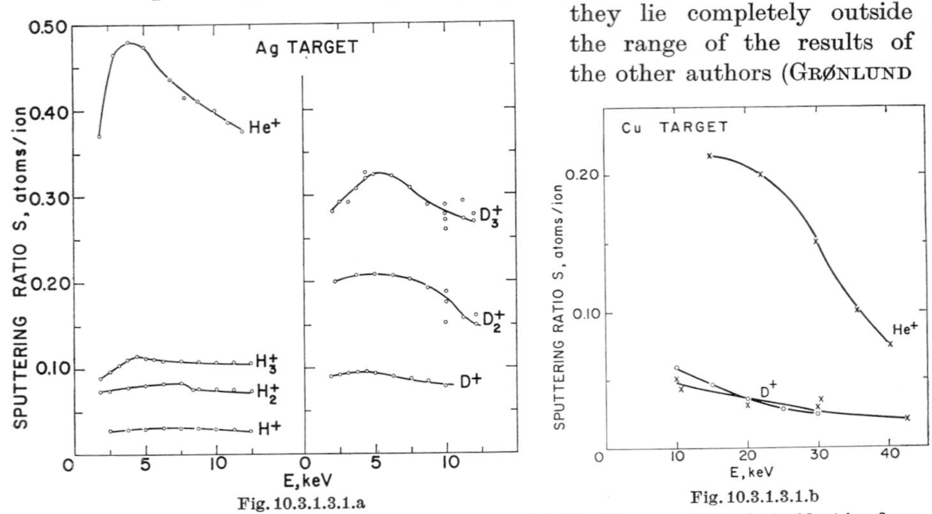

Fig. 10.3.1.3.1.a

Fig. 10.3.1.3.1.b

Fig. 10.3.1.3.1.a. The dependence of the sputtering ratio S on the energy E of the incident ion for a silver target bombarded by atomic and molecular hydrogen and deuterium ions and by helium ions (GRØNLUND et al. [258])

Fig. 10.3.1.3.1.b. The dependence of the sputtering ratio S on the energy E of the incident ion for a copper target bombarded by helium and deuterium ions. Points ×, YONTS et al. [787]; points ○, GUSEV et al. [275]

et al. [258] and WEHNER [771]). For example, in the case of 10-keV H_2^+ ions incident normally on a Cu target PLESHIVTSEV found a sputtering ratio of $S = 0.28—0.36$ atoms/ion; at 20 keV, $S = 0.65$ to 1.04 atoms/ion; and at 29 keV, $S = 0.40—0.57$ atoms/ion. Moreover, these values all lie far above those for the helium bombardment of a copper target, an extremely unlikely result. From the results shown in Figs. 10.3.1.3.1.a and 10.3.1.3.1.b one notices that the values of the function $S(E)$ for He^+ bombardment are always far above those for H_2^+ bombardment. More recent studies by WEHNER [771] on the bombardment of Ni targets with 9.3-keV ions gave the sputtering ratios $S(H_2^+) = (2.8 \pm 0.6) \times 10^{-2}$ atoms/ion and $S(H_3^+) = (4.5 \pm 0.4) \times 10^{-2}$ atoms/ion. These results agree at least in order of magnitude with the results of GRØNLUND and MOORE for Ag targets. This result is further evidence against accepting PLESHIVTSEV's values.

A comparison of the $S(E)$ curves when hard-sphere collisions predominate (e.g., Figs. 10.3.1.2.1.b and 10.3.1.2.1.c) with the curves (Figs. 10.3.1.3.1a and 10.3.1.3.1b.) shown here for weak screening indicates that the sputtering ratio S changes very little with ion energy in the weak-screening collision region. This can be understood from the fact that the displacement cross section σ_d in the energy range $E_A < E < E_B$ is independent of the ion energy, that is, $\sigma_d = \pi a^2$ with $a = a_0/(Z_1^{2/3} + Z_2^{2/3})^{1/2}$.

The $S(E)$ curves determined experimentally for the sputtering of Ag targets by various molecular ions can be compared with the values computed on the assumption that the energy of the molecular ion is divided equally among the atomic constituents so that the $S(E)$ of a molecular ion is just the sum of the $S(E)$'s of the component atoms. For example, the sputtering by D_2^+ at 8 keV is approximately double that of D^+ at 4 keV. Fig. 10.3.1.3.1.c shows that this sum rule is satisfied only approximately; at the lower energies the $S(E)$ constructed from those of the atomic components is usually below the $S(E)$ of the molecular ion, at higher ion energies the sum is above the value for the molecular ion. The figure shows a similar behavior for N_2^+ bombardment of Cu targets (ROL et al. [603]) in the energy range in which the collisions are principally of the hard-sphere type. There is, at present, no satisfactory explanation for this behavior. As will be discussed in more detail in Section 10.4.2.2, the theory developed by PEASE [550] for this energy range $E_A < E < E_B$ can predict the experimental values to within a factor of 2.

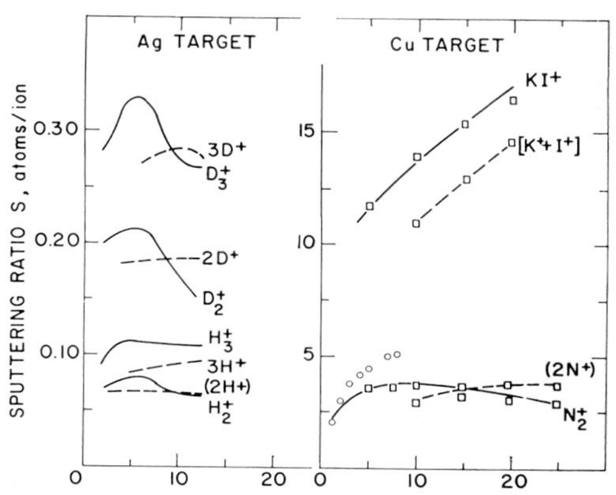

Fig. 10.3.1.3.1.c. The dependence of the sputtering ratio S on the incident-ion energy for Ag and Cu targets under bombardment by various molecular ions. The solid lines represent the experimentally determined $S(E)$ curves while the dotted lines have been calculated under the assumption that the energy of the molecular ions is divided equally among the atomic constituents so that the $S(E)$ of a molecular ion is the sum of the $S(E)$'s of the component atoms. The values for the Ag target according to GRØNLUND et al. [258] and the values for the Cu target according to ROL et al. [603]

10.3.1.3.2. Angular Distribution of Sputtered Particles

There are still only a few results concerning the angular distribution of sputtered target particles in the energy range $E_A < E < E_B$. O'BRIAN et al. [529] observed that the angular distribution of silver target atoms, sputtered by 9.25-keV H^+, H_2^+, and D^+ ions at normal incidence, deviates only slightly from a cosine distribution. Somewhat fewer particles are sputtered at small angles to the surface normal than are expected according to the cosine law and, conversely, at large angles to the surface normal there are more particles sputtered than for a cosine distribution. PLESHIVTSEV [*572*] found that the angular distribution of copper target atoms sputtered by 49-keV H_2^+ ions follows a Gaussian distribution rather than a cosine law. In contrast to these results, with 9.25-keV H_2^+ ions, more particles are sputtered in the direction of the normal than expected for a cosine distribution. This result agrees with the corresponding observations in the energy region $E < E_A$ (Section 10.3.1.2.2.)

10.3.1.4. Sputtering Yields in the Rutherford-Collision Region

Sputtering phenomena have scarcely been investigated in the energy region $E > E_B$, where RUTHERFORD scattering predominates. This is especially true in the case of „back sputtering", in which the target particles are ejected back from the surface struck by the incident beam. Thus, the sputtering studies of YONTS et al. [*787*] with D^+ ions on Cu targets (at 5—40 keV) extend so little into the energy range $E > E_B$ that the few data offer no adequate picture of the sputtering process in the RUTHERFORD-collision region. As an example of the order of magnitude of the results obtained by these authors, the sputtering ratio for 30-keV D^+ ions on copper is $S = 0.03$ atoms/ion. Further values can be read from Fig. 10.3.1.3.1.b.

PLESHIVTSEV [*572*] reported preliminary results for the sputtering of copper by deuterons in the energy range from 300 to 1,000 keV (curve *a* in Fig. 10.4.2.3.a). However his data disagree by 2—4 orders of magnitude with those reported by KAMINSKY [*361—363*] for the bombardment of a (100) plane of a copper monocrystal with normally incident deuterons in the range from 0.10—0.15 MeV and from 0.5 to 2.5 MeV. (The experimentally determined points are marked by open circles in Fig. 10.4.2.3.a.) The comparison of PLESHIVTSEV's results obtained for polycrystalline copper with those obtained by KAMINSKY for copper monocrystals is justified by the latter's observation that for the energy region considered the sputtering rate for such copper planes as (111), (110), and (100) differs by less than a factor of three. The sputtering rate decreases in the order $S_{(111)} > S_{(100)} > S_{(110)}$. KAMINSKY's data also show a decrease of the sputtering ratio S with increasing energy in qualitative agreement with theoretical predictions (calculated curves in Fig. 10.4.2.3.a are curve *b*: PEASE [*550*], curve *c*: THOMPSON [*720*], curve *d*: GOLDMAN et al. [*241, 242*], curve *e*: HARRISON [*292*]) but contrary to the findings of PLESHIVTSEV.

Unfortunately PLESHITVSEV did not state the experimental conditions under which he obtained his results. However, the high values of the sputtering ratio he obtained and their increase with increasing energy seem to indicate that an evaporation process was superimposed on the sputtering process. Such target evaporation becomes likely if the power transmitted by a high-energy beam at large ion currents is too great to be dissipated rapidly enough by a thin target. On the other hand, KAMINSKY's result that S decreases roughly as $\ln E/E$ may be expected since in RUTHERFORD scattering the total cross section σ_a varies inversely as the energy E and the mean recoil energy of the struck atom $\bar{E}$ varies as $\ln E$.For a more detailed discussion see Section 10.4.2.

The experiments of KAMINSKY [*363a, 363b*] clearly establish the existence of preferred ejection directions in the back sputtering of target particles for the RUTHERFORD collision region. (In back sputtering, the sputtered particles are ejected from the surface struck by the incident beam.) In contrast to the cosine distribution to be expected if the mechanism were evaporation of target material from a hot spot, as predicted by some theoretical treatments, a distinct spot pattern is observed (e.g., see Fig. 10.3.2.1.1.d). For normal incidence the deposit showed four main spots, corresponding to the [110] directions, and additional spots corresponding to the [112] and [100] crystal directions. The density distribution in an individual spot is Gaussian and not cosine. In order to obtain the above mentioned values of the sputtering ratio S for the (100) monocrystal planes of copper as well as those shown in Fig. 10.4.2.3. for a (100) plane of silver, it was necessary to integrate over the density distribution of all spots in the deposit. A more detailed discussion of the results on preferred ejection directions of metal monocrystals bombarded with ions in the RUTHERFORD collision region will be given in Section 10.3.2.1.

10.3.2. Sputtering of Neutral Target Atoms from Monocrystalline Targets

The first experimental evidence for the preferred ejection of sputtered particles along certain crystallographic directions of monocrystals of tungsten and silver by low-energy (125—150 eV) Hg^+ ion bombardment was found by WEHNER [*761*] in 1955. Results of this kind are of extreme interest since they offer insight into the microscopic mechanism by which the incident ion transfers momentum and energy to a crystal lattice.

SILSBEE [*653*] suggested that in a monocrystal regarded as rows of equal hard spheres, momentum may be focused along a line of atoms in a close-packed crystal direction. According to the model, which will be discussed in more detail in Section 10.4.2.1, momentum may be focused if the ratio of the sphere diameter d to the spacing D_{hkl} between the centers of the spheres along such a line (the subscripts h, k, l are MILLER indices) exceeds $\frac{1}{2}$ (Fig. 10.4.2.1.b). This is true, for instance, in the [110] direction of copper and silver monocrystals. The final result is a sequence of head-on collisions without any change of angle, and energy losses along the chain may be neglected. In this case the collisions do not displace lattice atoms from their original equilibrium positions and no mass transport occurs.

LEIBFRIED [*427*], THOMPSON [*720*], and GIBSON et al. [*235*] however, considered an energy attenuation in such focused collision sequences due to the interaction with atoms in adjacent close-packed lines. The reported calculated values (e.g., ref. [*235*], [*521a*]) for the energy loss per collision —for instance, for the [110] direction in copper—vary between $\frac{2}{3}$ and 1 eV. This model allows a good qualitative, and in some instances even semiquantitative, interpretation of the experimental results on preferred ejection directions of sputtered particles (see discussion in the following sections).

The mechanism of focusing by the formation of "crowdions", as suggested earlier by PANETH [*541*], contains some features similar to the SILSBEE mechanism. As a result of the introduction of an extra atom or ion into a close-packed line in a crystal at low temperatures, a "crowdion" may be formed. The atoms along the line need a certain relaxation time to accommodate the added member. The "crowdion" is "free" to move along the close-packed line and its mobility in this direction may be very high because the region of maximum compression along the line can move quickly without a large motion of the individual lattice atoms.

Another focusing mechanism, suggested by VINEYARD and his co-workers (ref. *235*, for an analytical treatment see also ref. *521a*) should be mentioned. According to this model, focused collision sequences may occur even along some of those close-packed crystallographic directions in monocrystals for which SILSBEE's focusing condition $D_{hkl}/d < 2$ is not fulfilled. For instance, in a fcc lattice an atom A_1 moving at a small angle θ_0 to the [100] direction (see Fig. 10.3.2.a) passes through a ring of four neighboring atoms B_1, B_2, B_3, and B_4 in a (100) plane, it undergoes a deflection ψ before hitting the next atom A_2 in the [100] direction. Such a ring of atoms in the (100) plane surrounding the [100] direction may be treated as a converging lens with a focal length given (for small values of θ_0) by $f_{hkl} \approx D_{hkl}(\theta_0/\psi)/\sqrt{2}$, in which $\psi = \theta_1 - \theta_1'$ (see ref. *521a*). This "lens focusing" mechanism is energetically less efficient than SILSBEE's hardsphere focusing and would be expected to become important only at higher energies. From an analytical treatment, NELSON and THOMPSON [*521a*] found that in a fcc lattice the effectiveness of atomic collision sequences decreases in the order [110], [111], [100]—a result which is in qualitative agreement with some of their experimental observations.

While the above-mentioned processes primarily involve momentum focusing without transporting atoms more than one lattice spacing, OGILVIE [*532*] has suggested a basically different process involving long-range transport of atoms. In this mechanism, energetic ions moving in metal monocrystals may move down a channel formed by properly placed atoms in the lattice. Computer calculations by ROBINSON et al. [*597a*], OEN et al. [*530a*], and BEELER et al. [*68a*] also predicted

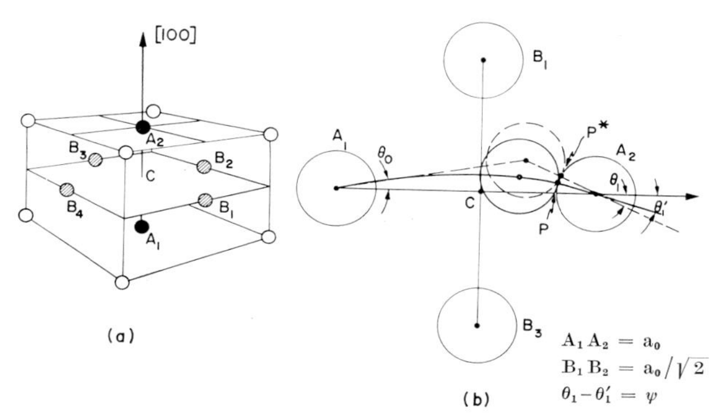

Fig. 10.3.2.a. Schematic illustration of the "lens-focusing" effect predicted by GIBSON et al. [*235*]. (a) Hard-sphere collision chain along the [100] direction in a fcc crystal. (b) Cross-sectional view of the [100] collision chain, illustrating the deflection of a lattice ion A_1, which moves at a small angle θ_0 to the [100] direction (in the plane $A_1B_1B_3$) and, passing between the neighboring atoms B_1B_3 undergoes a deflection $\psi = \theta_1 - \theta_1'$, before hitting the next atom A_2 at the point P. (P^* indicates the point of collision without deflection.) (Based on NELSON and THOMPSON [*521a*])

such channeling events; and analytical treatments of the same problem by LEHMAN and LEIBFRIED [*426b*] and by NELSON and THOMPSON [*521b*] reached similar conclusions.

Properly placed lattice atoms may form an effective potential well along the channel so that a moving particle is reflected back and forth across the channel in a series of glancing collisions with quasi-stationary lattice atoms as it passes through. A condition for channeling is that the spacing of lattice atoms across the width of a channel exceeds the distance of closest approach to the channel wall. In a fcc lattice, two type of channels are thought to exist [*521b*]: those formed in two dimensions between adjacent close-packed planes and those between rows of atoms in a particular direction. The first type of channel is most favorably formed between the (111) or (100) planes and the second type between the [110] or [100] directions. In their analytical treatment, NELSON and THOMPSON [*521b*] assumed that for motion of light ions the channel may be represented by a square-well potential in a good approximation and that the slowing down of the moving ion occurs predominantly by collisions with electrons. Such channeling mechanisms have been used to interpret not only experimental results on the range of high-energy ions in crystalline lattices (see for instance PIERCY et al.[*568b*] but also those on the preferential ejection of sputtered particles (e.g., NELSON and THOMPSON [*521c*]) or on the dependence of the sputtering ratio on the angle of incidence for metal monocrystals (e.g., FLUIT [*215a*]).

Since WEHNER's first investigations, atom ejection patterns have been studied in detail by several groups for various ions, metal targets, angles of incidence, and ion energies. The results for different crystal structures will be summarized below.

10.3.2.1. Atom Ejection Pattern for Face-Centered Cubic Metals

10.3.2.1.1. The (100) Surface

KOEDAM [*402*] bombarded a Cu (100) surface with noble-gas ions with energies below 200 eV and found four preferred directions for the sputtered particles. These directions corresponded to the four [110] directions in the lattice, which make angles of about 45° with the surface normal. WEHNER and ANDERSON [*16*] observed similar patterns of atom ejection from the (100) plane for Cu bombarded with Hg^+ and Ne^+, for Ni bombarded with Ne^+, for Al bombarded with A^+ and Ne^+, and for Au bombarded with Hg^+, A^+, and Ne^+ at energies up to 800 eV. Fig. 10.3.2.1.1.a shows the distribution of sputtered particles as observed by KOEDAM [*401, 402*] for the bombardment of a Cu (100) surface with normally incident Ne^+, A^+, Kr^+, and Xe^+ ions of 500-eV energy and also the stereographic projection of a fcc lattice on a plane parallel to a (100) lattice plane. In the case of Ne^+ and A^+ bombardment, one sees not only four spots corresponding to the [110] direction (in agreement with the theoretical predictions by SILSBEE) but also a center spot corresponding to the next-nearest-neighbor direction [100]. KOEDAM also observed such a center spot for Kr^+ bombardment if the ion energy was raised from 500 eV to 1000 eV, 1500 eV, and 2000 eV (Fig. 10.3.2.1.1.b). As the ion energy was increased, the density of the deposit at the center spot increased relatively more than that of the four side patches, a result which is also confirmed by other authors, as discussed later. At an ion energy of 2000 eV, more particles are ejected preferentially in the [100] direction than in any of the [110] directions for all ion species in KOEDAM's data.

On the basis of these results it is difficult to understand why ANDERSON and WEHNER [*16*] did not observe a detectable center spot in their ejection pattern for 800-eV Ne^+ ions on a Cu (100) surface, since the value of the ion energy is even higher than that (500-eV Ne^+) for which KOEDAM could already observe a center spot. Contrary to previous predictions the appearance of preferentially ejected particles along the

[110] directions could also be observed for incident ions of much higher energies. For example, YURASOVA et al. [*790*] reported such results for 8-keV A^+ ions bombarding a Cu (100) plane under normal incidence. By conducting such experiments also for different angles of incidence (45° and 60° with respect to the surface normal) the authors found the pattern of deposition to be analogous to those obtained at normal incidence. But in addition it was possible to observe a central spot corresponding to the [100] direction, since the incident beam now did not interfere with the deposition of the central spot deposit as it did in the case of normal incidence.

YURASOVA et al. compared the density of the sputtered copper in the individual patches by photometric measurements. The result was that with increasing energy of the bombarding ion the maximum density D_c in the central patch (corresponding to the [100] direction) increases relative to the maximum density D_s in the four side spots (corresponding to the [110] direction), in accordance with KOEDAM's [*401, 402*] data at lower ion energies. The ratio $r = D_c/D_s$ more than doubles,

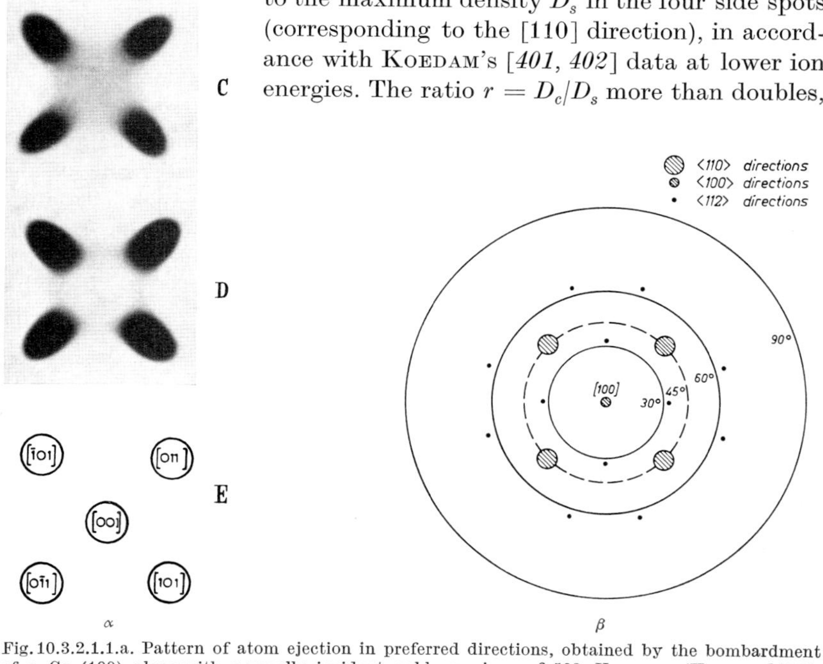

Fig. 10.3.2.1.1.a. Pattern of atom ejection in preferred directions, obtained by the bombardment of a Cu (100) plane with normally incident noble-gas ions of 500 eV energy (KOEDAM [*402*]). Diagram α: A, neon ions; B, argon ions; C, krypton ions; D, xenon ions; E, schematic diagram of the ejection pattern. Diagram β: stereographic projection of a fcc lattice on a plane parallel to a (100) lattice plane. The directions located on the 90° circle are not indicated

increasing from 1.5 to 3.75, as the energies of the A^+ ions incident at 60° increase from 6 to 50 keV (Fig. 10.3.2.1.1.c).

Ejection patterns similar to those found by A^+ bombardment were also observed for sputtering by the much lighter H_2^+ ion. The distribution of sputtered particles in each individual patch (the central patch and each of the four side patches) was determined photometrically. In each spot the density distribution follows a cosine law within an accuracy of $1-2^0/_0$. This result is also confirmed by the data of PEROVIC [*555*] who bombarded the (100) planes in single crystals of Cu and Pb with A^+ ions in the energy range from 10 to 25 keV.

In experiments by KAMINSKY [*362, 363*' *363a, 363b*], the (100) planes of copper and silver monocrystals were bombarded with protons, deuterons, and He^+ ions. The angles of incidence ranged from 0° to 45° from the normal and the energies from 0.10 to 0.15 MeV (mass-analyzed beam from a COCKCROFT-WALTON generator) and from 0.50 to 2.0 MeV (mass-analyzed beam from a VAN DE GRAAF). In this energy range, the energy of an incident ion is high enough that the collision can be treated in terms of the COULOMB repulsion between the nuclei (RUTHER-

A

B

C

D

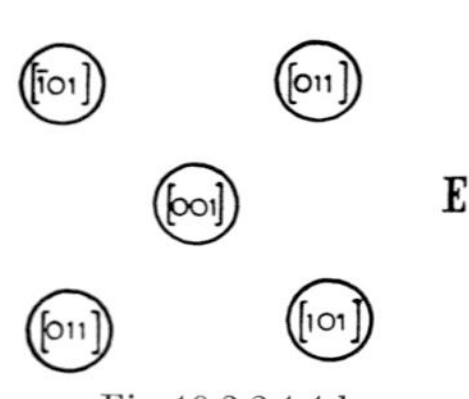

E

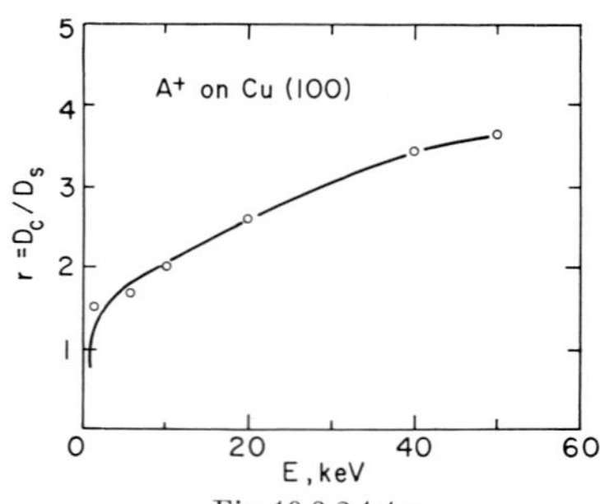

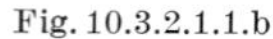

Fig. 10.3.2.1.1.b

Fig. 10.3.2.1.1.c

Fig. 10.3.2.1.1.b. Pattern of atom ejection in preferred directions, obtained by the bombardment of a Cu (100) plane with normally incident Kr^+ ions with various energies: A, 500 eV; B, 1000 eV; C, 1500 eV; D, 2000 eV; E, schematic diagram of the ejection pattern (KOEDAM [*402*])

Fig. 10.3.2.1.1.c. The ratio $r = D_c/D_s$, the maximum density D_c of the central patch of the deposit of sputtered copper [corresponding to the (100) direction] divided by the maximum density D_s of the four side spots of the deposit [corresponding to the (110) directions] (see, for example, Fig. 10.3.2.1.1.b), plotted as a function of the energy of the incident argon ion. The density measurements have been made photometrically (YURASOVA et al. [*790*])

FORD collision). This energy region presents several features of interest, as discussed in Section 10.4.2 and in reference [*362*]. One of the more significant aspects is that the incident ions at this energy have a mean free path of several thousand angstroms in the metal. This is 2—3 orders of magnitude larger than in any previous experiment on the preferred directions of back sputtering, since the corresponding measurements by other authors were conducted in the hard-sphere collision region. This in turn means that most of the target particles are displaced as primary and secondary knock-ons far below the surface layers. For the case in which preferred ejection directions can be observed, this allows a better check of the predicted ranges of the focused collision sequences, which may be based on the SILSBEE mechanism, the lens-focusing mechanism, or the channeling mechanism. It also allows a good check of other predictions about high-energy sputtering—e.g., to see if the ejected target particles follow the cosine distribution that would be expected from an evaporation process at a hot spot at the surface (Section 10.4.1) or if the sputtered particles are ejected in random directions since the displaced target particles have to undergo many collisions before reaching the surface layers.

The experimental results [*363, 363a*] established the existence of preferred ejection directions in the back sputtering of target particles for the RUTHERFORD collision region. For instance for 125 keV D^+ ions incident on the (100) plane of copper (angle of incidence 31°) and silver monocrystals (angle of incidence 45°), the spot pattern obtained is shown in Fig. 10.3.2.1.1.d. The four main spots, corresponding to the [110] directions, were conspicuous at all the angles of incidence. Additional spots corresponding to the [100] and [112] crystal directions were also observed throughout the investigated range of angles of incidence, but their densities were smaller than those corresponding to the [110] directions. With increasing angle of incidence, the deposits corresponding to the [100] and [112] directions became less dense. Optical transmission measurements of the deposits revealed that for ions incident at 45° approximately 75% of the sputtered material is in the four main spots ([110], $[1\bar{1}0]$, [101], and $[\bar{1}01]$ spots) and about 60% for ions incident at 25°.

Another significant observation was the increasing asymmetry in the densities of the four spots as the angle of incidence α increased. This seems to indicate an anisotropy in the momentum distribution of the primary recoil atoms, in contradiction to the interpretation of experimental results for the transmission sputtering of gold by proton bombardment in the RUTHERFORD collision region, as given by NELSON and THOMPSON [*521a*]. For instance, as the direction of incidence approaches the [110] direction ($\alpha \rightarrow 45°$), the density of the spot corresponding to the $[1\bar{1}0]$ direction increases relative to the densities of the [101] and $[\bar{1}01]$

spots, as shown in Fig. 10.3.2.1.1.d for $\theta = 31°$. At $\alpha = 45°$, the density of the [$1\bar{1}0$] spot is about 1.8 times that of the other two. The method used seems to offer a new tool to study the momentum distribution of primary recoil atoms.

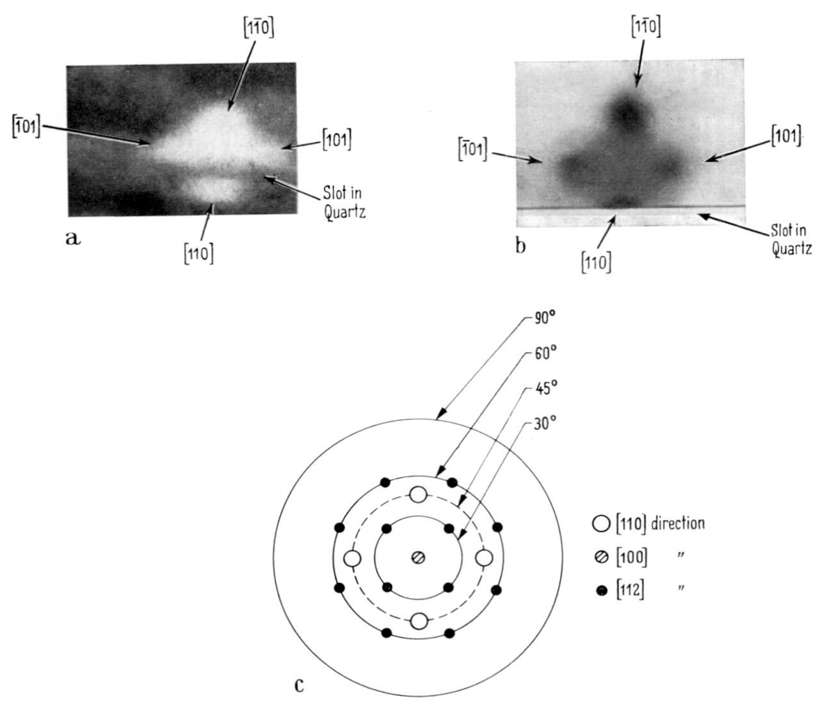

Fig. 10.3.2.1.1.d. Patterns of preferential ejection directions for sputtered particles. (a) Autoradiogram of deposit ejected from a Cu (100) surface bombarded with 125-keV deuterons incident at 31°. (b) Photograph of the deposit ejected from a Ag (100) surface bombarded with 125-keV deuterons incident at 45°. In both (*a*) and (*b*) most of the deposit is in the four [110] spots, the [$1\bar{1}0$] spot being the densest while the [110] spot is partically hidden by the slot in the quartz collector. Additional spots corresponding to the [112] and [100] directions are observed by microphotometry. (c) Stereographic projection of a face-centered cubic lattice onto a plane parallel to a [100] lattice plane. The figure shows only the [110], [100], and [112], directions (KAMINSKY [*363a, 363b*])

The sputtering ratio depends strongly and in a complicated way on the angle of incidence. Instead of increasing monotonically with increasing α, it drops at certain angles. For example, near $\alpha = 45°$ it drops to about the value of 0° but the intensities in the deposits of corresponding spots are different for the two angles. Optical transmission measurements (Fig. 10.3.2.1.1.e) show that the distribution in an individual spot is Gaussian, not cosine, and nearly independent of the angle of incidence. This result indicates that the recently suggested mechanism of sputtering by heated spikes [*721a*] is of no major importance in these experiments.

The interpretation of the observed spot pattern in the [110] directions has been based mainly on the action of a channeling mechanism and to a lesser extent on SILBSEE's mechanism. The spots corresponding to the [100] and [112] directions, however, can not be explained by the SILSBEE focusing mechanism since in these directions the spacings D_{112} and D_{100} are too large to fulfill the focusing condition $D_{hkl}/d < 2$. However, such spots may also appear as

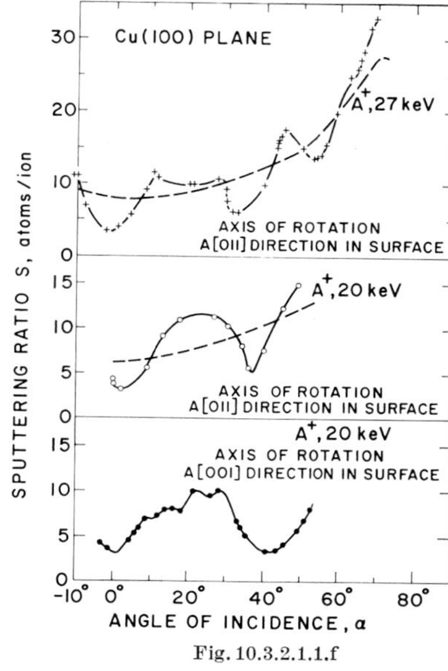

Fig. 10.3.2.1.1.e Fig. 10.3.2.1.1.f

Fig. 10.3.2.1.1.e. Density distribution of the deposit in an individual [110] spot. The center of the spot has been taken as $\theta = 0°$ and the edge of the spot (deposit density zero) as $\theta = 90°$. The particles in the deposit were sputtered from a Cu (100) plane by bombardment with 125-keV D^+ ions at normal incidence. The dotted line indicates a calculated normalized cosine distribution (KAMINSKY [*363b*])

Fig. 10.3.2.1.1.f. Variation of the sputtering ratio S with the angle of incidence α for a Cu (100) plane bombarded by argon ions. The data for 27-keV argon-ion bombardment have been obtained by MOLCHANOV et al. [*488*]; the dotted line represents $S(\alpha)$ for polycrystalline Cu, while the curve marked by (+) gives $S(\alpha)$ for a (100) crystal plane of Cu. The data for 20-keV argon bombardment (Cu crystal turned around a [011] direction in the surface) have been obtained by ROL et al. [*605*]; the dotted line represents $S(\alpha)$ for polycrystalline Cu, while the curve marked by (0) gives $S(\alpha)$ for a Cu (100) crystal plane. The data for 20-keV argon ion bombardment (Cu crystal turned around a [001] direction in surface) of a (100) plane of a Cu monocrystal marked by (•), have been obtained by FLUIT et al. [*215b*]

a result of the channeling mechanism and to a lesser extent (in view of the smaller ranges) to the lens focusing mechanism. For the channeling mechanism, the effectiveness in focusing decreases in the order [100], [112]—in qualitative agreement with the experimental observations. The experimental material available so far is insufficient to give a quantitative measure of the relative contributions of the two mechanisms to the observed pattern*.

* It should be noted, however, that while in KAMINSKY's sputtering experiments in the RUTHERFORD collision region the observed deposit thickness within

The variation of the sputtering ratio $S(\alpha)$ as a function of the angle of incidence α on a Cu (100) surface has been investigated in the hard-sphere collision region by several authors. ROL et al. [*605*] directed 20-keV A^+ ions onto a surface which was parallel to the (100) plane to within 3° and varied the angle of incidence by rotating the surface about a [011] axis in the surface. MOLCHANOV et al. [*488*] used 27-keV A^+ ions and a [011] axis in the surface; and FLUIT et al. [*215b*] used 20-keV Ne^+, A^+, and Kr^+ ions and a [010] axis in the surface. As can be seen in Fig. 10.3.2.1.1.f, the data of ROL et al. and of MOLCHANOV et al., which are both taken by turning the copper monocrystal around a [011] direction in the surface, agree qualitatively.

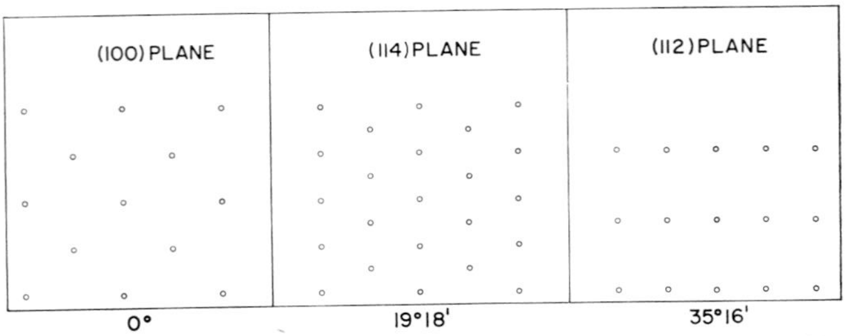

(100) PLANE (114) PLANE (112) PLANE

0° 19°18' 35°16'

Fig. 10.3.2.1.1.g. Schematic diagram showing the density of lattice points. A fcc crystal is projected on a plane normal to the beam (ROL et al. [*605*])

The sputtering ratio reached minima (at angles of about 0°, 35°, and 55° to the surface normal) when the direction of the ion beam coincided with the direction of one of the principal crystallographic directions [100], [112], and [111]. ROL et al. related the minima in the sputtering ratio to the small density of lattice points projected on a plane normal to the particular beam direction. As demonstrated in Fig. 10.3.2.1.1.g, the lattice is highly transparent for angles of about 0° and 35°, for example. On the other hand, a large density of projected lattice points corresponds to a high sputtering ratio; for instance the sputtering ratio reaches a maximum value at an angle of incidence of about 19°, at which the beam is incident along a [114] direction. (For a corresponding projection of lattice points, see Fig. 10.3.2.1.1.g.) The variation of the function $S(\alpha)$ with α has to be related also to the above mentioned channeling mechanism; the larger the range of the incident ion along such a channel, the smaller becomes the probability that displaced lattice

one spot decreases in the order [110], [100], [112], NELSON and THOMPSON [*521a*] predicted in their treatment that the effectiveness of focused collision sequences in fcc metals decreases in the order [110], [111], [100]. This result may be taken as indication for the dominant action of the channeling mechanism in KAMINSKY's experiments.

particles can reach the surface layers and the smaller is the sputtering ratio.

The experimental observations by FLUIT et al. [*215b*] are given in Fig. 10.3.2.1.1.f (20 keV A^+ ions on a Cu (100) plane, axis of rotation now a [001] direction). The sputtering yield shows minima at angles of incidence of about 0°, 11°, 18°, 27°, and 45° corresponding to the [010], [015], [013], [012], and [011] directions—in agreement with the above interpretation. The experimentally observed curves of $S(\alpha)$ presented by MOLCHANOV et al. [*488*] for 27-keV A^+ on Ni (100) and by FLUIT et al. [*215b*] for 20-keV Ne^+ on Cu (100) have characteristics similar to those shown in Fig. 10.3.2.1.1.f.

NELSON and THOMPSON [*521a*] investigated the directions of preferential ejection of particles sputtered from a copper (100) plane by 10-keV A^+ ions ($\alpha \approx 20°$) and from the (100) plane of silver and gold monocrystals with 10-keV A^+ and Xe^+ ions ($\alpha \approx 20°$). The observed spot pattern reveals preferential ejection in the [110] directions (four spots) in the [100] direction (one spot), and in the [112] directions (four spots). Oblique incidence results in an asymmetric deposit with the highest intensity on the side away from the incident beam. A similar asymmetry was observed by KAMINSKY [*362*] for the RUTHERFORD collision region.

Although the deposit patterns for the different ion-metal combinations show a certain similarity, the authors observed that the four spots corresponding to the [110] directions became less intense relative to the other on changing from A^+ to Xe^+ ions. They observed also that the intensity of the spot corresponding to the [100] direction decreased when a copper target was replaced by a silver one, and again when silver was replaced by gold. In the interpretation of their results, the authors invoked the SILSBEE mechanism for the appearance of the spots in the [110] directions and the lens focusing mechanism (for which they give an analytical treatment) for the spots in the [100] and [112] directions.

KOPITZKI and STIER [*405c*] investigated the mean kinetic energy $\bar{E}$ of the particles sputtered in the [110], [100], and [111] directions from a Cu (100) plane under bombardment with A^+ and Xe^+ ions with energies in the range from about 22 to 60 keV. They observed that $\bar{E}$ had the smallest value for the direction of densest packing [110] (A^+: $\bar{E}_{110} \approx 40$ eV; Xe^+: $\bar{E}_{110} \approx 35$ eV) and increased for the [100] direction (A^+: $\bar{E}_{100} \approx 50$ to 58 eV; Xe^+: $\bar{E}_{100} \approx 42$ eV) and the [111] direction (A^+: $\bar{E}_{111} \approx 100$ eV to 135 eV; Xe^+: 100 eV to 130 eV) over the investigated range of incident-ion energies. This agrees with the theoretical prediction ([*521a*], also Section 10.4.2.1) that the focusing energy* would have the smallest value

* The kinetic energy of a lattice atom moving in a lattice has to be smaller than the focusing energy in order to be focused by a particular focusing mechanism (e.g., by SILSBEE focusing or lens focusing).

for the direction of densest packing. For particle ejection along the [110] and [100] directions, the value of $\bar{E}$ remained nearly independent of the incident-ion energy over the investigated range from about 27—60 keV. Also the bombardment of the other crystal planes, such as the (110) and the (111) planes of copper, did not change the value of $\bar{E}$ significantly. In an analytical treatment, the authors express the mean kinetic energy $\bar{E}$ of the sputtered particles in terms of the focusing energy of a collision sequence in a particular crystallographic direction and the heat of evaporation. They find that the value of $\bar{E}$ does not depend explicitly on the mass and energy of the incident ion.

10.3.2.1.2. The (110) Surface

The angular distribution of material sputtered from a (110) plane of a face-centered cubic metal surface has been investigated by KOEDAM [*402*] for a Cu crystal bombarded with normally-incident Ne^+, A^+, and Kr^+ ions with energies ranging from 100—1500 eV, and by ANDERSON and WEHNER [*17*] for Cu and Ni surfaces bombarded with normally-incident Hg^+, Ne^+, and A^+ ions, for Au surfaces with Hg^+ and Ne^+ ions, and for Al targets with Ne^+ ions.

Fig. 10.3.2.1.2.a shows a pattern of preferential directions of ejection as found by KOEDAM for a Cu (110) plane bombarded with normally incident Kr^+ ions at 100 eV and 500 eV. For comparison, a stereographic projection of a fcc lattice onto a plane parallel to a (110) lattice plane is also shown. At low ion energies (100 eV, pattern A) the ejection pattern shows a center spot and four rather diffuse patches at an angle of about 60° to the surface normal. All five spots correspond to the [110] directions. (KOEDAM found that raising the target temperature to 500° C caused no marked change in the ejection pattern.) Similar results are observed in the experiments of ANDERSON et al. [*17*].

Both authors (KOEDAM and ANDERSON et al. agree that the density of the center spot increases with increasing energy (as shown also by comparing patterns A and B in Fig. 10.3.2.1.2.a). They also state that for certain cases (such as the 600-eV Ne^+ ions on Cu (110) surfaces, or the 400- and 800-eV Hg^+ ions on Ni (110) which were studied by ANDERSON et al., the pattern of the central patch seems to indicate the appearance of two additional spots at 45° to the surface normal. These correspond to the [100] direction, but are overlapped by the spot corresponding to the normal to the [110] surface. The lowest energy at which ANDERSON et al. were still able to observe a center spot ranged from 75 to 150 eV for all but one of the systems investigated. The exception was Ne^+ on Au, for which the spot could be observed below 50 eV.

For Hg^+ ion bombardment of an Au (110) plane, ANDERSON et al. [*17*] observed bands in the pattern deposited at about 30° to the surface normal. The authors related these bands, which seemed to consist of double spots, to the [411], [41$\bar{1}$], [141], and [14$\bar{1}$] directions, which are nearest-neighbor directions of crystals twinned to the (111) and (11$\bar{1}$) twinning planes. Additional spots corresponding to the [114] and [1$\bar{1}$4] directions could also be observed near 68° to the surface normal. However, such spots could not be observed for Ni (110) surfaces.

The preferred ejection directions at considerably higher incident-ion energies (but still in the hard-sphere collision region) were studied by NELSON and THOMPSON [*521a*, *521c*]. They bombarded the (110) planes of silver and gold monocrystals with 10-keV A^+ and Xe^+ ions and the (110) planes of copper monocrystals with 10-keV A^+

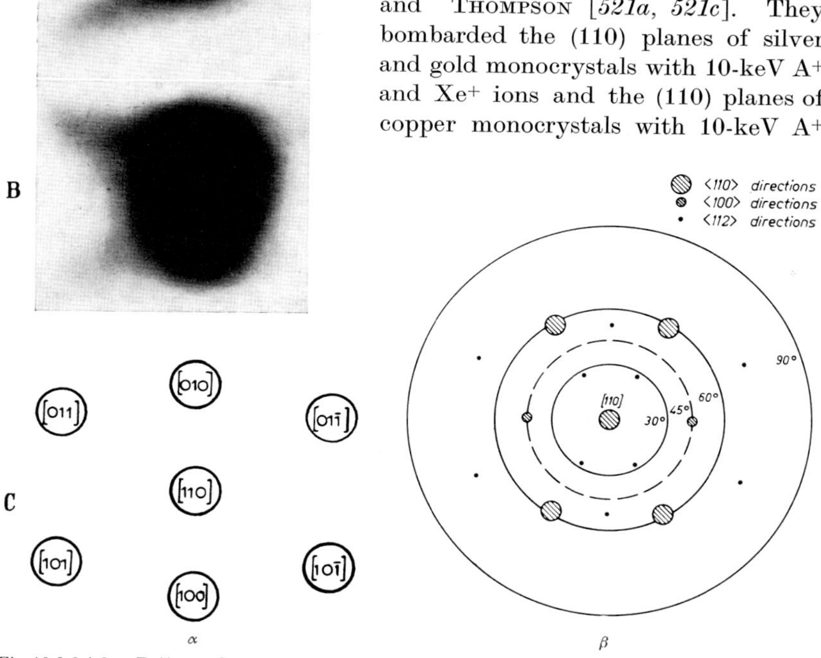

Fig. 10.3.2.1.2.a. Pattern of preferred ejection directions for atoms sputtered by the bombardment of a Cu (110) plane with normally incident Kr^+ ions with energies of 100 eV (curve A) and 500 eV (curve B). Curve C is a schematic ejection pattern (diagram α). Diagram β shows the stereographic projection of a fcc lattice onto a plane parallel to a (110) lattice plane (KOEDAM [*402*])

ions only, the angle of incidence being 20°. They also bombarded the Cu (110) planes with 25-, 50-, and 75-keV A^+ ions at normal incidence. They observed a spot pattern in the deposits of sputtered particles. For instance in the later experiments with the Cu (110) planes (ref. [*521c*]), they observed a strongly perferred ejection in the [110] directions and less strong preference in the [100] and [111] directions. In addition, they observed faint streaks joining the [110] spots. According to the authors, these streaks seem to correspond to the intersection of the (111) planes with the collector plane (which was kept parallel to the (110) plane of the crystal). The fact that the intensities of these streaks increase relative to those of the [110] spots with increasing incident-ion energy is taken by the authors as evidence that the focusing mechanism here is channeling between the widely spaced (111) planes. To what extent the spots corresponding to the [110] directions can be attributed to the action of such focusing mechanisms as channeling, lens focusing, and SILSBEE's mechanism cannot be clearly decided by these experiments.

FLUIT et al. [*215b*] studied the dependence of the sputtering ratio S on the angle of incidence α for a (110) plane of a copper monocrystal bombarded by 20-keV A^+ and Kr^+ ions. The axis of rotation for the crystal was the [001] direction in the surface. They observed sputtering minima at angles of incidence of 0°, 14°, 18°, 27°, and 45° angles at which the incident ions penetrated the lattice along the [011], [035], [012], [013], and [010] directions, respectively. As discussed in Section 10.3.2.1.1, such a behaviour of the function $S(\alpha)$ may be expected in view of the high transparency of the projected lattice points for these angles α and the more favorable channeling of the incident ions along these directions in the lattice.

As explained in Section 10.3.2.1.1 and reference [*405c*], measurements of particles sputtered in the [110] and [111] directions from a copper monocrystal bombarded with A^+ ions with energies in the range from about 25—60 keV indicated no marked difference between the mean kinetic energy $\bar{E}$ of particles sputtered from the (110) plane and the $\bar{E}$ from the (100) plane. For emission in the [110] direction, the observed energy has an almost constant value $\bar{E}_{110} \approx 30$ eV over the entire investigated range of incident-ion energies; but for the [111] direction the energy increases slightly (from $\bar{E}_{111} \approx 83$ eV to 93 eV) as the incident-ion energy increases from about 25—60 keV. The fact that the value $\bar{E}_{110}$ is smaller than the value $\bar{E}_{111}$ may be understood on the basis that the focusing energy in the [110] direction is smaller than in the [111] direction (see discussion in Section 10.4.2.1).

10.3.2.1.3. The (111) Surface

KOEDAM [*402*] investigated the angular distribution of material sputtered from a Cu (111) surface by normally incident Ne^+, A^+, and Kr^+ ions with energies ranging from 100—1500 eV. Fig. 10.3.2.1.3.a shows the patterns observed for a Cu (111) surface bombarded with Kr^+ ions at energies of (A) 250 eV, (B) 500 eV, (C) 1000 eV, and (D) 1500 eV. A stereographic projection (E) of a fcc lattice onto a plane parallel to a (111) plane is shown for comparison. A pattern of six preferential ejection directions can be recognized in the figure for ion energies of 250—1000 eV. These directions make an angle of about 30° with the surface normal, and their intersections with a plane parallel to the (111) plane lie on the vertices of a regular hexagon. Three of the six patches correspond to the [114] directions. However, the preferred ejection along the [114] direction disappears with increasing ion energy, as may be seen by comparison of pattern A (250 eV) and pattern D (1500 eV). A center spot corresponding to the [111] direction could not be observed for the energy range investigated. KOEDAM [*402*] obtained patterns qualitatively similar to these for Kr^+ ion bombardment also for bombardment by Ne^+ and A^+ ions.

ANDERSON and WEHNER [*17*] bombarded a Ni (111) plane with 600-eV Hg^+ ions and a Cu (111) plane with both 400-eV and 300-eV ions. In their ejection patterns they resolved one spot corresponding to the [110] direction and another which presumably corresponded to a [111] direction. Apparently they did not observe KOEDAM's

A

B

C

D

[110] [101]

E

[011]

α

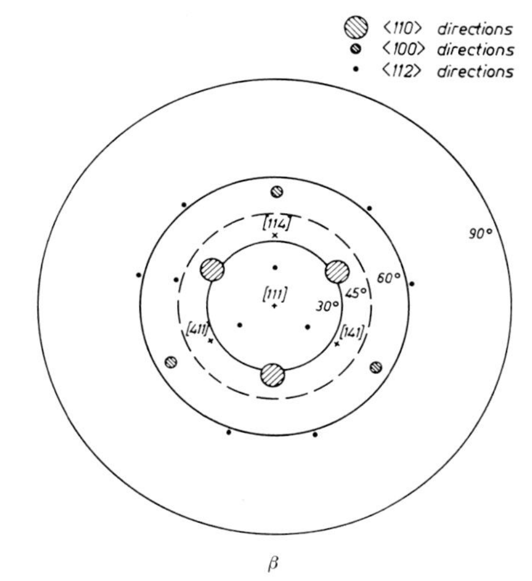

β

Fig. 10.3.2.1.3.a. Pattern of preferred ejection directions (part α) for atoms sputtered by the bombardment of a Cu (111) plane with normally incident Kr^+ ions with energies of 250 eV (pattern A), 500 eV (pattern B), 1000 eV (pattern C), and 1500 eV (pattern D). Diagram E is a schematic ejection pattern. Diagram β shows the stereographic projection of a fcc lattice onto a plane parallel to a (111) lattice plane (KOEDAM [*402*]) hexagonal pattern. No center spot corresponding to the [111] direction was observed, in agreement with KOEDAM's findings.

The preferred directions of ejection of sputtered particles at higher incident-ion energies have been studied for instance by NELSON and by NELSON et al. [*521, 521a, 521d, 521e*], who bombarded the (111) planes of Cu, Ag, and Au monocrystals with 10-keV A^+ or Xe^+ ions at an angle of incidence of 20° (ref. [*521a*]). They observed that particles were preferentially ejected, predominantly in the [110] directions and interpreted these results on the basis of SILSBEE's mechanism (section 10.3.2). Bombarding the (111) plane of an Al monocrystal with 50-keV A^+ ions at normal incidence, the same authors [*521d*] observed preferred ejection in the [110], [111], and [100] directions. Most of the sputtered particles were deposited in a central spot corresponding to ejection in the [111] direction and in three symmetrically placed spots corresponding to ejection in the [110] directions. The density of the [111] spot is higher than for a [110] spot. This result confirms earlier observations [*521a, 521e*] that the [111] ejection for Cu, Ag, and Au increases relative to the [110] ejection as the size of the atomic core becomes smaller in comparison with the spacing of lattice atoms.

They also found that the normalized density distribution of a [100] spot is broader and less peaked toward its maximum than the one for a [110] spot. This indicates that angular deviations from perfect focusing, caused for instance by thermal vibrations of lattice atoms, are greater in the [100] direction than in the [110] direction. This seems reasonable since focusing sequences in the [100] direction can be caused by the lens focusing and are therefore more sensitive to thermal vibration than those in the [100] direction (produced by SILSBEE's mechanism), because the vibration distorts the focusing rings in addition to displacing the atoms in the focusing line.

In other experiments by NELSON et al. [*521a*], a gold foil, which had two principal grain orientations with (111) planes in the surface was bombarded with 0.3-MeV protons. The particles leaving the other surface of the foil ("transmission sputtering" or "forward sputtering") were observed to be preferentially ejected along the [110] directions. NELSON [*521f*] extended his studies of the sputtering of metals to such ionic crystals as UO_2 (fluorite-type lattice $R_U : R_O > 0.732$) which has importance as a nuclear fuel. He bombarded the (111) plane of a UO_2 monocrystal with 50-keV Xe^+ ions and observed preferential ejection in the [110], [100], and [111] directions. The interpretation for the focusing in the [110] direction is based on SILSBEE's mechanism (Section 10.3.2); the focusing in the [100] and [111] directions is explained by the lens focusing mechanism (Section 10.3.2). The "focusing lens" for focusing in the [100] direction, for instance, is formed by a central section consisting of four uranium atoms in the fcc positions with a ring of four oxygen atoms on either side. The presence of the oxygen atoms along the [100] axis enhances the focusing action in comparison with a simple fcc lattice of uranium. The focusing in the [111] direction on the other hand is somewhat hindered by the presence of the oxygen atoms along the focusing direction.

ALMEN and BRUCE [*9*] measured a sputtering ratio S for 45-keV Kr^+ ions bombarding a Cu (111) surface at various angles of incidence α. Fig. 10.3.2.1.3.b shows the curve $S(\alpha')$ for a Cu monocrystal turned around

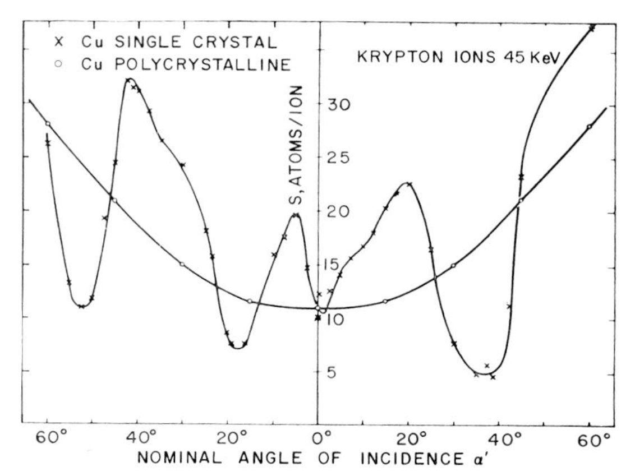

Fig. 10.3.2.1.3.b. The dependence of the sputtering ratio S on the angle of incidence α for a copper single crystal turned around a (110) axis in the (111) plane (points ×). The (111) direction is 3° from zero. The $S(\alpha)$ curve for a polycrystalline Cu surface is marked by ○ (ALMEN et al. [*9*])

a [110] axis in the (111) plane. The (111) plane was parallel to the surface within 2°. This curve shows sputtering minima corresponding to "transparent" directions (Section 10.3.2.1.1) at angles of 35.3° (corresponding to the [110] direction), 0° (corresponding to the [111] direction), 19.5° (corresponding to the [112] direction), and 54.7° (corresponding to the [100] direction). The authors point out that these are also the directions in which they were able to collect maximum amounts of Kr^+ ions in the lattice.

According to KOEDAM, a temperature rise of the target to 500° C did not cause a marked change in the ejection pattern. However, NELSON [*521*], who bombarded an Au (111) surface with 43.5-keV A^+ ions, observed a marked broadening of the three spots corresponding to the [110] direction when the target temperature was raised from 100° C to 900° C (see Fig. 10.3.2.1.3.c). NELSON et al. [*521g*] discussed three mechanisms which might cause the scattering of such focused collision sequences. These mechanisms, the first two of which had already been briefly considered by SILSBEE [*653*], are the following.

1. The lateral component of momentum of a lattice particle in its thermal motion may deflect the trajectory of an atom involved in a focusing collision sequence and moving with an energy E. However the angular deviation θ caused in this way is proportional to $(kT/E)^{1/2}$ and has very small values ($\theta < 0.03$).

2. The lattice atoms may vibrate from their equilibrium positions and cause an angular deviation $\theta = \delta/2\,R$, where δ is the relative displacement of adjacent atoms and R is the radius of a lattice atom.

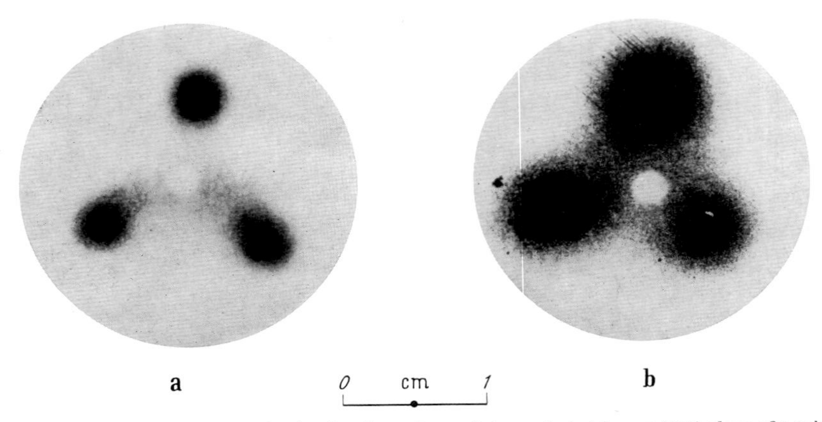

Fig. 10.3.2.1.3.c. Autoradiographs showing the pattern of atoms ejected from a (111) plane of a gold monocrystal under bombardment of 43-keV A^+ ions. The regular pattern of three spots corresponds to preferential ejection about the (100) axes (NELSON [*521*] and [*521a*]). a) Temperature of gold monocrystal kept at 370° K. b) Temperature of gold monocrystal kept at 1170° K

The angular deviation may reach values of the order of $\theta < 0.2$, since according to LINDEMANN's law $\delta \sim 0.4\,R$ at the melting point of the material.

3. The third mechanism for angular deviation of a collision sequence becomes important when such a sequence is caused by a "lens focusing" mechanism. If one considers, for example, the focusing in a [100] direction in a fcc lattice (Fig. 10.3.2.a) the ring of four B atoms surrounding the focusing line will be distorted by lattice vibrations, and cause a perturbation of the focusing action of the "lens" (formed by the four B atoms). The angular deviation caused by this effect is appreciable (order of magnitude 0.1) but smaller than in the case discussed under 2.

On the basis of the scattering effect mentioned under 2., NELSON et al. [*521g*] treated the observed broadening of the [110] spots analytically, using a vibration amplitude typical of surface atoms rather than those of lattice atoms in the bulk.

10.3.2.2. Atom Ejection for Body-Centered Cubic Metals

10.3.2.2.1. The (100) Surface

WEHNER et al. [*771*] and ANDERSON [*17a*] studied the angular distribution of material sputtered from the (100) surface of V, Fe, Nb, Mo, Ta, and W targets bombarded with Hg^+, Xe^+, A^+, and Ne^+ ions at lower energies ranging from 100—800 eV. For Fe, W, and Nb, the observed ejection directions correspond almost exactly to [111] directions for all ions in the above-mentioned energy range. These directions are the nearest-neighbor directions in bcc crystals, in which focusing-chain collisions (SILSBEE [*653*]) are possible. They make an angle of 54.8° with respect to the surface normal. Above a certain onset energy, which should not be considered as a sharp threshold energy, a center spot corresponding to the [100] direction appears. These onset energies depend on the incident ion, varying from 100—250 eV for bombardment by Ne ions to 500—800 eV for Hg ions, but are nearly independent of the mass of the target atom.

Another interesting observation is that the pattern of preferential ejection directions from Mo (100) surfaces differs significantly from those observed for the (100) planes of the other bcc metals investigated. WEHNER et al. [771] found no single spots corresponding to the [100] directions, but four spots surrounding each pole of the [111] direction. They suggest that a possible occurrence of (110) facets on the (100) surface might cause such a "splitting" effect.

Studies of preferred ejection directions from tungsten and molybdenum monocrystals at higher incident-ion energies (yet within the hard-sphere-collision region $E < E_A$) were conducted by NELSON [*521h*]. For instance, he cut a tungsten crystal in such a way that the [111] and [100] directions made equal angles with the surface normal and bombarded this surface with 50-keV A^+ ions. The deposit revealed the spot pattern corresponding to preferential ejection in the [111], [100], and [110] directions. The amount of tungsten in a particular spot decreased in the order [111], [100], and [110]. The ejection in the [111] and [100] directions, which are the directions with the more densely packed atoms, can be interpreted by hard-sphere collision sequences [SILSBEE mechanism, Section (10.3.2)]. On the other hand, the ejection in the [110] directions can be related to the "lens focusing" mechanism (Section 10.3.2.). Using such a model, NELSON performed calculations for the focusing energies and ranges of the [110] collision sequences.

According to NELSON, the general nature of the spot pattern obtained from the corresponding molybdenum plane was identical to the one from tungsten, except that the relative intensities of the different spots were different.

10.3.2.2.2. The (110) Surface

WEHNER [*771*] and ANDERSON [*17a*] studied the ejection pattern of sputtered material from (110) surfaces of Fe, V, Ta, Mo, and W targets bombarded by Hg^+ ions and/or in some cases, by Ne^+ and A^+ ions. They observed two spots corresponding almost exactly to the [100] directions for all cases investigated, but did not detect any center spot corresponding to the [110] direction. Moreover, no spots corresponding to the [111] were observed; but in some instances two spots, one on each side of the $(1\bar{1}0)$ plane, were found. The author [*771*] suggested that such a splitting might result from surface atoms that sit in positions with three nearest neighbors. [A surface atom in its usual position on a (110) plane has only two nearest neighbors.]

Contrary to these results at lower incident-ion energies, NELSON [*521h*] observed that bombarding the (110) planes of wolfram and molybdenum monocrystals with 50-keV A^+ ions produced both the spots corresponding to the [100] directions and also those corresponding to the [111] and [110] directions.

10.3.2.2.3. The (111) Surface

Results for the preferred ejection of atoms sputtered from Fe and V (111) surfaces under Hg^+ ion bombardment in the energy range of approximately 150—600 eV have been reported by WEHNER [*771*]. Spots corresponding to the [111] directions were observed at about 60° with respect to the surface normal (instead of the predicted 70.5°), and spots corresponding to the [100] directions were found at 52° (instead of 54.8°). A center deposit was also noted. The patterns are qualitatively the same for both metals.

ANDERSON [*17a*] bombarded a Fe (111) surface with A^+ ions of energies up to 800 eV and collected the particles on a spherical collector which allowed a somewhat better resolution of the deposit pattern than the previously used cylindrical collector [*771*]. He observed that the spot pattern obtained was similar to the one found for the Hg^+ ion bombardment, but noted an additional feature. The center spot, instead of being round had the shape of a three-pointed star with the points aimed between the [100] spots. Such a structure may be expected if one attributes preferential ejection of particles to the existence of three (110) planes which are inclined only 35° from the target surface [the (111) plane] and may even exist as small facets in the surface.

NELSON [*521h*] who bombarded the (111) planes of wolfram and molybdenum monocrystals with 50-eV A^+ ions (hard-sphere collision region, $E < E_A$), has observed a similar pattern in the central spot. Three faint streaks pointed in the directions of the intersections between the three (100) planes and the (111) surface plane. In addition, the [111] spot is surrounded by three subsidiary spots which may result from particle ejection at an angle of 10° to the [111] directions in the (110) planes. The three [100] spots seem to be surrounded by four poorly resolved spots, which may correspond to the ejection of particles at an angle of about 4° to the main direction in the (100) planes. The emission of the particles in the [111] and [100] directions is interpreted on the basis of the SILSBEE focusing mechanism for hardsphere collision sequences (Section 10.3.2).

10.3.3. Sputtering of Target Particles in an Excited Neutral or Ionized State from Polycrystalline and Monocrystalline Targets

When a metal surface is bombarded by ions, some of the sputtered target particles may leave the surface in an excited neutral or ionized state. Studies of the emission spectrum of ejected target atoms in excited states have been conducted by STUART and WEHNER [*690*], FLUIT et al. [*215c*], and KISTEMAKER and SNOEK [*385a*]. STUART and WEHNER monitored the emission lines of the target material by a monochromator and a photomultiplier as detector. They related the spectral-line intensities, characteristic for the sputtered target materials, to the sputtering yields; by measuring the line intensity as a function of the incident-ion energy they were able to extend such measurements to very low incident-ion energies and thereby to determine sputtering thresholds (the results are discussed in section 10.3.1.1).

FLUIT et al. [*215c*] and later KISTEMAKER and SNOEK [*385a*] studied the optical spectra emitted by the sputtered particles, in particular of those particles sputtered in a metastable state. They bombarded the (100) plane of a Cu monocrystal and polycrystalline copper and aluminum surfaces with 10—15-keV Ne^+, A^+, and Cu^+ ions and measured the emission spectra with a grating monochromator (bandwidth approximately 20 Å) and a photomultiplier as detector. In some instances they also used a prism spectrograph with photographic recording. While bombarding Cu targets KISTEMAKER and SNOEK observed for instance the Cu I and Cu II spectra and the emission spectra of the ionized gases used for the bombardment. They were able to relate the intensity of a spectral line of the target material to the sputtering yield; for instance they observed that the intensity of the Cu I resonance line (3247 Å) was nearly twice as strong in the case of a Cu (100) plane bombarded wits 15-keV A^+ ions as for bombardment with 15-keV Ne^+ ions. This agrees with the ratio of the particular sputtering yields determined by other methods (see Fig. 10.3.1.2.1.f).

Moreover they [*215c, 385a*] studied the intensity of the Cu I resonance line (3247 Å) as a function of the angle of incidence of an A^+ or

Ne^+ ion beam (15-keV A^+, Ne^+; 10-keV A^+) striking a (100) plane of a Cu monocrystal with its axis of rotation parallel to a [100] axis in the surface. They observed a very pronounced minimum of the line intensity at an angle of incidence $\alpha = 45°$, a result which agrees with the behavior of the function $S(\alpha)$ (see Fig. 10.3.2.1.1.d) and again confirms the close relationship between the sputtering ratio S and the intensity of a particular spectral line. Conducting such angle of incidence measurements, KISTEMAKER and SNOEK observed that the excited sputtered target atoms did not de-excite by photon emission while they were very close to the target surface.

There are still no conclusive results on the emission of metastable target atoms from metal surfaces under ion bombardment. For instance the results obtained by FLUIT, FRIEDMAN, VAN ECK, SNOECK and KISTEMAKER [*215c*] for the A^+ ion bombardment of monocrystalline and polycrystalline targets were later questioned by two of the original authors (KISTEMAKER and SNOEK [*385a*]). In the original paper the authors [*215c*] tried to detect the metastable particles by measuring the secondary-electron current produced by the de-excitation at a copper detector surface. (A pulsed-beam technique was used to distinguish secondary electrons produced by metastable atoms from those produced by photons and to allow time-of-flight measurements to determine the velocity distribution of the metastable atoms.) However, later experiments (reference [*385a*]) with an improved detection system indicated that the observed signals were due to particles striking the Cu detector surface with considerably higher energy (order of several keV) than originally concluded (order of 10 eV, reference [*215c*]); the signals may have originated from fast reflected particles or highly energetic sputtered particles. How far metastable target atoms contributed to the observed signals could not be decided by the experiments.

A better identification of the processes causing secondary electron emission, such as the process of de-excitation of metastable atoms (see Chapter 13), the potential emission mechanism (slow incident ions or atoms, Chapter 12), or the kinetic emission mechanism (fast incident ions or atoms, Chapter 14) may be obtained by measurements of the energy distribution of the secondary electrons, as discussed in more detail in chapter 13.

Investigations of such aspects of secondary-ion emission contribute significantly to a better understanding of the sputtering mechanism as well as such processes as the reflection of primary ions, secondary-electron emission, and the mechanism of ionic pumping. At present such data, which will be summarized here, are rather scarce compared with those on the sputtering of neutral particles. In particular there is a serious lack of detailed systematic investigations of the dependence of secondary-ion emission on such critical surface features as physical, chemical, and induced surface inhomogeneities (Sections 1.1.1., 1.1.2, and 1.1.3).

10.3.3.1. Threshold Values for the Sputtering of Charged Particles

Values for the minimum primary-ion energy at which the production of sputtered target ions was still detectable were reported by BRADLEY [*90*] and by STANTON [*674*]. Both authors detected the sputtered ions with mass spectrometers. BRADLEY found a "threshold" value of 80±8 eV for the emission of Mo^+ ions in the case of A^+ ion bombardment of a polycrystalline Mo surface, and a value near 60 eV for such ions impinging on a polycrystalline Ta surface. STANTON observed, within the limits of the experimental errors, that the yield of Pb^+ ions sputtered by He^+ bombardment of a Pb target was proportional to the ion current of the incident beam, and that the sputtered target atoms were hardly detectable below primary-ion energies of 30 eV. These thresholds are considerably higher than those reported for the emission of neutral particles. However the results of both authors were influenced by a strong contamination of the surface (as evidenced, for instance, by the detection of MoC_2^+ and TaN^+ ions by BRADLEY [*90*]) and by the increasing defocusing of the incident-ion beam with decreasing ion energy. Reliable threshold values for the emission of secondary negative ions are not yet available.

10.3.3.2. Species of Sputtered Positive and Negative Ions

Several authors [*90, 91, 218, 323, 363a, 407a, 485, 486, 737*] have investigated the mass spectrum of the sputtered ions and have detected a variety of both atomic and molecular species. One of the more striking results observed (e.g., see ref. [*90, 91, 218, 363a, 407a, 485, 737*]) is that sputtered ions appear only in a singly-charged state, while the re-emitted incident ions have been found to be both in singly- and doubly-charged states. However the results, summarized below, indicate that the mass spectrum of sputtered particles depends substantially on the experimental conditions, i.e., on the pressure and composition of the residual gas, the degree of surface coverage by adsorbed particles, and the impurities imbedded in the target lattice. Therefore any intercomparison of results obtained by different authors has to be made with extreme caution. Furthermore the data obtained by most authors are rather qualitative, since they did not report the absolute intensities of the detected peaks of the different species, and in most cases also failed to give their relative intensities. Also there are only estimated values for the ratio of the number of sputtered ions to the number of neutral particles. BRADLEY [*91*], for example, estimated that the ratio of sputtered Pt^+ ions to sputtered Pt neutral particles was about 10^{-3} if the Pt target, bombarded with noble-gas ions, was held at room temperature, and the ratio increased slightly with increasing target temperature. He did not mention how this ratio changed with the energy of the primary ion or with the type and the degree of the surface coverage.

Honig [*323*] studied the mass spectra of sputtered positive and negative ions as well as neutral particles (ionized by electron impact for the mass spectrometric detection) by bombarding polycrystalline silver targets with Kr^+ and Xe^+ ions with energies between 200 and 400 eV and current densities between 2 and 20 $\mu A/cm^2$. The residual gas pressure was less than 1×10^{-7} mm Hg. Some of the major mass peaks of the observed sputtered positive and negative ions are listed in Table 10.3.3.2A. In addition Honig observed large currents of re-emitted primary ions: Kr^+, Kr^{++}, Xe^+, and Xe^{++}.

Table 10.3.3.2.A. *Charged particles sputtered from an Ag surface by Xe^+ ions with energies of 300—400 eV* (Honig [*322, 323*])

Positive ions	Ag^+, Ag_2^+, Ag_3^+, Ag_2O^+, Na^+, Mg^+, Al^+, S^+, K^+, Ca^+, Fe^+, Hg^+
Negative ions	AgH^-, Ag_2^-, Ag_2O^-, $Ag_2O_2^-$, O^-, OH^-, F^-, S^-, SH^-, SH_2^-, Cl^-

Bradley [*90*] investigated the mass spectrum of positive ions sputtered from Pt, Ta, and Mo surfaces by noble-gas ions with energies up to 1 keV. The residual gas pressure in his experiments was about 10^{-8} mm Hg (partial pressures: $CO_2 \approx 7 \times 10^{-10}$ mm Hg; $CO \approx 1 \times 10^{-8}$ mm Hg). Before each run he tried to clean the surfaces by flash-filament technique (Section 3.1). Table 10.3.3.2.B shows some of the major peaks

Table 10.3.3.2.B. *Particles sputtered from Pt, Ta, and Mo targets by noble-gas ions at energies of 100—1500 eV* (Bradley [*89*])

Target	Positive ions
Pt	Pt^+, PtO_3^+, PtO^+, Li^+, Na^+, K^+, Rb^+, Cs^+
Ta	Ta^+, TaN^+
Mo	Mo^+, $Mo_2O_3^+$, MoO_2^+, MoC_2^+

observed. He observed that the MoO_2^+ peak disappeared at surface temperatures of about 300° C, and the $Mo_2O_3^+$ peak at about 1000° C. The Mo^+ peak was approximately tripled when the oxides had been removed from the surface. A MoO_3^+ peak could not be observed despite the fact that it is actually the most intense peak in the evaporation of molybdenum oxides. The observed emission of alkali metal and alkaline earth ions is probably due to bulk impurities in the target material.

Fogel et al. [*218*] investigated the mass spectrum of ions which were likewise sputtered from Mo targets by noble-gas ions, but at higher primary-ion energies (5—40 keV) than in the experiments of BRADLEY. The targets were heat treated before each run and the residual gas pressure was kept at $3—5\times10^{-6}$ mm Hg. Table 10.3.3.2.C lists the major peaks of the observed sputtered negative and positive ions. One notices that the overwhelming majority of emitted particles have negative charge.

Table 10.3.3.2.C. *Particles sputtered from a polycrystalline Mo surface by noble-gas ions at energies of 5—40 keV* (FOGEL et al. [*218*])

Negative ions	H^-, MoO_2^-, MoO_3^-, $Mo_2O_3^-$, C^-, CH^-, OH^-, C_2^-, C_2H^-, O_2^-
Positive ions	H^+, Mo^+

Instead of bombarding a Mo surface by noble-gas ions, as was done in references [*218*] and [*90*], VEKSLER and SHUPPE [*737*] used Hg^+ ions with energies up to 600 eV. With the exception of the H^- ion and the negative molybdenum oxide ions, the species of negative ions they observed were the same as those found by FOGEL et al. [*218*].

KROHN [*407a*] studied the emission of negative ions from targets of polycrystalline Cu, Ag, Au, Be, Cd, Al, Sn, Ta, W, and Ni bombarded by a beam of 1.1-keV Cs^+ ions. A PAUL mass filter [*545a*] was used to determine the charge-to-mass ratio of the emitted ions and the measurements were made a total pressures from 10^{-6} to 10^{-5} mm Hg. In most cases the strongest mass peak observed was due to the atomic ion of the target material or one of its oxides. However, molecular ions of the target material were observed in many instances (e.g., some observed ions and, in parentheses, their relative intensities, were Cu^- (25), Cu_2^- (15), Cu_3^- (6); Ag^- (8), Ag_3^- (5); Al^- (2), Al_2^- (30), Al_3^- (10); Sn^- (1), Sn_2^- (12)). It is of interest to note that for aluminum and tin targets the Al_2^- and Sn_2^- yields were an order of magnitude greater than the corresponding atomic ion yields. KROHN also made the interesting observation that the yield of negative ions from a copper target increased whenever an auxiliary beam of neutral cesium was directed against the target. This indicates that the presence of cesium on the target (effective lowering of the work function—see Chapter 1 and ref. [*160*]) enhanced the negative-ion emission.

Preliminary results for the mass spectra of positive as well as neutral target particles (ionized by electron impact for the mass-spectrometric detection) sputtered from the (100) planes of Cu and Ag monocrystals under high-energy ion bombardment (0.15—0.25-MeV and 0.5 to 2.0-MeV H^+, D^+, and He^+ ions) have been obtained by KAMINSKY [*363a*]. The residual gas pressure was 4×10^{-8} mm Hg. The mass spectra of sputtered ions revealed singly—and traces of doubly—charged atomic ions of Cu and Ag for higher incident ion energies and angles of incidence.

Mitropan et al. [*485, 486*] studied the emission of negative ions from aluminum, stainless steel, and copper targets bombarded by hydrogen and deuterium ions with energies of 200—1000 keV. The targets were degreased and outgassed (at 900° C) before each run and the residual gas pressure was kept down to about 10^{-6} mm Hg. In the emission spectrum from a copper target, the authors observed negative ions such as H^-, C^-, O^-, OH^-, C_2^-, CH_2^-, and $C_2H_2^-$. They also studied the coefficient of secondary-ion emission, the total number of secondary ions (positive *or* negative) per incident ion. The emission coefficient K^- for negative ions decreases monotonically as the energy of the primary hydrogen ion increases. The values of K^- are around 10^{-3} at a primary-ion energy of 200 keV and decrease to about 10^{-4} at 1 MeV. The value of K^- was found to increase from Cu to stainless steel to aluminum. The authors found that outgassing the target considerably decreased the secondary negative-ion emission and eliminated the hydrogen peak. This result emphasizes the great influence of surface conditions on secondary-ion emission, even at very high primary-ion energies where one would have expected the ion-emission process (not necessarily identical with a sputtering process) to become a bulk effect rather than a surface effect (Section 10.4.2.3).

Leck et al. [*97, 591*] made a more detailed investigation of the way in which impurities trapped in the target lattice are emitted in a charged state when the target is heated after positive-ion bombardment of its surface. When tungsten targets were bombarded with positive ions of the noble gases and then heated (up to 1600° C), the thermal emission of ions of sodium, potassium, rubidium, and cesium could be observed.

The thermal emission of ions (mainly K^+) continued for a long time after bombardment of the surface had ceased, and the rate increased strongly with increasing energy of the bombarding ions. This behavior is mainly due to changes in the surface condition (e.g., removal of "outgassed" surface layers to expose the uncleaned metal beneath) as a result of the ion bombardment.

Other experiments in which noble gases are first trapped in a Ni target and then re-emitted in the course of positive-ion bombardment of the target surface have been conducted by Carmichael and Trendelenburg [*108*]. The results were discussed in connection with the mechanism of ionic pumping (Section 3.2).

10.3.3.3. Dependence on the Energy of the Primary Ion

The coefficient of secondary-ion emission as a function of the velocity or energy of the primary ion has been investigated by several authors [*218, 738, 542*]. Fig. 10.3.3.3.a shows the dependence of this coefficient K on the mean velocity V of the primary ion for the case of a polycrystalline

Mo target bombarded with Ne^+, A^+, and Kr^+, as observed by FOGEL et al. [*218*]. The velocities of the incident Ne^+ ions range from about $V = 2.1 \times 10^7$ cm/sec (corresponding to 5 keV) to approximately 5.5×10^7 cm/sec (corresponding to 30 keV). For A^+ ions, the velocities range from about 1.8×10^7 cm/sec (corresponding to 8 keV) to approximately 4.5×10^7 cm/sec (corresponding to 50 keV); and for Kr^+ ions the range was from about 1.1×10^7 cm/sec (corresponding to 5 keV) to approximately 2×10^7 cm/sec (~ 17 keV). The curves $K(V)$ for a Mo target bombarded by Ne^+ ions and A^+ ions show a maximum near 3.5×10^7 cm/sec, which corresponds to about 12 keV for the case of Ne^+ and approximately 24 keV for A^+. This behavior is similar to that observed for the function $S(E)$ for sputtered neutral particles.

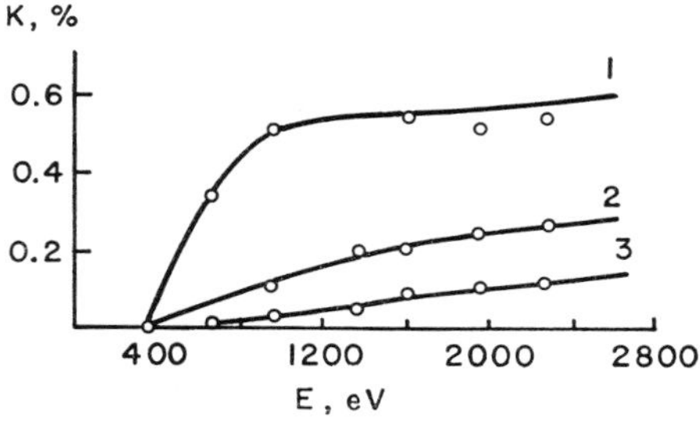

Fig. 10.3.3.3.a

Fig. 10.3.3.3.b

Fig. 10.3.3.3.a. The dependence of the secondary-ion-emission coefficient K on the mean velocity of the primary noble-gas ion for the cases of Ne^+, A^+, and Kr^+ ion bombardment of a Mo target (probably predominantly Mo^+ emission) (FOGEL et al. [*218*])

Fig. 10.3.3.3.b. The dependence of the secondary-ion-emission coefficient K on the incident-ion energy E for sputtered Ta^+ ions (curve *1*) at a Ta target temperature of 1293° K, and for sputtered Ni^+ ions at a Ni target temperature of 1240° K (curve *2*) and 1293° K (curve *3*), as obtained by VEKSLER et al. [*738*] for polycrystalline Ta and Ni targets bombarded by Cs^+ ions

Fig. 10.3.3.3.b shows the curves of $K(E)$ for sputtered Ta^+ ions [curve (*1*), at a target temperature of 1760° K] and Ni^+ ions [curve (*2*), at 1240° K; curve (*3*), at 1293° K] obtained by VEKSLER et al. [*738*] for polycrystalline Ta and Ni targets bombarded with Cs^+ ions with energies E ranging from 400 to 2800 eV. Here the secondary ions have been detected at an angle of 60° with respect to the primary-ion beam. It is of interest to note that the yield of Ta^+ ions is higher than for Ni^+ ions, while the opposite order is observed for the yield of sputtered neutral particles. This difference is not explained by the difference in target temperature, since the authors observed this same order over a broad temperature range (from room temperature up to 1293° K for Ni and 1760° K for Ta).

Panin [542] investigated the dependence of the total secondary-ion-emission coefficient R on the incident-ion energy. This coefficient R is the ratio of the total ion current leaving the surface to the incident ion current. The ions leaving the surface may include scattered primary ions as well as sputtered ions. Targets of Mo, Zr, and graphite were bombarded with H^+, H_2^+, H_3^+, He^+, C^+, N^+ Cl^+, Z^+, and Mo^+ ions in the energy range 10—100 keV. The author did not specify the exact angle of incidence used, but the experimental arrangement shown in his Fig. 2 seems to indicate an angle of incidence near 45° to the surface normal.

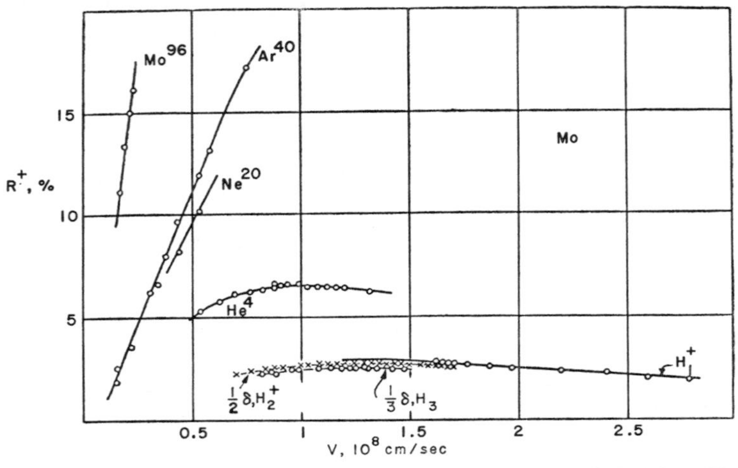

Fig. 10.3.3.3.c. The dependence of the total secondary-ion-emission coefficient R on the velocity V of the incident ion for the case of a Mo target (Panin [542])

Fig. 10.3.3.3.c shows the dependence of R^+ on the velocity of the incident ion for the case of a Mo target. The curves $R^+(V)$ are very similar for the other targets investigated. One notices that the values of R^+, the secondary-ion-emission coefficient for positive ions, are in the ratio 1:2:3 for bombardment by H_1^+, H_2^+, and H_3^+ ions, respectively. A similar behavior was found for the yields of sputtered neutral particles from bombardment by molecular ions (Fig. 10.3.1.3.1.c). However, one should keep in mind that the present values of R represent not just the number of sputtered ions but also include scattered primary ions.

10.3.3.4. The Effect of Target Temperature

Present experimental data indicate that the sputtering of charged particles differs most markedly from that of neutral particles in its sensitivity to target temperature. While the sputtering yields for neutral particles show only a slight temperature dependence (Section 10.3.1.2.3), the yields of secondary ions vary greatly. In the case of Pt targets, for instance BRADLEY et al. [*91*] observed a change of more than a factor of 10 between room temperature and 1500° C.

The results summarized later in this section seem to indicate that the rapid changes in the emission of secondary ions as the target is heated are partly due to such processes as desorption of gas layers adsorbed on the surface, diffusion and desorption of bulk impurities, and desorption of surface compounds formed by reactions of target atoms with surface contaminants. At higher target temperatures, the onset of thermal emission of electrons and ions masks the secondary emission of negative ions. As a result of this complexity of processes taking place when the target is heated, a detailed interpretation of the data on secondary-ion emission as a function of target temperature becomes rather difficult. This is particularly true because of

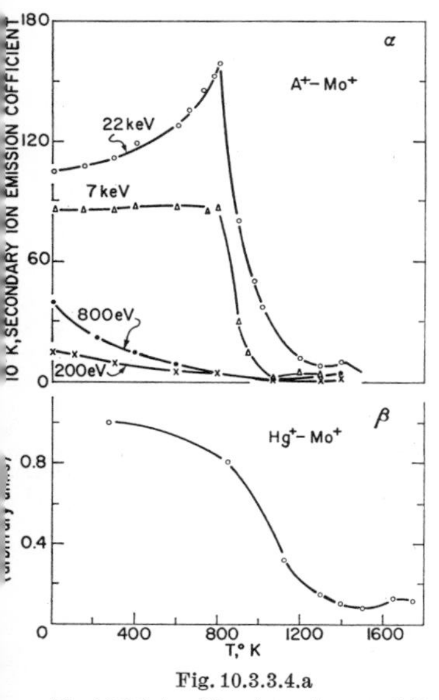

Fig. 10.3.3.4.a

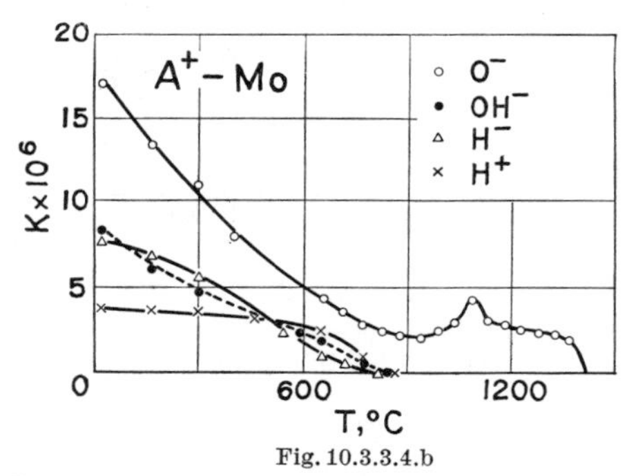

Fig. 10.3.3.4.b

Fig. 10.3.3.4.a. The dependence of the secondary-ion-emission coefficient K on the target temperature T. In the upper diagram, the two curves for 22-keV and 7-keV argon ions incident on a Mo target represent the data of FOGEL et al. [*217*]. The other two curves for 800-eV and 200-eV argon ions incident on Mo represent the results of BRADLEY [*90*]. The lower diagram shows the data of VEKSLER [*739*] for the emission of Mo^+ ions under Hg^+ ion bombardment of a Mo target at 2 keV

Fig. 10.3.3.4.b. The target-temperature dependence of the secondary-ion-emission coefficient K for emitted O^-, OH^-, H^-, and H^+ ions for the case of A^+ ion bombardment (22 keV) of a Mo target (FOGEL et al. [*217*])

the scarcity of experimental results now available. Also, the results obtained by different authors can be compared only with extreme caution, since different experimental conditions (such as surface cleanness, pressure of residual gas, etc.) strongly influence the results even for the same primary ion and target material.

As can be seen in Figs. 10.3.3.4.a and 10.3.3.4.b, the temperature coefficient of the secondary-ion emission can be either positive or negative. Curve α of Fig. 10.3.3.4.a shows the secondary-ion-emission coefficient K plotted as a function of the target temperature for a polycrystalline Mo target that has been bombarded by A^+ ions (FOGEL et al. [217], BRADLEY [90]) with energies ranging from 200 eV to 22 keV. Two curves in Fig. 10.3.3.4.a (α) represent the data of FOGEL et al. for secondary emission of Mo^+ ions as a result of a primary A^+ ion bombardment at ion energies of 22 keV and 7 keV, respectively, and the other curves show corresponding results (BRADLEY [90]) for A^+ ions at energies of 800 eV and 200 eV, respectively. Fig. 10.3.3.4.a (β) shows the data of VEKSLER [739] for the emission of Mo^+ ions under Hg^+ ion bombardment at 2 keV. From Fig. 10.3.3.4.a (α) it becomes apparent that the shape of the curve $K(T)$ depends less on the type of the incident ion than on its energy.

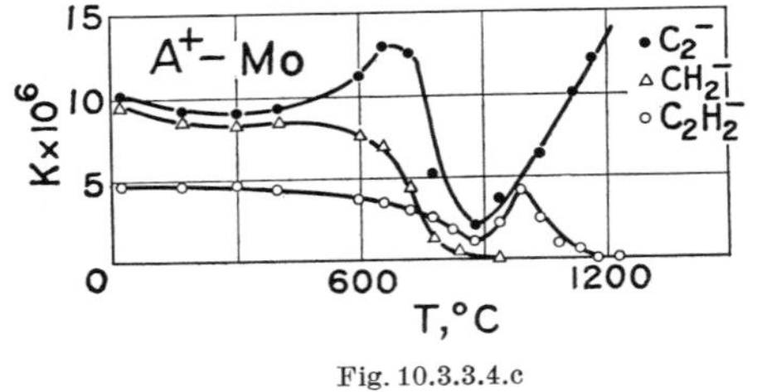

Fig. 10.3.3.4.c

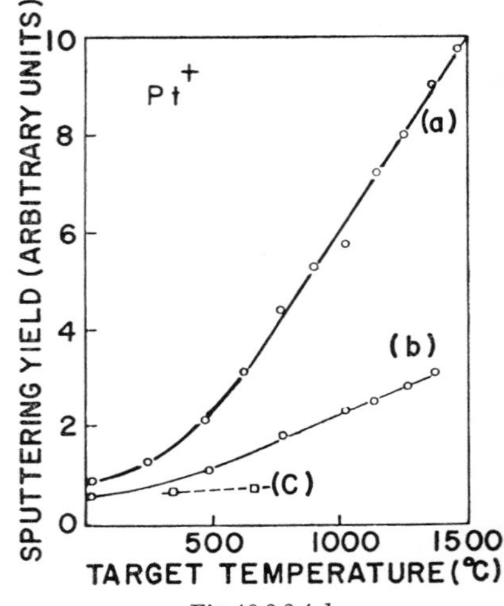

Fig. 10.3.3.4.d

Fig. 10.3.3.4.c. The dependence of the secondary-ion emission-coefficient K on the target temperature for the case in which C_2^-, CH_2^-, and $C_2H_2^-$ ions were ejected from a Mo target bombarded with 22-keV A^+ ions (FOGEL et al. [217])

Fig. 10.3.3.4.d. The dependence of the yield of sputtered Pt^+ ions on the target temperature for the cases of 870-eV Xe^+ ions (curve a) and of 1-keV A^+ ions (curve b) bombarding a Pt target (BRADLEY [91]). Curve c shows the results of WEHNER [771] for Pt^+ ion emission resulting from 300-eV Hg^+ ion bombardment of a Pt surface

Fig. 10.3.3.4.b shows values of the secondary-ion-emission coefficient K for emitted O^-, OH^-, H^-, and H^+ ions at different temperatures of the Mo target, which has been bombarded by A^+ ions with an energy of 22 keV. One sees that the emission of H^-, OH^-, and H^+ ions ceases when the temperature reaches roughly 850°—900° C. At this temperature the emission of O^- ions reaches a minimum and the emission of Mo^+ ions begins to fall off considerably, as can be seen in Fig. 10.3.3.4.a. As can be seen in Fig. 10.3.3.4.b, the emission of O^- ions ceases completely when the temperature exceeds 1400° C, whereas FOGEL et al. [217] report that the emission of negative ions of the molybdenum oxides (MoO_3^-, $Mo_2O_3^-$, MoO_2^-) already ceases above 1200° C. The CH_2^- emission ceases above about 850°—900° C (Fig. 10.3.3.4.c) while the emission of $C_2H_2^-$ and C_2^- ions shows a minimum in this temperature region. As the Mo target temperature increases further, however, the $C_2H_2^-$ emission again increases rapidly. These results suggest that certain chemical reactions at the metal surface may have to be invoked in order to explain these complex phenomena.

Bradley et al. [*91*] studied the yield of sputtered Pt^+ ions as a function of target temperature in the range from room temperature up to 1500° C by bombarding a Pt target with 870-eV Xe^+ ions (curve α) and with 1-keV A^+ ions (curve β). Also shown in curve c in Fig. 10.3.3.4.d are the results of Wehner [*771*] for the bombardment of Pt with 300-eV Hg^+ ions. As can be seen in Fig. 10.3.3.4.d, the curves (a) and (b) for the two primary ions indicate a larger positive temperature coefficient of the secondary Pt^+ ion emission over the temperature range investigated than has been observed for the corresponding yields for neutral particles (Section 10.3.1.2.3). Provided that the Pt surface was flashed at 1300° C before each run, no hysteresis in the function $K(T)$ could be observed.

Stanton [*674*], who bombarded a Be target with 1-keV ions of unknown composition from the background gas, observed a hysteresis in the secondary-ion currents of Be^+ when the target temperature was varied. An increase in the target temperature from 30° to 270° C led to a rapid increase in the number of secondary Be^+ ions. As the target was cooled, however, the ion yield decreased only slightly so that the yield at room temperature was approximately 5 times as large after a run as it was at the beginning.

Veksler et al. [*738*] investigated the function $K(T)$ for sputtered Ta^+ and Ni^+ ions when 2.62-keV primary Cs^+ ions bombarded Ta and Ni targets in the temperature range from room temperature to 1240° K for the Ni targets and to approximately 2000° K for the Ta targets. The temperature coefficient of the secondary-ion emission is positive for the Ta^+ and Ni^+ ions, except that it becomes negative at very high temperatures ($T > 1500°$ K) of the Ta target. For target temperatures between 900° and 1000° K for Ta, and 1000 to 1150° K for Ni, the ion yield rises steeply. This again may indicate a decrease in the surface contamination.

More systematic studies of the temperature dependence of secondary-ion-emission, with closer control of the surface conditions and of the pressure and composition of the residual gas, are necessary for a better understanding of the way in which gas adsorbed on the target surfaces influences the process of charged-particle emission.

10.3.3.5. Energy Distribution of Sputtered Ions

Measurements of the energy distribution of sputtered ions allow a better insight into the actual emission mechanism and may serve as a check of suggested sputtering mechanisms (e.g., the localized-hot-spot model with thermal emission of ions or a collison model). Unfortunately, the data now available are very scarce and, in most cases, were obtained only for gas-covered surfaces.

HONIG [*323*] studied the energy distribution of Ge^+ and Na^+ ions sputtered from a (111) plane of a Ge single crystal by Kr^+ ions in the energy range from 100 to 400 eV. The maximum of the distribution curve occurs at an energy of the order of 2 eV for both species, and the half width of the distribution curve is 2.5 eV; but he observed ions with energies even up to 20 eV.

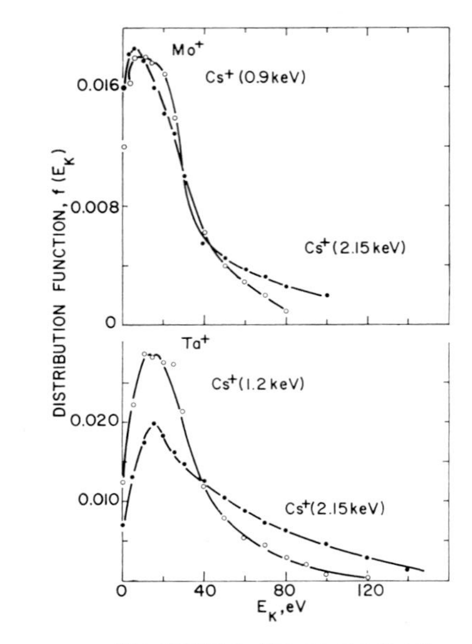

Fig. 10.3.3.5.a. The energy distribution $f(E_K)$ for Mo^+ and Ta^+ ions sputtered from Mo and Ta targets by Cs^+ ions of various energies (VEKSLER [*739*])

BRADLEY [*90*] made a mass spectrometric measurement of the energy distribution of Mo^+ ions sputtered from a molybdenum surface by 500-eV A^+ ions. The peak of this curve was in the neighborhood of 3—4 eV and its half width was about 5 eV. STANTON [*674*] found that the energy distribution of Be^+ ions sputtered from a Be target by 1-keV Ne^+ ions had a maximum at about 1 eV and a half width of approximately 5 eV. STANTON reported that these results remained the same for different primary ions such as A^+, Kr^+, Xe^+, N_2^+, and CO^+. However, the surface conditions in the experiments of all three authors are unknown, and in addition, the results have, in all probability, been influenced by the variation in the resolution and transmission of the mass spectrometer over the range of energies covered by the ion distribution.

In the studies of VEKSLER [*739*], experimental conditions such as the degree of surface coverage and the resolution and transmission of the mass spectrometer were better defined. He investigated the energy spectra of Cs^+, Mo^+, and Ta^+ ions sputtered from polycrystalline Mo and Ta targets (at about 1600° to 1800° K) by Cs^+ ions with energies ranging from 900 to 2150 eV. Fig. 10.3.3.5.a shows the energy distribution $f(E_K)$ for sputtered Mo^+ and Ta^+ ions for different energies of the primary Cs^+ ions (900 eV, 1200 eV, and 2150 eV). The half width of the curve $f(E_K)$ is found to be 30—35 eV for Mo^+ ions and 35—50 eV for Ta^+ ions. These half widths are several times as large as those observed in the work summarized above [*323, 90, 674*], and the maxima of $f(E_K)$ fall at higher ion energies (5—20 eV). As the energy of the primary ions increases, the maximum of the function $f(E_K)$ seems to shift to lower energies, an effect which is more pronounced in the case of Mo^+ ions than for Ta^+ ions. The explanation of this shift may be that more energetic primary ions penetrate more deeply into the target lattice, so that atoms sputtered from these deeper sites lose more energy by collisions in the lattice before they escape from the surface.

Veksler [*739*] pointed out that the half width of the function $f(E_K)$ was approximately the same for the sputtered Mo^+ ions as for the scattered Cs^+ primary ions. This observation indicates a relationship between the processes of sputtering and scattering, as will be discussed in Chapter 11. In view of the observed high values of the sputtered ion energies, it becomes apparent that target evaporation is a very unlikely explanation of the sputtering process.

10.3.4. Etching of Surfaces by Sputtering

The etching of crystal surfaces by ion bombardment was observed a long time ago (e.g., Stark et al. [*675*], Feitknecht [*209*], Baum [*56*]) but it is only recently that this field has attracted enormous attention and has been the subject of numerous investigations [*107, 133, 32, 46, 296, 297, 531, 532, 539, 555, 752a, 789, 790*]. In comparison with the conventional chemical or electrolytic method, the ionic method of etching has the great advantage of being applicable to practically all metals and alloys. Moreover, it allows the etching process to be closely controlled by varying the angle of incidence of the ion beam, the target temperature, the primary-ion energy, and the current density. It is therefore more versatile than chemical etching for certain metallographic applications. In the following, the most pertinent data for polycrystalline and single-crystal surfaces will be briefly discussed.

10.3.4.1. Etching of Polycrystalline Metal Surfaces

When a polycrystalline target is bombarded with ions, the grain boundaries in the surface become more apparent (Haymann [*296, 297*]), as has been confirmed for such different targets as Ta, Pt (Wehner [*762*]), Ag (Ogilvie et al. [*532*]), U, Th, and their alloys (Armstrong et al. [*32*]), uranium oxide, aluminum oxide (Bierlein et al. [*73*]), and Cu (Balarin et al. [*46*], Yurasova [*789*]). Wehner [*762*], who bombarded a Ta target with low-energy Hg ions, shows a micrograph of a region where three differently oriented grains join. Another rather common feature of ionically etched polycrystalline surfaces is the etch hillocks (Wehner, Balarin et al. [*46*], Cunningham et al. [*133*]). Fig. 10.3.4.1.a, an electron micrograph of a Pt target which has been bombarded by 130-eV Hg^+ ions under normal incidence, shows such etch hillocks

(Wehner [*762*]). Cunningham et al. [*133*] showed that if the angle of incidence of 8-keV A^+ ions varied between 60° and 75° with respect to the surface normal, the bombarded gold and aluminium surfaces con-

Fig. 10.3.4.1.a. Electron micrograph of a Pt target which had been bombarded with 130-eV Hg^+ ions at normal incidence (Wehner [*762*])

sisted of hillocks arranged in rows parallel to the direction of the ion beam. Fig. 10.3.4.1.b, an electron micrograph of such a bombarded polycrystalline gold surface (angle of incidence $\alpha = 70°$), shows a surface covered with grooves oriented in the beam direction. Balarin

Fig. 10.3.4.1.b. Electron micrograph of polycrystalline gold bombarded with 8-keV A^+ ions for 120 min at an angle of incidence of 70° (CUNNINGHAM et al. [*133*])

and HILBERT [*46*], who bombarded a polycrystalline Cu surface with 40-keV $^{24}Mg^+$ and $^{25}Mg^+$ ions at an angle of incidence of 70°, confirmed these results.

10.3.4.2. Etching of Single-Crystal Surfaces

The appearance of elongated grooves has been observed on single-crystal surfaces after bombardment with ions at high as well as low

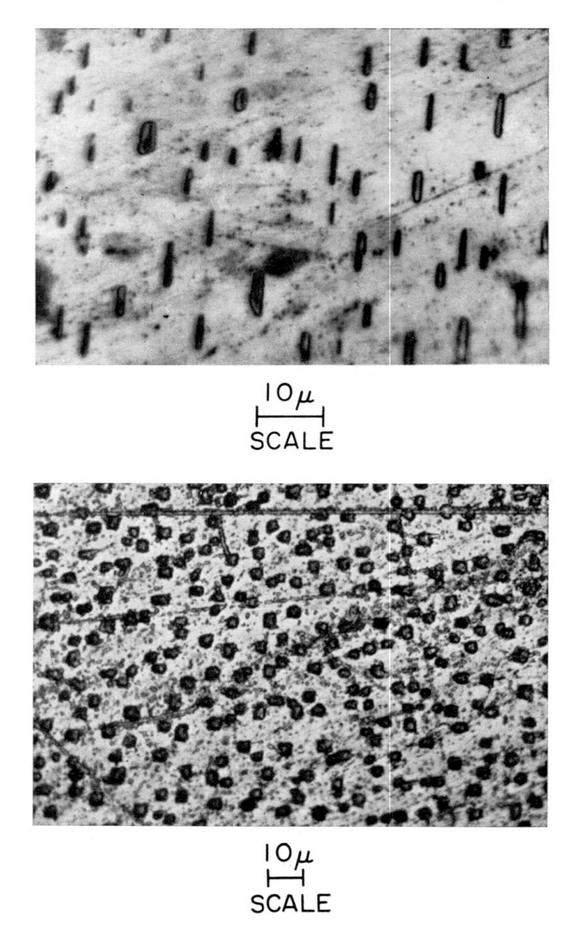

Fig. 10.3.4.2.a. Electron micrographs of a Cu (110) plane bombarded by 800-keV D^+ ions (Rutherford collision region) at an angle of incidence $\alpha = 45°$ for nearly 110 μA-hr (upper micrograph) and for normal incidence for nearly 150 μA-hr (lower micrograph). The elongated grooves which are noticeable in the upper micrograph ($\alpha = 45°$) are oriented along the beam direction (Kaminsky [*361*])

energies and incident at an oblique angle. After a high-energy ion bombardment (0.8-MeV D^+ ions—Rutherford collision region) of a (110) plane of copper at an angle of incidence $\alpha = 45°$, Kaminsky [*361*] observed the appearance of elongated grooves oriented along the direction of the beam, as may be seen in Fig. 10.3.4.2.a. (For comparison, the same surface is shown after D^+ ion bombardment at $\alpha = 0°$.) After a low-energy ion bombardment (8-keV A^+ ions—hard-sphere collision region) of a (0001) plane of a zinc crystal CUNNINGHAM et al. [*133*] also reported the appearance of elongated grooves oriented along the direction of the beam. Fig. 10.3.4.2.b shows the etch structure on an aluminum (100)

Fig. 10.3.4.2.b. Electron micrograph of an Al (100) plane bombarded with 8-keV A^+ ions for 85 min at an angle of incidence of 70° (CUNNINGHAM et al. [*133*]). The black arrow indicates the direction of the incident beam

plane after bombardment with 8-keV A^+ ions incident at 80°. One notices large arrow-shaped patterns, the ionic attack being from the wider ends of the hillocks towards their points (see black arrow in Fig. 10.3.4.2.b). This might be taken as visual evidence that the sputtering process is governed by the transfer of momentum (vectorial process) from the bombarding ions to the solid and not simply by the transfer of energy. Another well-established feature in the ionic etching of single-crystal surfaces is that the ionic attack is greatest in close-packed crystal planes, a result which is related to the observation of

Fig. 10.3.4.2.c. Electron micrograph of (110) planes of a silver single crystal, which are approximately parallel to the macroscopic surface and have been bombarded for some hours with 150-eV Hg^+ ions (WEHNER [*760*])

preferred ejection directions of sputtered particles (Sections 10.3.1.2.7, 10.3.2.1, 10.3.2.2).

Fig. 10.3.4.2.c shows an electron micrograph (WEHNER [*760*]) of the etch structure on a (110) plane of a silver crystal after several hours of bombardment with 150-eV Hg ions. The most material is removed from those zones in which the (111) planes coincide with the target surface and the least material is removed where the (110) planes are parallel to the surface. Corresponding results are obtained when a silver (111) plane, which is etched in such a way that the three neighboring (110) planes spaced 60° apart are developed (WEHNER [*762*]), is bombarded under similar conditions.

The etch structure of single-crystal surfaces becomes more pronounced at lower ion energies, as was shown by YURASOVA [*789*] for the case of Kr^+ ion bombardment of a copper (110) plane with ion energies of 300 eV and 6 keV.

The ion bombardment of a surface not only removes target material but may disturb the orientation of the crystal lattice near the surface. This effect has been observed by several investigators [*202, 203, 204, 531, 675, 729*]. TRILLAT [*729*] observed that a single-crystal film of gold is progressively disoriented by 12-keV ions until the single-crystal film is converted to a random polycrystalline one. OGILVIE [*531*] observed that when the (110) and (100) planes of a silver crystal were bombarded by A^+ ions in the energy range of 12 eV to 4000 eV, the surface became partly covered by a layer of crystallites with linear dimensions of about 100 Å. The crystallite face orientations are usually related to the original orientation by rotation about the [112] axis lying in or near the surface plane. OGILVIE suggests that the net current of interstitial atoms diffusing away from the primary dislocation causes the tilting of the lattice blocks.

An indirect confirmation of the production of such dislocation lines in the target by ion bombardment was obtained by SOSNOVSKY [*669*]. He investigated the decomposition of formic acid by the catalytic action of silver single-crystal surfaces, after the target surfaces had been bombarded by A^+ ions. The catalytic activity of the surface was altered considerably by the bombardment. Since the surfaces remained smooth, the changes were attributed to an increase in the number of dislocation lines intersecting the surfaces.

10.3.5. Structural Changes in Pure Metals or Alloys as a Result of Ion Bombardment

The composition changes of alloys and structural changes of pure metals by an ion bombardment of the target surfaces has been investigated by several authors [*33, 110, 214, 236, 290, 413, 533*]. When a binary alloy is bombarded with positive ions of an inert gas, in many instances one of the metals is initially removed more rapidly than the other. Consequently, a thin layer of alloy of a new and uniform composition is formed on the surface (GILLAM [*236*]). ASADA et al. [*33*] found that ion bombardment of Cu alloyed with very small amounts of Au sputtered the gold faster than the copper so that the surface was enriched in copper. When Cu_3Au alloys were bombarded with A^+ ions at energies up to 5 keV, however, OGILVIE [*533*] observed that the copper is removed more readily than gold. Since the ratio of sputtered Cu to sputtered Au atoms is more than 3:1, the surface layers became enriched in gold. After such a new surface layer has been formed, subsequent bombardment under the same conditions gradually eroded the surface but left the composition and the thickness of the layer unchanged. However, the thickness of such an altered layer depends on the energy and size of the incident ion (e.g., the thickness of the layer in a Cu_3Au target is 30 Å for 80-eV A^+ ions and 40 Å for 4000-eV A^+ ions). OGILVIE also observed that the orientation of the crystal planes in the altered layer is nearly the same as that in its substrate, though some damage in the form of disorientation is present. FISHER et al. [*214*] observed that when the alloy consisting of 70% brass, 30% stainless steel was sputtered by krypton ions the sputtered material had the same composition as the original alloy. HANAU [*290*] found similar results for aluminum alloys.

10.4. Theoretical Treatments of the Sputtering Process

Several models have been suggested to describe sputtering processes. Some of the more basic mechanisms on which such models rest are (*a*) a local evaporation process of the target material, either as a result of heating by the incident ion at the point of impact (the "hot-spot" model) or by some energetic primary recoil atoms of the lattice (the "heated spike" model) and (*b*) a momentum transfer from the incident particle to an atom of the target lattice which in turn may transfer momentum in successive collisions and eventually may cause a cascade of atomic collisions of other lattice atoms (the collision model).

10.4.1. "Hot-Spot" Model—"Heated Spike" Model

In the "hot-spot" theory, due originally to VON HIPPEL et al. [*315, 316*] and developed later by TOWNES [*725*], the energy of the impinging particle raises the temperature of a small region (of atomic dimensions) around the point of impact to such a high temperature that sputtering takes place by an evaporation process. The derivation of the relationship between the number of sputtered particles and the energy transferred to the target surface is based on macroscopic concepts of temperature and heat flow. As pointed out by PEASE [*550*], the legitimacy of such a model is very doubtful unless the hot spot has linear dimensions larger than the mean free path of electrons and photons in the solid and a duration longer than the period of thermal oscillations ($\approx 10^{-13}$ sec). This theory can be rejected for metal targets, however, on the basis of the experimental evidence presented in the preceding sections. For example, a decrease of the sputtering yield with increasing energy in the RUTHERFORD-collision region has been observed (KAMINSKY [*361, 363*]), contrary to the predictions of the hot-spot theory. In the very low and the high energy ranges, the angular distribution of the sputtered particles is not represented by a cosine law (see Sections 10.3.1.2.7 and 10.3.1.4) as the thermal-evaporation theory would predict. Also the strong dependence of the sputtering process on the angle of incidence cannot be explained by an evaporation process. Finally, the energy distribution of the sputtered particles deviates from the Maxwellian distribution expected for an evaporation mechanism and the mean energies of the sputtered particles are orders of magnitude larger than thermal energies. However, for such cases as the bombardment of thin target foils by high-energy ions, the emission of target particles may occur by the superposition of an evaporation and a sputtering process.

The hot-spot theory has been discussed in more detail by FRANCIS [*222*] and by WEHNER [*760*].

Another target evaporation process by "heated spikes" has been invoked by THOMPSON and NELSON [*721a*] to interpret their results on the energy spectrum of sputtered particles ejected from a polycrystalline gold target under A^+ and Xe^+ ion bombardment at an angle of incidence of 70°. They observed a low-energy peak at 0.15 ± 0.03 eV for 43-keV Xe^+ ion bombardment and 0.20 ± 0.05 eV for 42.5-keV A^+ ion bombardment and found that in the cases of Xe^+ the peak accounts for $12^0/_0$ of all sputtered particles and for $4^0/_0$ in the case of A^+ ion bombardment. The authors based the theoretical treatment of their results on the concept of localized heated spikes.

Such localized heated spikes had been considered earlier by SEITZ and KOEHLER [*649*] and by BRINKMAN [*94a*] who named them "thermal spikes" or "displacement spikes", respectively, according to whether or not displaced atoms accompanied the heating effect. According to the model, an energetic particle (e.g., a primary recoil atom of the lattice) moving through a lattice will rapidly transfer energy to its neighbors which then may become highly excited. In a macroscopic approach it is conceived that the cooling off of a group of excited lattice atoms may be treated by the classical theory of heat conduction. In metals the ordinary heat conductivity is mainly the sum of contributions from the lattice (ion-ion collisions) and the conduction electrons (electron-electron collisions).

THOMPSON and NELSON [*721a*] assumed that the coupling between lattice ions and electrons is sufficiently loose so that the excitation in the spike, starting entirely in the lattice, will not be transferred to the conduction electrons during the important phase of the spike. In an approximation, the spike is therefore assumed to cool only by the thermal lattice conductivity (ion-ion collisions). According to the classical theory of heat conduction, a hot region of linear dimensions r cools in a characteristic time τr^2, where $\tau = c\varrho/4\varkappa$, c being the specific heat, ϱ the mass density, and $\varkappa$ the thermal conductivity.

The authors estimated the decay time of a spike under the additional crude assumption that a metal region disordered locally by a thermal spike acts thermally as an insulator (they chose $\varkappa = 10^{-2}$ cal cm^{-1} sec^{-1} deg^{-1}, as for porcelain rather than $\varkappa = 0.75$ cal cm^{-1} sec^{-1} deg^{-1} as for Au). Their estimated value $\tau r^2 \approx 10^{-11}$ sec for the decay time of the spike was obtained by choosing $r = 10^{-6}$ cm, a value taken to represent the order of magnitude of the maximum range of a collision sequence in a monocrystalline gold target under experimental conditions similar to theirs (although they actually used a polycrystalline target). Since this rough estimate of the decay time of the spike is approximately 100 periods of thermal oscillation, the authors consider the application of such a "local heating" model to be legitimate. It should be noted, however, that the choices of the values of $\varkappa$ and r are arbitrary. Some other equally likely choice of $\varkappa$ and r values (e.g., $\varkappa = 0.50$ cal cm^{-1}sec^{-1}deg^{-1} and $r = 5 \cdot 10^{-7}$ cm) reduces the decay time of a spike to the order of magnitude of the period of thermal oscillation.

In order to determine the number of evaporated atoms per incident ion in the energy interval dE at E (the differential sputtering ratio for the thermal emission), additional assumptions had to be made. For instance, the spike was assumed to be a sphere of radius r over which the "temperature" T_1 is constant; the spike radius was assumed in a first approximation to be constant and independent of the energy of the moving particle (i.e., independent of the energy of the primary recoil atom); and the flux of atoms evaporationg from a surface area at "temperature" T_1 was calculated by MAXWELL-BOLTZMANN statistics. The authors derived an expression for the differential sputtering ratio for thermal emission; and by fitting the calculated values to the observed results, values for the spike temperature T_1 have been obtained. For the Xe^+ bombardment, for instance, the best fit to the observed peak at 0.15 eV was obtained with a spike radius $r = 110$ Å; for this energy and spike radius, they obtained $T_1 = 1750 \pm 300°$ K.

In view of the rough estimates and numerous assumptions made, the quantitative reliability of the theory seems questionable. However, certain qualitative features seem suggestive. THOMPSON and NELSON point out that in their experiments the thermal mechanism of sputtering contributed far fewer sputtered particles than did the mechanism of focused collision sequences; but the thermal mechanism should become more significant for sputtering by high-energy heavy-ion bombardment. In contrast, in his experimental observations on sputtering in the RUTHERFORD collision region (ions with high initial energy and low mass, e.g., 2-MeV D^+), KAMINSKY [*363b*] found that the density distribution of sputtered material indicated no detectable contribution from thermally ejected particles. That is, he found that the distribution in an individual spot was Gaussian rather than cosine as mentioned in Section 10.3.2.1.1.

10.4.2. Collision Models

The concept of momentum transfer from the incident particle to the lattice atom was suggested by LAMAR and COMPTON [*415*]. Since then, a number of theories have been derived on the basis of such a concept of of momentum transfer. The first types of collision to be discussed are those occurring for incident ions in the nonrelativistic velocity range.

The interaction potential between the incident particle of mass m and a "stationary" lattice atom of mass M is the screened COULOMB potential between the two nuclei. This potential has the form

$$V(r) = \frac{Z_1 Z_2}{r} e^2 \exp\left(\frac{-r}{a}\right), \qquad (10.4.2\text{–}1)$$

where r is the separation of the two nuclear charges, a is the screening radius of the orbital electrons, and Z_1 and Z_2 are the atomic numbers of the incident particle and the target atom, respectively (BOHR [*85*]). For the radius a, which can be chosen to fit the THOMAS-FERMI field, BOHR suggested the relation.

$$a = \frac{a_0}{(Z_1^{2/3} + Z_2^{2/3})^{1/2}}, \qquad (10.4.2\text{–}2)$$

in which a_0 is the BOHR radius of the hydrogen atom ($\approx 0.57 \times 10^{-8}$ cm). Another important scattering parameter is the collision diameter b. It is the distance of nearest approach of the two nuclei in a head-on collision in the absence of screening (RUTHERFORD scattering). The parameter b is given by the relation

$$b = \frac{2 Z_1 Z_2 e^2}{\mu V^2}, \qquad (10.4.2\text{–}3)$$

where $\mu = mM/(m + M)$ is the reduced mass and V is the velocity of the incident particle.

Depending on the energy of the incident particle, three types of interaction need to be considered. (1) Particles with high energies interact through the Coulombic repulsions of their nuclear charges (RUTHERFORD collisions). (2) At medium energies the electron clouds cause a partial screening of the positively charged nuclei and the collisions are treated as weakly-screened COULOMB collisions. (3) At low energies there is little penetration of the electron clouds and the collisions are approximately of the hard-sphere type.

In addition, the principal mode of energy loss for fast-moving charged particles in a solid is by collisions raising electrons to excited states. According to SEITZ [*648*, *649*] this mode of energy loss exceeds all others by a large factor ($\approx 10^3$) when the kinetic energy E of the incident particle is greater than a limiting energy E_c and is zero when $E < E_c$.

In metals, inelastic collisions become infrequent when the atomic velocities become small compared with the velocities of the electrons of the FERMI level. The limiting energy E_c can be approximately represented by

$$E_c = \frac{1}{16} \frac{m}{m_e} W_i \,, \qquad (10.4.2\text{–}4)$$

where W_i is the FERMI energy of the free electrons and m_e is the electron mass. As illustrations, the limiting energies E_c for bombardment of a silver target by various ions have been calculated to be 0.7 keV for H^+, 1.3 keV for D^+, 2.6 keV for He^+, 9.1 keV for N^+, 10.4 keV for O^+, and 13.1 keV for Ne^+.

Finally, for slow-moving ions the effect of trapping in the lattice structure of the target and the conversion of translational energy into the vibrational energy of the solid structure become important.

RUTHERFORD *Collisions*

The laws of classical RUTHERFORD scattering will apply down to small angles θ if the relationships

$$\frac{b}{a} \ll 1 \quad \text{and} \quad b \gg \frac{\hbar}{\mu V}$$

are satisfied. Such RUTHERFORD collisions predominate when the energy E of the incident ion is greater than a limiting energy E_B given by the relation (e.g., KINCHIN and PEASE [*379*])

$$E_B = 4 E_R^2 Z_1^2 Z_2^2 (Z_1^{2/3} + Z_2^{2/3}) \frac{m}{M} \frac{1}{E_d} \,, \qquad (10.4.2\text{–}5)$$

where E_R is the RYDBERG energy for hydrogen (13.68 eV) and E_d is the energy required to displace an atom from its lattice site. For many metals E_d lies between 20 and 25 eV (e.g., for Ag, $E_d = 21$ eV; for Cu, $E_d = 25$ eV). Calculated values for E_B for different bombarding particles on various targets are given in Table 10.4.2.A.

For the energy range $E > E_B$, BOHR [*85*] expresses the cross section σ_d for a RUTHERFORD collision in which an energy greater than E_d is transferred as

$$\sigma_d = 4 \frac{m}{M} Z_1^2 Z_2^2 E_R^2 \left(\frac{1 - E_d}{\Lambda E} \right) \frac{\pi a_0^2}{E_d E} \,, \qquad (10.4.2\text{–}6)$$

where

$$\Lambda = \frac{4\,m\,M}{(m+M)^2}\,.$$

Because the term $(1 - E_d/\Lambda E)$ is close to unity in the energy range considered, the mean recoil energy $\bar{E}$ of the struck atom is

$$\bar{E} = E_d \ln \frac{E_{\max}}{E_d} \quad \text{for} \quad E_{\max} \gg E_d\,, \qquad (10.4.2\text{–}7)$$

with

$$E_{\max} = \frac{4\,m\,M}{(m+M)^2}\,E\,. \qquad (10.4.2\text{–}8)$$

Weak-Screening Collisions

If the kinetic energy E of the incident particle lies within the energy range $E_A < E < E_B$, the screening of the nuclei by the orbital electrons becomes important. The energy limit E_A is given by the relation (Bohr [85])

$$E_A = 2\,E_R Z_1 Z_2 (Z_1^{2/3} + Z_2^{2/3})^{1/2}\,\frac{m+M}{M}\,. \qquad (10.4.2\text{–}9)$$

Values of E_A for different combinations of incident particle and target are given in Table 10.4.2.A.

Table 10.4.2.A. *Limiting energies for ionic collisions. (Hardsphere region; $E < E_A$; weak-screening region, $E_A < E < E_B$;* Rutherford*-collision region $E > E_B$)*

Bombarding ion	Silver target		Copper target	
	E_A, keV	E_B, keV	E_A, keV	E_B, keV
H^+	4.8	10.2	2.6	4.1
H_2^+	9.9	85.0	5.4	34.8
D^+	4.9	20.4	2.6	8.2
He^+	10.1	160.0	5.6	69.1
N^+	41.1	8.26×10^3	24.4	3.52×10^3
O^+	48.5	12.58×10^3	29.0	5.39×10^3
Ne^+	63.7	25.70×10^3	39.0	11.13×10^3
A^+	140.5	186×10^3	93.4	82.6×10^3
Xe^+	796	748×10^3	632	353×10^3

The displacement cross section in this energy range is roughly constant and is given by $\sigma = \pi a^2$, where a is given by relation (10.4.2–2). The mean recoil energy of the struck atom is

$$\bar{E} = E_d \left(\frac{E_B}{E}\right) \ln\left(1 + \frac{4\,E^2}{E_A{}^2}\right). \qquad (10.4.2\text{–}10)$$

Hard-Sphere Collisions

The collisions are approximately of the hard-sphere type if the conditions $b/a \gg 1$ and $a \gg \lambda$, with $\lambda = \hbar/\mu V$ are fulfilled. The collision cross section for the energy range $E < E_A$ is given by

$$\sigma_d = \pi R^2, \qquad (10.4.2\text{–}11)$$

where R is the distance of closest approach under hard-sphere conditions and is obtained by solution of

$$E = \frac{Z_1 Z_2 e^2}{R} \exp\left(-\frac{R}{a}\right). \qquad (10.4.2\text{–}12)$$

The solution can be expressed in the form

$$R = \frac{a}{(Z_1^{2/3} + Z_2^{2/3})^{1/2}} \ln\left(\frac{Z_1 Z_2 e^2 (m + M)}{4\pi R E M}\right). \qquad (10.4.2\text{–}13)$$

The calculated values of R for a number of target materials bombarded by noble-gas ions of different energy can be read from a graph published by ALMEN and BRUCE [*10*]. The cross sections given by Eqs. (10.4.2–11) and (10.4.2–12) should be considered only as rough estimates, because effects such as closed-shell repulsion are neglected. In the hard-sphere approximation, the displacement cross section σ_d is given by PEASE [*550*] as

$$\sigma_d = (1 - E_d/E_{\max})\, \pi R^2, \qquad (10.4.2\text{–}14)$$

where $E_{\max}$ (the maximum energy that can be imparted to the stationary atom) is given by relation (10.4.2–8). The energy distribution of the struck atoms is assumed to be uniform in the energy range $0 < E \leq E_{\max}$ so that the mean energy is

$$\bar{E} = \tfrac{1}{2} (E_{\max} + E_d)\,. \qquad (10.4.2\text{–}15)$$

10.4.2.1. Collision Theories for the Energy Range $E < E_A$ (Hard-Sphere Region)

KEYWELL [*375*] modified a neutron cooling theory so that it could be used to calculate how the incident particle transfers its energy in successive collisions to the target atoms. As shown schematically in Fig. 10.4.2.1.a, the incident particle penetrates the surface and travels its mean free path before it strikes an atom. If the kinetic energy E of the incident particle is greater than the displacement energy E_d, it will displace a target atom from its lattice site as a "primary knock-on". The displaced atom, in turn, collides with neighboring lattice atoms and produces secondary displacements. The swiftly recoiling target atoms continue this process until the successive collisions reduce their energy, as well as the energy of the impinging particle, to a value below the displacement energy E_d.

KEYWELL assumed that the collisions are of the hard-sphere type and that the number of sputtered atoms could be related to the number of displaced atoms. He neglected the fact that in the energy range $E_c < E < E_A$ the incident ion loses most of its energy to electronic excitation. He also considered the successive collisions of the lattice atoms to be uncorrelated and neglected an influence of the ordered crystal structure. For the energy region $E < E_A$ (actually valid only for $E < E_c$), he expresses n_d, the number of atoms displaced as a result of the nth collision, in the form*

$$n_d = \left(\frac{\bar{E}_n}{E_d}\right)^{1/2}, \quad (10.4.2.1\text{–}1)$$

where $\bar{E}_n$ is the average energy of the lattice atom displaced by the nth collision of the primary atom. From neutron cooling theory, KEYWELL derived the relation

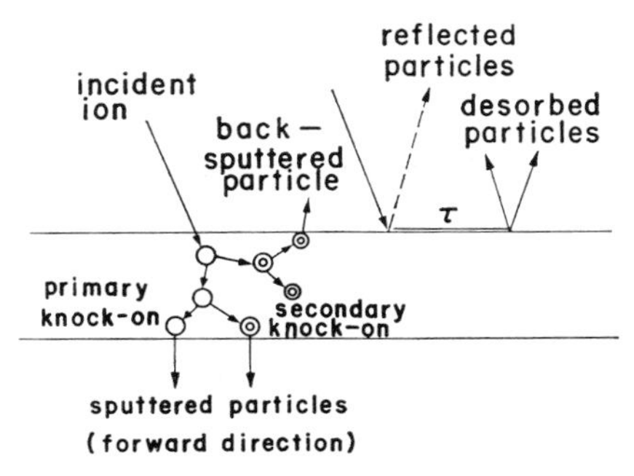

Fig. 10.4.2.1.a. Schematic diagram of some of the possible ionic impact phenomena on metal surfaces. (The time τ is the mean adsorption lifetime of the incident particle on the metal surface)

$$\bar{E}_n = \alpha E \exp\left[(1-n)\xi\right], \quad (10.4.2.1\text{—}2)$$

with

$$\alpha = \frac{2\,m\,M}{(m+M)^2} \text{ and } \xi = 1 - \frac{(m-M)^2}{2\,m\,M} \ln\left[\frac{m+M}{m-M}\right].$$

The total number N of displaced atoms, found by summing over all collisions, is

$$N = \sum n_d = \left(\frac{\alpha E}{E_d}\right)^{1/2} \sum_{n=1}^{n_c} \exp\left[-\tfrac{1}{2}(n-1)\xi\right], \quad (10.4.2.1\text{–}3)$$

with $n_c = \xi^{-1} \ln(\alpha E/E_d)$.

KEYWELL took account of the probability that an atom displaced from a distance x below the surface might be captured in the lattice before it could escape from the surface. This probability assumed to be proportional to $\exp(-\beta x)$. The depth x_n from which an atom starts as a result of the nth collision of the incident particle is $x_n = k\sqrt{n}$, the distance the incident particle penetrates in the course of n collisions along a random walk. Finally he considers that when the mass m of the incident ion is less than the mass M of the target atom, there exists a certain probability f_p that the incident particle penetrates into the target and a probability f_r that it rebounds on the first collision.

* It should be noted that KEYWELL uses the erroneous relation $n_d = \sqrt{\bar{E}_n/E_d}$ instead of the $\frac{1}{2}\bar{E}_n/E_d$ given by SEITZ. However, for ions incident at very low energies the difference between these expressions is not too large.

From such considerations, KEYWELL derived an expression for th sputtering ratio S, namely,

$$S = \left(\frac{\alpha E}{E_d}\right)^{1/2} \sum_{n=1}^{n_c} (f_p + 1.32\,\delta f_r) \exp\left[-(n-1)\frac{\xi}{2} - k n^{1/2}\right], \quad (10.4.2.1\text{–}4)$$

where

$$\delta = \begin{cases} 0 & \text{for} \quad n \neq 1 \\ 1 & \text{for} \quad n = 1\,. \end{cases}$$

The parameter k cannot be obtained experimentally and must be adjusted to fit the data. Values of the probabilities f_p and f_r can only be estimated roughly.

According to Eq. (10.4.2.1–4) the sputtering ratio S varies as $E^{1/2}$ in first approximation. For a certain energy range within the limits $E < E_A$, this is in agreement with some experimental results (Section 10.3.1.2.1). However, MOORE et al. [*529*] claimed that their experimental data for the sputtering of silver targets by 10-keV hydrogen ions could not be interpreted by the expression derived by KEYWELL [*375*]. This, however, is not surprising since the kinetic energy E of the incident protons in the experiment of MOORE et al. is near the energy limit E_B. This is on the limit of the RUTHERFORD-collision region where KEYWELL's theory is not applicable. Yet at very low ion energies (just above the sputtering threshold, where the theory should be valid), some experimental values (Section 10.3.1.1) of the sputtering ratio vary as the square of the kinetic energy instead of the square root. Such an E^2 dependence was predicted theoretically by LANGSBERG [*419*] but this law is not applicable in general. Adjoining this region in which the sputtering ratio is a quadratic function of energy, LANGSBERG finds a region in which the energy dependence is linear. This has been confirmed experimentally for several systems.

Another theoretical approach based on an even simpler collision model than the ones mentioned above, was given by ROL, FLUIT, and KISTEMAKER [*604*]. They assumed that only the first collision of the incident particle with the target atoms near the surface contributes to the sputtering process and neglected contributions due to the successive collisions of the displaced particles. The chance for a collision process near the surface is inversely proportional to the mean free path $\lambda(E)$, and the energy transferred in the first collision is proportional to the maximum energy $E_{\max}$ [Eq. (10.4.2–8)]. Thus, in first approximation, ROL et al. write

$$S \propto \frac{1}{\lambda(E)}\, E_{\max}\,, \quad (10.4.2.1\text{–}5)$$

so that

$$S = \frac{K}{\lambda(E)} \frac{m M}{(m + M)} E, \qquad (10.4.2.1\text{–}6)$$

where K is a constant which can be adjusted to fit the data and λ is given by

$$\lambda = \frac{1}{\pi R^2 n_0}. \qquad (10.4.2.1\text{–}7)$$

Here n_0 is the number of target atoms per unit volume and, in the hard-sphere model assumed by the authors, R is given by Eq. (10.4.2–13). The model used here also fails to take account of the effects of electronic excitation in the energy range $E_c < E < E_A$ and of anisotropy of the energy transfer within the lattice of a monocrystal. Despite the rather severe approximations made in the model, the authors show that the theoretical predictions are in fair agreement with their experimental results (Rol et al. [*604*]) for the sputtering ratio of copper bombarded with ions of several elements (i.e., N^+, Ne^+, A^+, Zn^+, D^+, and Ti^+) in the energy range from 5 to 20 keV. Almen and Bruce [*10*], who compared their experimental data with the values calculated from Rol's theory, found discrepancies up to 20%.

All the preceding theories neglected the effect of directional correlation between successive collisions, which is introduced by the ordered structure of the crystal. As mentioned in Section 10.3.2, Silsbee [*653*] has pointed out that in a monocrystal regarded as rows of equal hard spheres, momentum may be focused along a line of atoms in a close-packed crystal direction.

Fig. 10.4.2.1.b shows schematically such a line of hard and equal spheres along the x axis, with D_{hkl} as the spacing between the centers of the spheres along such a line (the subcripts h, k, l are Miller indices). The diameter d of a sphere is taken equal to the distance of closest approach of the atoms in a head-on collision. If two spheres collide at the collision point P, the knocked-on second sphere moves in the direction PC making an angle θ_1 with the x axis since momentum is transferred only in the direction of the radius vector between the centers of the two spheres. The incident atom rebounds perpendicular to the direction PC. For small values of the angle θ_0 between the x axis and the momentum of the incident atom, the relation between θ_1 and θ_0 can be shown to be

$$\theta_1 = \theta_0 \left(\frac{D_{hkl}}{d} - 1 \right). \qquad (10.4.2.1\text{–}8)$$

Silsbee focusing occurs whenever $f = \theta_{n+1}/\theta_n < 1$ or, in other words, whenever $D_{hkl}/d < 2$; the successive angles then become monotonically smaller. The final result is a sequence of head-on collisions without any change of angle, and energy losses along the chain may be neglected. In this case the collisions do not displace the lattice atoms from their original equilibrium positions and no mass transport occurs.

In the case of a face-centered cubic Cu monocrystal, SILSBEE's focusing condition $f < 1$ is fulfilled only for the most densely packed crystal directions, the [110] directions in which the experiments show the strongest sputtering effect (see Section 10.3.2.1.1). A focusing energy E_f, below which SILSBEE focusing occurs, can be defined as the

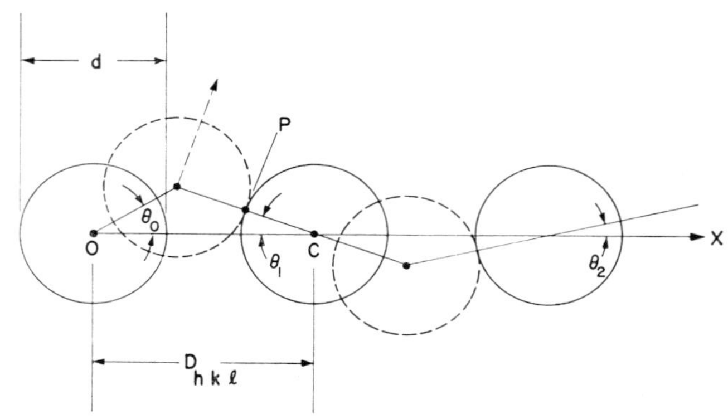

Fig. 10.4.2.1.b. Schematic illustration of SILSBEE's [*653*] mechanism for the momentum transfer along a line of hard and equal spheres. The distance D_{hkl} is the spacing between the centers of the spheres along the particular axis (subscripts h, k, l, are MILLER indices) and d is the diameter of a sphere

energy at which $f = \theta_{n+1}/\theta_n = 1$ and $D_{hkl}/d = 2$. If a BORN-MAYER repulsive potential energy $E_\varphi = A \exp(-r/a)$ is assumed to characterize the interaction between two ion cores in the lattice, then their diameter (taken as their distance of closest approach in a head-on collision) is given by

$$d = a \ln(2A/E) ; \tag{10.4.2.1–9}$$

and the focusing energy in the [110] direction is

$$E_f = 2A \exp(-D_{110}/2a) . \tag{10.4.2.1–10}$$

Using $A = 20$ keV and $a = D_{110}/13$ from data on the compressibility of copper (ref. [*330a*]), THOMPSON [*720*] obtained $E_f = 60$ eV for the focusing energy for the [110] direction in a Cu monocrystal, while the machine calculations of GIBSON et al. [*235*] led to only $E_f = 30$ eV.

However, as SILSBEE [*653*], LEIBFRIED [*427, 428*], GIBSON et al. [*235*], and THOMPSON [*720*] pointed out, the dominant energy attenuation in such focused collision sequences is due to interaction with atoms in adjacent close-packed lines. By moving up to collide with its neighbor, for instance along the [110] direction in a fcc crystal, the atom shortens the distance between itself and four atoms in neighboring lines. The resulting increase in the repulsive potential leads to an energy loss ΔE, which for the [110] direction in a fcc lattice can be represented by the expression (LEIBFRIED [*427*]),

$$\left(\frac{\Delta E}{E_f}\right)_{110} = 2\left[\exp - \left(\frac{3 D_{110}}{8 a}\right) - \exp\left(- \frac{D_{110}}{2 a}\right)\right]. \quad (10.4.2.1–11)$$

Reported values for the energy loss vary between $\frac{2}{3}$ and 1 eV per collision [*235, 720*]. The number n of collisions in a collision sequence initiated by an atom with energy E_f is given by $n = E_f/\Delta E$. This would be between 60 and 90 collisions for the Cu [110] direction and corresponds to a maximum range of about 206 Å to 309 Å. As already mentioned in section 10.3.2.1.3, the thermal vibration of lattice atoms about their equilibrium positions may cause the scattering of such focused collision sequences and effectively increase the energy loss per collision. NELSON et al. [*521g*] analytically treated the experimentally observed broadening of the [110] spots with increasing target temperature, using a vibration amplitude typical of surface atoms rather than those of lattice atoms in the bulk. When the angle of deviation is θ, the angle between successive collisions is 2θ because the momentum vector oscillates from side to side of the focusing axis. Thus, for small values of θ, one sees from the triangle of momenta that $\sin\theta \approx \theta \approx \sqrt{2 M \Delta E}/\sqrt{2 M E}$ so that the transverse momentum generated in each collision $2\theta\sqrt{2 M E}$. The energy loss due to this scattering effect is therefore $4\theta^2 E$ and the relative energy loss $(\Delta E/E_f)_{\text{therm}} = 4\theta^2 E/E_f$ would have to be added to the right-hand side of Eq. (10.4.2.1–11) to obtain the total energy loss per collision.

As briefly discussed in Section 10.3.2, another focusing mechanism has to be invoked in order to explain such experimental observations as the preferred ejection directions along the [100] and [111] directions of a fcc lattice, in which the spacings D_{100} and D_{111} are so large that SILSBEE's focusing condition $D_{hkl}/d < 2$ is not fulfilled. By machine calculations based on a model representing a copper lattice, VINEYARD et al. [*235*] obtained results indicating that when an atom A_1 moving at a small angle θ_0 to the [100] direction passes through a ring of four atoms B_1, B_2, B_3, and B_4 in a (100) plane (Fig. 10.3.2.a), it will undergo a deflection before hitting the next atom A_2 in the [100] direction. Whenever the angle θ_1' which the deflected particle makes with the [100] direction is smaller than the angle θ_0, focusing has occurred.

An analytical treatment given by NELSON and THOMPSON [*521a*] indicated, for a collision sequence along a line in the [100] direction, a surrounding ring of atoms $B_1 B_2 B_3 B_4$ in the (100) plane has a focusing effect formally similar to that of a converging lens in optics. They find that when an atom moving with velocity V in a straight line collides with a stationary lattice atom, it transfers a momentum

$$P = \frac{A D}{a V} \exp\left(-\frac{D}{4a}\sqrt{1 + \frac{16 b^2}{D^2}}\right), \qquad (10.4.2.1\text{–}12)$$

where the impact parameter b is in the range $D/\sqrt{3} \leqslant b < D/\sqrt{2}$, and A, a, and D have the meanings explained above.

As the atom A_1 passes through the ring of four B atoms, four glancing collisions occur simultaneously. If one makes the simplifying assumption that the recoil motion is confined to a (100) plane (the plane of $B_1B_3A_1$ (see Fig. 10.3.2.a), then momenta P_1 and P_3 are transferred in collisions with impact parameters b_1 and b_3. Then the other pair of collisions lead to equal and roughly opposite momenta transfers (P_2 and P_4) and do not cause any deflection. The net momentum change of the moving atom A_1 is therefore $(P_1 - P_3)$ leading to a deflection Ψ*) of the trajectory towards the [100] axis, with $\Psi = (P_1 - P_3)/MV$. For small angles θ_0 one obtains $b_1 = (D/\sqrt{2})\,(1 - \theta_0)$ and $b_3 = D\sqrt{2}\,(1 + \theta_1')$. Neglecting small quantities, Nelson and Thompson find

$$\Psi = \theta_0 \frac{2}{3} \frac{A a^2}{E D^2} \exp\,(-\,3D/4a)\,. \qquad (10.4.2.1\text{–}13)$$

Since Ψ is proportional to θ_0, the ring of atoms is considered as a lens with a focal length $f = D/\sqrt{2}\,(\theta_0/\Psi)$. For the [100] direction in a fcc lattice, this becomes

$$f_{100} = \frac{3}{2\sqrt{2}} \frac{E}{A} \frac{a^2}{D} \exp \frac{3D}{4a}\,. \qquad (10.4.2.1\text{–}14)$$

The focusing condition $\theta_1' < \theta_0$ (or, in other words, $4 f_{100} < D_{100}\sqrt{2}$) leads to a focus focusing energy E_{100} below which focusing along the [100] axis occurs. This energy is

$$E_{100} = \frac{A}{3} \frac{D^2}{a^2} \exp\left(-\frac{3D}{4a}\right). \qquad (10.4.2.1\text{–}15)$$

For copper, Nelson and Thompson calculated a focusing energy $E_{100} = 65$ eV and an energy loss per collision of about 3.4 eV. (They assumed a constant energy loss per collision.) These results indicate that the "lens focusing" mechanism is energetically less efficient than

---

*) At the intersection of a line connection the center of A_1 (see Fig. 10.3.2.a) with the center of the colliding particle (not A_2) the angle Ψ is formed by a line connecting this intersection with the center of the colliding particle and another line, which connects the center of A_2 and point P and extends to the intersection. The difference between the angle Ψ and ψ (shown in Fig. 10.3.2.a) is not large if θ_0 is small. The quantity a_0 in Fig. 10.3.2.a corresponds to the quantity $D/\sqrt{2}$.

SILSBEE's hard-sphere focusing. The calculations of NELSON and THOMPSON indicate that in a fcc lattice the effectiveness of atomic collision sequences in different directions decreases in the order [110], [111], [100]—a result which is in qualitative agreement with some low-energy sputtering observations (NELSON and THOMPSON [*521a*]). For additional theoretical treatments of focused collision sequences in fcc and bcc lattices, the reader is referred to articles by NELSON [*521h*], DEDERICHS and LEIBFRIED [*142a*], LEHMANN and LEIBFRIED [*426b*], and OEN et al. [*530a*].

For a discussion of focusing mechanisms other than the "lens" focusing mechanism (such as the channeling mechanism or the crowdion motion) the reader is referred to Section 10.3.2.

THOMPSON [720] has attempted to develop a quantitative sputtering theory which takes account of the emission of target particles in preferred directions and is supposed to be valid in a certain energy range of the hard-sphere collision region ($E < E_A$). However, he made several simplifying assumptions—e.g., he assumed that the range of the incident particle is considerably longer than the range of a focusing collision sequence in a given crystallographic direction (although the actual treatment is restricted to the emission in the [110] directions of a fcc lattice, it could be extended to other directions and structures) and that $Z_1 < Z_2/5$, where Z_1, Z_2 are the atomic numbers of incident particle and lattice atom, respectively. These assumptions restrict the application of the theory to light incident ions with energies closer to the energy limit E_A. THOMPSON's actual expression for the sputtering ratio S indicates that for $E \gg E_f$, where E_f is the focusing energy for a given crystallographic direction, S is nearly independent of the energy E of the incident particle. A typical estimated value of E_f is 60 eV for the [110] direction in Cu (ref. *521a*). This predicted behavior of the function $S(E)$ for the region $E < E_A$ is in agreement only with certain experimental results on sputtering by ions with very low mass (e.g., H^+, H_2, and D^+ on Ag) but disagrees with other results on sputtering such as those obtained by D_3^+ or He^+ ion bombardment of a Ag target, systems for which the theory should also be valid. One notices for instance that in Fig. 10.3.1.3.1.a the function $S(E)$ depends strongly on the incident-ion energy and runs through a maximum near $E = 5$ keV for He^+ on Ag ($E_A = 10.1$ keV), contrary to the theoretical expectations.

10.4.2.2. Collision Theories for the Energy Range $E_A < E < E_B$ (Weak-Screening Region)

PEASE [*550*] calculated values of the function $S(E)$ on the basis of Eq. (10.4.2.3–5) (the derivation of which will be given in more detail in Section 10.4.2.3) for a silver target bombarded by He^+, D^+, and H^+ ions with energies in the range $E_A < E < E_B$. By inserting the proper values for the displacement cross section σ_d in this energy region and the mean energy $\bar{E}$ of the struck atom [Eq. (10.4.2–10)] as well as the displacement energy E_d (21 eV for Ag) and the heat of sublimation E_s (3 eV for Ag), he obtained the values shown as dashed lines in Fig. (10.4.2.2.a). The values are compared with those determined experimentally by O'BRIAN et al. [*529*], which are also shown as the solid curves in Fig. 10.4.2.2.a. The theoretical values are low by a factor of 1.5 to 2. This discrepancy is not surprising, however, since Eq. (10.4.2.3–1) is more accurate for the region $E > E_B$ than for the region $E_A < E < E_B$, where, for instance, the energy spectrum of the primary knock-ons begins to change significantly with E. The general shape of the $S(E)$ curve seems however to be represented rather well by the equation. Since the displacement cross section σ_d is roughly constant in the energy region considered, the function $S(E)$ shows only slight changes with energy.

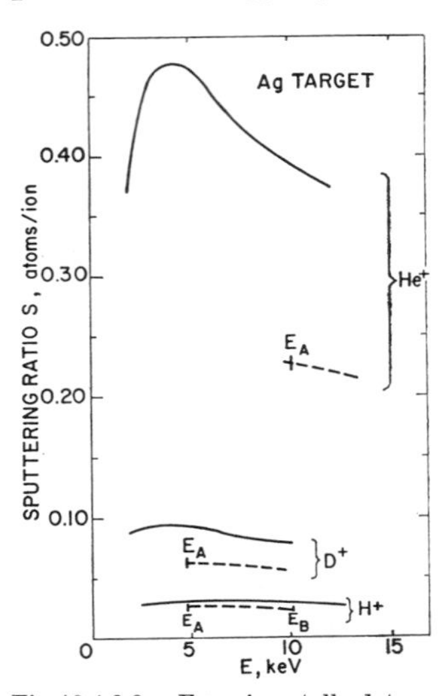

Fig. 10.4.2.2.a. Experimentally determined $S(E)$ values for He^+, D^+, and H^+ ion bombardment of a silver target (O'BRIAN et al. [*529*]) (solid lines) compared with theoretical values calculated by PEASE [*550*] (dashed lines). E_A and E_B indicate the limiting energies given by Eqs. (10.4.2–9) and (10.4.2–5)

THOMPSON [*720*] calculated sputtering ratios for a Cu target bombarded by D^+ ions in the energy range $E_A < E < E_B$ by interpolation of theoretical curves he had obtained for the energy region $E < E_A$ (discussed in the previous section) and the region $E > E_B$ (Section 10.4.3.2) for the corresponding system. The agreement between the interpolated curves for $E_A < E < E_B$ (several curves have been obtained for various BORN-MAYER potentials chosen) and the experimentally observed one by YONTS et al. [*787*] is rather poor. This result is not surprising since the interpolation method used is rather crude and the sputtering ratios calculated for $E > E_B$ depend strongly on the right choice of the parameters A and D/a of the BORN-MAYER potential for the Cu-Cu repulsion (see page 228).

10.4.2.3. Collision Theories for the Energy Range $E > E_B$ (RUTHERFORD-Collision Region)

As the energy of the bombarding ion increases above the energy limit E_B, the incident particle undergoes Coulomb collisions with the nuclei of the target atoms along its path and displaces them from their lattice sites as primary knock-ons. The primary knock-ons in turn may suffer hard-sphere collisions with neighboring lattice atoms and produce secondary displacements if their energy exceeds the displacement energy of a lattice atom (e.g., 25 eV for Cu, 21 eV for Ag) and/or they may transfer momentum along focusing collision sequences (which in turn may intersect the target surface and knock off target particles) and/or they may be channeled along "open" channels in the lattice (by the channeling mechanism discussed in Section 10.3.2) and reach the surface and escape. The cascades of collisions causing displacements continue until the energies of the displaced atoms fall to about the displacement energy.

For the energy region $E > E_B$, several simplifying features emerge in certain processes involved in the sputtering mechanism. For instance the energy spectrum of the primary knock-ons becomes a very insensitive function of the energy and the type of the incident particle. Furthermore, if the mean energy [defined in Eq. (10.4.2.–7)] of the primary knock-ons is much smaller than the maximum transferable energy $E_{\max}$ then the primary knock-ons are displaced normal to the direction of incidence. The mean free path λ of the incident particle increases greatly (e.g., Harrison [*292*] found that for 500 keV deuterons in Cu, $\lambda = 0.8 \times 10^{-4}$ cm) while the knock-on particles maintain an approximately constant mean free path ($\approx 10^{-7}$ cm). This increase of the mean free path λ of the incident ion reduces the fraction of the primary and secondary knock-ons that have a chance to reach the surface and escape; on the other hand, larger values of λ offer a better chance to study the ranges of focussed collision sequences in less perturbed surface layers of the target.

Several authors [*241*, *242*, *292*, *550*, *572*] have tried, on the basis of different concepts, to relate the number of sputtered atoms to the number of such displaced target atoms. Certain simplifying assumptions are commonly made in these theories. For instance, the anisotropy of the energy transfer within the lattice is neglected, as is the electronic excitation as a principal mode of energy loss for fast-moving charged particles in a solid. As long as $\bar{E} < E_c$, this effect will be important only for the energy loss of the incident particle.

Goldman and Simon [*241*] assumed that the number of primary knock-ons per incident particle is inversely proportional to the mean free path λ of the incident particle. The subsequent secondary collisions were treated as hard-sphere interactions and the sputtering ratio S was assumed to be proportional to the average number $\bar{\nu}$ of displaced secondary atoms. Under simplifying assumptions, corrections were made for displaced target atoms getting trapped in the lattice. These authors expressed the sputtering ratio S as

$$S = \frac{k\bar{\nu}}{\lambda} = \frac{0.17\,\bar{\nu}}{\sigma_d \lambda}, \qquad (10.4.2.3\text{–}1)$$

where $\sigma_d = \pi R^2$ is the cross section for hard-sphere collisions. By inserting the proper expressions for λ and $\bar{\nu}$ into Eq. (10.4.2.3–1) as described by SEITZ and KOEHLER [*649*] for the energy region considered, the authors obtain

$$S = 0.17\,\bar{\nu}\,\frac{Z_1^2 Z_2^2 e^4}{E E_d R^2}\,\frac{m}{M}\,\frac{1}{\cos\alpha}, \qquad (10.4.2.3\text{–}2)$$

where

$$\bar{\nu} = \left[0.885 + 0.561 \ln\left(\frac{x+1}{4}\right)\right]\frac{x+1}{x}, \qquad (10.4.2.3\text{–}3)$$

with

$$x + 1 = \frac{4\,m M}{(m+M)^2}\,\frac{E}{E_d}.$$

Here α is the angle of incidence and the other symbols are the same as in Eqs. (10.4.2–5) and (10.4.2–6). At high energies, x becomes large compared to unity and the sputtering ratio depends on the energy through the relation

$$S \propto \frac{m \ln E}{M E \cos\alpha}. \qquad (10.4.2.3\text{–}4)$$

Values for $S(E)$, calculated from Eq. (10.4.2.3–4) for a copper target bombarded by deuterons, are plotted as curve d in Fig. 10.4.2.3.a.

Applying the same model, GOLDMAN, HARRISON, and COVEYOU [*242*] performed a MONTE CARLO calculation and obtained the results shown as x's in curve e in Fig. 10.4.2.3.a. A comparison between curve d and these x-marked values shows a qualitative agreement for the function $S(E)$, but quantitatively, the latter values are smaller than those in curve d over the whole energy range.

HARRISON [*292*] used a modified model to derive an analytical expression which includes the equation of GOLDMAN and SIMON as a special case. The sputtering ratios calculated from HARRISON's theory are also shown on curve e (solid line) of Fig. 10.4.2.3.a.

According to PEASE [*550*], the sputtering ratio S is determined by three factors: the effective collision area available within a layer ($S \propto \sigma_d n^{2/3}$), the number of layers contributing to sputtering ($S \propto 1 + N^{1/2}$), and the total number of displaced atoms per primary knock-on ($S \propto \frac{1}{2}\bar{E}/2E_d$). Thus the sputtering ratio for normal incidence and for $2E_d < \bar{E} < E_{\max}$ is given by

$$S = \sigma_d n^{2/3}(1 + N^{1/2})\,\tfrac{1}{4}\,\frac{\bar{E}}{E_d}. \qquad (10.4.2.3\text{–}5)$$

Here σ_d is the displacement cross section for the energy region considered, n is the number of target atoms per unit volume, and N is the number of collisions the primary knock-ons make by hard-sphere collisions with other atoms in slowing down to an energy E_s corresponding to the heat of sublimation of the target.

If binary collisions are assumed in a random-walk calculation (CHANDRASEKHAR [*111*]), the total number of atomic layers (including the surface layer) contributing to sputtering is

$$1 + \sqrt{N} = 1 + \sqrt{\frac{\ln(\bar{E}/E_s)}{\ln 2}} . \qquad (10.4.2.3\text{–}5a)$$

By inserting this into Eq. (10.4.2.3–4) one gets

$$S = \sigma_d n^{2/3} \left[1 + \sqrt{\frac{\ln(\bar{E}/E_s)}{\ln 2}}\right] \cdot \frac{1}{4} \frac{\bar{E}}{E_d} . \qquad (10.4.2.3\text{–}6)$$

On the basis of Eq. (10.4.2.3–6) PEASE calculated the sputtering ratio S of silver targets bombarded by protons, deuterons, and helium ions. The curves b in Fig. 10.4.2.3.a represent such calculated values for deuterons bombarding copper and silver. (Values for Ag were taken from Fig. 1 of PEASE's article.) One notices that the values calculated by PEASE are larger than those calculated by the other authors [*241*, *242*, *292*] and also larger than the experimentally determined values. However, in such a comparison of the experimental and theoretical values one should keep in mind that the theoretical values have been calculated for polycrystalline surfaces while the experimental values have been determined for the (100) planes of monocrystalline copper and silver. (As discussed on Page 185 the ratios between the S values for polycrystalline and monocrystalline surfaces of the same material are expected to be smaller than a factor of three for the region $E > E_b$). In addition, in view of the crude assumptions made in the different theoretical treatments (neglecting for instance the action of focusing and channeling mechanisms) any close quantitative agreement between the theoretical values and those determined experimentally for monocrystals would seem rather fortuitous. Qualitatively, however, the theories seem to predict correctly that the sputtering ratio S decreases with increasing energy E nearly as ln E/E, and indicate the right trend for the variation of S with the target material (e.g., $S_{\mathrm{Ag}\,(100)} > S_{\mathrm{Cu}\,(100)}$).

In contrast to the random-collision models adopted in the above mentioned theories, THOMPSON [*720*] considered in his theoretical treatment that recoiling atoms may transfer momentum to atoms in close-packed crystal directions in focused collision sequences. In his model, the number of sputtered particles per incident ion depends mainly on the energy spectrum of the primary recoil atoms and the number of sequences produced by a primary recoil atom and traveling in a particular crystallographic direction. He restricted his actual calculations to fcc metals; a brief discussion of some simplifying assumptions made in the theory has been given earlier in Section 10.4.2.1. For the energy range $E > E_b$, THOMPSON calculated the function $S(E)$ for the system D^+ on Cu for three different sets of parameters A and D/a of the BORN-MAYER repulsive potential $[E_\varphi = A \exp(-r/a)]$ for copper. The calculated sputtering constants depend critically on the choice of the parameters A and D/a. The curve c in Fig. 10.4.2.3.a

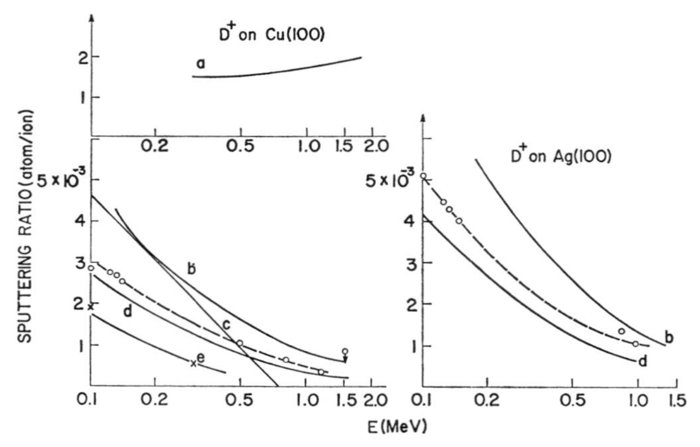

Fig. 10.4.2.3.a. Comparison of calculated sputtering ratios with those determined experimentally by bombarding the (100) planes of Cu and Ag monocrystals with deuterons at normal incidence for the energy region $E > E_B$ (RUTHERFORD collision region). Curve *a* represents the $S(E)$ values reported by PLESHIVTSEV [*572*]. Curves *b, c, d,* and *e* represent the $S(E)$ curves calculated by PEASE [*550*], THOMPSON [*720*], GOLDMAN and SIMON [*241*], and GOLDMAN et al. [*242*], respectively. The points *x* in curve *e* have been calculated by HARRISON [*292*]. The experimentally determined values marked by (○) ,and connected by a dotted line, were obtained by KAMINSKY [*361, 363*]

represents the values of $S(E)$, —calculated for the parameter values $A = 90$ keV and $D/a = 17$—, which come at least moderately close to the experimental values determined by KAMINSKY [*363b*]. The other calculated $S(E)$ curves, including the one calculated with the more commonly accepted parameter values $A = 22$ keV and $D/a = 13$ (see ref. [*235*], [*330c*]), give S values which are nearly one order of magnitude higher than the experimentally determined ones. In addition, one notices that the calculated function $S(E)$ in curve c drops off more sharply with increasing energy E than the experimentally determined function $S(E)$. The calculated function varies nearly as E^{-1}, while the experimental one seems to vary nearly as $\ln E/E$ in agreement with the theoretical predictions by PEASE [*550*], GOLDMAN et al. [*241, 242*], and HARRISON [*292*].

Contrary to the predictions made in the preceding theories, PLESHIVTSEV [*572*] claimed that the sputtering ratio increases with increasing energy in the energy region $E > E_B$. He tried to correlate the sputtering ratio with the total number of displaced particles, a number which will actually increase with increasing energy. However, as mentioned above, only a fraction of the total number of displaced particles will have a chance to reach the surface and to escape. PLESHIVTSEV's values of the sputtering ratios are therefore larger than those predicted by other authors. PLESHIVTSEV's values for the sputtering of copper by deuterons are given in curve *a* in Fig. 10.4.2.3.a. The values in curve *a* are 3 to 4 orders of magnitude higher than the calculated curves *b*, *c*, and *e* and they increase with increasing energy.

The experimentally-determined sputtering ratios (dotted line) (KAMINSKY [*361, 363*]) disagree with those reported by PLESHIVTSEV by three to four orders of magnitude. The data also show a decrease of the sputtering ratio with increasing ion energy, in qualitative agreement with theoretical predictions [*241, 242, 292, 550, 720*] but contrary to the findings of PLESHIVTSEV.

11. Ion Scattering from Metal Surfaces

11.1. Definitions: Ion-Reflection Coefficient and Secondary-Emission Coefficient

An important process during ion bombardment of a metal target is the possible reflection of the impinging ions from the target surface. A summary of the older results has been given by MASSEY and BURHOP [*461*]. The investigation of this phenomenon requires careful consideration of other processes which occur simultaneously. These include secondary emission of ions (e.g., reflected ions, sputtered ions, desorbed ions), electrons, and neutral particles, and surface effects (e.g., formation of surface layers by the incident particles being adsorbed at the surface and/or trapped in the target lattice [at the higher incident-particle energies], and ionic etching of the surface), which can have a significant effect on the measurements of ion reflection. In particular, it is necessary to distinguish the secondary-emission coefficient R from the ion-reflection coefficient ϱ concepts which are unfortunately used interchangeably by numerous authors (as explained in Section 11.2).

Fig. 11.1.a is a schematic illustration of the various particle currents (the ion current I^i, the electron current I^e, and the current I^m of metastable atoms) which can occur at target and collector when a primary-ion current I_1^i bombards the target. It has been assumed here that there is no potential difference between target and collector and that the target can be at high temperature (see for example, Section 11.3.6). In Fig. 11.1.a, I_2^i is the current of primary ions reflected from the target. This current includes both the elastically reflected ions (whose average energy is equal to that of the primary ions) and inelastically reflected ions (whose average energy is less than that of the primary ions). The available results on the energy distribution of the reflected ions (Section 11.3 5) indicate that the elastically reflected ions make a negligible contribution to I_2^i when the primary ions are normally incident on the target, but they become appreciable at the larger angles of incidence. The current I_d^i of thermally desorbed primary ions becomes significant at high target temperatures (see Chapter 8 on Surface Ionization). The characteristic current of the target particles sputtered as ions is denoted by I_s^i. This current may become important in certain cases, e.g., when noble gas ions at high energies are incident at large angles. (This contribution is discussed in more detail in Chapter 10.) Other particle currents shown in Fig. 11.1.a are the current I_2^e of secondary electrons ejected by the primary beam, the current of thermal electrons I_{th}^e at higher target temperatures (see Section 11.3.6), the current of reflected metastable atoms I_2^m, and the electron currents I_{3m}^e, I_{3i}^e, and I_{3e}^e of electrons ejected from the collector by impinging metastable atoms, ions, and electrons, respectively.

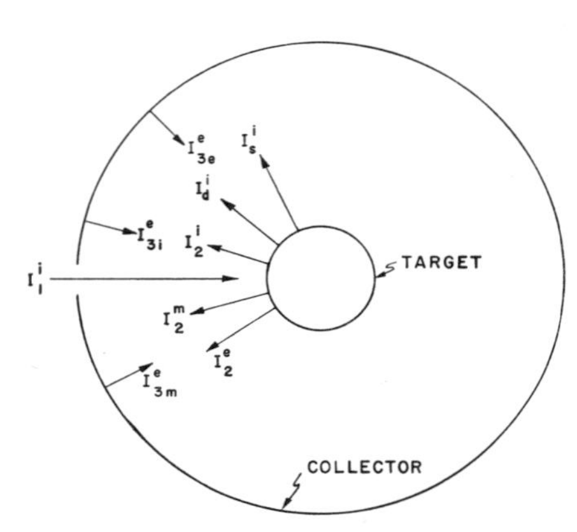

Fig. 11.1.a. Schematic illustration of various particle currents (ion currents I^i, electron currents I^e, and currents of metastable atoms I^m) which may occur at target and collector when a primary-ion current I_1^i strikes the target. The symbols used have the following meanings: I_2^i is the current of reflected primary ions, I_d^i the current of thermally desorbed primary ions, I_s^i the current of target particles sputtered as ions, I_2^m the current of reflected metastable atoms, I_2^e the current of secondary electrons emitted by ion impact, and I_{3e}^e, I_{3i}^e, I_{3m}^e are currents of electrons ejected from the collector by impinging electrons, ions, and metastable atoms, respectively.

Applying a suitably chosen voltage between collector and target can reduce the secondary-electron emission I_2^e and I_{th}^e. Furthermore, the contribution I_{th}^i of thermally emitted ions and the current I_s^i of target particles sputtered as ions can be made negligible by a suitable choice of experimental conditions. (The appropriate conditions are described in Sections 11.2 and 14.2.) Thus, in first approximation, the measured total currents I_T at the target and I_C at the collector are expressible

(HAGSTRUM [*289*]) as

$$I_T = I_1^i - I_2^i - I_{3m}^e \tag{11.1–1}$$

and

$$I_C = I_2^i + I_3^e + I_{3m}^e . \tag{11.1–2}$$

Many authors have determined the ratio of collector current to target current, i.e.,

$$R = I_C/I_T . \tag{11.1–2a}$$

Although this has often been called the "reflection coefficient", it is evident from Eqs. (11.1–1) and (11.1–2) that the quantity R is not characteristic of the actual reflection process but rather is determined by other secondary and tertiary effects. The quantity R will therefore be referred to in what follows as the "secondary-emission coefficient", while the term "reflection coefficient" will be restricted to the ratio

$$\varrho = I_2^i/I_1^i . \tag{11.1–3}$$

For the special case represented by Eqs. (11.1–1) and (11.1–2), the quantity ϱ is related to R by the expression

$$R = \varrho + \gamma_2 \varrho + \gamma_m \varrho_m . \tag{11.1–4}$$

Here, as defined by HAGSTRUM, $\gamma_2 = I_3^e/I_2^i$ is a secondary-electron coefficient, $\gamma_m = I_{3m}^e/I_2^m$ is the emission coefficient for AUGER electrons, and $\varrho_m = I_2^m/I_2^i$ is a conversion coefficient for the transformation of ions into metastable atoms. A determination of the quantity ϱ, of interest in this chapter, requires additional experiments to eliminate such effects as those represented by the second and third terms in Eq. (11.1–4) or to determine them separately. Some of the methods developed for this purpose are described briefly in the following section.

11.2. Experimental Techniques

In his investigations of the reflection of noble gas ions from metal surfaces, HAGSTRUM [*289*] measured the currents I_T at the target and I_C at the collector. For the case in which a spherical collector surrounding the target was at negative potential (e.g., at —40 eV relative to the target), these currents can be reproduced in first approximation by Eqs. (11.1–1) and (11.1–2). In order to find the quantity $\varrho = I_2^i/I_1^i$, the author applied a magnetic field of 60 G in the target region perpendicular to the plane of the paper in Fig. 11.1.a. By this means he was able to suppress the secondary-electron current and to determine the quantity ϱ by

$$\varrho = I_C/(I_C + I_T) = I_2^i/I_1^i = R_B . \tag{11.2–1}$$

The second unknown in Eq. (11.1–4), namely, the secondary electron-emission coefficient γ_2, could be determined if the potential V_{TC} between target and collector and the potential V_{SC} between ion source and collector are so selected that the primary ions (constituting I_1^i) are repelled by the target and impinge on the collector. Then in first approximation γ_2 can be determined from the measured quantities I_T and I_C by

$$\gamma_2 = \left(\frac{I_T}{I_T + I_C}\right) \quad \text{for} \quad V_{TC} > V_{SC}. \tag{11.2–2}$$

The third unknown in Eq. (11.1–4), the secondary-electron-emission coefficient γ_m, could not be determined directly. Nevertheless, HAGSTRUM [*282, 288*] assumes on the basis of his experimental results that the values of γ_2 and γ_m are approximately equal. In this way the various components of the quantity R in Eq. (11.1–4) are determined.

ARIFOV et al. [*25, 27, 30*] bombarded a metal surface at high temperature with alkali metal ions and employed a technique of ion beam modulation to distinguish between the thermally desorbed primary ions I_d^i from the surface, the reflected ions I_2^i, the desorbed ions I_x^i coming from the target impurities, the secondary electrons I_2^e, and the impinging primary ions. Briefly, the method may be described as follows. An accelerated primary-ion beam is passed through an electrical condenser, where its intensity is modulated by a square-wave generator with a frequency of 500—1000 cps, and then impinges on the target. The secondary-emission current from the target is detected by a collector, whose potential relative to the target is modulated by a sawtooth pulse generator having a frequency of 25 cps. This current is fed into the vertical amplifier of an oscilloscope whose horizontal sweep is synchronized with the sawtooth generator.

If the process of secondary emission is instantaneous, each pulse in the primary-ion beam gives a corresponding pulse in the secondary beam and results in a deflection of the oscilloscope beam. Since the frequency of the square-wave pulse is made several times as great as the sawtooth frequencies, several secondary-current pulses occur during a single sweep.

The horizontal part of the square wave, corresponding to the position of the beam at the instant when the primary current is not striking the target, is taken as the zero line that separates the secondary-ion current from the secondary-electron current. The beam-modulation method can also be used to measure the components whose emission is delayed with respect to the primary-ion-beam pulse. These are the fraction of primary ions that were adsorbed on the target surface and then after a certain residence time on the surface (corresponding to their mean adsorption lifetime τ_i as discussed in Sections 8.2 and 8.3.1.5) desorb partially as ions.

When the primary pulse is cut off, these surface-ionization currents will decay exponentially and thus distort the square-wave pattern of the modulated signal on the oscilloscope. This provides a way of distinguishing between the almost instantaneous process of ion scattering and the delayed process of ion desorption. However, at high target temperatures the mean adsorption lifetimes become so small that such a distinction becomes impossible. It should be noted that none of the authors [*25, 27, 30*] have published data on the mean adsorption lifetime of the incident beam particles on the target surface.

In the majority of the investigations of "ion reflection", the measuring techniques were inadequate for a determination of the ion-reflection coefficient ϱ. Only the secondary-emission coefficient R could be found by the methods used, the commonest being to change the polarity of the target and the collector potentials in order to distinguish between secondary electrons and secondary ions (cf. references [*217, 529, 557*]).

11.3. Experimental Ion-Reflection and Secondary-Emission Coefficients

11.3.1. Dependence on the Incident Ion Energy

11.3.1.1. Incident Noble-Gas Ions

When positive noble-gas ions are incident normally on a metal surface, the ion-reflection coefficient ϱ has been found to be very small and practically independent of the kinetic energy of the incident ion. Fig. 11.3.1.1.a illustrates such results obtained by HAGSTRUM [*289*] for the function $\varrho(E)$ for He^+, Ne^+, and A^+ ions reflected from a cleaned wolfram surface. The values of ϱ range from 10^{-4} to 10^{-3}*. For He^+ on cleaned Mo surfaces, the function $\varrho(E)$ was found to be nearly independent of the incident-ion energy, but the absolute values of ϱ were smaller ($8 \times 10^{-4} < \varrho < 1 \times 10^{-3}$) than the corresponding ones for W surfaces. (See the corresponding results for R in the article by FOGEL et al. [*217*]). PETROV [*557*] investigated the reflection of He^+ ions from a Ta target at much higher incident-ion energies E (in the range 1—30 keV) and also reported that the reflection was nearly independent of E. However, he observed a larger fraction of reflected ions ($\approx 5\%$) than HAGSTRUM did. The results of BRADLEY et al. [*91*] for the reflection of Xe^+ on Pt surfaces also indicate that the reflection is independent of E in the investigated range (100—900 eV). However the observed strong pressure

* It should be noted that the values of ϱ shown in Fig. 5 in HAGSTRUM's paper [*289*] are erroneously plotted too high by a factor of ten, as is the value of ϱ for He^+ on W reported by HAGSTRUM in an earlier paper [Phys. Rev. **93**, 652 (1954)]. The values of ϱ listed in Tables II and III of his paper [*289*] are correct [private communication].

dependence of the Xe^+ "reflection", and also the observed small values of the mean energy of the "reflected" Xe^+ ions, indicate that a possible sputtering effect of physisorbed Xe atoms is superimposed in these experiments.

In contrast to the behavior of the function $\varrho(E)$, the functions $\varrho_m(E)$, $\gamma_2(E)$, and $\gamma_{2m}(E)$ increase with increasing energy E. Figs. 11.3.1.1b and 11.3.1.1.c reproduce the values found by HAGSTRUM for the functions $\varrho_m(E)$ and $\gamma_2(E)$ for He^+, Ne^+, and A^+ ions on clean wolfram surfaces. In this connection, it is of interest to point out that if one plots the co-

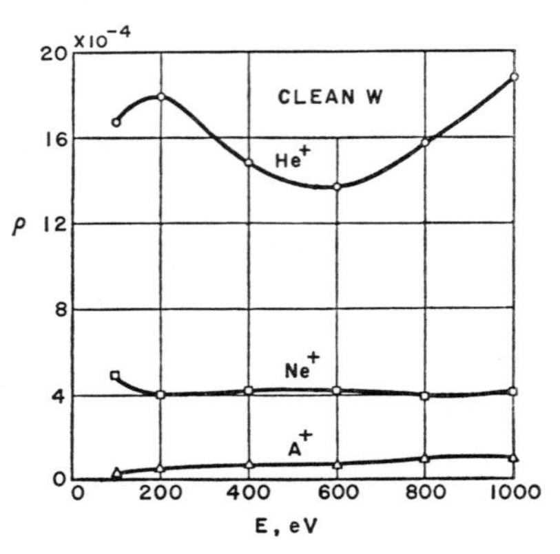

Fig. 11.3.1.1.a

Fig. 11.3.1.1.b

Fig. 11.3.1.1.a. The dependence of the ion-reflection coefficient ϱ on the kinetic energy E of the incident ion for the case of noble-gas ions reflected from a clean polycrystalline wolfram surface (HAGSTRUM [*289*])

Fig. 11.3.1.1.b. The conversion coefficient ϱ_m for the transformation of an ion into a metastable atom as a function of the kinetic energy E of the incident ion for the case of noble-gas ions striking a clean wolfram surface (HAGSTRUM [*289*])

efficient ϱ_m against the velocity V of the primary ions for different noble-gas ions (e.g., for the case of He^+ and A^+ bombardment of W surfaces) all of the values lie on the same curve (HAGSTRUM [*284*]). An explanation which has been given by HAGSTRUM will be discussed in Chapter 12. The behavior of the functions $\varrho_m(E)$ and $\gamma_2(E)$, on the assumption that $\gamma_2 \approx \gamma_{2m}$ (HAGSTRUM [*282, 288, 289*]), explains, for instance, the observation by HAGSTRUM [*289*] that for the case of the bombardment of a clean wolfram surface by He^+ ions with energies ranging from 100 eV to 1 keV the function $R(E)$ increases with increasing energy E. Qualitatively, a similar behavior of the function $R(E)$ (as defined in Eq. 11.1–2a) for the corresponding energy region has been observed by HEALEA et al. [*300, 301*], bombarding heated nickel surfaces with He^+, Ne^+, A^+, and hydrogen ions (see Fig. 11.3.1.1.d). The absolute values of R found by HAGSTRUM for He^+ on clean wolfram are two orders of magnitude below those observed by HEALEA et al. for He^+

on gas-covered nickel. In such a comparison, however, one should keep in mind that the experimental conditions were different in the two cases. Since the values of R (influenced by the values of such quantities as ϱ, $\varrho_m, \gamma_2, \gamma_m$) depends for instance very strongly on surface conditions, such discrepancies are not surprising.

The secondary-ion-emission coefficient R as a function of the energy E or velocity V of the primary ions at considerably higher energies (or velocities) was studied by FOGEL et al. [*217, 218*] for different primary ions (among them such ions as Kr^+, A^+, Ne^+, and He^+ with energies

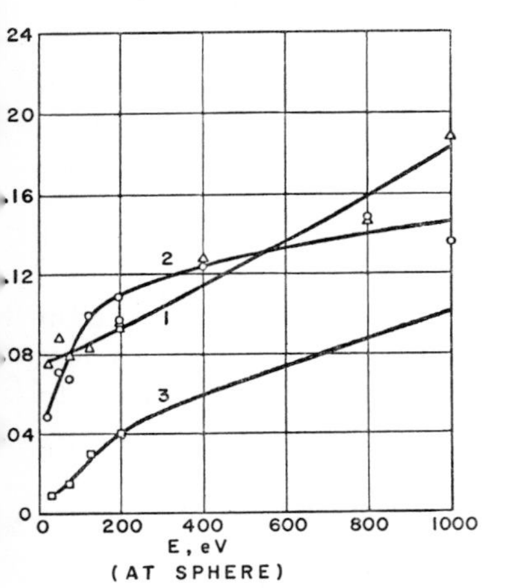

Fig. 11.3.1.1.c

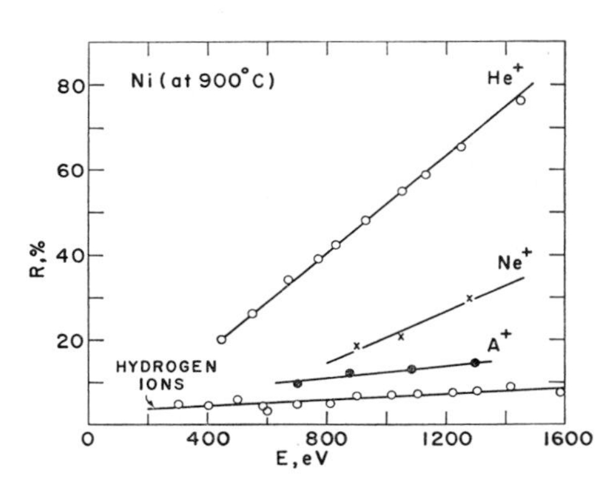

Fig. 11.3.1.1.d

Fig. 11.3.1.1.c. The coefficient γ_2 for the emission of secondary electrons from the collector as a result of the impact of reflected primary ions I_2^f as a function of the kinetic energy E of such ions. The curves *1*, *2*, and *3* represent such results for the cases of He^+, Ne^+ and A^+ ions, respectively, striking a clean wolfram surface (HAGSRTUM [*289*])

Fig. 11.3.1.1.d. The dependence of the secondary-ion-emission coefficient R on the kinetic energy E of the incident ion, for noble-gas ions striking a Ni target held at 900° C (HEALEA et al. [*300, 301*])

ranging from approximately 10—40 keV) on Mo, Ta, W, Cu, and Fe surfaces; by PANIN [*542*] for different ions (among them A^+, Ne^+, and He^+ ions with energies varying from 10—100 keV) incident at an angle* $\alpha = 45°$ with respect to the surface normal on such targets as Mo, Zr, and C; and by WALTHER and HINTENBERGER [*751a*] for He^+, Ne^+, A^+, Kr^+, and Xe^+ ions incident at an angle $\alpha = 45°$ on gold, copper, and graphite targets with energies ranging from 100 eV to 30 keV (the emitted secondary ions were detected under 0° and 45° with respect to the surface normal). The data obtained by the above-mentioned authors [*217, 218, 542, 751a*] for Ne^+, A^+, and Kr^+ ion bombardment ($E > 10$ keV) of different target materials show in qualitative agreement that the

* This angle has not been specified by PANIN [*542*] but has been estimated from a schematic figure showing his experimental arrangement.

function $R(E)$ increases with increasing energy [the results obtained by FOGEL et al. and PANIN for $R(V)$ have to be converted to the function $R(E)$]; for the case of He^+ ion bombardment of different target materials only a very slight dependence of R on E has been observed. According to WALTHER and HINTENBERGER the values of $R(E)$ vary with the type of the incident ion for all three target materials investigated (and for the energy region $4\,\text{keV} \leqslant E \leqslant 30\,\text{keV}$) in the order $R_{Xe} > R_{Kr} > R_A > R_{Ne} > R_{He}$, if the secondary ions were detected at $\theta = 0°$ with respect to the surface normal. Such a trend may be expected if secondary ions are dominantly produced by a sputtering process. If the angle of detection θ of secondary ions differs significantly from 0°, a change in the above-mentioned order of $R(E)$ values occurs. The data of PANIN and FOGEL et al. indicate that for $E > 20\,\text{keV}$ the order of the $R(E)$ values becomes $R_{Ne} > R_A > R_{He}$; WALTHER and HINTENBERGER mention also a change in the order of $R(E)$ values for $\theta = 45°$ (this is specular to the incident beam direction).

The observed changes in the order of $R(E)$ values for $\theta \neq 0°$ and in particular the relative increase of the R values for the lighter ions may be explained by a relative increase of the contribution of reflected ions to the value of R. (Notice, that the ϱ values are in the order $\varrho_{He} > \varrho_{Ne} > \varrho_A > \varrho_{Kr} > \varrho_{Xe}$.)

WALTHER and HINTENBERGER reported that the function $R(E)$ (determined for $\theta = 0°$), measured for Ne^+, A^+, Kr^+, and Xe^+ bombardment of gold, copper, and graphite targets, passed through a maximum at approximately $E = 1.6 \pm 0.4\,\text{keV}$. The value of this maximum of the $R(E)$ curve decreases in the order of the target materials gold, copper, and graphite and in the order of the incident ions Xe > Kr > A > Ne. For H^+ ion bombardment, no maximum has been observed. According to the investigators, the position of the maximum seemed to be independent of the target material, the type of the incident ions, the value of the incident-ion current, and the residual gas pressure. There is at present no ready explanation for the occurrence of this maximum in the $R(E)$ curve.

The absolute values of $R(E)$ observed by WALTHER and HINTENBERGER for the various noble gas ions incident on gold, copper, and graphite targets are nearly two orders of magnitude smaller than those reported by the other authors (refs. [*217, 218, 542*]) for the different noble gas ions incident on a Mo target. This may be explained in part by the fact that in the experiments of WALTHER and HINTENBERGER the effective solid angle for the detection of secondary ions* is nearly two orders

* A quantitative correction for the difference in the effective solid angles in these experiments would necessitate a better and quantitative knowledge of the angular distribution of the particles which are emitted by the different mechanisms

of magnitude smaller than in the experiments of PANIN and FOGEL et al. In addition, since other experimental conditions reported by these authors are also quite different—e.g., the reported vacuum in the target region during the run has been $2-3\times10^{-8}$ mm Hg (PANIN [*542*]), $2-3\times10^{-7}$ mm Hg FOGEL et al. [*217*]), $1-6\times10^{-6}$ mm Hg (WALTHER and HINTENBERGER [*751a*])—such large discrepancies are not surprising. In summary, the available experimental material indicates that although the reflection coefficient ϱ for noble-gas ions (and probably also for other ions for which $I \gg \varphi$) is practically independent of the primary energy, the secondary-emission coefficient R increases with increasing ion energy for Ne^+, A^+, Kr^+, and Xe^+ ions over a large energy range.

No experimental data seem to be available for the reflection coefficient ϱ_0 for neutral noble-gas atoms reflected from clean metal surfaces [in analogy to the quantity ϱ of Eq. (11.1–3) we define $\varrho_0 = I_2^0/I_1^0$, where I_1^0 and I_2^0 are the neutral particle currents of the incident and reflected beam, respectively]. Only some qualitative information on the reflection of neutral atoms from metal surfaces has been obtained so far. MEDVED* used two different methods to investigate the reflection of neutral neon and argon atoms from a flashed Mo surface and an uncleaned Pt surface. The atoms were incident with various energies, ranging from 0.5 to 2.5 keV. In one method the secondary and tertiary electron currents I_2^e and I_3^e produced by the primary and secondary neutral beam particles were monitored; in the other method the energy transfer of the primary and secondary beam particles (the later detected only in the direction of specular reflection) to the uncleaned surfaces of two Pt-disk thermocouples was measured. From the data MEDVED inferred that there is a strong preference for noble-gas atoms to be ejected as neutrals rather than ions. He estimated that for 750-eV neutral argon atoms incident on a flashed Mo surface, approximately 20% are reflected; while for argon ions incident on a clean W surface, only approximately 0.02% are reflected (HAGSTRUM [*289*]). (For a Mo surface one may expect even a smaller value of ϱ, as indicated by HAGSTRUM's results for He^+ on W and Mo.) In addition MEDVED concluded from the data that below an incident-particle energy of approximately 1.5 keV there is a monotonic increase in the number of reflected neutral particles with decreasing particle energy. (A similar behavior has been observed for the reflection of K^+ ions from clean Mo (see Fig. 11.3.1.2.a), BRUNNEÉ [*102*]). These conclusions, however, should be considered suggestive rather than

---

(e.g. sputtering, elastic and inelastic scattering, desorption) and contribute to the complexity of the secondary ion emission coefficient.

* The communication published by MEDVED [*467a*] is based in part on results reported earlier in the Proceedings of the 23rd Physical Electronics Conference (March 1963) by A. COMEAUX and D. B. MEDVED.

conclusive in view of several uncertainties involved in the experimental methods and conditions. For example, the voltage signals (V_1^i and V_2^i produced by the two thermocouples during bombardment by the primary and secondary beams of ions and V_1^0 and V_2^0 from the beams of neutral particles) depend not only on the number of particles (n_1^i, n_2^i, n_1^0, n_2^0) reaching the thermocouple surface, but also on their energy distribution [$f_1(E_i)$, $f_2(E_i)$, $f_1(E_0)$, $f_2(E_0)$] and their thermal accommodation at the surface (see Chapter 6, Section 6.3.1.1). In view of the complex dependence of V on n, $f(E)$, and α, and since the values of most of these quantities are insufficiently known [in particular n_1^0, n_2^0, $f_2(E_i)$, $f_1(E_0)$, $f_2(E_0)$] neither their ratios $r_i = V_2^i[n_2^i, \alpha_2^i, f_2(E_i)]/V_1^i[n_1^i(\alpha_1^i, f_1(E_i)]$, nor $r_0 = V_2^0[n_2^0, \alpha_2^0, f_2(E_0)]/V_1^0[n_1^0, \alpha_1^0, f_1(E_0)]$, nor $R = r_0/r_i$ can be accurately related to the reflection properties of noble-gas ions and neutrals from metal surfaces and the reflection coefficients $\varrho_i = n_2^i/n_1^i$ and $\varrho_0 = n_2^0/n_1^0$. (A further source of uncertainty is the assumption that the neutral particles detected in the specular direction by the thermocouple have been produced by reflection rather than by sputtering or desorption of particles in this direction; there is also an uncertainty in the value of the quantity $\gamma_2^0 = I_3^e/I_2^0$ and the function $\gamma_2^0(E)$; furthermore in the case of Pt, the surfaces have not been clean.)

11.3.1.2. Incident Alkali Ions

The experimental investigation of the reflection of alkali ions from metal surfaces at low target temperatures is more difficult than for noble-gas ions because of the enhanced adsorption of the incident beam (as discussed in Chapters 4 and 5 on "Chemisorption") and the build-up of surface layers. In order to obtain data representative of the reflection of ions from clean metal surfaces rather than from adsorbed gas layers, the surface must be cleaned by such special techniques as flashing, sputtering, etc. (Chapter 3), and any appreciable surface coverage by the incident beam and the rest gas during the run has to be avoided.

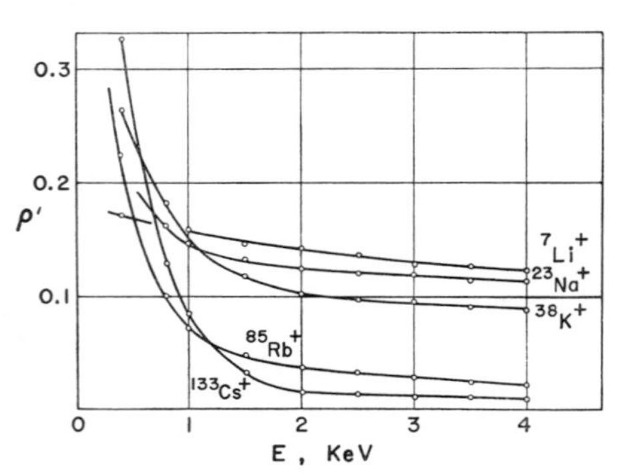

Fig. 11.3.1.2.a. The dependence of the ion-reflection coefficient ϱ' on the kinetic energy E of the incident ion for a clean Mo surface bombarded by the indicated isotopes of alkali metals (BRUNNEÉ [*102*])

In preparing the surface for his reflection measurements BRUNNEÉ [*102*], for example, used the flash-filament method (residual gas pressure $\approx 5 \times 10^{-8}$ mm Hg) and made a measurement 10—15 sec after each flashing. He did not state how much the target had cooled in such a short time. Fig. 11.3.1.2.a illustrates the results obtained by BRUNNEÉ for the reflection of alkali ions from a cleaned Mo surface in the energy range from 0.4 to 4 keV. The coefficient ϱ' plotted here against the primary-ion energy is a secondary-emission coefficient which is corrected for the secondary and tertiary electron currents I_2^e and I_3^e but not for neutralized or sputtered ions. In the following the reflection coefficient will be represented by ϱ' whenever it seems appropriate to neglect the influence of ion neutralization or sputtering.

Fig. 11.3.1.2.a reveals the interesting result that, as in the reflection of noble-gas ions, the values of ϱ' are nearly independent of energy except at low energies. This result is in qualitative agreement with the findings of EREMEEV [*188*] for the reflection of Li^+ and K^+ ions from Ta and W surfaces (which had to be kept at a temperature above 1000°K in the case of K^+ reflection) in the broader energy range of 1—13 keV and those of PETROV [*558*] for K^+ ions on W surfaces ($\varrho' \approx 30\%$ for 1 keV $< E < 5$ keV). The reflection coefficients reported by VEKSLER [*739*] for Cs^+ ions on Ta ($\varrho' \approx 6\%$ at 900 eV) and on Mo ($\varrho' \approx 4\%$ at 2 keV) and by ZANDBERG [*793*] for Pt surfaces bombarded by 780-eV ions of Cs^+ ($\varrho' \approx 3\%$), Rb^+ ($\varrho' \approx 9\%$), K^+ ($\varrho' \approx 15\%$), Na^+ ($\varrho' \approx 17\%$), and Li^+ ($\varrho' \approx 17\%$) agree within a factor of three with those of BRUNNEÉ. In view of the fact that here the values of ϱ' have been obtained under different experimental conditions and are being compared for different target materials (for a more detailed discussion, see Section 11.3.3), the agreement is rather good.

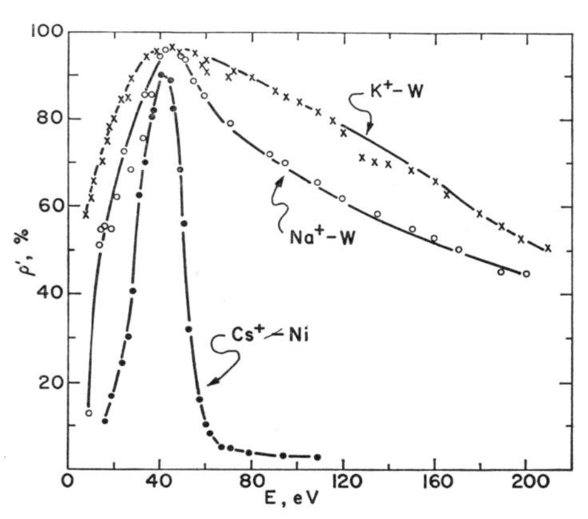

Fig. 11.3.1.2.b. The dependence of the partially-corrected ion-reflection coefficient ϱ' on the kinetic energy E of the incident ion in the low-energy region, for alkali ions striking Ni and W surfaces (VEKSLER [*739*])

At energies $E < 1$ keV, BRUNNEÉ [*102*] found that the reflection coefficient for Mo surfaces increases rapidly with decreasing energy, as shown in Fig. 11.3.1.2.a. Similar observations have been made by ARIFOV et al. [*24, 30*] for the case of reflection of Na^+ and K^+ ions from cleaned W and Ta surfaces and Cs^+ ions from a cleaned Ni surface. As illustrated in Fig. 11.3.1.2.b, VEKSLER [*739*] actually observed that with decreasing energy E the function $\varrho'(E)$ passes through a maximum (reaching a value of $\varrho'_{max} \approx 95\%$ at 44 eV for Na^+ on W, $\varrho'_{max} \approx 96\%$ at

43 eV for K^+ on W, and $\varrho'_{\max} \approx 90\%$ at 42 eV for Cs^+ on Ni). At present this increase of $\varrho'(E)$ with decreasing E is not well understood. One suggested explanation is that at low energies the incident ion penetrates the target lattice so little that the usual mode of reflection becomes a single collision between incident ion and target atom.

The reflection coefficient ϱ' for both clean and gas-covered surfaces bombarded by alkali ions is nearly independent of the primary-ion energy at the higher energies, as was the case for noble-gas ions. In contrast, the secondary-ion-emission coefficient R rises with increasing energy, as found for instance by Ploch [*573*], Paetow and Walcher [*540*], and Flaks [*215, 216*]. Ploch observed a rather strong increase of the function $R(E)$ for Li^+ ions bombarding gas-covered surfaces of Pt ($2\% < R < 30\%$ für 1 keV $< E <$ 7 keV), Mo ($4\% < R < 12\%$ for 1 keV $< E <$ 6 keV), and Cu ($3\% < R < 6.5\%$ for 0.4 keV $< E$ $<$ 1.7 keV). For degassed surfaces the changes of the function $R(E)$ are smaller (e.g., for Li^+ on Pt, $23\% < R < 29\%$ for 1 keV $< E <$ 7 keV) and for Li^+ on Mo, $R(E)$ even decreases slightly with increasing energy. In the case of Cs^+ ions on gas-covered W surfaces. Paetow and Walcher observed the same behavior of $R(E)$ as was found by Ploch. Results for $R(E)$ for singly- and multiply-charged potassium ions (K^+, K^{++}, K^{+++}) bombarding gas-covered Pt surfaces have been reported by Flaks [*215, 216*]. The $R(E)$ curves rise with increasing energy and level off at energies above about 25 keV. At 20 keV Flaks reports that $R \approx 49\%$ for K^+, $R \approx 62\%$ for K^{++}, and $R \approx 81\%$ for K^{+++}.

For a discussion of older results, the reader is referred to the summaries given by Massey and Burhop [*461*] and by Brunneé [*102*].

11.3.1.3. Incident Protons, Deuterons, and Various Other Atomic and Molecular Ions

The secondary-emission coefficient R and the reflection coefficient ϱ' depend only very slightly on the primary-ion energy when metal surfaces (e.g., Mo, Zr, Ta, W, Cu, Ag) are bombarded with protons and molecular ions such as H_2^+ and H_3^+ in the energy range 1—40 keV [*217, 429, 529, 542, 557*]. Fogel et al. [*217*] observed the secondary-emission coefficient R and reflection coefficient ϱ' for protons bombarding Mo, Ta, W, Cu, and Fe surfaces in the energy range 5—40 keV. For proton bombardment of a Mo surface, Fogel et al. [*217*] observed only a very slight decrease of the values of R and ϱ' with increasing velocity V of the incident ion ($0.6\% > R > 0.5\%$ and $2.2\% > \varrho' > 1\%$ for 9×10^7 cm/sec $< V < 2.2 \times 10^8$ cm/sec corresponding to 4 keV $< E$ $<$ 25 keV). Similar results for H^+ on Mo have been reported by Panin [*542*] ($2.5\% > \varrho' > 2.3\%$ for 1.2×10^8 cm/sec $< V <$ 2.7×10^8 cm/sec corresponding to 7 keV $< E <$ 35 keV). For proton bombardment of a Zr target, he reported $3\% > \varrho > 1.5\%$ for energies ranging from 6 to 35 keV. The secondary-emission coefficient R for metal surfaces such as Mo and Zr (Panin or Ta (Petrov [*557*]) is found to be almost twice as great for H_2^+ bombardment, and almost three times as great for H_3^+ bombardment, as for protons of the same velocity. This indicates that the molecular ion dissociates at the surface in the process of being reflected. Petrov pointed out that for H_2^- bombardment of a heated Ta surface (1600° K) the secondary-emission coefficient had the nearly-constant value $R \approx 6\%$ over the incident-ion energy range from 1 to 30 keV but that the value of R at room temperature is 1.3—1.5 times this value. This behavior is opposite to that commonly observed (Section 11.3.6).

Fogel et al. [*217, 218*] investigated the magnitudes of R and ϱ when a Mo surface was bombarded by positive and negative ions of the same element, for instance H^+ and H^- ions. They observed that at an incident-ion energy of 22 keV the values of R are practically the same [$R(H^+) \approx 0.44\%$, $R(H^-) \approx 0.45\%$] and the values of ϱ' differ only slightly [$\varrho'(H^+) \approx 1.56\%$, $\varrho'(H^-) \approx 2.03\%$].

Although the coefficients R and ϱ' are almost independent of the incident-ion energy E for primary H^+, H_2^+, and H_3^+ ions, they depend more markedly on E for such ions as N^+, Mo^+, O^+, and Cd^+. In all cases investigated so far, namely, C^+, N^+ on Mo, N^+ on Zr (Panin Cd^+ on W (Petrov [*557*]), and O^+ on Mo (Fogel et al. [*217, 218*]), the curves $R(E)$ and $\varrho'(E)$ or $R(V)$ and $\varrho'(V)$ rise with increasing primary-ion energy or velocity. For example, Panin reported that for the N^+ and Mo^+ bombardment of a Mo surface, $4.6\% < R(N^+) < 8\%$ for 3.75×10^7 cm/sec $< V < 7.5 \times 10^7$ cm/sec (corresponding to 10 keV $< E <$ 60 keV) and $10\% < R(Mo^+) < 16\%$ for 1.3×10^8 cm/sec $< V$ $< 2.5 \times 10^8$ cm/sec (corresponding to 10 keV $< E <$ 30 keV).

The dependence of the slope of the curves $R(V)$ and $\varrho(V)$ on such parameters as the target temperature, degree of surface contamination, angle of incidence, etc., will be discussed in the following sessions.

11.3.2. Dependence on the Mass and Electronic Configuration of the Incident Ion

The ion reflection coefficient ϱ and the only partly corrected reflection coefficient ϱ' (defined in Section 11.3.1) decrease with increasing mass m of the primary ion for certain ranges of incident-ion energies E, as incidated by the experimental results for noble-gas ions over the complete investigated range of energies (100—1000 eV, see Fig. 11.3.1.1.a) and for alkali ions for energies E larger than approximately 1.3 keV [see $\varrho'(E)$ values in Fig. 11.3.1.2.a].

In order to study the influence of the electronic configuration of the incident ion on the ion reflection, BRUNNEÉ [*102*] investigated the isotopic effect on the reflection coefficient ϱ' for isotopic ion pairs such as $^6Li^+$ and $^7Li^+$, $^{39}K^+$ and $^{41}K^+$, and $^{85}Rb^+$ and $^{87}Rb^+$, and determined the ratio $[\Delta\varrho'(m, Z)/\Delta m]_{Z=\mathrm{const}}$, where Z is the atomic number and m the ion mass. Fig. 11.3.2.a illustrates BRUNNEÉ's results for the behavior of the function $\varrho'(m)$ for the investigated alkali isotopes. The slope of the curve $\varrho'(m)$ for the alkali ions coincides with the value $\Delta\varrho'/\Delta m$ for each isotopic ion pair within the experimental error [as indicated by the possible range of slopes shown for each isotopic pair in Fig. (11.3.2.a)]. This indicates that the value of the ion reflection coefficient ϱ' is determined mainly by the mass of the incident ion and to a lesser extent by its electronic configuration. These results are supported by PLOCH's [*573*] data for the isotopic effect on the reflection of the Li ion, which indicate that the reflection coefficient ϱ' depends predominantly on the ion mass. As additional support, one may consider the work of O'BRIAN et al. [*529*] who observed that the value of ϱ' for H^+ was remarkably different from that for D^+ when these ions bombarded a bright silver surface [$\varrho'(H^+) \approx 12\%$, $\varrho(D^+) \approx 7\%$ for an ion energy of 9.25 keV].

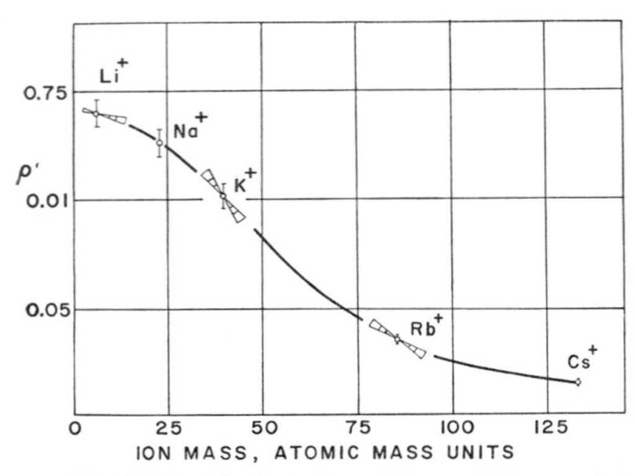

Fig. 11.3.2.a. The dependence of the partially corrected ion-reflection coefficient ϱ' on the mass m of the incident ion. The curve illustrates BRUNNEÉ's results for the function $\varrho(m)$ for such isotopic ion pairs as $^6Li^+$ and $^7Li^+$, $^{39}K^+$ and $^{41}K^+$, and $^{85}Rb^+$ and $^{87}Rb^+$ incident on a clean Mo surface. The experimental error in the value $\Delta\varrho/\Delta m$ is indicated by the possible range of slopes shown for each isotopic pair. Also shown are the $\varrho(m)$ values for $^{23}Na^+$ and $^{133}Cs^+$ ions (BRUNNEÉ [*102*])

Contrary to these conclusions MEDVED [*467a*] invoked a significant influence of the electronic configuration of the incident particle on the reflection process to explain his observation of a marked difference in the reflection characteristics of A^+ ions and neutral argon atoms incident on a flashed Mo surface on the one hand and a similarity in the reflection characteristics on K^+ ions and neutral argon atoms incident on a flashed Mo surface on the other. According to MEDVED the observed difference in the reflection properties may be related to AUGER neutralization processes, which can occur for A^+ ions but not for K^+ ions and neutral argon atoms. However, in view of the more qualitative nature of the experimental results (see discussion in Section 11.3.1.1) and the fact that MEDVED did not actually determine the quantity $\varrho(E)$ or $\varrho'(E)$ the conclusions do not appear to be sufficiently substantiated yet and more quantitative material is needed.

11.3.3. Dependence on the Mass and Work Function of the Target Material

The reflection coefficient ϱ and the secondary-emission coefficient R depend, significantly on the mass of the target material. For clean target surfaces, the values of ϱ (or ϱ')* and R increase as the mass of the target atom increases. For the case in which the reflection process consists of a single collision between the incident ion of mass m and a target atom of mass M at very low ion energies, Brunneé [*102*] has derived the relation

$$\varrho \approx \left(1 - \frac{m}{M}\right) \quad \text{for} \quad m < M\,. \tag{11.3.3–1}$$

The experimental results for $\varrho(M)$ [particularly the relative values $\varrho(M_1)/\varrho(M_2)$] are represented rather well by this relationship, as can be seen from the data of various authors. Hagstrum [*289*] observed that for clean W and Mo surfaces bombarded by He^+ ions with energies of 0.1—1.0 keV the reflection coefficients were $\varrho_W = 0.17\,^0/_0$ and $\varrho_{Mo} \approx 0.09\,^0/_0$ so that the relative value is $(\varrho_W/\varrho_{Mo})_{exper} = 1.8_8$. This agrees within a factor of two with the theoretical ratio $(\varrho_W/\varrho_{Mo})_{theor} = 1.2_5$ calculated from Eq. (11.3.3–1). An increase of ϱ' and R with target mass has been confirmed by data such as those of Panin [*542*] who found for 55-keV A^+ on Mo, Zr, and C that $\varrho'_{Mo} \approx 11.5\,^0/_0$, $\varrho'_{Zr} \approx 10.8\,^0/_0$, and $\varrho'_C \approx 4.6\,^0/_0$; of Ploch [*573*] who bombarded degassed Pt and Mo surfaces with 2-keV Li^+ ions and obtained $R_{Pt} \approx 23.5\,^0/_0$ and $R_{Mo} \approx 18\,^0/_0$; of Gehrtsen [*233*] whose bombardment of various surfaces by hydrogen ions showed that $R_{Pt} > R_{Ag} > R_{Ni} \approx R_{Cu}$ and $R_{Al} > R_{Mg} > R_{Be}$; and of Eremeev et al. [*188, 189, 190*] who found, for example, that for W and Ta bombarded with K^+ in the energy range 2 keV $\leqslant E \leqslant$ 6 keV the values were $R_W \approx 20.5\,^0/_0$ and $R_{Ta} \approx 19.8\,^0/_0$.

11.3.4. Dependence on the Angle of Incidence

At present no experimental results are available concerning the dependence of the reflection coefficient ϱ on the angle of incidence α of the primary-ion beam. There are, however, data on the behavior of the functions $R(\alpha)$ and $\varrho'(\alpha)$ for polycrystalline targets by several authors (Read [*583*], Gurney [*270*], Sawyer [*627*], Woodcock [*785*], Longacre [*443*], Eremeev et al. [186] and Mitropan et al. [*486*]) and for a monocrystalline target by Nelson and Thompson [*521b*]. Such investigations were carried out with alkali ions incident on surfaces of

* The quantity ϱ' has been defined in Section 11.3.1.

polycrystalline Pt (references [*270, 583, 627, 785*]), Ta (reference [*186*]) and Na (references [*627, 785, 443*]) and with 50-keV protons and noble-gas ions incident on a Cu (110) surface (reference [*521b*]). Because of the poor vacuum conditions and the insufficient cleanliness of the surfaces prevailing in most of these experiments, the results obtained are characteristic only of gas-covered surfaces.

The results of several authors [*186, 270, 443, 486, 583, 627, 785*] on the behavior of the functions $R(\alpha)$ and $\varrho'(\alpha)$ agree qualitatively that the values of R and ϱ' increase with increasing angles of incidence α (measured from the normal). Furthermore, a sizeable fraction of the secondary "emitted" ions are ejected as if reflected, i.e., the maximum of the angular distribution $f(\beta)$ of these scattered particles is in the neighborhood of the angle for specular reflection. Read [*583*], Gurney [*270*], Sawyer [*627*], and Eremeev et al. [*186*] are in agreement, however, that the position of this maximum depends on the energy of the incident ion in such a way that with increasing energy (for $E > 100$ eV) the value β_{max} of this maximum shifts toward smaller ejection angles. As an example of this, Fig. 11.3.4.a shows the results observed by Read, who bombarded Pt surfaces with Li^+ ions. In this figure, one sees that at the lower energies of incidence the curve of β_{max} as a function of E goes through a maximum value, a result that was observed also by Sawyer [*627*] in the reflection of Li^+ from Ni surfaces. Longacre [*443*] who investigated the energy distribution of the secondary ions as a function of the total angle of deviation $\varphi = 2\pi - (\alpha + \beta)$ for the case of Li^+ ions bombarding a Ni surface, found that the ratio $r = E'/E$ of the most probable energy E' of the reflected ions to the energy E of the incident ions decreased with increasing values of φ (cf. Section 11.3.5).

One should also mention the results of Mitropan et al. [*486*] who investigated the secondary-emission coefficient for positive and negative ions as a function of the angle of incidence of the primary-ion beam impinging with an energy of 50 keV on surfaces such as Cu, stainless steel, aluminium, and Be. The results incidate that for Cu and stainless steel targets the number of scattered protons is larger than the number of secondary negative ions at all grazing angles $\gamma = 90° - \alpha$. For Al and Be targets, the number of secondary negative ions (negative R-values) exceeds the number of scattered protons at angles $\gamma > 30° - 40°$, as shown in Fig. 11.3.4.b. It is of interest to note that when a 50-keV proton beam strikes a Cu target at an angle of incidence α, the secondary-emission coefficient increases from $R \approx 0.1\%$ at $\alpha = 0°$ to $R \approx 1.7\%$ at $\alpha = 65°$—an increase of a full order of magnitude.

Studies of the secondary-ion emission from copper monocrystal surfaces were conducted by Nelson and Thompson [*521b*]. For various angles of incidence α, they bombarded with 50-keV H^+, He^+, Ne^+, and

Xe^+ ions the (110) plane of a copper monocrystal, which could be turned around an axis parallel to the [110] direction. The current of secondary ions was measured at a collector which enclosed the target entirely (except a hole for entrance of the primary beam) and was biased to -120 V with respect to the target. Rotating the crystal around the [110] axis (keeping a particular angle of incidence α constant) they observed that the secondary-ion current passed through several minima which occurred whenever the primary beam was incident parallel to lower index planes [i.e., for $\alpha = 30°$ parallel to the (111) and (100) planes; for $\alpha = 45°$ parallel to the (111) planes; and for $\alpha = 60°$, parallel to the (110) and

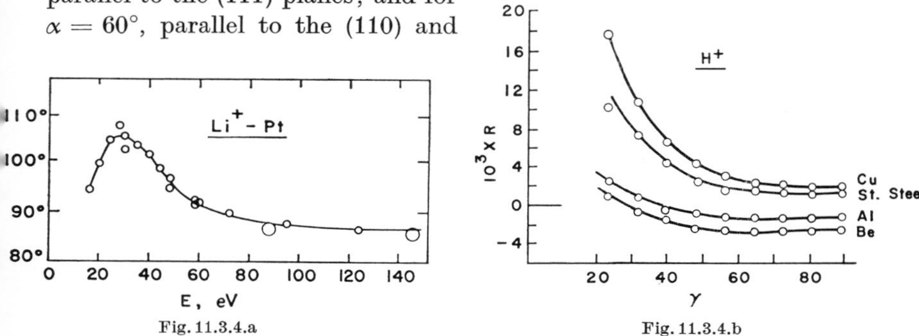

Fig. 11.3.4.a

Fig. 11.3.4.b

Fig. 11.3.4.a. Positions β_{max} of the maximum of the curve of the angular distribution function of scattered Li^+ ions, plotted against the kinetic energy E of the Li^+ ion incident on a Pt surface (READ [*583*])

Fig. 11.3.4.b. The dependence of the secondary-ion-emission coefficient R on the grazing angle γ (degrees) for the case of 50-keV protons striking Cu, stainless steel, Al, and Be targets (MITROPAN et al. [*486*])

(100) planes] or lower index directions (i.e., for $\alpha = 45°$, parallel to the [100] direction and for $\alpha = 60°$, parallel to the [110] direction). NELSON and THOMPSON interpret the appearance of such minima as the results of channeling of the incident ion between adjacent close-packed planes or down "open channels" along certain crystallographic directions (see discussion in Section 10.3.2) since hereby the chance of re-emission of the incident ion is greatly reduced. The observed magnitudes of the secondary-emission coefficient at these minima suggest that the efficiency of channeling in copper (fcc lattice) is the highest along a [110] direction and becomes successively smaller in the directions of the (111) and (100) planes.

11.3.5. Energy Distribution of Reflected and Secondary Ions

The energy distribution of primary ions reflected or re-emitted from metal surfaces has been investigated by several authors [*91, 102, 186, 188, 270, 443, 557, 627, 739, 740a, 751a, 793*] as a function of incident-ion energy E, angle of incidence α, and the masses M and m of the target atom and incident ion, respectively, Let E_2 be the energy of a reflected or reemitted ion and let $(E_2)_{max}$ be the limiting energy above which the distribution curve $f(E_2)$ is zero. For normal incidence with $m < M$, EREMEEV et al. [*186, 188*], ZANDBERG [*793*], BRUNNEÉ [*102*], and VEKSLER [*739*] report that $(E_2)_{max}$ increases linearly with increasing energy E of the incident ion, and also increases as the mass of the incident ion decreases. This result may be considered as support of a model postulating elastic collisions between incident ions and individual *free* atoms of the target.

From the classical laws of elastic collision, one finds the relationship (EREMEEV [*186, 188*], BRUNNEÉ [*102*])

$$(E_2)_{max} = \frac{M - m}{M + m} E = \mu' E . \tag{11.3.5–1}$$

The cutoff energy $(E_2)_{min}$ at the low end of the distribution curve is given by

$$(E_2)_{min} = \left(\frac{M - m}{M + m}\right)^2 E = \mu'^2 E . \tag{11.3.5–2}$$

From this elastic-collision model, it also follows that the energy distribution function is

$$f(E_2) = \frac{dN}{dE_2} = \begin{cases} \text{constant for } (E_2)_{min} < E_2 < (E_2)_{max} , \\ 0 \qquad \text{for } E_2 < (E_2)_{min} \text{ and } E_2 > (E_2)_{max} . \end{cases} \tag{11.3.5–3}$$

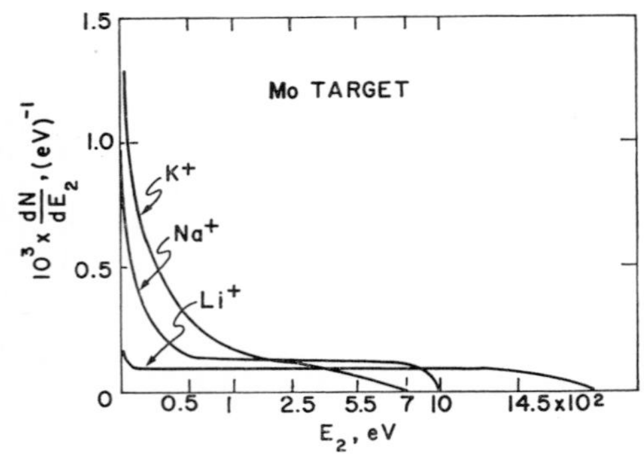

Fig. 11.3.5.a. The energy distribution dN/dE_2 of Li^+, Na^+, and K^+ ions reflected from a cleaned Mo surface. The primary ions were incident normally at an energy of 2 keV (BRUNNEÉ [*102*])

Fig. 11.3.5.a illustrates BRUNNEÉ's [*102*] results for the energy distribution of Li^+, Na^+, and K^+ ions reflected from a cleaned Mo surface when the primary ions are incident normally at an energy of 2 keV. For secondary-ion energies $E_2 > 50$ eV, the distribution curves are rather level and the cutoff value $(E_2)_{max}$ agrees well with the value calculated from Eq. (11.3.5–1). Table 11.3.5.A compares the calculated values of μ''s with the experimental values obtained from the results of EREMEEV [*186, 188*] and VEKSLER [*739*] by use of Eq. (11.3.5–1). The good agreement between calculated and observed values of μ''s indicates that the model of elastic collisions between an incident ion and a "free" lattice atom describes the observed phenomenon rather well. However, at the lower cutoff value $(E_2)_{min}$ calculated from Eq. (11.3.5–2) the curves do not drop to zero but extend considerably into the region between $(E_2)_{min}$ and 0. Similar results [$f(E_2) > 0$ for $E < (E_2)_{min}$] have been obtained by VEKSLER [*739, 740*] for Cs^+ ions reflected from Ta, W, and Mo surfaces; by ZANDBERG [*793*] for Na^+, K^+, and Rb^+ ions reflected from W and Pt surfaces; and by EREMEEV for primary Li^+ and K^+ ions re-emitted from Ta surfaces. The assumption of several collisions in

Table 11.3.5.A. *Comparison of calculated values of $\mu' = (M - m)/(M + m)$ with values which have been obtained from the experimental results of Eremeev [186, 188], BRUNNEÉ [102] and Veksler [739]*

Target	Incident ion	μ'		$(E_2)_{max}$ for $E = 2$ keV		Reference
		Exper.	Theor.	Exper. keV	Theor. keV	
Mo	Li^+	0.90	0.88	1.8	1.7	[*102*]
	Na^+	0.50	0.61	1.0	1.2	
	K^+	0.32	0.40	0.65	0.85	
	Rb^+	0.12	0.06	0.25	0.12	
Ta	Li^+	0.82	0.92			[*186, 188*]
	K^+	0.60	0.64			
Ta	Cs^+	0.13	0.16			[*739*]

sequence between an incident ion and the lattice atoms (BRUNNEÉ [*102*], and PETROV [*557*]) leads to the result that $f(E_2) > 0$ for $E < (E_2)_{min}$ and thus offers a possible explanation for the observed results; but it does not greatly change the distribution $f(E_2)$ in the range $(E_2)_{min} < E < (E_2)_{max}$. In contrast to these results for the values $(E_2)_{min}$ and $(E_2)_{max}$ obtained by the above-mentioned authors for the reflection and re-emission of alkaline ions from metal surfaces, the values of $(E_2)_{max}$ and $(E_2)_{min}$ observed by WALTHER and HINTENBERGER [*751a*] in their experiments* on the reflection of noble-gas ions from metal surfaces disagree strongly with the values predicted by Eqs. (11.3.5–1) and (11.3.5–2). This result is surprising since one would not expect the crude model of elastic collisions between an ion and a free target atom to be less applicable in the case of incident noble-gas ions than for alkaline ions. In the experiments of WALTHER and HINTENBERGER different targets (probably gas-covered since the vacuum in the target region was worse than 1×10^{-6} mm Hg during the run) such as Au, Pt, W, Ta, Ag, Cu, Fe, Al, and C were bombarded with various noble-gas ions with different energies ($E = 5$, 10, and 15 keV) at an angle of incidence of 45°.

* In a private communication the author has been informed that more recently performed experiments indicate that the original results regarding the value of $(E_2)_{min}$ have been influenced by a systematical error, which leads to a cutoff of the energy distribution at the low—energy side.

The re-emitted noble-gas ions were detected at an angle of 45° with respect to the normal to the target surface (i.e., at the angle of specular reflection) with a parabola mass spectrograph, which allows the determination of both the charge-to-mass ratio e/m of the re-emitted and reflected ions and their energy distribution. WALTHER and HINTENBERGER observe that the experimentally determined cutoff $(E_2)_{max}$ varies according to the relationship

$$(E_2)_{max} = \frac{M^2 + m^2}{(M + m)^2} E \,, \qquad (11.3.5\text{–}4)$$

instead of according to Eq. (11.3.5–1) as had been found by the other authors [*102, 186, 188, 739, 793*]. They were not able to relate this result to a particular collision model for the incident ion and target atoms and offered only an empirical relationship to describe the function $E_2(M, m, E)$. More experimental material on the energy distribution of noble-gas ions reflected and re-emitted from metal surfaces preferably obtained by different experimental techniques is needed to verify these findings of WALTHER and HINTENBERGER and to exclude the possibility that these results may have also been influenced by systematic experimental errors, as in the case of their reported $(E_2)_{min}$-values (see footnote on page 255).

We turn now to a discussion of the case $m > M$, e.g., the reflection of Cs^+ ions from Mo surfaces as investigated by BRUNNEÉ [*102*], VEKSLER [*739, 740, 740a*], and PETROV [*557*]. In this case, the single-elastic-collision model [Eq. (11.3.3.–1)] predicts no reflection of ions*; but reflected primary ions were actually observed. Fig. 11.3.5.b shows

* It should be mentioned that such a behavior has actually been reported by WALTHER and HINTENBERGER [*751a*] for the reflection or re-emission of noble-gas ions from metal surfaces at the angle of specular reflection. They observed that the target materials Au, Pt, W, and Ta reflected all noble-gas ions (He^+—Xe^+), Ag reflected only He^+—Kr^+ ions, and Cu and Fe reflected only He^+—A^+ ions. However, such a close agreement between the predictions made from a single-elastic-collision model and these results is surprising since their other results for the function $E_2(E)$ [in particular, the values $(E_2)_{max}$] cannot be interpreted by such a model and because several other authors [*102, 557, 739, 740a*] actually observed the reflection of ions for the case $m > M$, although one should keep in mind that they investigated alkali ions and not noble-gas ions. Actually the appearance of reflected ions for the case $m > M$ is not surprising if one considers that the incident ion may come in contact with more than one target atom in one collision process (e.g., whenever the cross section of the incident ion becomes larger than that of the target atom). This process may be described as an increase in the effective mass M_{eff} of the target atom, so that one obtains the condition $m < M_{eff} > M$. PETROV [*558*] and VEKSLER [*740a*] have used the assumption $M_{eff} > M$ in the interpretation of their results. For the case of the reflection of 1-keV Cs^+ ions from a Mo surface (scattered at an angle of 90° with respect to the incident beam), PETROV [*558*] derived a value of $M_{eff} = 179$ atomic mass units, while $M_{Mo} = 96$ atomic mass units.

BRUNNEÉ's results for the reflection of Rb^+ ions ($m < M$) and for Cs^+ ions ($E = 2$ keV) on Mo ($m > M$). For the later case, more than 80% of the reflected ions have energies smaller than 25 eV, a result which agrees qualitatively with the observations of PETROV [*557*]. In contrast to the case $m < M$, for which experimental observations confirm the theoretical prediction that the value of $(E_2)_{max}$ increases linearly with increasing primary-ion energy E, for $m > M$ it increases more slowly with E. Fig. 11.3.5.c illustrates PETROV's [*558*] results for $(E_2)_{max}$ as a function of E for the reflection of Cs^+ on Mo. The shape of the function $(E_2)_{max}(E)$ can be interpreted on the basis of a simultaneous collision between an incident ion and several lattice atoms, as pointed out by PETROV [*558*]. BRUNNEÉ's data for the reflection of Cs^+ on Mo at 2 keV fit the extrapolated curve of $(E_2)_{max}$ as a function of E, and VEKSLER's [*739*] data for the same system agree qualitatively.

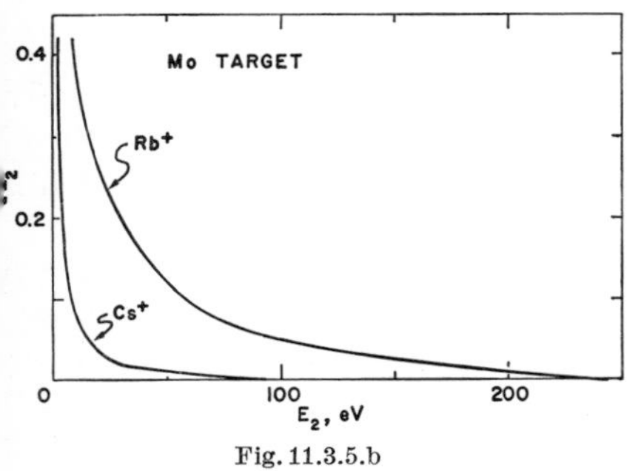

Fig. 11.3.5.b

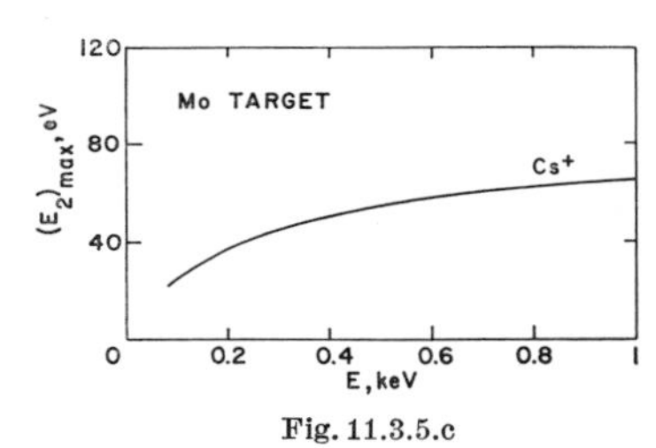

Fig. 11.3.5.c

Fig. 11.3.5.b. The energy distribution dN/dE_2 of Rb^+ and Cs^+ ions reflected from a cleaned Mo surface. The primary ions were incident normally at an energy of 2 keV (BRUNNEÉ [*102*])

Fig. 11.3.5.c. The dependence of the upper-limit energy $(E_2)_{max}$ [above which the distribution curve $f(E_2)$ is zero] on the energy E of the incident ion, for the case of Cs^+ ion bombardment of a Mo target (PETROV [*557*])

The same model of a simultaneous collision between one incident ion and several lattice atoms has also been used by VEKSLER [*740a*] to explain some of his data on the behavior of the energy distribution dN/dE_2 as a function of E for the case $m < M$ as well as $m > M$. He investigated the energy distribution of secondary ions emitted from a hot molybdenum surface [$T = 1400-1450°$ K] during the bombardment by Rb^+ or Cs^+ ions with energies ranging from 15—250 eV and with an angle of incidence $\alpha = 60°$. For primary-ion energies E larger than 60—80 eV, he observed two maxima in the distribution curve dN/dE_2: one at a low value of E_2 (near about 3—5 eV), the other at a larger value of E_2. Only the maximum at the higher value of E_2 can be considered as being caused primarily by scattered primary ions. The observation that the position of this maximum shifts only slowly to higher values of E_2 as the primary-particle energy E increases may be taken as evidence that (for the systems investigated) the scattering of the primary ion is not caused by a collision between an incident ion and the single lattice atom but by collisions between the incident ion and several lattice atoms (i.e., the effect may be described as an increase of the "effective" mass of the target atom; see footnote earlier in this section).

Bradley et al. [*91*] who bombarded a Pt surface with Xe ions ($m < M$), observed that at an ion energy $E \approx 300$ eV the re-emitted Xe^+ ions had an average energy of less than 1 eV and an energy spread of the same order of magnitude. On the basis of this result, they assume that the Xe^+ ions are sputtered from surface layers rather than reflected.

Older experiments on ion reflection, conducted by Gurney [*270*], Longacre [*443*], and Sawyer [*627*] for lower primary-ion energies and different angles of incidence, are representative of gas-covered surfaces because of the inadequate vacuum conditions. The results of these authors indicate that, for a given primary-ion energy E, the mean energy of the re-emitted ion beam increases approximately linearly with increasing angle of incidence (as in the example of K^+ on Pt, Gurney [*270*]).

The foregoing discussion was confined to the reflection or re-emission of *positive* ions, but the few results that follow apply to the energy distribution of re-emitted negative ions from a surface being bombarded by positive ions. Mitropan et al. [*485*] bombarded copper, stainless steel, and aluminum surfaces with beams of purified positive ions (H_1^+, H_2^+, D_1^+, D_2^+) at energies ranging from 200 to 1000 keV and observed the formation of various negative ions such as H_1^-. They reported that the energies of the re-emitted negative ions did not exceed 10 eV. Levine [*434*], who bombarded a tungsten surface with positive hydrogen ions at an energy of 1.24 keV, observed negative hydrogen ions emitted with energies ranging from 0 to 550 eV.

The energy distribution of reflected ions will be discussed further in connection with the ion-reflection theories of von Roos [*611*] and Vapnik et al. [*733*] in Section 11.4.

11.3.6. Dependence on the Target Temperature and on Surface Contamination

Several authors [*24, 188, 189, 190, 192, 217, 540, 558*] have observed that the target temperature influences the reflection and secondary emission of incident ions from metal surfaces. Petrov [*558*], Arifov et al. [*24*], and Eremeev et al. [*188, 189, 190, 192*] all reported that the reflection of alkali ions from metal surfaces such as W, Ta, and Mo is nearly independent of target temperature for temperatures above about 1200° K in the range of primary-ion energies investigated ($E \approx 0.1-20$ keV). For lower target temperatures, however, marked changes in the functions $\varrho'(T)$ or $R(T)$ have been observed in the case of K^+ ion bombardment of a W surface at $E = 2.2$ and 2.7 keV (Eremeev [189, 190] and at $E = 560$ eV (Arifov et al. [24]), and of a Ta surface (Eremeev [189, 190]), as illustrated in Fig. 11.3.6.a. One notices a sharp rise in the function $\varrho'(T)$ at a target temperature of about 800° K for $E \approx 3$ keV (Eremeev [189, 190]) and at about 1100° K for $E = 560$ eV (Arifov [24]). This rise might be partially explained by an increase of thermally evaporated primary ions (surface ionization) with increasing target temperatures, an interpretation which is supported by

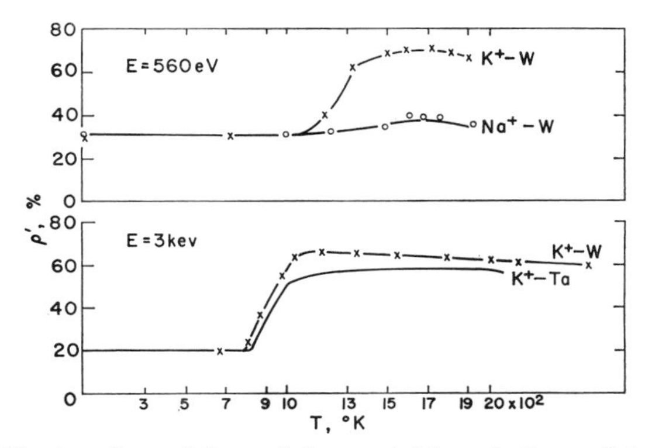

Fig. 11.3.6.a. The dependence of the partially-corrected ion-reflection coefficient ϱ' on the target temperature T, as reported for alkali ions incident on a W surface at 560 eV by Arifov et al. [24], and for potassium ions incident on W and Ta surfaces near 3 keV by Eremeev [189, 190]

the observation of Arifov et al. [24] that with rising temperature the energy distribution of the secondary particles shows an increase in the number of particles with thermal energies—as would be expected if surface ionization became important.

For the reflection of Li^+ ions from a W and Ta surface, Eremeev [188] and Eremeev et al. [189, 190] observed that the partially corrected secondary-emission coefficient R' remains practically unchanged over the investigated temperature range (from room temperature to 2250° K for Ta and from room temperature to 2500° K for W, $R' \approx 20\%$) as can be seen in Fig. 11.3.6.b. A similar result for the function $\varrho'(T)$ has been reported by Arifov et al. [24] for Na^+ ions reflected from W (Fig. 11.3.6.a).

For protons with energies $E > 10$ keV incident on Mo surfaces, Fogel et al. [217] observed that the partially corrected reflection coefficient ϱ' increased rather sharply from $\varrho' \approx 1.6\%$ at 600° C to $\varrho' \approx 3.8\%$ at 1000° C. However, Petrov [558] reported that for H_2^+ ions reflected from a Mo surface, the value of ϱ' decreased by a factor of 1.3—1.5 when the target temperature was raised from room temperature to about 1600° K.

For the reflection of noble-gas ions, FOGEL et al. [*217*] reported that the reflection coefficient ϱ' for a Mo target had a positive temperature coefficient, while BRADLEY et al. [*91a*] observed a negative temperature coefficient for the case of a Cu target.

EREMEEV and YUR'EV [*192*] investigated the dependence of the function $R'(T)$ on the presence of oxide films in the case of K^+ and Li^+ ions reflected from W and Ta surfaces. As can be seen in Fig. 11.3.6.b, the presence of an oxide film on the surface of a metal bombarded with K^+ ions increases the reflection factor, in agreement with the findings of PAETOW and WALCHER [*540*] for Cs^+ ions reflected from an oxidized W surface. As can be seen in Fig. 11.3.6.b, at high target temperatures ($T > 1500°$ K) the value of R' for K^+ ions reflected from an oxidized Ta

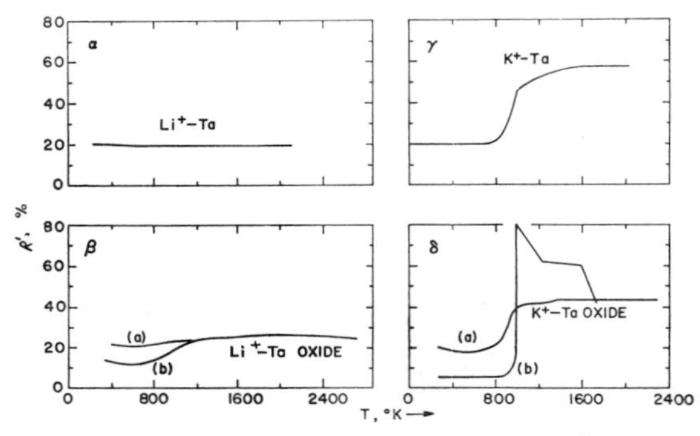

Fig. 11.3.6.b. The dependence of the secondary ion-emission coefficient R' on the target temperature T for the case of Li^+ and K^+ ions striking a Ta and a Ta oxide surface (primary ions in the energy range 1—40 keV) (EREMEEV [*188*] and EREMEEV et al. [*189, 190*])

surface approached the value for a clean Ta surface. It is of interest to note [curves (*a*) and (*b*) of graph δ of Fig. 11.3.6.b] that the relatively sharp peak of the function $R'(T)$ at about 1000° K for K^+ reflected from oxidized Ta disappears after annealing the target at 2200° K (curve *a*).

The available results on surface contamination as an influence affecting ion reflection are still few and rather controversial. HAGSTRUM [*289*] observed that the value of ϱ for 200-eV He^+ ions decreased from about 0.18% for clean tungsten to about 0.05% for contaminated tungsten, and similarly for He^+ and for Ne^+ on Mo FOGEL et al. [*217*] reported that ϱ' increased when the surface was cleaned. However, for the reflection of alkali ions from surfaces covered to an unknown degree with contaminant of unknown composition, ARIFOV et al. [*24*] found for Cs^+ on Ni and EREMEEV observed for K^+ on liquid Sn that the re-emission of the incident ion decreased as the surface became cleaner. However, the dominant factor in the latter cases seems to be the work function of the system comprising adsorbed atom and metal surface. This can be seen from the results of PAETOW and WALCHER [540] who studied the reflection of Cs^+ ions from clean and oxidized tungsten and from tungsten with a monolayer of adsorbed caesium. They found that the value of ϱ' decreases as the work function of the surface decreases. For example, for Cs^+ incident at $E = 600$ eV,

$$\varrho'_{\mathrm{W-O}} \approx 1.7\% > \varrho'_{\mathrm{W}} \approx 1.4\% > \varrho'_{\mathrm{W-Cs}} \approx 0.9\%,$$

$$\varphi_{\mathrm{W-O}} \approx 6.3 \text{ eV} > \varphi_{\mathrm{W}} \approx 4.5 \text{ eV} > \varphi_{\mathrm{W-Cs}} \approx 1.8 \text{ eV}.$$

More quantitative results on the dependence of the reflection coefficient ϱ on the degree of surface coverage θ are not yet available.

11.4. Theoretical Treatment of Ion Scattering from Metal Surfaces

Theoretical treatments of the scattering of ions from metal surfaces were attempted by von ROOS [611] and VAPNIK et al. [733]. ROOS treated the ion reflection phenomenon as a collision process between the incident ions and free lattice atoms, i.e., he neglected the binding of the lattice atoms so that the lattice was regarded as a gas. In a further simplification he used the BORN approximation for the calculation of the interaction cross section between the incident ion and the lattice atom. In this he assumed that the interaction potential between a lattice atom and an incident ion is of the form

$$V(r) = \frac{Z_{\mathrm{eff}}}{r} \exp\left(-\frac{r}{r_0}\right), \tag{11.4–1}$$

where the screening radius r_0 and the effective charge Z_{eff} are two parameters to be fitted by the experimental results, and r is the distance between the centers of the ion and the lattice atom. It should be noted, however, that the application of the BORN approximation to such calculations is not valid (VAPNIK [733]) in the region of low incident-ion energies. Only when the ion energies E are much greater than the binding of the lattice atom do the collisions show the same angular distribution and energy distribution as for collisions between free particles so that the BORN approximation becomes applicable.

The use of such a free-particle-collision model allows VON ROOS to describe the reflection process by applying the BOLTZMANN equation

$$\frac{dN(r, V, t)}{dt} = -n\sigma_s(V)\, V N(r, V, t) + n \int dV' f(V', V)\, V' \sigma_s(V') N(r, V', t), \tag{11.4–2}$$

which represents the time rate of change of the distribution function $N(r, V)$ of the incident ions moving among the "free" lattice atoms. Here $\sigma_s(V)$ is the cross section for the scattering process as a function of incident-ion velocity V, $f(V', V)$ is the probability for the change of the ion velocity from V' to V in a single collision, and n is the number of collision centers per unit volume. The first term on the right-hand side of Eq. (11.4–2) represents the rate at which particles leave the space element $(\vec{r}, \vec{V})$, and the second term gives the rate at which particles are scattered into the space element $(\vec{r}, \vec{V})$.

For the case of a stationary state, VON ROOS solved the BOLTZMANN equation approximately by expanding $N(z, K, \theta)$ in a series of spherical harmonics $P_n(\cos\theta)$. Here z is the penetration depht of an incident ion, $K = MV/\hbar$ is a characteristic wavenumber of the incident ion, and θ is the angle between the velocity V and the normal to the surface. By including only terms that are either independent of θ or porportional to $\cos\theta$ in the expansion of N, he was able to express ϱ as a function of E in the form

$$\varrho = \varrho_0(\mu) - \varrho_1(\mu)\,\frac{M r_0^2}{\hbar^2}\,E\,. \tag{11.4–3}$$

Here the term $\varrho_0(\mu)$ represents a function depending only on the mass ratio $\mu = M/m$, where M is the mass of the target atom and m is that of the incident ion. The factor $\varrho_1(\mu)$ is a complicated integral not given explicitly by the author. The other symbols have the meanings established for Eqs. (11.4–1) and (11.4–2).

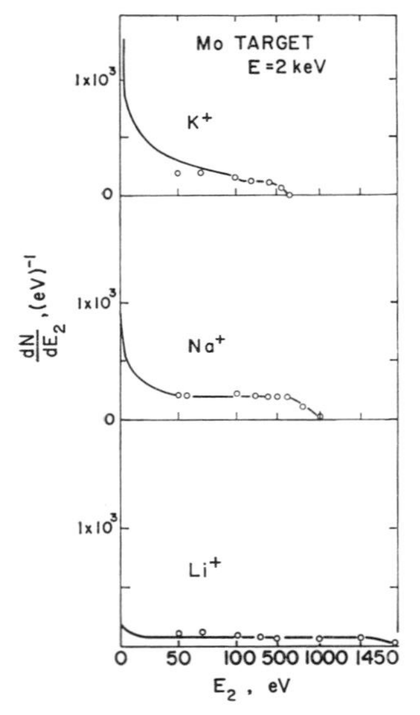

Fig. 11.4.a. Comparison of the experimental data for the energy distribution function $f(E_2) = dN/dE_2$ of reflected ions (solid curves, according to data of BRUNNEÉ [102]) and the calculated values (shown as circles, according to VON ROOS [*611*])

According to Eq. (11.4–3), the reflection coefficient ϱ decreases linearly with increasing ion energy E. For higher ion energies ($E > 2$ keV) this is in qualitative agreement with the experimental results for Li^+, Na^+, and K^+ ions striking a Mo surface (BRUNNEÉ [*102*]). The deviations of the function $\varrho(E)$ at lower energies from the behavior predicted by Eq. (11.4–3), in particular the sharp rise in the region $E < 1$ keV, cannot be explained by VON ROOS's theory. However, the rate at which $\varrho(E)$ increases with decreasing E might be expected to exceed that indicated by Eq. (11.4–3) as soon as the scattering distribution changes from isotropy in the center-of-mass system to isotropy in the laboratory system, i.e., when E becomes small enough that the momentum begins to be delivered to the lattice instead of to individual atoms so that the binding between lattice atoms must be taken into account. On the basis of Eq. (11.4–3), von Roos calculated values for the reflection coefficient ϱ for K^+, Rb^+, and Cs^+ ions incident on a Mo surface for $E \approx 2$ keV. The values obtained are 3—4 times the experimental ones reported by Brunneé [*102*]. On the other hand, he also calculated the energy distribution $N(E)$ of the reflected ions. In view of the drastic approximations made in the theory, the agreement between his calculated values (shown as circles in Fig. 11.4.a) for the function $N(E)$ and the experimental data (solid curve), as for instance reported by Brunneé for Li^+, Na^+, and K^+ on a Mo surface, might be somewhat fortuitous.

The theoretical treatment of the ion reflection given by Vapnik et al. [*733*] is basically similar to that of von Roos. However, the former differ from the latter in their method of calculating the collision cross sections σ_s. According to their results, the function $\varrho(E)$ is independent of the ion energy E.

12. Neutralization of Ions on Metal Surfaces (Potential Emission of Secondary Electrons)

12.1. Introductory Remarks and Definitions

When an ion or an excited atom approaches a metal surface, the ion may be neutralized and the excited atom may be de-excited or undergo resonance ionization. These processes are of particular interest because of the role they play in secondary-electron-emission phenomena, in gas-discharge phenomena, in the interaction of plasmas with electrode surfaces, and in the surface-ionization mechanism. The evidence available so far from experimental data for slow incident ions leads to the conclusion that the electronic transitions involved in the processes mentioned above are almost independent of the kinetic energy of the incident particle but are governed by its potential energy of excitation.

The following discussion will be limited to four types of electronic transition which an incident ion or excited atom may undergo at a metal surface. These are resonance neutralization, resonance ionization, Auger de-excitation, and Auger neutralization.

Oliphant and Moon [*537*] first suggested the possibility of resonance neutralization of an ion at a metal surface by the mechanism of metal electron tunneling through the potential barrier between the surface and the adion (see Chapter 1) to fill an excited state of the incident particle. Massey [*459, 460*], Shekhter [*650*], and Cobas and Lamb [*116*] investigated theoretically the two-step process of resonance neutralization followed by secondary-electron emission in an Auger de-excitation of the excited atom.

If one considers the system comprising the incident ion A^+ and all metal electrons Ne_m^-, the resonance neutralization processes may be represented by

$$A^+ + Ne_m^- \to A^* + (N-1)\, e_m^-\,, \tag{12.1–1}$$

where A^* is the excited atom formed and N is the total number of electrons in the metal.

If the shift from one energy level to another in the particle as it arrives at the surface is neglected, the process (12.1–1) can occur whenever the condition $e\overline{\varphi} < E_{M1} < W_a$ is fulfilled. Here $e\overline{\varphi}$ is the average work function of the metal, W_a is the "outer work function" (see Chapter 1) and is equal to the total depth of the potential well representing the metal (Fig. 12.1.a, curve α), and E_{M1} is the energy* of the metal electron relative to the FERMI level W_i. The potential diagram shown in Fig. 12.1.a. curve α, illustrates the resonance-neutralization process. The solid arrow indicates the electronic transition involved; a dashed arrow represents the reverse process of resonance ionization of an excited atom at a metal surface, the process $A^* \to A^+ + e_m^-$. For noble gases, for instance, resonance ionization evidently is possible only if the atom is excited and the condition $e\overline{\varphi} < E_{M1}$ is fulfilled.

The excited atom A formed in the process (12.1–1) may in turn undergo AUGER de-excitation near the metal surface if the condition $eI > e\overline{\varphi}$ is fulfilled, eI being the ionization energy of the incident particle. The AUGER de-excitation process, which will be discussed in more detail in Chapter 13, may be written as

$$A^* + Ne_m^- \to A + e^- + (N-1)\, e_m^-\,. \tag{12.1–2}$$

Here e^- is the secondary electron ejected from the metal surface. Fig. 12.1.a, curve β, shows the potential diagram for the Auger de-excitation of a metastable helium atom (2^3S) on a Mo surface. The electron-exchange process is indicated by the full lines. Whenever the condition $E^* < e\overline{\varphi}$ is fulfilled, where E^* is the excitation energy, secondary electrons are emitted. From the diagram in Fig. 12.1.a, curve β, one can see that the maximum kinetic energy $(E_e)_{\max}$ of the emitted secondary electron is given by

$$(E_e)_{\max} = E^* - e\overline{\varphi} \tag{12.1–3}$$

and the minimum energy by

$$(E_e)_{\min} = E^* - W_a\,,\quad (E_e)_{\min} = 0 \text{ if } E^* < W_a\,. \tag{12.1–4}$$

HAGSTRUM [282] pointed out that while a metastable atom can undergo AUGER de-excitation whenever $eI > e\overline{\varphi}$, resonance ionization followed

* The quantity E_{M1} is identical with the quantity E_{m1} shown in Fig. 12.1.a, and correspondingly $E_{M2} \equiv E_{m2}$.

by AUGER neutralization is the more probable process (for a given mean distance of nearest approach r) if $eI - E^* < e\overline{\varphi}$. Consequently it is difficult to distinguish secondary electrons ejected by metastable atoms

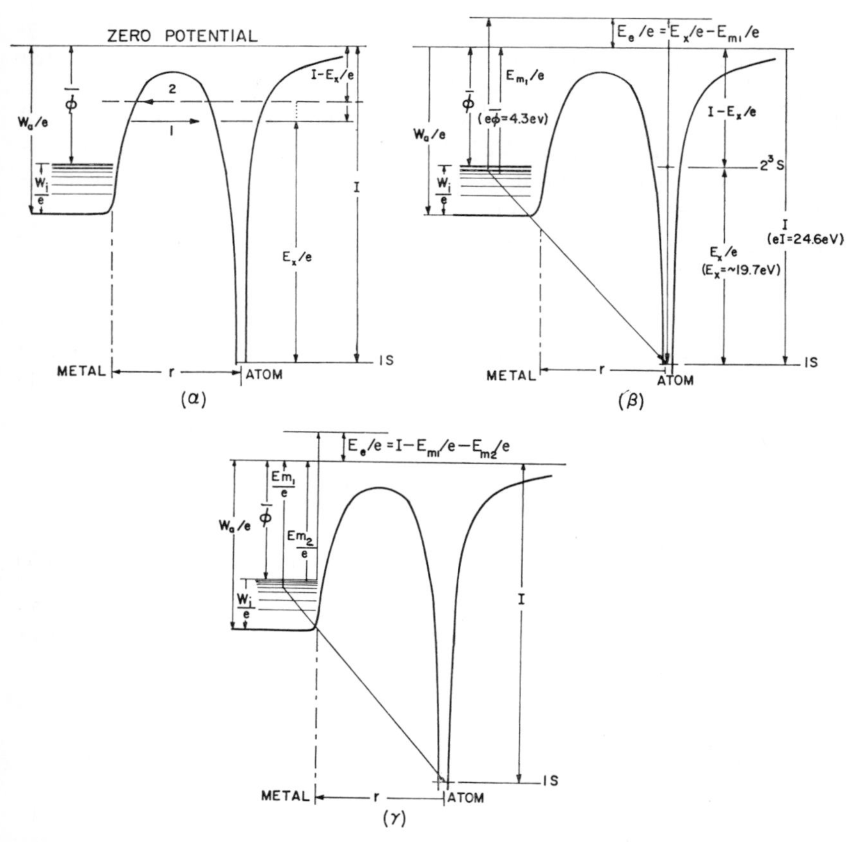

Fig. 12.1.a. Diagram α: Schematic illustration of the resonance neutralization of an ion (transition *1*) or resonance ionization of an excited atom (transition *2*) at a metal surface. Diagram β: Schematic illustration of the AUGER de-excitation of a metastable helium atom (2^3S) on a Mo surface. The electron-exchange process is indicated by the full lines. Diagram γ: Schematic illustration of the AUGER neutralization of an ion at a metal surface. The notations used in diagrams α, β, and γ are defined in section 12.1. (After HAGSTRUM [*282*])

from those ejected by the incident ions. However, the distinction can be made by observing the maximum and minimum energies of the ejected electrons [Eqs. (12.1–3), (12.1–4), (12.1–6), and (12.1–7)].

The broadening of the energy levels as the particle nears the surface reduces the likelihood of the two-stage resonance-neutralization process (resonance neutralization followed by AUGER de-excitation). For instance HAGSTRUM [*282, 286*], who bombarded clean surfaces of a refractory metal with noble-gas ions, found that AUGER neutralization is more probable than the two-stage resonance neutralization for slow primary ions ($E < 10$ eV).

SHEKHTER [*650*] suggested and treated theoretically the one-stage process of secondary-electron emission by AUGER neutralization in which the ion is neutralized directly to the ground state. This process has been the subject of further extensive experimental study and theoretical interpretation (HAGSTRUM [*282*], TAKEISHI [*701*]). The AUGER neutralization process can occur if $eI > e\overline{\varphi}$, although the resonance-neutralization process becomes more probable at a given r if the condition $e\overline{\varphi} < eI < W_a$ is fulfilled. The process of AUGER neutralization will be written as

$$A^+ + N e_m^- \rightarrow A + e^- + (N - 2)\, e_m^- . \qquad (12.1\text{–}5)$$

The potential diagram (Fig. 12.1.a, curve γ) is a schematic diagram of the AUGER neutralization process, the full lines indicating the electronic transitions. The limiting kinetic energies of the emitted secondary electrons are seen to be

$$(E_e)_{\max} = eI - 2e\varphi \qquad (12.1\text{–}6)$$

and

$$(E_e)_{\min} = eI - 2W_a,\ (E_e)_{\min} = 0 \text{ if } eI < 2W_a . \qquad (12.1\text{–}7)$$

As will be discussed in more detail in the following sections, most noble-gas ions with small kinetic energy ($E \lesssim 10$ eV) are neutralized by this process on atomically clean refractory-metal surfaces.

12.2 Auger Neutralization; Resonance Neutralization

12.2.1. Experimental Methods

Ion neutralization and AUGER de-excitation of excited atoms have been studied experimentally only in a few rare cases. In the following discussion, those methods that allow one to draw conclusions about the neutralization on the basis of measurements of secondary-electron emission will be distinguished from those that neglect thermionic and secondary-electron emission and use the primary-ion current I_1, the secondary-ion current I_2, and their difference $I_1 - I_2$ to determine the neutralization rate.

The older work, such as that of PENNING [*551, 552*], OLIPHANT [*535, 536*], ROSTAGNI [*615*], and D'ANS, DARIOS and MALASPINA [*135*], will not be discussed since it was done entirely without definite knowledge of the condition of the target surface or of the composition, charge, and energy of the incident ions. Instead the reader is referred to the survey of such older data by MASSEY and BURHOP [*461*] and to the critical discussion by HAGSTRUM [*286*]. Most of the available results of the neutralization process (and also of the de-excitation process, Chapter 13) have been obtained by the study of secondary-electron emission. Especially noteworthy are the more modern investigations of HAGSTRUM [e.g. *284, 285, 286, 287, 288*] in which the experimental conditions were carefully controlled. In his studies of secondary-electron emission when metal surfaces were bombarded with low-energy ions, the primary ions produced by electron impact passed through a magnetic mass analyzer and then were focused by an electrostatic lens onto a ribbon target, as shown in the schematic diagram of Fig. 12.2.1.a. The kinetic

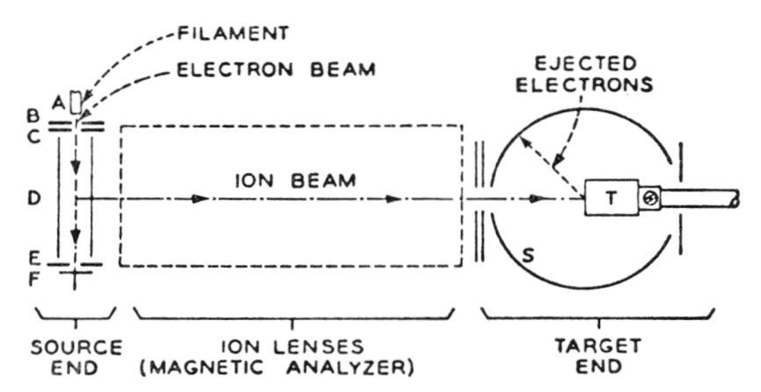

Fig. 12.2.1.a. Schematic representation of the experimental arrangement used by HAGSTRUM [*284*] in his studies of the potential emission of secondary electrons

energy of the ion beam could be varied from 10 to 1000 eV, and the total energy spread of the beam was less than 1 eV. The target was surrounded by a spherical collector, large compared to the size of the flat target. However the electric field distribution around a flat target inside such a spherical collector may still be considered unfavorable for the determination of the energy distribution of the secondary electrons. The electric field acts differently on different velocity components, and as a result the apparent energy of some electrons is too low, as pointed out by HAGSTRUM himself. However, all his studies have been conducted with it.

HAGSTRUM also had to consider the effects of the fringing magnetic field in the neighborhood of his target (e.g., 45 gauss in his experiments with Xe^+ ions). He tried to balance the fringing field with a field applied by an auxiliary magnet straddling the tube in the region of the target. With an uncompensated fringing field, the electron energy distribution is shifted to regions of lower electron energies. His experimental data indicate that the fringing field could not be balanced out uniformly over the volume of the collector, since for a 1-eV electron ($rB = 3.4$ cm-gauss for a 1-eV electron) the residual field must be reduced considerably below 1 gauss to make the effect undetectable. His measurements of the total secondary-electron yield γ as a function of the kinetic energy of the incident ion for different incident ions and target surfaces were conducted by maintaining the collector at only a small positive voltage relative to the target ($V_{tc} \approx 2-5$ V). According to the author, these measurements were reproducible to within 5%. The kinetic-energy distribution of the secondary electrons was determined by applying retarding potentials between the target and the electron collector. Corrections for contact potentials were made by the method of retarding the thermoelectrons from the heated target. The target surfaces used were cleaned thoroughly by the flash-filament technique and the vacuum in the target region was in most cases of the order of 4×10^{-9} mm Hg.

Similar ion-beam techniques (but without the use of a mass analyzer) have been employed by Parker [*544*], Arifov and Rakhimov [*26*], and Oliphant [*536*]. Parker produced an ion beam in a Finkelstein [*213*] type ion source (gas ionization by electron impact) with an energy spread of the order of 2 eV. This beam struck either cleaned metal surfaces or surfaces with different types of adsorbed gas layers, and he determined the secondary-electron emission. In the experiment of Arifov and Rakhimov also, an ion beam struck the metal target; but its intensity was modulated by a rectangular pulse. The secondary-electron current emitted from the target reached a collector, whose potential was modulated by a sawtooth voltage so that the authors could distinguish between the contributions from secondary-ion and electron currents. The primary-ion beams used in the experiments of Oliphant undoubtedly contained singly- and doubly-charged ions as well as metastable atoms in proportions which could be functions of the ion-beam energy. Also, because of the inadequate vacuum techniques applied in his experiments, the data refer to contaminated surfaces only.

Investigations of the secondary-electron emission from metal surfaces under bombardment by ions with low energies have also been conducted with methods other than the ion-beam technique. For instance, Varney [*735*] and Molnar [*490*] used a pulsed Townsend discharge. In such experiments a parallel-plate Townsend discharge triggered by a flash of ultraviolet on the cathode gives a current to the anode which depends strongly on the total number γ of secondary electrons ejected by the ions incident on the cathode. The values of γ are measured as a function of ε/p, the ratio of field strength to gas pressure. The total secondary-electron-emission coefficient γ has been related to the first Townsend coefficient for ionization by electrons. The main disadvantages of the method are the uncertainty in the energy spread of the incident ions and in their composition as well as the certainty of gas coverage on the target surface.

Other ways of determining γ indirectly have been used in measurements of the normal cathode fall in glow discharges (TAKEISHI [*702*]) and in measurements of breakdown voltages (HAGSTRUM [*286*]).

As already mentioned, ion neutralization can be studied by measuring the currents of primary and secondary ions as well as by measuring currents of secondary electrons. ARIFOV and KHADZHIMUKHAMEDOV [*29*] used the double-modulation technique (already mentioned in reference [*26*]) to determine the various secondary-electron currents resulting from interaction of the primary-ion beam with the metal surface), and deduced the neutralization rate from the difference between the currents of primary and secondary ions. They neglected the contribution of secondary electrons (except for thermionic electrons) to the measured secondary-ion currents.

12.2.2. Experimental Results

12.2.2.1. Singly-Charged Ions

12.2.2.1.1. Secondary-Electron Yield γ and its Dependence on the Energy E of the Incident Ion

Fig. 12.2.2.1.1.a shows the experimental data for the total secondary-electron yield γ for Mo surfaces bombarded by various noble-gas ions with energies ranging from 10 to 1000 eV (HAGSTRUM [*285, 286*]). The electron beam producing the ions had energies of 100, 33, 28, and 22 eV in the ionization region of the source. Since the last three energies are just below the threshold value for the formation of metastable ions, the values presented in Fig. 12.2.2.1.1.a are not affected by a detectable amount of metastable ions, in contrast to the values reported previously for the same systems by HAGSTRUM [*277, 281*].

The general characteristics of the $\gamma(E)$ curves in Fig. 12.2.2.1.1.a for molybdenum are the same as those for tungsten. The absolute values of γ, however, are greater for molybdenum than for tungsten. This effect is partly due to the difference in work function ($e\bar{\varphi}_{Mo} = 4.3$ eV; $e\bar{\varphi}_W = 4.52$ eV), since a decrease in φ will increase the electron escape probability and result in a larger value of γ. However, other

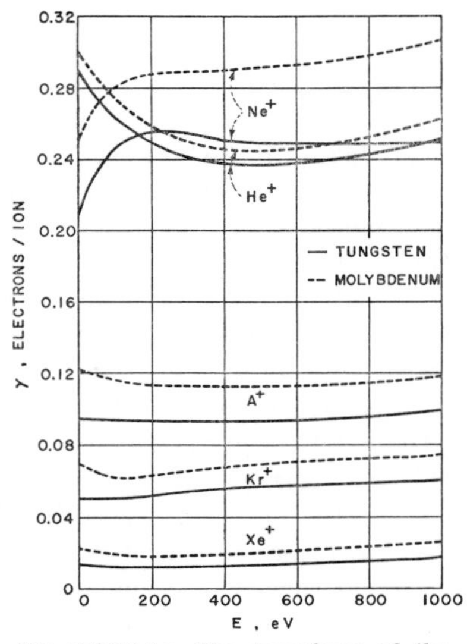

Fig. 12.2.2.1.1.a. The dependence of the secondary-electron yield γ on the kinetic energy E of the incident ion for the case of various noble-gas ions incident on atomically clean wolfram and molyb- denum surfaces (HAGSTRUM [*285, 286*])

factors also appear to contribute to such a behavior. For example, as the difference between the ionization energy eI of the incident particle and the work function $e\overline{\varphi}$ of the incident particle decreases, the ratio $\gamma(\mathrm{Mo})/\gamma(\mathrm{W})$ increases from 1.04 for He^+ to 1.69 for Xe^+, as can be seen in Table 12.2.2.1.1.A.

Table 12.2.2.1.1.A. *Total secondary-electron yields (electrons per incident ion) for 10-eV noble-gas ions on various metal surfaces*

Incident ions	$E = 10$ eV			Value of E not indicated	$E = 10$ eV		
	γ(Ta)	γ(Mo) [286]	γ(W) [286]	γ(Ni) [702]	γ(Pt) [544]	$\frac{\gamma(\mathrm{Mo})}{\gamma(\mathrm{W})}$	$eI - e\overline{\varphi}$ (Mo target)
	$e\overline{\varphi}=4.15$ eV	$e\overline{\varphi}=4.27$ eV	$e\overline{\varphi}=4.52$ eV	$e\overline{\varphi}=5.0$ eV	$e\overline{\varphi}=6.3$ eV		
He^+	0.14 [277]	0.300	0.290	0.159		1.04	20.3
Ne^+		0.254	0.213	0.128		1.19	17.3
A^+	0.007 [544]	0.122	0.095	0.050	0.021	1.29	11.5
Kr^+		0.069	0.050			1.38	9.7
Xe^+		0.022	0.013	0.005		1.69	7.8

Hagstrum's result for the function $\gamma(E)$ for He^+ and A^+ ions incident on cleaned Mo surfaces have been confirmed qualitatively by Mahadevan et al. [452a] and Medved et al. [467b]. The authors bombarded a thoroughly cleaned Mo surface with He^+ and A^+ ions in their ground state over the energy region from about 100 eV to 2500 eV. Their results (ref. 452a) for He^+ on Mo agree rather well with those of Hagstrum for energies up to 500 eV; above this energy they observe a steeper rise of the curve $\gamma(E)$ than Hagstrum. This rise of the function $\gamma(E)$ above approximately 500 eV has been related to the contribution of electrons emitted by the kinetic ejection mechanism increasing relative to the potential emission. The observed onset value near 500 eV is consistent with a model for the kinetic emission of electrons for the system He^+ on Mo, as suggested by Parilis et al. [543] (see Chapter 14, Section 14.4). The values of $\gamma(E)$ observed for the system A^+ on Mo by Hagstrum over the energy region from about 100 eV to 1000 eV are approximately 25% higher than those observed by Mahadevan et al. and Medved et al. for the corresponding energy region. In addition, Mevded et al. observe that the curve $\gamma(E)$ remains practically constant ($\gamma \approx 0.074$ electrons/ion) over the energy region from 100 eV to 700 eV and then shows a steep increase, while Hagstrum's $\gamma(E)$ curve passes through a broad minimum (between approximately 200 eV—600 eV, see Fig. 12.2.2.1.1.a) and shows only a slight increase with increasing ion energy E above approximately 600 eV. Again Medved et al. relate the observed sharp rise of the function $\gamma(E)$ above approximately 700 eV to an increase of the contribution of electrons emitted by the kinetic ejection mechanism relative to the potential ejection mechanism. The observed onset energy, at about 700 eV for the sharp rise in the function $\gamma(E)$, is in good agreement with the threshold value for the kinetic emission of electrons as predicted by the theory of PARILIS et al.

As can be seen in Fig. 12.2.2.1.1.a, values of γ increase as the mass of the incident ion decreases. This trend is in agreement with HAGSTRUM's theory of AUGER neutralization, as will be discussed in Section 12.2.3. The initial drop in the $\gamma(E)$ curve for He^+ ions as the ion energy increases is also explained by the theory based on AUGER neutralization. As the He^+ ion approaches the metal surface with increasing velocity, γ drops because the effective ionization potential is reduced and the energy states are broadened since (on the average) the AUGER neutralization occurs nearer to the surface.

HAGSTRUM's theory of AUGER neutralization cannot, however, account for the experimental observation that γ increases above approximately 500 eV. As mentioned before, here another mechanism of secondary-electron emission, the kinetic ejection (see Chapter 14) starts to become the dominant emission mechanism. This seems to be supported also by the marked change in the energy distribution of the secondary electrons as the ion energy increases.

For the case of Ne^+ ions, the rapid rise of γ with increasing ion energies is attributed to a portion of the Ne^+ ions being resonance neutralized and de-excited by an AUGER de-excitation process. Such a mechanism would lead to higher yields than would the AUGER neutralization process. This interpretation is also supported by the corresponding data for the energy distribution of the secondary electrons. The $\gamma(E)$ curves for the heavier noble-gas ions (A^+, Kr^+, Xe^+), do not change much with the energy of the primary ion except for a slight drop at low ion energies. These results can be reasonably well interpreted by an AUGER neutralization process.

As will be discussed in more detail in Section 12.2.2.1.3, the value of γ is extremely sensitive to even a small fraction of a monolayer of contamination on the target. HAGSTRUM, for instance, showed that the presence of a monolayer of adsorbed gas (unknown composition on a W surface) decreased $\gamma(He^+)$ from 0.29 to 0.18 electrons per incident ion. MAHADEVAN et al. [*452a*] reported that the value of γ for the case of 600-eV A^+ ions incident on a Mo surface covered by a monolayer of gas (unknown composition) is 25% smaller than for the case of a clean Mo surface.

More results on secondary-electron emission from Mo surfaces under bombardment by noble-gas ions have also been reported by ARIFOV and

RAKHIMOV [*26*] and by ARIFOV et al. [*30a*]. Attempts to study the emission of secondary electrons under the impact of neutral noble-gas atoms on a cleaned Mo surface have been conducted by MEDVED et al. [*467b*] and by COMEAUX and MEDVED [*119a*] for helium and argon atoms with energies ranging from 500 eV to 2.5 keV and by ARIFOV et al. [*30a*] for argon atoms in the energy range from 0.2 to 2.0 keV. Since the emission of the secondary electrons by the impact of neutral atoms results from the kinetic emission mechanism, the discussion of the results will be given in Chapter 14, Section 14.3.1.1.1.

Besides such studies of the neutralization of noble gases on metal surfaces, the only others are a very few investigations of the neutralization of positive alkali ions on such targets as W, Mo, and Ta as a function of the target temperature and the primary-ion energy [*29*, *498*]. ARIFOV and KHADZHIMUKHAMEDOV [*29*] used the double-modulation method, described briefly in Section 12.2.1, to measure the primary- and secondary-ion currents. They determined the integral neutralization coefficient R_0 on the assumption that, because of the conservation of matter, the coefficient of secondary-ion emission R_Σ plus the integral coefficient of neutralization must be equal to 100%, i.e.,

$$R_\Sigma + R_0 = 100\,\%\,, \qquad (12.2.2.1.1\text{–}1)$$

with

$$R_\Sigma = \frac{I_s + I_e + I_d}{I_1}\,, \qquad (12.2.2.1.1\text{–}2)$$

where I_s is the current of scattered ions, I_e is the current of evaporated ions, I_d is the current of diffused ions, and I_1 is the current of primary ions in the beam. The values of R_Σ were not corrected for the contributions from secondary- and tertiary-electron currents (Chapter 14), except for a correction for thermoelectrons at the higher target temperatures. In view of the results discussed in Chapter 14, neglecting such electron currents does not seem justified in general, and throws some doubt on the accuracy of the reported values of R_0. The authors investigated the neutralization of 560-eV Cs^+ ions incident on a tungsten surface at target temperatures ranging from 300° to 1700° K. Fig. 12.2.2.1.1.b shows the curves $R_0(T)$, $R_\Sigma(T)$, and $R_s(T) = I_s/I_1$ for this case. The neutralization coefficient has a constant value of about 77% throughout the temperature interval from 300° to 1000° K. In the temperature range from 1000° to 1300° K the coefficient R_0 drops off abruptly to zero because of the increasing influence of the surface-ionization effect. This means that, for target temperatures higher than 1300° K all primary Cs^+ ions leave the surface as ions (as may be expected from the discussion in Chapter 8).

Results similar to those shown in Fig. 12.2.2.1.1.b were obtained with Rb^+ ions on W. However, in this case the value of R_0 was 20% at target temperatures above 1200° K, instead of dropping to zero.

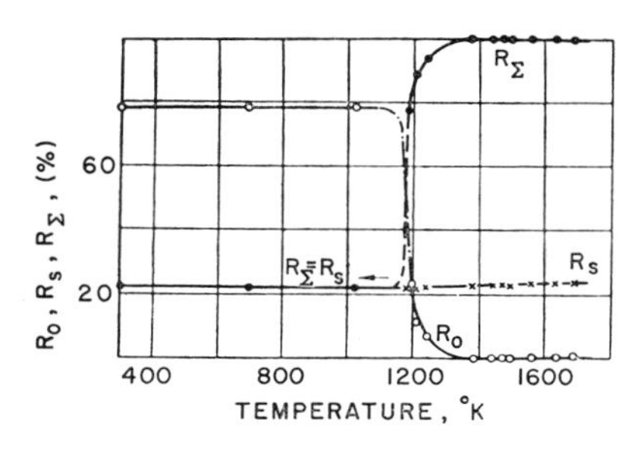

Fig. 12.2.2.1.1.b

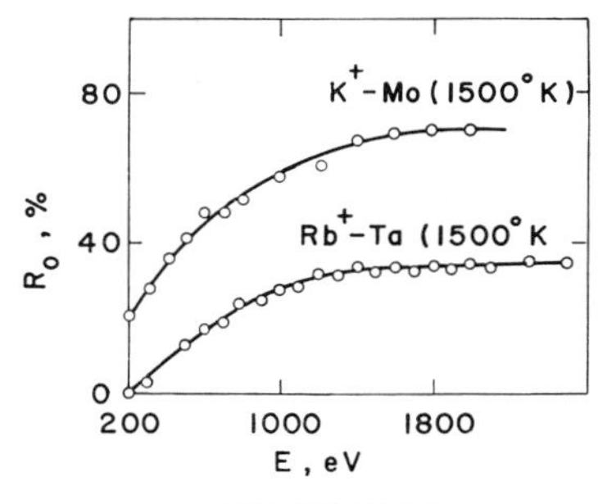

Fig. 12.2.2.1.1.c

Fig. 12.2.2.1.1.b. The dependence of the integral neutralization coefficient R_0, the partially-corrected secondary-ion-emission coefficient R_Σ, and the partially-corrected ion-reflection coefficient R_s on the target temperature for the case of 560-eV Cs^+ ions incident on a wolfram surface (Arifov et al. [*29*])

Fig. 12.2.2.1.1.c. The dependence of the integral neutralization coefficient R_0 on the kinetic energy E of the incident ion for the case of K^+ ions striking a Mo target and Rb^+ ions striking a Ta target. The target temperature is kept at at $\approx 1500°$ K (Arifov et al. [*29*])

The authors also investigated the coefficient R_0 as a function of the energy of K^+ and Rb^+ ions incident on a molybdenum surface, of Rb^+ ions incident on a tantalum surface, and of Na^+ ions on a tungsten surface. The investigated range of primary-ion energies extended from 200 to 2500 eV and the target temperature was kept at $\approx 1500°$ K. Fig. 12.2.2.1.1.c shows such curves for $R_0(E)$ for K^+ ions on a Mo surface and Rb^+ ions on a Ta surface. One sees that with increasing primary-ion energy the values of R_0 increase and tend to saturation at the higher energies. Some experimental values of R_0 for different target materials and for different ions incident with energies of 1200 eV and for a target temperature of $\approx 1500°$ K are listed in Table 12.2.2.1.1.B (Arifov et al. [*29*]). The values of R_0 listed therein decrease with decreasing values of $(eI - e\overline{\varphi})$, a result which is in qualitative agreement with the findings for the noble gases (Table 12.2.2.1.1.A) since a decrease in γ is related to a decrease in the neutralization rate.

Table 12.2.2.1.1.B. *Experimental values of the integral neutralization coefficient R_0 with $E = 1200$ eV and $T \approx 1500°$ K* (Arifov et al. [*29*])

Incident ion	Target	R_0 (%)	$eI - e\overline{\varphi}$ (eV)
Na^+	W	76	0.57
K^+	Mo	62	0.17
Rb^+	Mo	35	0
Rb^+	Ta	30	−0.05
Cs^+	W	0	−0.63

12.2.2.1.2. Energy Distribution of Secondary Electrons and its Dependence on the Incident-Ion Energy

The energy distribution $f(E_e)$ of emitted secondary electrons has been determined from retarding-potential measurements by use of the relationship (HAGSTRUM [*286*])

$$f(E_e) = \frac{d\,i}{d\,V_{ct}}\,. \qquad (12.2.2.1.2\text{–}1)$$

Here $i = I_c/(I_t - I_c)$, where I_c is the collector current and I_t is the target current; and V_{ct} is the electron-retarding potential applied between collector and target. The scale for the energy E_e of the secondary electrons was determined from the V_{ct} scale by correcting for the measured

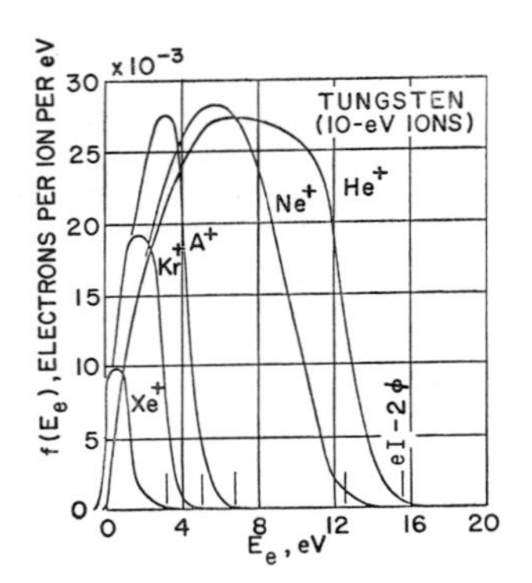

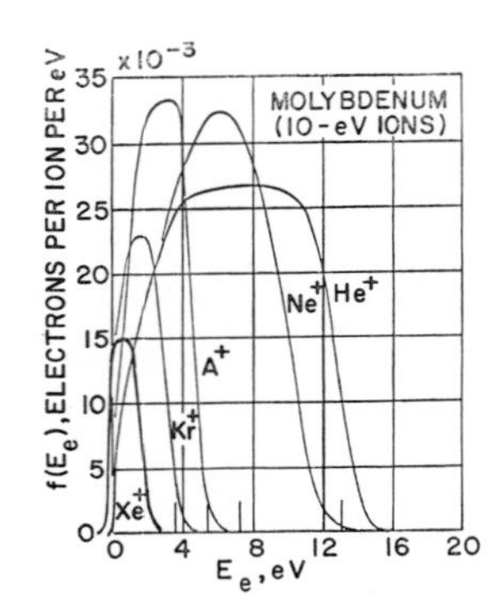

Fig. 12.2.2.1.2.a. Energy distribution function $f(E_e)$ for secondary electrons emitted from atomically clean wolfram and molybdenum surfaces under bombardment by singly-charged noble-gas ions of 10-eV energy. The vertical lines on the abscissa indicate the energy $(E_e)_{\max} = eI - 2e\varphi$ (HAGSTRUM [*285, 286*])

contact potential. The experimentally obtained curve of $di/d\,V_{ct}$ vs V_{ct} [corresponding to $f(E_e)$ vs E_e] has been smoothed by weighting eight neighboring points equally (as suggested by L. E. KAPLAN [*367*]). Fig. 12.2.2.1.2.a shows such smoothed $f(E_e)$ curves obtained by HAGSTRUM [*281, 286*] for thoroughly cleaned tungsten and molybdenum surfaces which have been bombarded with 10-eV ions of various noble gases. In these experiments the data were not influenced by admixtures of ions in a metastable state or metastable atoms, as were the previously published data by HAGSTRUM [*277, 285*] for the same systems. Comparison of the $f(E_e)$ curve for a Mo target with that for W reveals a remarkable similarity in the shapes of the distributions (Fig. 12.2.2.1.2.a). One also notices the important result that the ion neutralization in these cases is due to an AUGER neutralization process, since the upper limit of the kinetic energy $(E_e)_{\max}$ of the secondary electrons (indicated by vertical lines in Fig. 12.2.2.1.2.a) agrees well with the theoretical limit $eI - 2e\overline{\varphi}$

for such a process [Eq. (12.1–6)]. However, contrary to the prediction [Eq. (12.1–7)] that the value of the lower limit $(E_e)_{min}$ of the kinetic energy is positive, the experimental values of $(E_e)_{min}$ extend to negative values—the more so the heavier the incident ion. This phenomenon has been related to the presence of an uncompensated magnetic fringing field (caused by the field of the magnetic analyzer) in the region of the target and collector, as already discussed in Section 12.2.1.

Fig. 12.2.2.1.2.b shows $f(E_e)$ curves for the case of Mo and W targets which have been bombarded with noble-gas ions at the higher energy of

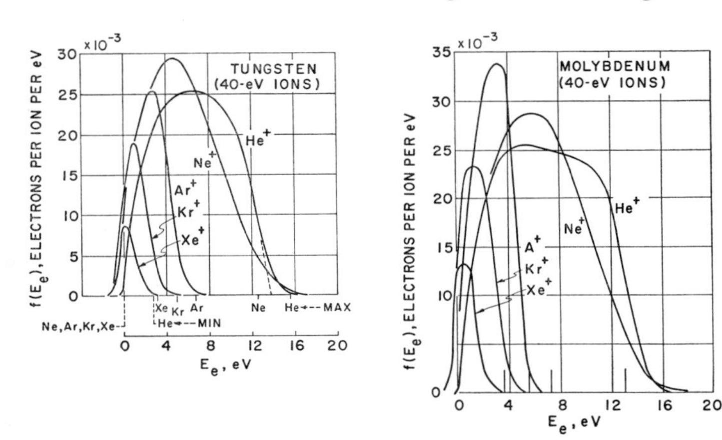

Fig. 12.2.2.1.2.b. Energy distribution $f(E_e)$ of secondary electrons emitted from atomically clean wolfram and molybdenum surfaces under bombardment by singly-charged noble-gas ions of 40-eV energy. The vertical lines on the abscissa indicate the energy $(E_e)_{max} = eI - 2e\varphi$ (HAGSTRUM [285, 286])

40 eV. While the upper limit $(E_e)_{max}$ for the ions He^+, A^+, Kr^+, and Xe^+ again agrees rather well with the theoretical limit $eI - 2e\overline{\varphi}$ expected for an AUGER neutralization process, there is a remarkable deviation in the case of the Ne^+ ion. Here the observed value of $(E_e)_{max}$ considerably exceeds the theoretical one. It is thought that at these energies the AUGER neutralization process is significantly supplemented by a competing two-stage neutralization process (resonance neutralization followed by AUGER de-excitation). Such a two-stage process would indeed lead to a higher value $(E_e)_{max} = eI - e\overline{\varphi}$ of the upper energy limit. Such an interpretation is supported by the sharp rise observed in the graph of $\gamma(E)$ in the range of ion energies considered, as already pointed out in Section 12.2.2.1.1.

In order to exhibit the influence of the primary-ion energy on the distribution function $f(E_e)$ over the broader energy range from 10 eV to 1 keV, $f(E_e)$ curves for He^+ ions bombarding a Mo target are shown in Fig. 12.2.2.1.2.c. One notices that as the energy of the primary ions increases from 10 to 40 to 100 and to 200 eV, the maxima of the distribution curves $f(E_e)$ are reduced and the high-energy tails of the distributions extend to higher values of E_e. This behavior can be explained by the theory of AUGER neutralization (Section 12.2.1) if one also considers the broadening of the energy levels as the incident particle approaches the metal surface. For the higher primary-ion energies (600 and 1000 eV) the behavior of the distribution function can no longer be explained by an AUGER neutralization process alone. Both the appearance of a peak in the $f(E_e)$ curve near zero electron energy and also the considerable extension of the high-energy tail of the distribution indicate the additional action of a non-AUGER process of electron ejection. A discussion of a different secondary-electron-emission mechanism (kinetic ejection) will be given in Chapter 14.

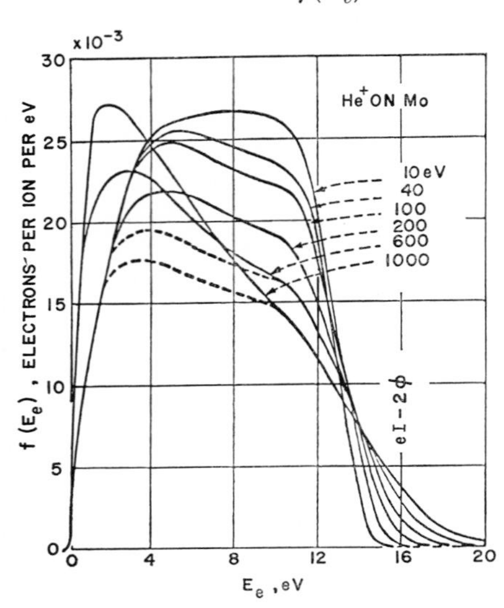

Fig. 12.2.2.1.2.c. Energy distribution $f(E_e)$ of secondary electrons emitted from a molybdenum surface under bombardment by He^+ ions of various energies (10—1000 eV). The dashed lines separate the distributions at 600- and 1000-eV ion energy into parts attributable to AUGER (under the dashed line) and non-AUGER processes of electron ejection (HAGSTRUM [*286*])

12.2.2.1.3. Influence of Surface Contamination on the Yield and Energy Distribution of the Secondary Electrons

While the influence of surface contamination on the yield γ and energy distribution $f(E_e)$ of the ejected electrons has been recognized by several authors [*277, 281, 286, 452a, 490, 544, 735*], there have been no extensive studies of the subject. Three important results can be obtained from the available data. The first one, indicated for instance by the data of HAGSTRUM [*277, 281, 286*] and PARKER [*544*], is that a surface coverage of adsorbed gases leads to a decrease in the value of the total yield of secondary electrons, the decrease being greater for higher degrees of coverage and for metals with higher work functions. HAGSTRUM [*286*] found that one monolayer of adsorbed background gas (unknown composition) on a tungsten surface dropped the value of γ for He^+ from 0.29 to 0.18 electrons per ion. For a 10% coverage, the value of γ dropped to 0.25. PARKER [*544*] observed that the values of γ for cleaned Pt or Ta surfaces bombarded with low-energy A^+ ions (particularly for the region $E < 35$ eV) were higher than those when these surfaces were treated with such gases as H_2, N_2, or O_2. Fig. 12.2.2.1.3.a shows the results of PARKER for cleaned and for gas-covered Pt surfaces bombarded with A^+ ions. One notices that the values of γ decrease for the different surfaces in the sequence $\gamma_{Pt} > \gamma_{Pt/H} > \gamma_{Pt/N} > \gamma_{Pt/O}$, where γ_{Pt} is for a degassed Pt surface and the others are for Pt coated with an adsorbed layer of the indicated gas. This result might be expected, since the work functions of the particular surfaces increase in sequence $e\bar{\varphi}_{Pt}$ (4.72 eV, [*169*]) $< e\bar{\varphi}_{Pt/H}$ ($\approx$4.95 eV, [*480*]) $< e\bar{\varphi}_{Pt/O}$ (5.90 eV, [*237, 528*]). For the A^+ ion bombardment of cleaned Ta surfaces and those treated with the same gases (H_2, N_2, and O_2), PARKER observed qualitatively similar results in the trend of γ. However, one should notice that for the clean metal surfaces $\gamma_{Pt} > \gamma_{Ta}$, despite the fact that $e\bar{\varphi}_{Pt} > e\bar{\varphi}_{Ta}$.

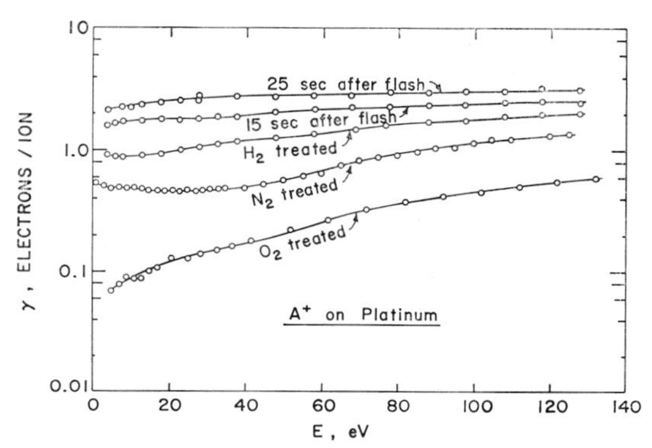

Fig. 12.2.2.1.3.a. The dependence of the secondary-electron yield γ on the kinetic energy E of the incident ion for the case of A^+ ions striking both degassed and gas-covered Pt surfaces (PARKER [*544*])

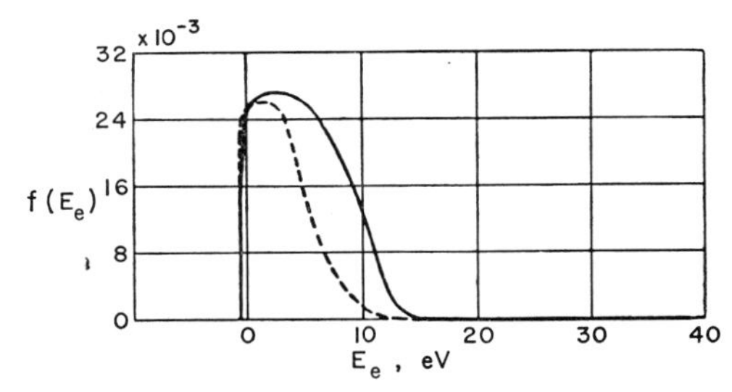

Fig. 12.2.2.1.3.b. Energy distributions $f(E_e)$ of secondary electrons emitted when 40-eV He^+ ions bombarded a clean Mo surface (solid line) and a Mo surface covered with a layer of adsorbed background gas of unspecified composition (dashed line) (HAGSTRUM [*277*])

Fig. 12.2.2.1.3.a also illustrates the second result: γ-values increase faster with increasing primary-ion energy in the case of gas-covered surfaces than for the clean surface. It is possible that this increase in γ for the gas-covered surface and the fact that at higher ion energies these values seem to approach the value of the clean metal surface is due to progressive surface cleaning with increasing ion energy (sputtering process).

Fig. 12.2.2.1.3.b shows the energy distribution $f(E_e)$ for secondary electrons ejected by 40-eV He^+ ions from a clean Mo surface (solid line) and from a Mo surface covered with a layer of adsorbed background gas (unspecified composition), as obtained by HAGSTRUM [*277*]. This figure illustrates the third result: a layer of adsorbed gas atoms on the metal surface seems to reduce the number of fast secondary electrons relative to the number of slow secondary electrons.

12.2.2.2. Multiply-Charged Ions

12.2.2.2.1. Secondary Electron Yield γ and its Dependence on the Energy E of the Incident Ion

Experimental data obtained by HAGSTRUM [*281*] for the total secondary-electron yield γ for singly- and multiply-charged noble-gas ions incident on cleaned wolfram surfaces are shown as a function of the energy of the incident ion in Fig. 12.2.2.2.1.a. Two interesting results can be noticed. One is the increase of the value of γ with increasing ionic charge and hence with total ionization energy. If one forms the ratios $\gamma_r = \gamma^{(n+1)+}/\gamma^{n+}$ of electron yields and $I_r = I^{(n+1)+}/I^{n+}$ of total ionization energies for the case of ions differing only by one unit charge, one notices (Table 12.2.2.2.1.A) that the values of γ_r are in fair agreement with those of I_r for He, Ne, and the more highly charged ions of A, Kr, and Xe. However, the difference between γ_r and I_r becomes rather large for the less highly charged ions of the heavier gases.

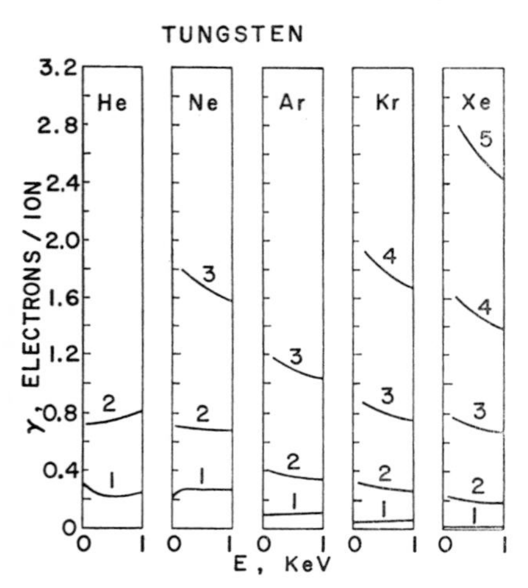

Fig. 12.2.2.2.1.a. The dependence of the secondary-electron yield γ on the kinetic energy of singly- and multiply-charged noble-gas ions incident on atomically clean wolfram. (The charge of the ions is indicated on each curve.) (After HAGSRTUM [*281*])

This may be understood if one assumes with HAGSTRUM that multiply-charged ions are neutralized in a series of approximately iso-energetic steps. The energy released per step for the lighter atoms and the more

Table 12.2.2.2.1.A. *Ratios of secondary-electron yields γ_r and total ionization energies I_r for noble-gas ions incident on a wolfram surface* (HAGSTRUM [*281*])

$\frac{(n+1)^+}{n^+}$	$\frac{2^+}{1^+}$		$\frac{3^+}{2^+}$		$\frac{4^+}{3^+}$		$\frac{5^+}{4^+}$	
Atom	γ_r	I_r	γ_r	I_r	γ_r	I_r	γ_r	I_r
He	2.6	3.2						
Ne	3.1	2.9	2.5	2.0				
A	4.3	2.8	3.0	2.0				
Kr	6.8	2.8	2.9	2.0	2.2	1.7		
Xe	12.0	2.8	3.5	1.9	2.0	1.8	1.8	1.5

highly charged heavier ions is sufficient to excite all or nearly all electrons to energies above the surface barrier of the metal. Therefore, the ratio of electron escape probabilities P_r (as defined in Section 12.2.3.2) is in this case nearly unity. Since $\gamma_r \approx P_r I_r$, this means that $\gamma_r \approx I_r$. For the heavier singly-charged ions A^+, Kr^+, Xe^+, on the other hand, only a fraction of the metal electrons are excited to a high enough energy to escape. The result is that, for instance, the ratio $P_r = P^{2+}/P^+ \gg 1$. This explains why the values of γ_r exceed those of I_r in such cases. This is most conspicuous in the case of Xe, which has the lowest ionization energy.

Besides the investigation of secondary-electron emission by multiply-charged ions incident on a wolfram surface, HAGSTRUM [*286*] also studied the neutralization of doubly-charged noble gas ions on a Mo surface. The $\gamma(E)$ curve is similar to the one observed for the wolfram surface.

The neutralization process is thought to consist of three steps: (1) resonance neutralization of the doubly-charged ion to an excited state of the singly-charged ion, (2) AUGER de-excitation of the excited singly-charged ion to the ground state of the ion, and (3) AUGER neutralization of the singly-charged ion to the ground state of the atom. Secondary electrons are emitted only in the second and third stages. The yield γ_{mi} of secondary electrons per incident ion in the second stage considerably exceeds the yield γ for the third stage, as is evident from Table 12.2.2.2.1.B. The second important result, which can be obtained

Table 12.2.2.2.1.B. *Yield γ_{mi} of secondary electrons in the process of Auger de-excitation of singly-charged ions in excited states and the yield γ from Auger neutralization of singly-charged ions for noble gases at 10 eV energy on a Mo surface* (HAGSTRUM [*286*]). *The total yield (electrons per incident ion) is $\gamma_t = \gamma_{mi} + \gamma$*

Atom	$(\gamma_t)_{exp}$	γ_{exp}	$(\gamma_{mi})_{exp} = (\gamma_t - \gamma)_{exp}$	$(\gamma_{mi})_{calc}$*
He	0.81	0.30	0.51	
Ne	0.68	0.25	0.43	
A	0.42	0.12	0.30	
Kr	0.39	0.07	0.32	
Xe	0.30	0.02	0.28	0.23

* The method of calculation is described in Section 12.2.3.

from the data in Fig. 12.2.2.2.1.a, is that the value of $d\gamma/dE$ for each of the multiply-charged ions becomes increasingly negative with increasing ion energy, except in the case of He^{2+}. This effect is attributed to the broadening of the energy states in both the metal and the incident particle as the steps of neutralization occur closer to the metal, the same explanation as already given in the case of $\gamma(He^+)$. The behavior of $\gamma(He^{2+})$ remains unexplained.

12.2.2.2.2. Energy Distribution of Secondary Electrons

The energy distribution $f(E_e)$ of secondary electrons emitted from a wolfram surface by 200-eV krypton ions in different charge states (Kr^+, Kr^{2+}, Kr^{3+}, Kr^{4+}) is shown in Fig. 12.2.2.2.2.a (HAGSTRUM [*281*]). One notices that the maximum height of the distribution curves $f(E_e)$ depends strongly on the charge state of the ion and, in particular, increase with increasing ionic charge. On the other hand, the mean energy of the emitted electrons is nearly independent of the ionic charge.

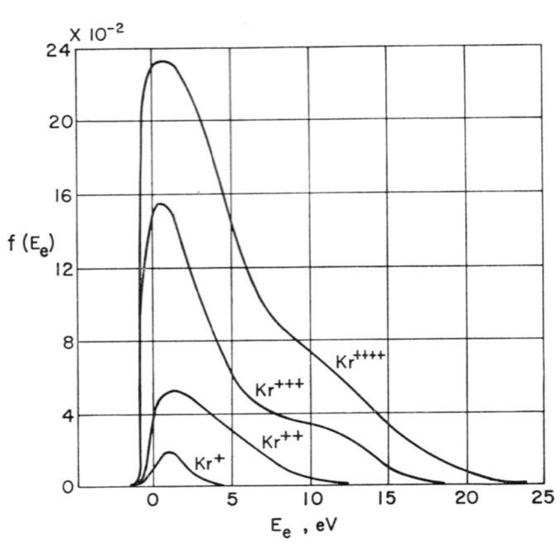

Fig. 12.2.2.2.2.a. Energy distribution $f(E_e)$ of secondary electrons emitted from a molybdenum surface under impact of 200-eV krypton ions in different charge states (HAGSTRUM [*281*])

The dependence of the function $f(E_e)$ on the type of the incident ion is illustrated in Fig. 12.2.2.2.2.b, in which the energy distribution $f(E_e)$ is shown for the case of secondary electrons emitted from a Mo surface under bombardment by 40-eV doubly-charged noble-gas ions (HAGSTRUM [*286*]). One notices that the maximum of the $f(E_e)$ curves appears to be nearly independent of the type of the incident ion and show only a slight drop toward heavier ions. The drop is thought to be the result of the lower ionization energy of the heavier ions and their consequent inability to cause the escape of every metal

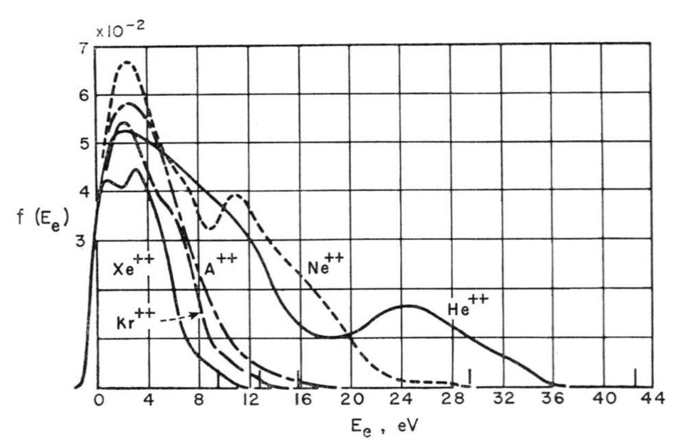

Fig. 12.2.2.2.2.b. Energy distribution $f(E_e)$ of secondary electrons emitted from a molybdenum surface under impact by 40-eV doubly-charged noble-gas ions. The vertical lines indicate the theoretical value for the maximum kinetic energy $(E_e)_{\max} = eI_2 - eI_1 - \frac{1}{2} W_i - 2e\varphi$ for the case of doubly-charged ions (HAGSTRUM [*286*])

electron they excite. HAGSTRUM [*281*] observed that this effect becomes less apparent the higher the ionic charge, as should be expected.

Finally one should notice that three observed values for the upper energy limit $(E_e)_{\max}$ of the function $f(E_e)$ indicate the validity of the model of a three-stage neutralization process of the doubly-charged ion. Of the three stages involved (resonance neutralization of the doubly-charged ion to an excited state of the singly-charged ion, AUGER de-excitation of the excited singly-charged ion to the ground state of the ion, and AUGER neutralization of the singly-charged ion to the ground state of the atom), the last two stages determine the upper energy limit $(E_e)_{\max}$, which is given by

$$(E_e)_{\max} = eI(2+) - eI(1+) - \tfrac{1}{2} W_i - 2e\overline{\varphi}\,. \quad (12.2.2.2.2\text{–}1)$$

Here $eI(2+)$ and $eI(1+)$ are the ionization energies of the doubly- and singly-charged ions, W_i is the FERMI level, and $e\overline{\varphi}$ is the mean work function of the metal surface. The term $\frac{1}{2} W_i$ implies that on the average the tunneling metal electron comes from the middle of the conduction band. The values of $(E_e)_{\max}$ calculated according to Eq. (12.2.2.2.2–1) are indicated by vertical lines in Fig. 12.2.2.2.2.b. The agreement with the observed values is rather good.

12.2.2.2.3. Influence of Surface Contamination on the Electron Yield and the Electron Distribution

There are still very few experimental data on the influence of surface contamination on the secondary-electron emission in the process of neutralization of multiply-charged ions. Fig. 12.2.2.2.3.a (HAGSTRUM [*277*]) illustrates the effect of surface contamination on the total electron yield γ as a function of the ion energy of singly- and doubly-charged helium ions (He^+, curves *1* and *2*; He^{2+}, curves *3* and *4*) incident on a Mo surface.

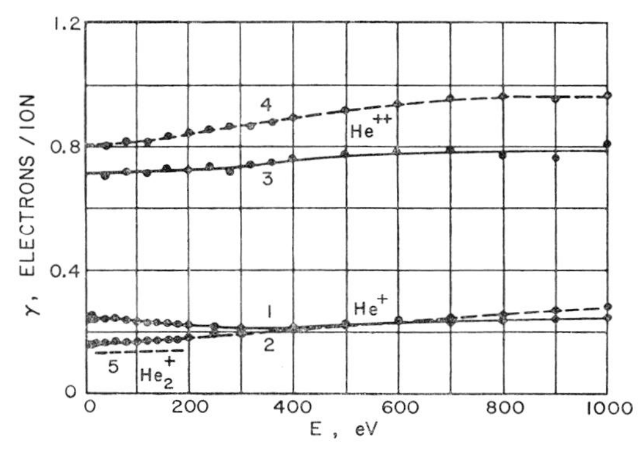

Fig. 12.2.2.2.3.a. The dependence of the secondary-electron yield γ on the kinetic energy E of the incident ion beam for the case of singly- and doubly-charged helium ions striking a clean Mo surface (curves *1* and *3*) and a gas-covered Mo surface (curves *2* and *4*) (HAGSTRUM [*277*])

For incident-ion energies below 400 eV one notices the interesting result that while γ is higher for clean surfaces (curve *1*) than for gas-covered ones (curve *2*) bombarded by singly-charged He ions (as is also the case for the other singly-charged noble-gas ions, as discussed in Section 12.2.2.1.3), the opposite is true of doubly-charged ions. Over the whole energy range investigated (10—1000 eV) the values of γ for the gas-covered surfaces (curve *3*) are higher than those for the clean surface (curve *4*). This behavior remains unexplained. Additional studies, particularly of the dependence on target temperature, are highly desirable.

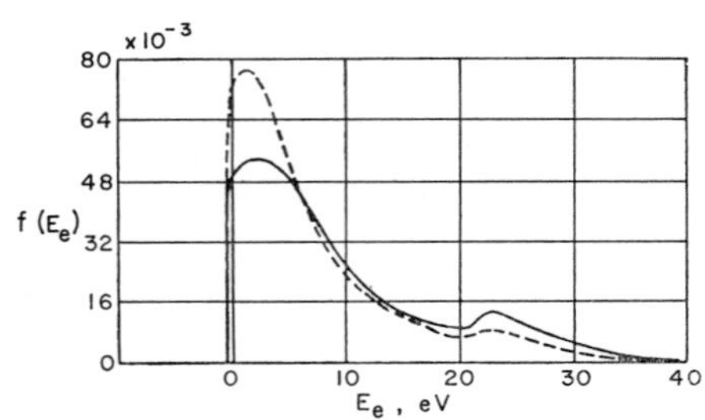

Fig. 12.2.2.2.3.b. The energy distribution $f(E_e)$ of secondary electrons emitted from a clean Mo surface (solid line) and from a gas-covered Mo surface (dotted line) under the impact of 80-eV He^{+++} ions (HAGSTRUM [*277*])

The influence of surface contamination on the distribution of secondary electron energies is illustrated in Fig. 12.2.2.2.3.b. Here the distribution function $f(E_e)$ is plotted as a function of the energy E_e of the secondary electrons for the case of electrons emitted under bombardment of clean and gas-covered Mo surfaces by He^{2+} ions at 80 eV energy. One notices that, as in the case of singly-charged ions (Section 12.2.2.1.3), adsorbed gas layers on the target reduce the number of fast electrons relative to the number of slow electrons. On the other hand, the presence of the gas layer on the surface does not shift the onset of the distribution function near $E_e = 0$. This behavior, particularly the structure observed in the distribution function, has not yet been satisfactorily explained.

12.2.3. Theory of Auger Neutralization

12.2.3.1. General Definitions of the Total Transition Rate and Related Probability Functions

The electronic transitions in the process of AUGER neutralization of an ion at a metal surface involve two metal electrons from the conduction band. One metal electron neutralizes the incident ion directly to the ground state of the atom; then the energy released in this transition causes another metal electron to be ejected from the metal surface (see potential diagram in Fig. 12.1.a, curve γ). There are two concepts to describe this process of neutralization and electron emission; but in the non-relativistic limit they lead to the same result, as pointed out by BURHOP [*104a*]. In the approach used by PROBST [*578a*] in his treatment of AUGER neutralization, the radiation field set up by the first metal electron as it falls to the ground state of the atom is treated as the perturbation that excites the second metal electron. In the other approach, discussed for instance by HAGSTRUM [*282*], the Coulomb interaction between the two participating metal electrons is the perturbation. The following brief discussion will be limited to the later point of view.

As long as the relative velocity between incident ion and metal is much smaller than the orbital velocities of atomic electrons, the neutralization process can be treated as quasi-stationary by a time-independent perturbation theory. The transititon probability per unit time for the process which occurs when the ion is at a distance r from the surface and which results in an excited metal electron with energy E_k with velocity vector lying in the element of solid angle $d\Omega$ is given by

$$P_x(r)\,d\Omega = \frac{2\pi}{\hbar}\,|H_i^f|^2 \varrho(E_t)\,d\Omega\,, \qquad (12.2.3.1.–1)$$

where $\varrho(E_t)$ is the electron density of the final states at total energy E_t and H_i^f is the matrix element

$$H_i^f = \int\int \psi_f^*(e_1)\,\psi_f^*(e_2)\,\Delta E\,\psi_i(e_1)\,\psi_i(e_2)\,d\tau_1\,d\tau_2\,. \qquad (12.2.3.1–2)$$

Here $\psi_f^*(e_1)$ and $\psi_f^*(e_2)$ are the complex conjugates of the eigenfunctions $\psi_f(e_1)$ and $\psi_f(e_2)$, where these functions and $\psi_i(e_1)$, $\psi_i(e_2)$ are the Schrödinger spatial eigenfunctions of the final and initial states of the two electrons (e_1, e_2) participating in the Auger neutralization process and ΔE is the perturbation energy. So far no valid attempt has been made to evalute the matrix element H_i^f directly. However, as will be discussed in the following sections, Hagstrum [*282*] developed an approximate method of evaluating the matrix element as a function of the distance r from the surface to the incident particle, and of the angle between the surface normal and the velocity of the excited electron.

The probability $P_n(r, v)\,dr$ that an incident ion having a constant velocity v will undergo neutralization in the region between r and $r + dr$ from the surface can be represented by the product of the probabilities of two events. The first one, the probability that the incident ion starting from an infinite distance and moving toward the metal surface will reach the distance r without undergoing transition, is $\left\{1 - \int_r^\infty [1 - P_n(r, v)]\,dr\right\}$. The second, the probability that the ion is neutralized in the time $dt = dr/v$, is equal to $(dr/v)\,R(r)$. As shown by Cobas and Lamb [*116*] the probability $P_n(r, v)$ then is given by

$$P_n(r, v) = \frac{R(r)}{v}\left[1 - \int_r^\infty P_n(r, v)\,dr\right]. \qquad (12.2.3.1–3)$$

This integral equation can be solved for $P_n(r, v)$ by inserting the total transition rate $R(r)$. The total transition rate for all elemental processes occurring when the ion is at a distance r from the surface is given by

$$R(r) = \int\int\int\int P_x'(r)\,d\Omega\,dE_k \times \delta(E_{M1} + E_{M2} + eI - W_a - E_k) \times N_c(E_{M1})\,N_c(E_{M2})\,dE_{M1}\,dE_{M2}\,. \qquad (12.2.3.1–4)$$

Here $P'_x(r)\,d\Omega$ is the transition probability already defined in equation (12.2.3.1–1), except that the density $\varphi(E_t)$ of final states has been approximated by the density $N(E_k)$ of final states available to the excited electron at energy E_k inside the metal; $N_c(E_{M1})$ and $N_c(E_{M2})$ are the densities of states in the conduction band in the neighborhood of the electron energies E_{M1} and E_{M2}. The DIRAC δ function in the energy assures that the energy condition for the AUGER neutralization process is fulfilled, namely $E_{M1} + E_{M2} = E_k + W_a - eI$, as may be understood from the potential diagram in Fig. 12.1.a.

As already mentioned, the direct evaluation of the matrix element, which according to Eq. (12.2.3.1–1) determines P'_x, is too difficult. Therefore HAGSTRUM [278] suggested an approximate method for evaluating $P'_x(r)$. In his treatment, the transition rate $P'_x(r)\,d\Omega$ was written as

$$P'_x(r)\,d\Omega = F(E_{M1}, E_{M2})\,N(E_k)\,P_\Omega(\theta, E_k)\,d\Omega\,. \qquad (12.2.3.1\text{–}5)$$

Here $F(E_{M1}, E_{M2})$ is a function of the initial-state energies only and $P_\Omega(\theta, E_k)\,d\Omega$ is the probability that the excited electron at energy E_k has its velocity vector lying in the solid angle $d\Omega$ at direction θ. HAGSTRUM assumes that the matrix element is independent of the azimuth angle φ.

In order to determine the energy distribution function $f_i(E_k)$ of electrons excited in the AUGER neutralization process, as well as the energy distribution $f(E_k)$ of emitted secondary electrons and the total secondary-electron yield γ, an additional probability function has to be defined. Let $P_k(E_k, r)\,dE_k$ be the probability that the excited electron produced in a neutralization process at a distance r will have energy in the range dE_k at E_k. According to HAGSTRUM, the probability $P_k(E_k, r)$ is given by the expression

$$P_k(E_k, r) = C_1 N(E_k) \int\!\!\int N_c(E_{M1})\, N_c(E_{M2})$$
$$\times\,\delta\,[E_k - E_{M1} - E_{M2} + W_a - eI(r)]\,dE_{M1}\,dE_{M2}\,. \qquad (12.2.3.1\text{–}6)$$

Here $C_1 = [g(2\,mc)^{3/2}/2\,\pi^2\hbar^3]\exp(-W_i/kT)$, where g is a degeneracy factor (degeneracy of spin excluded), m is the electron mass, c is the velocity of light, and W_i is the energy difference between the bottom of the conduction band and the FERMI level. The other notations in Eq. (12.2.3.1–6) are the same as those used previously.

Then the function $f_i(E_i)$ for the energy distribution of electrons excited in the process of AUGER neutralization can be expressed as

$$f_i(E_k) = \int_0^\infty P_n(r, v)\,P_k(E_k, r)\,dr\,. \qquad (12.2.3.1\text{–}7)$$

Here it has been assumed that all significant contributions to $f_i(E_k)$ occur while the ion is moving inward with velocity v toward the metal. The energy distribution $f(E_k)$ of electrons emitted from the metal surface is then given by

$$f(E_k) = f_i(E_k) \int_0^{2\pi} \int_0^{\theta_c} P_\Omega(\theta, E_k) \sin\theta \, d\theta \, d\varphi \,. \qquad (12.2.3.1\text{–}8)$$

Here $\theta_c(E_k)$ is the maximum value of θ for which excited electrons can escape over the surface barrier. In the following, the energy distribution will be written as a function of $E_e = E_k - W_a$ (see Fig. 12.1.a), the kinetic energy of the emitted electron.

Finally, integrating $f(E_e)$ over E_e gives the total electron yield γ for the AUGER neutralization process in the form

$$\gamma = \int_0^\infty f(E_e) \, dE_e \,. \qquad (12.2.3.1\text{–}9)$$

In order to obtain quantitative expressions for the energy distribution function $f(E_e)$ and the yield γ, further approximations have to be made in evaluating the probability functions that determine the functions $f(E_k)$ and γ [Eqs. (12.2.3.1–7) to (12.2.3.1–9)]. This will be discussed in the following sections.

12.2.3.2. Total Transition Rate and the Probability for Auger Neutralization

Since the electronic transitions in the AUGER neutralization process occur only over a limited range of r, HAGSTRUM [*278*] assumes that the total transition rate, which specifies the dependence of the matrix element on r, follows the exponential law

$$R(r) = A \exp(-ar) \,. \qquad (12.2.3.2\text{–}1)$$

Here A and a are two parameters determined by fitting the theory to the experimental results. In the theory of SHEKHTER [*650*] and COBAS and LAMB [*116*] in which a similar exponential law is used, the factor A is represented by a polynominal in r.

HAGSTRUM found on the basis of his experimental results that the value of the parameter a most probably lies in the range 2×10^{-8} cm^{-1} $< a < 5\times10^{-8}$ cm^{-1}. For the case of He^+ ions incident on wolfram, he found $A = 7.1\times10^{16}$ sec^{-1} for $a = 2\times10^8$ cm^{-1} and $A = 1.5\times10^{20}$ sec^{-1} for $a = 5\times10^8$ cm^{-1}.

Substituting expression (12.2.3.2–1) into (12.2.3.1–3), COBAS and LAMB obtained

$$P_n(r, v) = \frac{A}{v} \exp\left[-\frac{A}{av} e^{-ar} - ar\right]. \qquad (12.2.3.2\text{–}2)$$

The function $P_n(r, v)$ passes through a maximum value at

$$r_m = \frac{1}{a} \ln \left(\frac{A}{a v} \right) . \tag{12.2.3.2–3}$$

Here the velocity v of the incident particle is assumed to be constant. The value of the function P_n at $r = r_m$ is $P_n(r_m) = 0.368\, a$ and is independent of the velocity v. If one assumes that the exponential law of Eq. (12.2.3.2–1) holds to $r = 0$, then from Eq. (12.2.3.2–2) it follows that

$$\int_0^\infty P_n(r, v)\, dr = 1 - \exp(-A/av) = 1 - \exp[-\exp(a r_m)] , \tag{12.2.3.2–4}$$

and

$$\lim_{v \to 0} \int_0^\infty P_n(r, v)\, dr = 1 . \tag{12.2.3.2–5}$$

This means that for the lower ion velocities the value of r_m is large enough to make the probability $P_n(r, v)\,dr$ of AUGER neutralization nearly unity.

On the basis of this result and the use of the additional approximation that the probability function $P_k(E_k, r)$ is taken at r_m [where $P_n(r, v)$ has its maximum], the energy distribution $f_i(E_k)$ appearing in Eq. (12.2.3.1–7) can be expressed as

$$f_i(E_k) = P_k(E_k, r_m) \int_0^\infty P_n(r, v)\, dr , \tag{12.2.3.2–6}$$

which with the aid of Eq. (12.2.3.2–5) becomes

$$f_i(E_k) = P_k(E_k, r_m) . \tag{12.2.3.2–7}$$

Values of $f_i(E_k)$ can be calculated from Eqs. (12.2.3.1–6) and (12.2.3.2–7). Before Eq. (12.2.3.1–8) can be used to calculate the energy distribution $f(E_k)$ of secondary electrons, it is still necessary to derive an expression for the probability $P_\Omega(\theta, E_k)\, d\Omega$ that an excited electron of energy E_k will leave the metal surface within the solid angle $d\Omega$.

12.2.3.3. The Probability of Emission of an Excited Electron from the Metal Surface

One should distinguish two processes which allow electrons excited in an AUGER neutralization process, to leave the metal surface. On the one hand, a certain fraction of the originally excited electrons have a large enough component of momentum p_n normal to the metal surface to pass over the surface barrier W_a (e.g., $p_n^2/2\, m_e \geqslant W_a$) and escape from the metal surface. Such a process has been considered by HAGSTRUM [*282*] in his treatment of AUGER neutralization. On the other hand, PROBST [*578a*] pointed out that this is not the only possible way for electrons to escape from a metal surface. A large fraction of the originally excited electrons* cannot pass over the surface barrier W_a directly (since $p_n^2/2m_e \leqslant W_a$); but there is the possibility that interactions with other metal electrons may enable them to escape from the surface. The following discussion will start with the first mentioned possibility for the escape of excited metal electrons, as considered in the treatment of HAGSTRUM.

Let the probability for the escape of an excited electron from the metal surface be defined as

$$P_e(E_k) = \int_0^{2\pi}\int_0^{\theta_c} P_\Omega(\theta, E_k) \sin\theta \, d\theta \, d\varphi , \qquad (12.2.3.3\text{–}1)$$

where θ_c is the maximum value of the angle θ for which an electron can escape over the surface barrier, as already mentioned in Section 12.2.3.1. This critical angle is given by

$$\theta_c = \cos^{-1}\sqrt{W_a/E_k} . \qquad (12.2.3.3\text{–}2)$$

Since the direct determination of the function $P_\Omega(\theta, E_k)$ by the evaluation of the matrix element H_f^i [see Eqs. (12.2.3.1–2) to (12.2.3.1–5)] is rather difficult, HAGSTRUM suggested that as a first approximation the angular distribution of the excited electrons may be taken as isotropic. Then $P_\Omega(E_k, \theta)$ is a constant equal to $\frac{1}{4}\pi$. Substituting this into equation (12.2.3.3–1) yields

$$P_e(E_k) = \begin{cases} \frac{1}{2}[1 - \sqrt{W_a/E_k}] & \text{for } E_k > W_a , \\ 0 & \text{for } E_k < W_a . \end{cases} \qquad (12.2.3.3\text{–}3)$$

However the experimental results (for instance, the absolute value of the yield γ) indicate that P_Ω is not isotropic, but that the value $P_{\Omega 1}$ at angles $\theta < \theta_c$ is larger than the value $P_{\Omega 2}$ at angles $\theta > \theta_c$. HAGSTRUM defined a parameter $f^2 = P_{\Omega 1}/P_{\Omega 2}$ whose value he obtained by fitting the yield data. (For example, for 40-eV He^+ ions on W, he found $f = 2.2$.) He considered furthermore that the function $P_\Omega(\theta, E_k)$ should be symmetrical about the origin because $H_i^f(\theta) = i\, H_i^f(\pi + \theta)$. Then in second approximation, after normalizing the function $P_\Omega d\Omega$, he was able to express $P_e(E_k)$ in the form

$$P_e(E_k) = \begin{cases} \frac{1}{2}\left[\dfrac{1 - \sqrt{W_a/E_k}}{1 - (1 - 1/f^2)\sqrt{W_a/E_k}}\right] & \text{for } E_k > W_a , \\ 0 & \text{for } E_k < W_a . \end{cases} \qquad (12.2.3.3\text{–}4)$$

* In the following they will be called primary AUGER electrons, while in the abstract by HAGSTRUM et al. [*289a*] they are called secondary AUGER electrons.

Hagstrum and also Takeishi [*701, 702*] have substituted Eq. (12.2.3.3–4) in Eqs. (12.2.3.1–8) and (12.2.3.1–9) to calculate $f(E_k)$ and the secondary-electron yield γ. One notices the important result that the expressions for $f(E_k)$ and γ derived in this way do not contain the kinetic energy of the incident ion. This means that the theory is unable to account for the observed slight energy dependence of these quantities. Graph α of Fig. 12.2.3.5.a shows $f(E_e)$ curves calculated by use of Eqs. (12.2.3.3–4), (12.2.3.2–7), and (12.2.3.1–8)*.

While Hagstrum's theory does not consider the possibility that additional electrons may be emitted by the interaction of primary Auger electrons with metal electrons, Probst [*578a*] included such an effect in his treatment of the potential emission of electrons. The energy distribution $f_i'(E_k)$ of those primary Auger electrons that cannot escape directly from the surface is given by

$$f_i'(E_k) = [1 - P_e(E_k)]\, f_i(E_k)\,. \qquad (12.2.3.3\text{–}5)$$

Probst assumed that the secondary metal electrons that are excited by the primary Auger electrons have the same energy distribution $f_s(E_k)$ as those secondary electrons that are produced in a metal under the impact of primary external electrons that are normally incident on the surface of the metal in which these electrons have the energy distribution $f_i'(E_k)$. He used several simplifying assumptions in an attempt to construct the energy distribution $f_s(E_e)$, or $f_s(E_k)$ since $E_k = E_e + W_a$, of the secondary electrons resulting from interactions of primary Auger electrons with metal electrons. He also sought to estimate their relative contribution γ_s to the total yield γ by use of Harrower's [*293a*] experimental results on the relative energy distributions of secondary electrons emitted from polycrystalline tungsten under the impact of 7-, 10-, and 20-eV primary electrons and the results of Morgulis et al. [*504a*] on γ_s and γ for low-energy electrons incident on polycrystalline tungsten. For the case of 100-eV He^+ ions incident on polycrystalline tungsten Probst estimated that nearly 50% of the emitted secondary electrons result from interactions between primary Auger electrons and metal electrons—a surprisingly large yield. It should be pointed out, however, that from their recent results on the emission of Auger electrons from a (111) plane of a Ni and Ge surface bombarded by singly-charged noble-gas ions with energies ranging from 4 to 100 eV Hagstrum and Takeishi [*289a*] inferred that the number of secondary electrons due to the interaction of primary Auger electrons with metal electrons are much less numerous (somewhere in the region of 1—6%) for these materials than estimated by Probst for tungsten.

* The energy distribution curves in Fig. 12.2.3.5a have been plotted as a function of $E_e = E_k - W_a$.

12.2.3.4. Effects that Broaden the Electron Energy Distribution

By making a further approximation in his theory of AUGER neutralization, HAGSTRUM [*282*] was able to include two effects that cause a broadening of the function $f_i(E_k)$. One is a variation in the ionization energy of the incident particle with changing distance of approach r. The other is the fact that the initial and final states of the process have finite lifetimes.

The changes of the ionization energy of the incident particle with changing distance r result from various interaction forces between metal surface and incident particle—such as the COULOMB image force of the ion at the surface, the VAN DER WAALS forces resulting from the polarizability of the particle, the possible presence of exchange forces, and the repulsive forces resulting from interpenetration of the electron clouds of the incident particle and target atoms.

HAGSTRUM assumed that the dependence of the ionization energy eI on the distance of approach r can be given approximately by

$$eI(r) = eI(r_m) + k(r - r_m), \tag{12.2.3.4–1}$$

where

$$k = \left. \frac{d(eI)}{dr} \right|_{r = r_m}.$$

The energy distribution $f_i(E_k)$ calculated by substituting Eq. (12.2.3.4–1) into equation (12.2.3.1–7) now becomes broader, the lower limit $(E_e)_{min}$ of the energy distribution shifting to lower energies, the upper limit $(E_e)_{max}$ to higher.

As already mentioned, the $f_i(E_k)$ distribution curves are also broadened by the fact that the initial and the final states of the process have finite lifetimes. If one assumes that only one process will occur in the range of r over which $P_n(r, v)$ differs from zero, the lifetime of the initial state is determined by the total transition ratio $R(r)$ [Eq. (12.2.3.1–4)]. The final state has a finite lifetime because the electrons emitted in the AUGER process leave holes in the filled portion of the conduction band, and these holes in turn are filled by secondary AUGER processes involving electrons lying higher in the band. LANDSBERG [*418*] estimated the broadening to be zero when the holes lie at the FERMI level and to be a maximum when the holes lie at the bottom of the conduction band. HAGSTRUM pointed out that the tailing of the $f_i(E_k)$ distribution at lower energies is caused by the finite lifetimes in both initial and final states, whereas that at the high-energy limit results from the lifetime of the initial state only. HAGSTRUM's theory included corrections for the broadening due to the lifetime of the initial states only.

The $f(E_e)$ distribution calculated for 40-keV noble-gas ions incident on a wolfram surface is shown in graph β of Fig. 12.2.3.5.a. This illustrates the type of distribution obtained when the calculations include corrections for the effects that broaden the distribution of electron energies.

12.2.3.5. Comparison of Experimental and Theoretical Values for the Energy Distribution and Yield of Secondary Electrons

Comparison of the general form of the calculated $f(E_e)$ distribution in graph β of Fig. 12.2.3.5.a and the experimentally observed ones in graph γ of Fig. 12.2.3.5.a for 40-eV noble-gas ions incident on a W target reveals a fair agreement, particularly for the high-energy tails. According to HAGSTRUM, the deficiency of slow electrons observed for He^+ results from the fact that $eI - 2W_a > 0$ at the values of r at which the AUGER transition occurs. The relative maxima of the $f(E_e)$ curves observed for He^+ and Ne^+ are not reproduced by the theory but, as already discussed in Section 12.2.3, this is probably due to the presence of a competing two-stage neutralization process (resonance neutralization followed by AUGER de-excitation) in the case of Ne^+.

In Table 12.2.3.5.A, values of the total electron yield γ, determined as areas under the $f(E_e)$ curves are compared with the experimental values for the case of 40-eV noble-gas ions incident on a wolfram surface.

Table 12.2.3.5.A. *Theoretical and experimental total electron yields γ (electrons per incident ion) for 40-eV ions on wolfram* (HAGSTRUM [282])

Bombarding ion	Theoretical yield γ	Experimental yield γ
He^+	0.279	0.282
Ne^+	0.220	0.232
A^+	0.050	0.097
Kr^+	0.027	0.048
Xe^+	0.012	0.017

A comparison of the calculated and observed values of γ for He^+ and Ne^+ shows fairly good agreement, in spite of the fact that in the case of Ne^+ the electron emission is caused not only by AUGER neutralization but probably also by AUGER de-excitation. For the A^+, Kr^+, and Xe^+ ions, on the other hand, the theoretical values of γ are clearly too low. Such deviations might be expected, however, since for these ions the calculation of γ values becomes much more sensitive to the specific choices made for $N_c(E_{M1})$, $N_c(E_{M2})$, and P_Ω than are the results for He^+ and Ne^+. As HAGSTRUM pointed out, this stems from the fact that for A^+, Kr^+, and Xe^+ only those electrons which lie in the high-energy tail of the $f_i(E_k)$ distribution have a chance of escaping from the surface, whereas for He^+ and Ne^+ practically the entire $f_i(E_k)$ distribution lies at $E_k > W_a$.

HAGSTRUM was also able to show that the $f(E_e)$ curves calculated for different incident-ion energies agreed qualitatively with the experimental results (Fig. 30 of article by HAGSTRUM [*282*]). TAKEISHI [*701*, *702*] calculated values of γ and $f(E_e)$ for noble-gas ions incident on BaO and Ni surfaces on the basis of HAGSTRUM's theory. His calculated values of γ are in fair agreement with the observed ones.

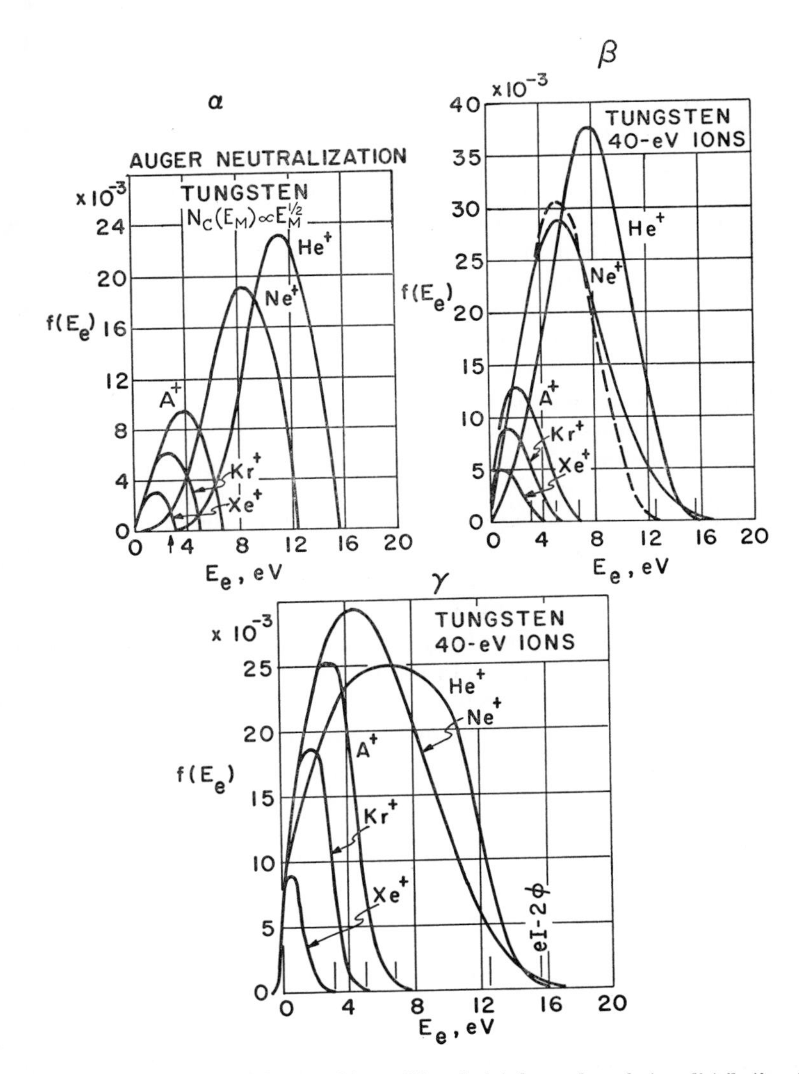

Fig. 12.2.3.5.a. Comparison of the general form of the calculated secondary-electron distributions $f(E_e)$ (graphs α and β) with the experimentally observed one (graph γ) for 40-eV singly-charged noble-gas ions incident on a clean W target. The curves in graph α have been calculated by use of Eqs. (12.2.3.3–4), (12.2.3.2–7), and (12.2.3.1–8). The curves in graph β have been calculated by use of Eqs. (12.2.3.4–1) and (12.2.3.1–7). (Both data and calculations are from HAGSTRUM [*282*])

13. De-excitation of Metastable Atoms and Ions on Metal Surfaces

13.1. Introductory Remarks and Description of Experimental Methods

The de-excitation of a *metastable atom* follows either a collision of the first kind (in which the metastable atom undergoes a transition into a higher energy level, after which it goes to the ground state by emission of radiation) or collisions of the second kind (in which the excitation energy is transmitted to the particle with which it collides and may cause the ejection of secondary electrons). The secondary electrons emitted in collision processes of the second kind may in turn result from two different processes. In one process the metastable atom will be converted into an ion (resonance ionization, Section 12.1) and then undergo Auger neutralization with electron emission (Section 12.1, Fig. 12.1.a, curve γ). In another process the metastable atom will undergo Auger de-excitation—also with electron emission (Section 12.1, Fig. 12.1.a, curve β).

The de-excitation of a *metastable ion* A^{+m} can consist in the Auger de-excitation process $A^{+m} \rightarrow A^{+}$ followed by Auger neutralization $A^{+} \rightarrow A$, processes which also lead to secondary-electron emission. It is therefore possible to study the de-excitation processes of metastable atoms and ions by measuring the electron yield γ and the energy distribution $f(E_e)$ of the secondary electrons.

It was Webb [*755*] who first showed that metastable mercury atoms striking a nickel surface caused the emission of secondary electrons. Since then the secondary-electron emission from a metal surface following impact by metastable atoms and ions has been the subject of extended research [*127, 156, 224, 268, 269, 284, 288, 373, 374, 416, 421, 468, 483, 527, 535, 667, 670, 755, 791*]. For a more detailed review of the older data, which apply almost entirely to gas-covered surfaces, the reader is referred to the summaries given by Massey and Burhop [*461*] and Granowski [*253*].

In the following, some techniques for producing metastable atoms and ions and a method of detecting them by the secondary-electron emission they cause in interacting with a metal surface will be discussed briefly.

Metastable atoms and ions have been produced by a number of methods, such as the method of optical excitation of atoms and ions (Langmuir [*421*], Kenty [*373, 374*], Spivak [*670*], Gurewitsch [*269*]), the gas-discharge method (Yuterhoeven et al. [*791*], Spivak [*670*]), the sputtering process [*215c, 385a*], a method based on the interaction of ions with the metal surface of an auxiliary electrode (Oliphant [*535*]), or finally the method of electron impact (Coulliete [*127*], Sonkin [*667*],

HAGSTRUM [*284, 288*], DORRESTEIN [*156*]). The first and the last methods in particular have been successfully used for relatively well controlled excitation of the gas atoms. HAGSTRUM, who applied the last method, used the experimental arrangement already shown in Fig. 12.2.1.a. The bombarding electrons were passed through the ionization chamber (*D* in Fig. 12.2.1.a) at right angles to the initial direction of the beam of metastable atoms or ions. The total acceleration of the electrons is given by the voltage V_{AC} between the cathode (*A* in Fig. 12.2.1.a) and the chamber electrode (*C* in Fig. 12.2.1.a).

In order to count only the secondary electrons emitted from metal surfaces being bombarded by metastable ions and atoms, it becomes necessary to discriminate against those electrons ejected by other effects. One possible contribution to the secondary-electron current comes from photoelectrons resulting from the radiation emitted in the de-excitation of metastable atoms and ions before they reach the target as has been pointed out by several authors (KENTY [*373, 374*], SPIVAK and REICHRUDEL [*670*], MESSENGER [*468*], FRANK and EINSPORN [*224*]).

In order to study the relative importance of the electron emission due to the impact of metastable atoms and that due to the adsorption of photons, SPIVAK and REICHRUDEL [*670*] and MESSENGER [*468*] placed plates of glass, quartz, or fluorite of known optical transmission between the excited gas and the target surface. These plates would allow the photons to reach the surface but not the metastable atoms or ions. They were able to show that the contribution of the photoelectrons to the total secondary-electron current could be neglected only for certain gas/metal systems.

Another unwanted source of secondary electrons is the interaction of unexcited singly-charged ions (which often are present in the incident beam of metastable atoms or ions) with the target surface. Several methods of detection have been used to distinguish between the contribution to the total secondary-electron current due to metastable particles and that due to the singly-charged ions in the ground state. NOVICK and COMMINS [*527*] modulated the metastable component in the incident beam (and hence the secondary-electron current) at 280 cps. The modulated electron current was fed into an electrometer, a wideband preamplifier, a narrowband amplifier, and lock-in detector. The authors obtained a signal-to-noise ratio of approximately 440. A different method used by HAGSTRUM [*284, 288*] is based on the fact that the electrons ejected by singly charged metastable ions are more numerous and also have higher kinetic energy than those ejected by singly-charged ions in the ground state. This method depends on the ability to detect the change in electron yield in response to the appearance of the metastable ions against the "background" of singly-charged ions in the ground state.

HAGSTRUM measured the electron yield as a function of the energy of the bombarding electrons in the ion source. If the electron energy is still below the ionization energy $eI(2+)$ for the doubly-charged ions, the observed secondary-electron yield may be written (HAGSTRUM [*284, 288*])

$$\gamma_{\text{obs}} = (1 - f_{im})\,\gamma_{i1} + f_{im}\gamma_{im}\,, \tag{13.1–1}$$

where f_{im} is the fraction of the incident beam that is in a metastable state, γ_{i1} the total electron yield* for incident singly-charged ions in the ground

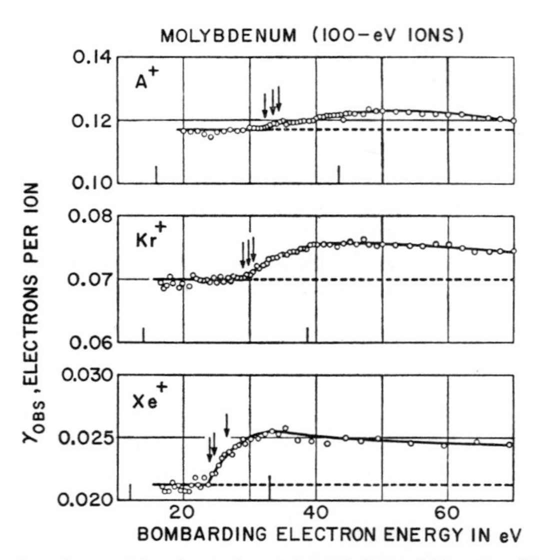

Fig. 13.1.a. The dependence of the observed secondary-electron yield γ_{obs} on the energy of the bombarding electrons used to ionize the noble-gas atoms for the case of 100-eV A^+, Kr^+ and Xe^+ions incident on atomically clean molybdenum. The arrows indicate the energies of metastable ionic states and the vertical lines on the abscissa at lower and higher electron energies indicate the first and the second ionization energies (eI_1 and eI_2) of the particular noble gas (HAGSTRUM [*284, 288*])

state, and γ_{im} the total yield per metastable ion. Fig. 13.1.a gives a set of data for γ_{obs} vs the bombarding-electron energy for A^+, Kr^+, and Xe^+ ions bombarding a Mo surface. The arrows in Fig. 13.1.a indicate the energies of metastable ionic states as well as the first and second ionization energies $eI(1+)$ and $eI(2+)$. One notices that the yield below the threshold value of metastable ion formation has a constant value equal to the quantity γ_{i1}. The value of γ_{im} could not be obtained experimentally so that f_{im} is also left undetermined. However, HAGSTRUM proposed that the

* Here the notation γ_{i1} is used instead of γ (as in the previous sections) to distinguish more clearly between the yields from singly-charged ions (γ_{i1}) and doubly-charged ions (γ_{i2}).

value of γ_{im} had to be nearly equal to the measured yield γ_{i2} for doubly-charged ions. This assumption rests upon the model in which a doubly-charged ion slowly approaching a metal surface will be partially neutralized to an excited state of the singly-charged ion before any AUGER-type process can occur. In case such a resonance transition occurs first, subsequent AUGER-type processes in which electrons are emitted will be very similar to those that would occur if the particle incident on the surface was initially a metastable singly-charged ion.

If Eq. (13.1.–1) is rewritten in the form

$$\frac{\gamma_{\text{obs}} - \gamma_{i1}}{\gamma_{i1}} = \frac{f_{im}(\gamma_{im} - \gamma_{i1})}{\gamma_{i1}} = f_{im}\gamma_r , \qquad (13.1\text{–}2)$$

then the left-hand side of the equation includes only the experimentally observable quantities γ_{obs} and γ_{i1}. On the assumption that $\gamma_{im} = \gamma_{i2}$, HAGSTRUM calculated values for the ratio γ_r for metastable singly-charged ions of the noble gases incident on Mo. Table 13.1.A lists such values of γ_r.

Table 13.1.A. *Calculated values* (HAGSTRUM [*284, 288*]) *of* γ_r

	He^{+m}	A^{+m}	Kr^{+m}	Xe^{+m}
$\frac{\gamma_{im} - \gamma_{i1}}{\gamma_{i1}}$	1.62	2.42	4.25	10.9

One notices in Table 13.1.A that the value of γ_r increases toward the heavier noble gases. This results from a reduction of the values of γ_{i1} for smaller values of the first ionization energy $eI(1+)$. As HAGSTRUM pointed out, the detection of metastable ions becomes more efficient as γ_{i1} becomes smaller since the emission of electrons by ions in the ground state becomes less important. On the basis of Eq. (13.1–2) HAGSTRUM also determined values of f_{im} for different bombarding-electron energies for the excitation of the neutral atom beam) and found that the maximum value of f_{im} for the different metastable noble-gas ions is nearly 2% and is independent of the target surface, as one should expect.

Finally, it should be mentioned that the de-excitation of *metastable* atoms colliding with a metal surface has also been detected by the emission of secondary electrons, as for instance in the work of DORRESTEIN [*156*], GREENE [*256*], SCHULZ and FOX [*643*], LAMB and RETHERFORD [*416*], and NOVICK and COMMINS [*527*]. DORRESTEIN [*156*] eliminated ionic components in his atomic beam by magnetic and electric deflecting fields; and by additional modulation of the incident beam he was able to distinguish between the contributions of the photo-electrons and the AUGER electrons to the total secondary-electron current.

13.2. Experimental Data for the Total Yield and Energy Distribution of Secondary Electrons from De-excitation of Metastable Atoms and Ions on Metal Surfaces

The total yield γ_m of secondary electrons emitted from a metal surface under the impact of metastable noble-gas atoms is close to the value of the total yield γ for singly-charged ions in the ground state. This may be understood since the metastable atom approaching the surface can be resonance ionized and subsequently AUGER neutralized if the energy levels are appropriate (Section 13.1). Thus the secondary-electron emission due to primary-ion bombardment should be almost completely indistinguishable from that due to ions formed by resonance ionization of a metastable atom within a critical distance from the surface. This view is supported by the comparison of experimentally observed values of γ_m and γ in Table 13.2.A.

Table 13.2.A. *Experimental values of the total electron yield γ_m for metastable atoms and γ_{i1} for the corresponding singly-charged noble gas ions. The yield γ_{i1} is expressed in electrons per ion*

Metastable atom	Metal surface (gas covered)	γ_m	γ_{i1}	Author
A^m	Ta	0.023	0.022*	MOLNAR [*490*]
A^m	Mo	0.065	0.071	MOLNAR [*490*]
He^m	Pt	0.24	0.29-Mo** 0.27–W	DORRESTEIN [*156*]

* The values of γ_{i1} have been observed for $\mathcal{E}/p = 117.2$ V/cm per mm Hg.

** Because no values of γ_{i1} are available for Pt surfaces, the values given here are those of HAGSTRUM [*286*] for Mo and W surfaces.

On the basis of his experimental results, SONKIN [*667*] estimated $\gamma_m \approx 0.5$ for the case of metastable mercury atoms bombarding a gas-covered tungsten surface. OLIPHANT [*535*], in his experimental studies of the secondary-electron emission due to bombardment of Mo, Ni, and Mg surfaces with metastable helium atoms, noted that $\gamma_m < 1$ and that the value did not vary significantly for the different target surfaces.

Experimental investigations of the energy distribution $f(E_e)$ of electrons emitted under the bombardment of surfaces by metastable atoms have been conducted by OLIPHANT [*535*], CHAUDHRI and KHAN [*112*], and GREENE [*256*]. Fig. 13.2.a shows the $f(E_e)$ curves reported by GREENE for the case of gas-covered Mo surfaces bombarded by metastable atoms of helium, neon, and argon. The upper energy limit $(E_e)_{\max}$ of the distribution curve is in closer agreement with the value $(E_e)_{\max} = eI_1 - 2e\overline{\varphi}$ (Section 12.2.2.1.2) characteristic of an AUGER neutralization process than with a value $E^* - e\overline{\varphi}$ characteristic of an AUGER de-excitation process. The agreement between the observed value of $(E_e)_{\max}$ and the theoretical value $(E_e)_{\max} = eI_1 - 2e\overline{\varphi}$ is even closer if one considers that the value of the work function may have been increased because of adsorbed gas layers on the surface. Furthermore, the value of the lower energy limit of the distribution $f(E_e)$ agrees better with the value $(E_e)_{\min} = eI_1 - 2W_a$ characteristic of the AUGER neutralization process than with the value $E^* - W_a$ expected for an AUGER de-excitation process. For the case of metastable helium atoms striking a Mo target, the lower energy limit for the first case is calculated to be $eI(1+) - 2W_a = 2.8$ eV; while for the second case the value is $E^* - W_a \approx 9.1$ eV, a value which considerably exceeds the observed one. That the observed value for the lower energy limit is even smaller than the value 2.8 eV is probably due to the presence of surface contamination, which tends to shift the distribution to lower energy values (see Section 12.2.2.1.3).

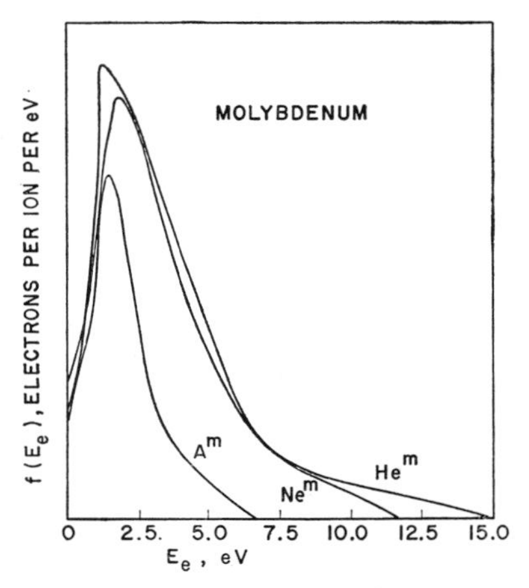

Fig. 13.2.a. Experimentally determined energy distributions $f(E_e)$ for metastable atoms of He, Ne, and A incident on a molybdenum surface which was probably covered with at least a monolayer of foreign gas (GREENE [*256*])

In conclusion, the data for γ_m and for the energy distribution $f(E_e)$ indicate that *metastable noble-gas atoms in interaction with metal surfaces* undergo resonance ionization with subsequent AUGER neutralization rather than AUGER de-excitation.

According to HAGSTRUM [*286*], the total electron yield γ_{im} obtained in the *de-excitation of metastable noble-gas ions on a metal surface* can be expected to be close to the value γ_2, the total electron yield for a doubly-charged ion. The reason for this is that the doubly-charged ion A^{++} may undergo the transition $A^{++} \rightarrow A^{+*}$ by resonance neutralization, in which case the initial state (A^{++} or A^{+*}) of the incident particle loses its importance for the electron emission. The energy distribution curve $f(E_e)$ for doubly-charged ions should therefore also be expected to be similar to that for metastable singly-charged ions.

Fig. 13.2.b shows experimentally observed energy distribution functions for secondary electrons emitted from a Mo target under bombardment with metastable singly-charged ions (curve *1*) and doubly-charged Xe ions (curve *2*) (HAGSTRUM [*284, 288*]). The distribution curve for the process $Xe^{++} \rightarrow Xe^+$ has been obtained as the difference between the measured $f(E_e)$ curve for Xe^{++} ions and the one for Xe^+ ions. The

$f(E_e)$ curve for the process $Xe^{+m} \to Xe^+$ has been obtained as the difference between $f(E_e)$ functions measured with and without metastable ions present in the beam and has been normalized to the same area as the curve for the process $Xe^{++} \to Xe^+$. Also shown is the distribution function for the transition $Xe^+ \to Xe$ corresponding to the final transition stage of each of the above-mentioned processes.

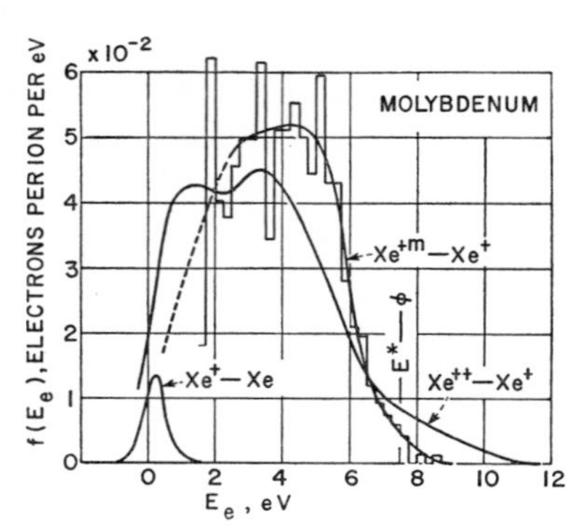

Fig. 13.2.b. The energy distributions $f(E_e)$ for secondary electrons emitted from a molybdenum surface under impact of singly-charged Xe^+ ions, metastable Xe^{+m} ions, and doubly-charged Xe^{++} ions. The curve labelled ($Xe^{++} \to Xe^+$) is obtained by subtracting the ($Xe^+ \to Xe$) curve from the $f(E_e)$ distribution curve measured for Xe^{++} ions. The curve ($Xe^+ \to Xe$) is that measured for Xe^+ ions and corresponds to the final stage in which the doubly-charged ion or the metastable ion is involved in the neutralization or de-excitation, respectively. The curve $Xe^{+m} \to Xe^+$ has been normalized to same area as the ($Xe^{++} \to Xe^+$) curve. (After HAGSTRUM [*284*])

Despite a certain similarity in the course of curves *1* and *2*, one notices that the observed upper energy limit for both curves lies somewhat higher than the value calculated by $(E_e)_{\max} = E^* - e\overline{\varphi} = 11.8 - 4.3 = 7.5$ eV. The high-energy tail of distribution curve *2* (transition $Xe^{++} \to Xe^+$) extends to even higher electron energies than the one for curve *1* (transition $Xe^{+m} \to Xe^+$). According to HAGSTRUM, this is probably due to a partial neutralization of the Xe^{++} ion into the singly-charged ion in excited levels that are higher than the metastable states. For instance, for the highest possible excited level, one would find the value for the upper energy limit to be $(E_e)_{\max} = eI(2+) - eI(1+) - 2e\overline{\varphi} = 21.1 - 8.5 = 12.6$ eV for Xe^{++} on Mo.

Analyzing the distribution curves, HAGSTRUM [*284, 286, 288*] was able to show that for metastable noble-gas ions incident on Mo and W targets the yield values γ_{im} were close to γ_2. Values for γ_2 are given in Section 12.2.2.2.1.

HAGSTRUM [*284, 288*] also investigated the influence of surface contamination (but with an unknown degree of coverage and unknown chemical composition of the gas layers) on the total yield γ_{im} for metastable ions. In his experiments with A^{+*}, Kr^{+*}, and Xe^{+*} ions incident on contamined W surfaces, he observed that the number of secondary electrons emitted was smaller by an order of magnitude than that for a clean surface. He also observed that in this case the yield γ for a singly-charged ion decreases more than does the yield γ_{im} for a metastable ion. This has the effect of increasing the detection sensitivity $[\approx (\gamma_{im} - \gamma_{i1})/\gamma_{i1}]$ for metastable ions.

13.3. Theoretical Aspects of the Auger De-excitation of Metastable Atoms on Metal Surfaces

The experimental data on the secondary-electron emission caused by the de-excitation of metastable noble-gas atoms on metal surfaces indicate the occurrence of a two-stage transition process involving a resonance-ionization process followed by an AUGER neutralization process—not an AUGER de-excitation process—as already discussed in Section 13.2. This process occurs at a definite distance of approach r. In a theoretical treatment, COBAS and LAMB [*116*] showed that in the case of metastable helium atoms (in a 2^3S state) with a velocity $v = 10^5$ cm/sec striking a Mo surface, the de-excitation process occurs with appreciable probability only for a distance of approach $r \lesssim 1$ Å. Since, on the other hand, metastable helium atoms are almost completely ionized at distances greater than 6 Å, it is clear that AUGER de-excitation of an approaching metastable atom makes only a very small contribution to secondary-electron emission when r is in the range greater than 6 Å.

However, it is probably possible to find combinations of incident atom and metal surface for which the distance of approach of the metastable atom will be small enough to allow the occurrence of the AUGER de-excitation process. It seems therefore appropriate to briefly outline HAGSTRUM's treatment of the process.

HAGSTRUM was able to use the same mathematical formalism that he developed for the AUGER neutralization process (Section 12.2.3) since he assumes that one of the two conduction-band electrons participating in the transition process has a constant energy $E_{M1} = W_a - [eI(1+) - E^*]$. The total transition rate for all the transition processes occurring at a distance r can be written in analogy to Eqs. (12.2.3.1–4) and (12.2.3.1–5) as

$$R^*(r) = \int\int\int F^*(E_{M2}) N(E_k) P^*_\Omega(\theta, E_k) \delta(E_{M2} + E^* - E_k) \times N_c(E_{M2}) dE_{M2} d\Omega\, dE_k . \quad (13.3\text{–}1)$$

Here the stared functions F^*, F^*_Ω, and R^* have the same definitions with respect to AUGER de-excitation as the unstared functions have for the AUGER neutralization process. HAGSTRUM assumes furthermore that the function F^* has a constant value C^* and that its dependence on E_{M2} is included in an "effective" $N_c(E_{M2})$ function. Then the transition rate $R^*(r)$ becomes

$$R^*(r) = C^* \int_{W_i}^{\infty} N(E_k) N_c(E_k - E^*) dE_k . \quad (13.3\text{–}2)$$

The other functions describing the AUGER de-excitation process, such as $P^*_k(E_k, r)$, $P^*_e(E_k)$, $f^*_i(E_k)$, and $f^*(E_k)$, are also identical in form with the corresponding functions for AUGER neutralization defined in Eqs. (12.2.3.1–6), (12.2.3.3–1), (12.2.3.1–7), and (12.2.3.1–8), respectively. The total yield γ_m of secondary electrons emitted in an AUGER de-excitation process is the integral of $f(E_e)$ over E_e in the analogy to the definition in equation (12.2.3.1–9).

HAGSTRUM used some further approximations to evaluate the functions $R^*(r)$, $P_e^*(E_k)$, and $P_\Omega^*(\theta, E_k)$ quantitively. He assumed that the total rate $R^*(r)$ follows the same exponential law [Eq. (12.2.3.2–1)] as in the case of AUGER neutralization and that the functions $P_e^*(E_k)$ and $P_\Omega^*(\theta, E_k)$ can be evaluated in the same way as for the AUGER neutralization process. Neglecting shifts in the atomic energy levels of the incident particle near the metal surface, HAGSTRUM determined the $f_i^*(E_k)$ function for $N_c(E_M) = K$ (valid for $0 < E_M < W_i$, where K is a constant). The $f^*(E_k)$ functions obtained from these by use of the $P_e^*(E_k)$ function for isotropic P_Ω^* allowed a crude computation of γ_m [analogous to Eq. (12.2.3.1–9)], some values of which are listed in Table 13.3.A. One notices that while one should expect $\gamma_m \approx \gamma_{i1}$, the calculated values of γ_m are smaller in comparison with the experimentally observed values of γ_{i1}.

Table 13.3.A. *Values of γ_{i1} and γ_m (electrons per ion) calculated for a wolfram surface, and experimentally determined values of γ_{i1} for 10-eV incident ions* (HAGSTRUM [282])

Incident particle	$\gamma_{i1,\,calc}$ $[N_c(E_M) \propto E_M^{1/2}]$	$\gamma_{i1,\,obs}$	$\gamma_{m,\,calc}$ $[N_c(E_M) \propto K]$
He	0.142	0.293	0.155
Ne	0.114	0.213	0.129
A	0.04	0.094	0.063
Kr	0.021	0.047	0.038
Xe	0.0065	0.018	0.017

However, as HAGSTRUM pointed out, a broadening of the $f_i^*(E_k)$ functions and an increase in the values of γ_m can be expected if account is taken of the shifts in the atomic energy levels as the incident particle nears the surface.

14. The Emission of Electrons from Metal Surfaces by Bombardment with Charged and Uncharged Particles (Kinetic Emission)

14.1. Introduction

As a sequel to the discussion of "potential emission" of secondary electrons in Chapter 12, the investigations of the "kinetic emission" of secondary electrons from metal surfaces under the bombardment of charged and uncharged particles will be reviewed in this chapter. For the case of bombardment by ions the ejection of electrons by the kinetic emission process occurs whenever $eI < e\varphi$, where eI is the ionization energy and $e\varphi$ is the work function of the metal, and the incident-ion energy exceeds a certain threshold value. Even for systems for which $eI > 2e\varphi$ so that there can be a yield γ_{pot} of secondary electrons from potential emission by an Auger neutralization process, the yield γ_{kin} of secondary electrons produced by kinetic emission starts to contribute significantly to the total yield $\gamma = \gamma_{\text{pot}} + \gamma_{\text{kin}}$ when the incident-ion energy exceeds a certain threshold E_{th}. One notices in Fig. 14.1.a that for the case of A^+ ions incident on a clean Mo surface the value of E_{th}

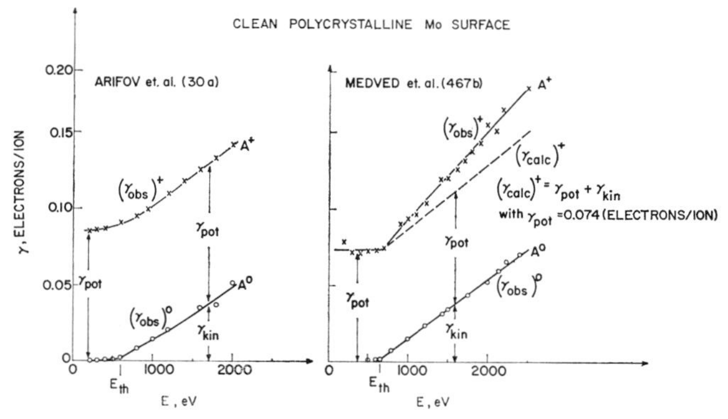

Fig. 14.1.a. The observed total secondary electron yield $(\gamma_{\text{obs}})^+$, and the yields γ_{pot} and γ_{kin} for secondary electrons emitted by the potential and the kinetic emission mechanisms, respectively, as functions of the kinetic energy E of the incident argon ions (A^+) and atoms (A^0) striking a clean polycristalline Mo surface (Arifov et al. [*30a*], Medved et al. [*467b*]). For incident-ion energies E exceeding a certain energy threshold value E_{th}, the total yield is $\gamma = \gamma_{\text{pot}} + \gamma_{\text{kin}}$; for $E < E_{\text{th}}$, the total yield is $\gamma = \gamma_{\text{pot}}$

varied between approximately 600 and 700 eV according to the results of Arifov et al. [*30a*] and Medved et al. [*467b*]. For the case of incident neutral particles in their ground state, no potential emission of electrons takes place; but above a certain energy threshold E_{th}, secondary electrons are emitted by the kinetic ejection process. This is demonstrated in Fig. 14.1.a for neutral argon atoms (A^0) incident on a clean Mo surface, with $E_{\text{th}} \approx 600-700$ eV according to the results of Arifov et al. and Medved et al. For a more detailed discussion of the curves in Fig. 14.1.a, the reader is referred to Section 14.3.1.1.1.

It is of interest to note that the "kinetic" electron emission occurs not only from metallic surfaces but also from the surfaces of ionic crystals (e.g., KCl, Lowzow et al. [*445*]), and from glass surfaces (Batanov [*54*]). This result rules out certain models for the kinetic emission (for instance, that the excitation of secondary electrons is restricted to the presence of free metal electrons). Also another long-lasting belief that the secondary-electron emission is necessarily connected with a gas coverage of the target surface and that the secondary-electron yield will be zero for a gas-free surface (see EREMEEV [*188*] and ARIFOV et al. [*23*]), has been disproved by most recent investigations of the electron emission from atomically pure metal surfaces.

At present, however, there is an unfortunate lack of systematic studies of the kinetic emission of secondary electrons from metal surfaces, in particular from atomically clean metal surfaces. This is rather surprising in view of the important role the kinetic electron emission plays in plasma physics, gas-discharge phenomena, and high-voltage breakdowns, and in such practical applications as the secondary-electron multiplier.

14.2. Experimental Methods

The experimental methods used to study the kinetic emission of secondary electrons are basically similar to those applied in investigations of the potential emission of secondary electrons, which have been briefly described in Section 12.2. The experimental arrangements used in most of the earlier work did not allow control of some of the important parameters that influence the emission of secondary electrons. For example, there was uncertainty about the degree of surface coverage (because of poor vacuum conditions), the composition of the primary-ion beam (i.e., in experiments using gas-discharge sources there might be atomic or molecular ions, a mixture of ions of different elements, and ions in different charge states), the energy of the primary-ion beam, the angle of incidence of the primary-ion beam, the reflection of primary ions, the formation of negative ions at the target, and the formation of tertiary electrons at the collector. Therefore, the following description does not include the older techniques such as a direct determination of the secondary-electron yield by the method of OLIPHANT [*536*], or an indirect determination from the measurement of the secondary TOWNSEND coefficient (HOLST et al. [*320*], KRUITHOFF et al. [*408*], HUXFORD [*335*]) or by a calorimeter method (GÜNTHERSCHULTZE et al. [*261, 262, 263*]). Instead, the reader is referred to the review articles of LITTLE [*440*], MASSEY and BURHOP [*461*], RÜCHARDT [*619*], GEIGER [*232*] and THOMSON et al. [*719*].

In more recent experiments, several authors (HILL et al. [*311*], AARSET et al. [*1*], MURDOCK et al. [*518*], HIGATSBERGER et al. [*308*], DUNAEV et al. [*168*], FLAKS [*216*], BRUNNEÉ [*102*], PETROV [*558*], and FOGEL et al. [*217*]) used mass-analyzed ion beams with a well defined energy to bombard target surfaces. Unfortunately, in nearly all cases in which such well-defined mass-analyzed ion beams have been used, the target conditions were rather poorly known because of inadequate surface preparations and poor vacua. BRUNNEÉ [*102*], however, who used the mass spectrometric arrangement shown in Fig. 14.2.a, was able to study the secondary-electron emission from nearly atomically-clean target surfaces, since he operated at a background pressure of about 5×10^{-8} mm Hg and took the readings within 15 sec after each target baking ("flash-filament" technique).

In other experimental setups with no provisions for mass analysis of the incident ion beam, several authors (e.g., PAETOW and WALCHER [*540*], ARIFOV et al. [*24, 28*]) tried at least to separate the neutral component

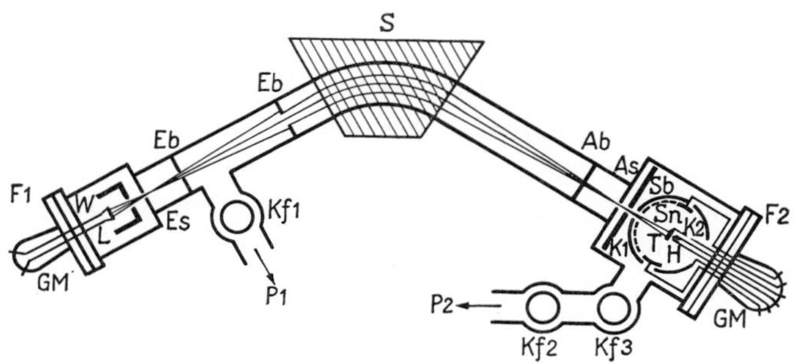

Fig. 14.2.a. Schematic representation of the experimental arrangement used by BRUNNEÉ [*102*] for his studies of secondary-electron emission and ion reflection from metal surfaces under bombardment with isotopic ions. The alkali ions are produced at W by surface ionization, accelerated, focused by the lens L, and enter the mass spectrometer through entrance slot E_s. After passing the collimating slits E_b, they enter the magnetic sector magnet S of the mass analyzer and then pass through the collimating slot A_b and the exit slit A_s and strike the target T. In the target chamber, K_1 and K_2 are two collectors and S_n is a grid. A suitable voltage between K_1 and S_n serves to suppress secondary electrons emitted from K_1 under ion impact. Ports P_1 and P_2 connect to mercury diffusion pumps and $Kf_{1,2,3}$ are liquid-nitrogen traps

of the beam from the ionic one by electric fields (e.g., by use of cylindrical condensers). PAETOW and WALCHER and ARIFOV and AYUKHANOV produced the primary-ion beam by the surface-ionization method (Chapter 8), and thereby restricted the possible impurities in the ion beam to only a few elements.

An accurate determination of the secondary-electron yield requires corrections for the possible reflection of a fraction of the primary-ion beam at the target, for the emission of negative ions from the target, and for possible tertiary electrons emitted from the collector and from metal walls (surrounding the target and collector) under ion and secondary-electron bombardment. At the collector K_1 (Fig. 14.2.a), BRUNNEÉ [*102*] detected the fraction of the primary-ion beam which had been reflected at the target. The secondary electrons emitted from the target were detected at the collector K_2. The fraction of tertiary electrons formed by the impact of reflected primary ions at the collector K_1 were suppressed by a suitable voltage between the grid S_n and collector K_1. Negative ions emitted from the target may have contributed to the secondary-electron current but were neglected.

Arifov et al. [*24, 28*] and Petrov [*558*] used beam-modulation techniques [primary-ion-beam modulation (Petrov), primary-ion-beam as well as secondary-electron-beam modulation (Arifov et al.)], which have already been described briefly in Chapter 12.2, to distinguish between the emitted secondary electrons and other secondary and tertiary emission currents. Finally Waters [*754*] and Tel'kovskii [*708*] applied weak magnetic fields in the collector region to separate the secondary electrons from the emitted negative ions.

14.3. Experimental Results

14.3.1. Dependence of Secondary-Electron Yield γ on the Energy or Velocity of an Incident Ion or Neutral Atom

A discussion of experimental results for the function $\gamma(E)$ demands first a distinction between possible contributions to the value of $\gamma(E)$ from potential emission and from kinetic emission of secondary electrons. As has already been discussed in Chapter 12, the adiabatic process of potential ejection of secondary electrons from atomically clean metal surfaces occurs only if the potential energy available for the neutralization of the incident ion at the metal surface exceeds twice the work function φ of the metal ($eI > 2e\varphi$, with $eI \equiv$ ionization energy of incident ion). If the potential emission is energetically possible, as for noble-gas ions incident on metal surfaces, this type of emission occurs at all incident-ion energies. But the electron yield γ, which is determined solely by the potential ejection mechanism, depends only very slightly on the kinetic energy of the incident ion (see corresponding results in Chapter 12, Section 12.2.2.1.1). In the event that the potential energy available in the interaction of the incident ions with the metal surface is less than twice the work function of the metal, as in the case of alkaline-ion bombardment of metal surfaces, the kinetic emission will be the only mechanism operative. In this case more significant changes in the $\gamma(E)$ curves are observed. The value of γ increases with increasing energy up to very high energies and then starts to flatten out and finally to drop, a behavior which suggests a close relationship between the formation of secondary electrons and the cross section σ_i for ionization of an atom by ion impact (see discussion in Section 14.4).

Whenever both mechanisms contribute to the electron emission, as in the case of bombardment of metals by noble-gas ions with medium energies, it is possible in an approximation to distinguish between the different contributions on the assumption that the yield γ_{pot} due to potential ejection remains at a constant value over the whole energy range and can be determined for incident-ion energies $E < E_{th}$, as indicated in Fig. 14.1.a for the system A^+ on Mo. Therefore values of $\gamma_{kin}(E)$ for ion energies $E > E_{th}$ will be obtained by the relationship $\gamma_{kin}(E) = \gamma(E) - \gamma_{pot}$, where γ is the measured total yield of secondary electrons. Values of γ_{kin} and γ determined in this way are shown in Figs. 14.1.a and 14.3.1.1.1.a.

In the following, results for γ as well as γ_{kin} will be discussed. In most cases they have been obtained for gas-covered surfaces and only in very few cases for clean metal surfaces.

14.3.1.1. Medium Energy Range

14.3.1.1.1. Atoms and Ions of Noble Gases

Table 14.3.1.1.1. A lists the systems which have been investigated recently by several authors [*28, 30a, 119a, 129, 308, 451a, 451b, 452, 452a, 467a, 557, 559, 560, 561, 562, 573, 709*] to study the functions $\gamma(E)$ or $\gamma(V)$ (where V is the ion velocity) for the bombardment of metal surfaces by atoms and ions of noble gases. The earlier work has been reviewed by Massey and Burhop [*461*], Francis and Jenkins [*223*], Little [*440*], and Granowski [*253*].

A good illustration of the relative contributions of secondary electrons, emitted by the potential- and/or the kinetic-emission mechanism to the total yield γ is given in Fig. 14.1.a, where $\gamma(E)$ curves are plotted for A^+ and A^0 incident normally on clean polycrystalline Mo surfaces. One notices that for the case of bombardment by neutral argon atoms A^0 the ejection of secondary electrons occur only above an energy threshold $E_{th} \approx 600-700$ eV, where the observed yield $[\gamma_{obs}(E)]$ is primarily due to the kinetic-emission mechanism, i.e., $(\gamma_{obs})^0 = \gamma_{kin}$. For the case of bombardment with argon ions A^+, one notices that the yield $[\gamma_{obs}(E)]^+$ increases with increasing incident-ion energy E above an energy threshold E_{th}, whose value coincides closely with the one observed for the bombardment with neutral argon atoms. This again indicates the onset of the kinetic emission of secondary electrons.

For ion energies $E < E_{th}$, the emission of secondary electrons is predominantly determined by the potential emission mechanism, i.e., $(\gamma_{obs})^+ = \gamma_{pot}$. It is of interest to note that Parilis and Kishinevskii [*543*] predicted theoretically a velocity threshold value for the kinetic emission of electrons by considering collisions between incident ions and lattice atoms (as will be explained in more detail in Section 14.4). For A^+ incident on Mo, the predicted velocity threshold is near 0.6×10^7 cm/sec. This corresponds to an energy threshold $(E_{th})_{theor} = 690$ eV, which is in good agreement with the observed threshold value. For the system He^+ on Mo, Comeaux and Medved [*119a*] and Hagstrum [*286*] determined

Table 14.3.1.1.1.A. *List of some systems of noble-gas atoms or ions on metal surfaces, as used by various authors to study the function* $\gamma(E)$

Target materials	Be	Al	Zr	Cu-Be Alloy	Ag-Mg Alloy	Ni	Ta	Zn	Fe	Co	Mo	W	Cu	Pt	Author
Incident particle	Ne^+													Ne^+	Ploch [573]
											A^+				Philbert [562]
							He^+ A^+					A^+			Petrov [557, 560]
						He^+ A^+									Petrov [559]
	A^+	A^+	A^+			A^+	A^+	A^+	A^+	A^+	A^+	A^+	A^+		Slodzian [658]
	A^+	A^+	A^+					A^+	A^+	A^+	A^+	A^+	A^+	A^+	Cousinié [129]
			A^+ A^{2+} A^{3+}								He^+ He^+ A^+ A^{2+} A^{3+} He^0 A^0				Tel'kovskii [709]
							Ne^+ A^+ Kr^+				Ne^+ A^+ Kr^+	Ne^+ A^+ Kr^+			Arifov [28]
											He^+ A^+				Mahadevan [452a]
											He^0 He^+ A^0 A^+				Comeaux [119a]

Incident particle															
											A^0 A^+				MEDVED [467b]
											A^0 A^+				ARIFOV [30a]
		A^+	A^+			A^+	A^+				Ne^+ A^+ Kr^+ Xe^+		Ne^+ A^+ Kr^+ Xe^+		MAGNUSON [451a]
													A^{+*}		MAGNUSON [451b]
		A^+	A^+			A^+	A^+						A^+		MAGNUSON [452]
				A^+	He^+ Ne^+ A^+ Kr^+ Xe^+										HIGATSBERGER [308]

* Three different planes [(111), (110), (100)] of a copper monocrystal were bombarded.

$\gamma(E)$ curves, which indicate an energy threshold value $E_{th} \approx 500$ eV. This value cannot be compared with the one calculated on the basis of the theory of PARILIS et al. since their treatment is only valid for systems satisfying the inequality $\frac{1}{4} < Z_1/Z_2 < 4$ (where Z_1 and Z_2 are the atomic numbers of the incident ion and the target atom, respectively). However, the observed threshold seems to be consistent with the values obtained by extrapolating the theoretical curves for other noble-gas ions.

In Fig. 14.1.a, one sees that for ion energies $E > E_{th}$ the values of $(\gamma_{obs})^+$ observed by ARIFOV et al. [30a] for incident argon ions agrees with the yield curve calculated under the assumption

$$[\gamma_{obs}(E)]^+ = \gamma_{pot} + \gamma_{kin}(E) \quad (14.3.1.1.1\text{–}1)$$

with

$\gamma_{pot} = (\gamma_{obs})^+$ for $E < E_{th}$

and

$$\gamma_{kin} = (\gamma_{obs})^0 .$$

The yield curve $(\gamma_{obs})^+$ in Fig. 14.1.a determined by MEDVED at al. [467b], however, indicates that

$$[\gamma_{\text{obs}}(E)]^+ > [\gamma_{\text{calc}}(E)]^+ = \gamma_{\text{pot}} + \gamma_{\text{kin}}(E)\,. \quad (14.3.1.1.1\text{–}2)$$

MEDVED et al. tried to interpret the discrepancy between $(\gamma_{\text{obs}})^+$ and $(\gamma_{\text{calc}})^+$ under the assumption that (in the energy region considered) neutral noble-gas atoms are more readily reflected from metal surfaces than are noble-gas ions, which could lead to the result $(\gamma_{\text{kin}})^0 < (\gamma_{\text{kin}})^+$. However, the assumption that noble-gas atoms reflect more readily from metal surfaces than noble-gas ions is based only on qualitative results on particle reflection, which cannot be considered as being conclusive (see critical reviews in Section 11.3.1.1). From the reported experimental conditions, it is difficult to judge the extent to which the discrepancy may be attributed to three different sources. (1) There may have been a difference in temperature. The target temperature was 1100° K in the experiments of ARIFOV et al.; MEDVED et al. did not state their target temperature but their first measurements were taken 15 sec after flashing the target to 2000° K. (2) There may have been a slight surface coverage (probably less than a monolayer) in the experiments of ARIFOV et al. [30*a*]. Their background pressure was 10^{-7} mm Hg, but they conducted the double-modulation measurements (Section 11.2) quickly after flashing the surface to 2200° K. (3) Possibly there were differences between the grain structures of the polycrystalline Mo target surfaces (due to some such cause as annealing during the flashing procedure). It should be mentioned that MAGNUSON and CARLSTON [*451a*] confirmed the results obtained by MEDVED et al. for the function $\gamma(E)$ for the system A^+ on polycrystalline Mo in the energy region from 500 to 2500 eV.

The functions $\gamma(E)$ and $\gamma_{\text{kin}}(E)$ increase linearly with increasing ion energy E for the case of atoms and ions of noble gases incident on metal surfaces in the investigated medium-energy region $E_{\text{th}} < E < 10$ keV, as has been reported for different ranges of incident-ion energies by ARIFOV et al. [*28*, *30a*], MAHADEVAN et al. [*452a*], COMMEAUX and MEDVED [*119a*], MEDVED et al. [*467b*], and MAGNUSON and CARLSTON [*451a*, *451b*]. The curves $\gamma(E)$ and $\gamma_{\text{kin}}(E)$ shown in Fig. 14.1.a for A^0 and A^+ bombardment of clean Mo surfaces (ref. [*30a*], [*467b*]) and in Fig. 14.3.1.1.1.a for He^+, Ne^+, A^+, and Kr^+ ion bombardment of a clean Mo surface (ref. [*28*]) illustrate such a behavior. The values of γ_{kin} in Fig. 14.3.1.1.1.a were determined by ARIFOV et al. under the assumption $\gamma_{\text{kin}}(E) = \gamma(E) - \gamma_{\text{pot}}$ for $E > E_{\text{th}}$. The experimentally observed dependence $\gamma(E) \propto E$ is in agreement with theoretical predictions made by PARILIS and KISHINEVSKII [*543*] (see Section 14.4). It is of interest to notice that in contrast to these results for the function $\gamma(E)$ for noble-gas atoms and ions, BRUNNEÉ [*102*] observed for alkali ions incident on atomically clean Mo surfaces for a corresponding energy region that the yield varied as $\gamma(E) \propto E^2$. A subsequent theoretical treatment by VON ROOS [*612*] led to the same energy dependence of the yield γ.

The experimental data available so far on the values of $\gamma(E)$ and $\gamma_{\text{kin}}(E)$ indicate a rather complicated dependence on the mass m and the atomic number Z of the incident ion (see also discussion in Section 14.3.3). For the bombardment of a polycrystalline Mo surface with the noble-gas ions He^+, Ne^+, A^+, and Kr^+ incident normally with energies in the medium range $E_{\text{th}} < E < 10\,\text{keV}$ the values of both γ and γ_{kin} increase with decreasing mass and atomic number of the incident ion (i.e., $\gamma_{\text{He}} > \gamma_{\text{Ne}} > \gamma_{\text{A}} > \gamma_{\text{Kr}}$) as has been observed for instance by ARIFOV et al. [*28*] (see Fig. 14.3.1.1.1.a) and MAGNUSON and CARLSTON [*415a*].

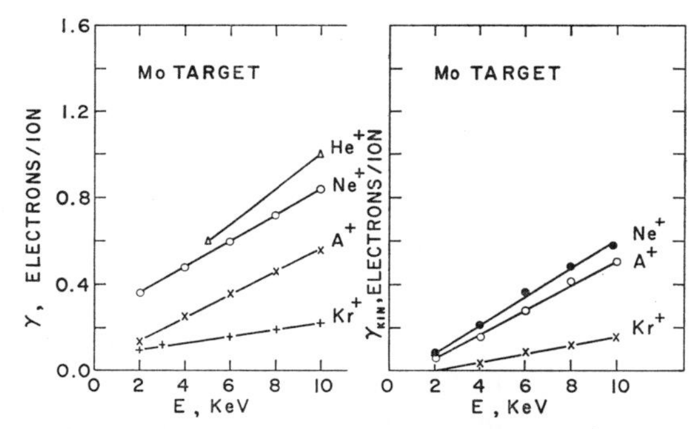

Fig. 14.3.1.1.1.a. The total secondary-electron yield γ and the yield γ_{kin} for secondary electrons emitted by the kinetic emission mechanism as functions of the kinetic energy of the incident ion for noble-gas ions striking a Mo surface (ARIFOV et al. [*28*]). The curve (He^+-Mo, points Δ) is based on data reported by TEL'KOVSKII [*707*]

However, for Xe^+ ion bombardment of such a Mo surface, MAGNUSON et al. observed that the $\gamma(E)$ curves for Xe and Kr crossed at energies near 6.5 keV above which $\gamma_{\text{Xe}} > \gamma_{\text{Kr}}$, and that they come to another cross-over point near 10 keV. In contrast to these observations made for a Mo surface, MAGNUSON and CARLSTON reported for the bombardment of a polycrystalline Cu surface by normally incident noble-gas ions Ne^+, A^+, Kr^+, and Xe^+ that the γ curves depend on the type of the incident ion such that $\gamma_{\text{Ne}} > \gamma_{\text{A}} > \gamma_{\text{Kr}} > \gamma_{\text{Xe}}$ for the energy region $500\,\text{eV} < E \lesssim 4.3$ to 4.8 keV (values taken from Fig. 7 of the article in [*415a*] but show the inverse dependence for energies above 4.3 to 4.8 keV namely $\gamma_{\text{Kr}} > \gamma_{\text{A}} > \gamma_{\text{Ne}}$. However, the $\gamma_{\text{Xe}}(E)$ curve does not cross any of the $\gamma(E)$ curves of the other noble-gas ions over the entire investigated medium-energy range $500\,\text{eV} < E < 10\,\text{keV}$. At present there is no theoretical treatment available that offers a sufficient explanation of the observed complicated behavior of the function $\gamma(E, m, Z)$.

It is of interest to note in the plots of γ versus the velocity V of the incident ion that for the Ne^+ and A^+ ion bombardment of a polycrystalline

Mo surface the $\gamma(V)$ curves come to a cross-over point at $V = 2.4 \times 10^7$ cm/sec (see Fig. 14.3.1.1.b) while the corresponding $\gamma(E)$ curves do not show any cross over. The yield values in Fig. 14.3.1.1.1.b for the lower and the higher velocity region have been obtained by different authors [*26, 28, 708*]. If the data of MAGNUSON and CARLSTON for A^+ and Kr^+ on Mo are plotted as functions $\gamma(V)$, the curves come to a cross-over point near $V = 1.55 \times 10^7$ cm/sec (a velocity above which $\gamma_{Kr} > \gamma_A$). As will be discussed in Section 14.3.1.2, this behavior $\gamma(V_1, m_1, Z_1) > \gamma(V_1, m_2, Z_2)$ with $m_1 > m_2$ and $Z_1 > Z_2$ becomes typical for the higher energy region.

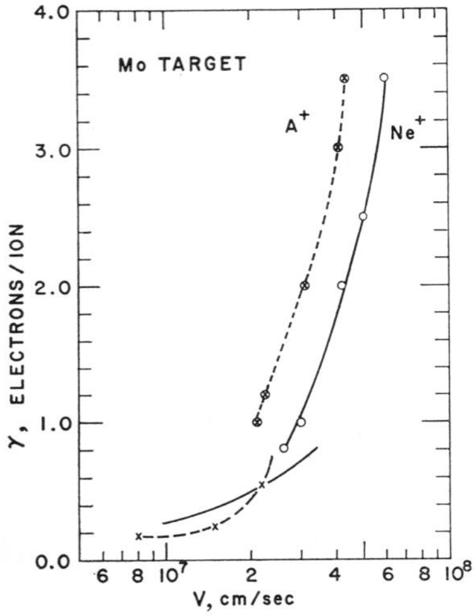

Fig. 14.3.1.1.1.b. The dependence of the secondary-electron yield γ on the velocity V of the incident ion for the case of Ne^+ and A^+ ions bombarding a Mo target. The γ values at lower velocities for Ne^+ (solid line) and A^+ (dashed line) were obtained by ARIFOV et al. [*28*]. The γ values at higher ion velocities (Ne^+, solid line; A^+, dashed line) were obtained by TEL'KOVSKII [*708*]

The complicated dependence of the functions $\gamma(E)$ and $\gamma(V)$ on the mass of the incident ion has not been taken sufficiently into consideration in many experiments, such as those in which the precise measurements of isotopic ratios rely on a comparison of secondary-electron currents (see discussion in Section 14.3.3).

The dependence of the secondary-electron yield on the type of target material is illustrated in Fig. 14.3.1.1.1.c for the case of A^+ ion bombardment of various degassed surfaces. One notices that the disagreement between the results of different authors for the same system is larger than the differences between the values of γ for different target materials. MAGNUSON and CARLSTON [*415a*], who investigated the emission of secondary electrons from such cleaned polycrystalline target surfaces as Cu, Ni, Al, Zr, Mo, and Ta under bombardment by normally incident A^+ ions, observed that ordinarily the targets with fcc structure (Cu, Ni, Al) yielded somewhat larger values of γ in the energy region 3 keV $< E <$ 10 keV ($\gamma_{Co} > \gamma_{Ni} > \gamma_{Al}$) than those with bcc structure (Zr, Mo, Ta) for the corresponding energy region ($\gamma_{Zr} > \gamma_{Mo} > \gamma_{Ta}$). However, the values of γ for the different investigated bcc and fcc target materials differ at most by 40% from each other.

In view of these data, a somewhat surprising result was obtained by MAGNUSON et al. [*415b, 452*] in experiments in which three different clean copper single-crystal planes [(111), (100), and (110)] were bombarded by A^+ ions in the energy range from 500 eV to 10 keV. As illustrated in Fig. 14.3.1.1.1.d, they observed rather large differences in the yield curves $\gamma(E)$ for the different crystal planes bombarded $[\gamma_{111}(E) > \gamma_{100}(E) > \gamma_{110}(E)$, for 1 keV $< E <$ 10 keV]. For example, at $E = 10$ keV the value of γ_{111} is nearly 240% larger than the value of γ_{110}. The authors suggest a simple geometrical model, relating the difference in the value of γ for each of the three surfaces to the difference in transmissivity* of the three

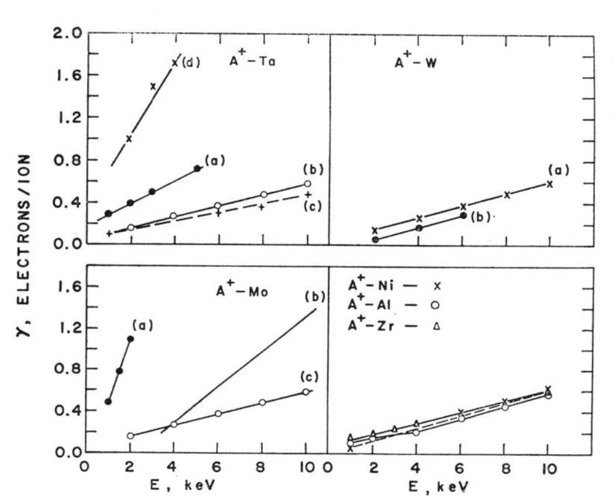

Fig. 14.3.1.1.1.c. The dependence of the secondary-electron yield γ on the kinetic energy of the incident ion for A^+ ions incident on various target materials. For A^+—Ta: curve *a*, PETROV [*560*]; curve *b*, ARIFOV et al. [*26, 28*]; curve *c*, MAGNUSON [*452*]; curve *d*, BERRY [*71*]. For A^+—Mo: curve *a*, PHILBERT [*562*]; curve *b*, TEL'KOVSKII [*708*]; curve *c*, ARIFOV et al. [*26, 28*]. For A^+—W: curve *a* ARIFOV et al. [*26, 28*]; curve *b*, PETROV [*560*]. The values for A^+ on Ni, Al, and Zr were obtained by MAGNUSON [*452*]

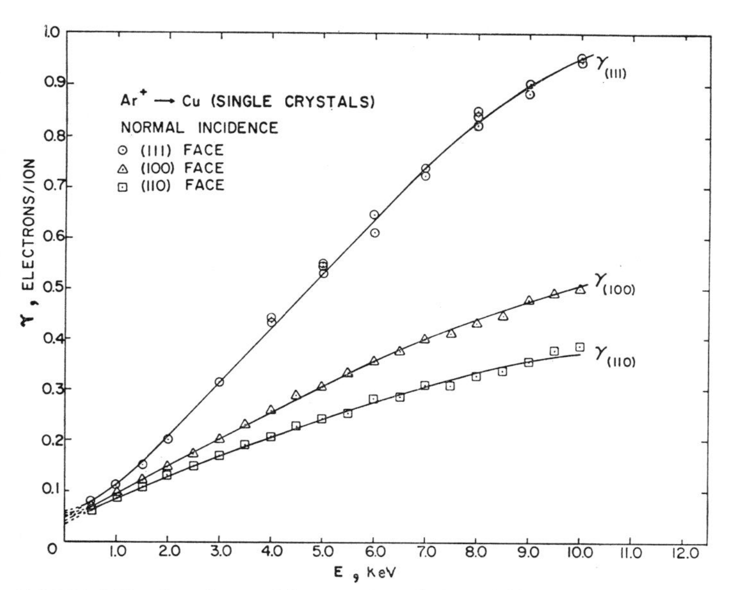

Fig. 14.3.1.1.1.d. The dependence of the secondary-electron yield γ on the kinetic energy of the incident ion for A^+ ions bombarding the (111), (100), and (110) planes of copper single crystals. (MAGNUSON et al. [*451b*] and MAGNUSON [*452*])

* The transmissivity is inversely proportional to the density of lattice points projected on a plane normal to the beam (see also the discussion in Section 10.3.2.1.1 and Fig. 10.3.2.1.1.g).

low-index planes, the transmissivity being the highest for the (110) plane and the lowest for the (111) plane. According to the model, the less transparent the lattice, the higher is the probability of incident-ion/lattice-atom collisions and therefore the higher the probability of secondary-electron production by ionization of lattice atoms by ion impact. One should keep in mind, however, that an additional effect, the channeling of the incident ion between adjacent close-packed crystallographic directions or close-packed planes [see Sections 10.3.2 and 10.4.2.1] is an additional cause for the observed order in the decrease of the yield-values. More experimental data on the emission of secondary electrons from single-crystal surfaces under ion impact are badly needed to verify the few existing results and to check the validity of the suggested model.

For the case of gas-covered surfaces COUSINIÉ et al. [*129*] and SLODZIAN [*658*] have observed a stronger dependence of γ on the target material. For instance for the bombardment of various surfaces with 30-keV A^+ ions, COUSINIE et al. observed that the values of γ for different target materials were in the order $\gamma_{Be} > \gamma_{Al} > \gamma_{Zr} > \gamma_{Zn} > \gamma_{Mn} > \gamma_{Co} > \gamma_{Pb} > \gamma_{Fe} > \gamma_{Ni} > \gamma_{Cu} > \gamma_{Au} > \gamma_{W} > \gamma_{Pt} > \gamma_{Mo}$. These observed differences in the values of γ for the various targets cannot be explained only by the difference in the work function φ of the surfaces as has been suggested by PAWLOW et al. [*549*] who proposed the relation $\gamma = A e^{-B\varphi}$, where A and B are constants, even if the work-function changes due to the gas adsorption are taken into consideration. (For corresponding results of such work function changes, see the review article by EBERHAGEN [*169*]).

14.3.1.1.2. Alkali or Alkaline Earth Metal Ions

For bombardment of metal surfaces by alkali or alkaline earth metal ions, only the kinetic energy of the ion is available to release secondary electrons, since in this case $eI \ll 2e\varphi$ (the symbols have the same meaning as in Section 14.2.1.1.1). The observed total yield γ then is equal to the kinetic yield γ_{kin}.

Most studies of the secondary-electron emission from metal surfaces under bombardment by alkali or alkaline earth metal ions have been conducted for gas-covered target surfaces, as for instance those investigations given in the references [*168, 188, 189, 192, 217, 252, 540, 573, 753, 754*]. Table 14.3.1.1.2.A, lists some of the ion/metal-surface systems investigated. Corresponding earlier work has been reviewed elsewhere [*223, 253, 440, 461*]. Data on the secondary-electron emission from degassed or even atomically-clean metal surfaces are very scarce. Table 14.3.1.1.2.B lists some of the ion/metal-surface systems investigated in recent work.

Table 14.3.1.1.2.A. *Ion/metal-surface systems used to study the function $\gamma(E)$ on gas-covered surfaces*

Target material	Be	Ni	Cu	Sn	Ta	W	Pt	Author
Incident ion						Cs		PAETOW [*540*]
	Li		Li				Li, K	PLOCH [*573*]
				Li	Li	Li		EREMEEV [*188*]
					Li K	Li K		EREMEEV [*189, 190*]
					Li K	Li K		EREMEEV [*192*]
		Na Ba Ca						DUNAEV [*168*]
							K	FLAKS [*215*]
						Li Cs		WATERS [*753, 754*]

Table 14.3.1.1.2.B. *Ion/metal-surface systems used to study the function $\gamma(E)$ on degassed surfaces*

Target material	Mo	Ta	W	Authors
Incident ion	Li, Na, K, Rb, Cs			BRUNNEÉ [*102*]
			Li, Cs	WATERS [*753, 754*]
		K	K	PETROV [*560*]
	Na, K	Na, K	Na, K	ARIFOV [*28*]

Fig. 14.3.1.1.2.a shows $\gamma(E)$ curves for singly-charged alkali ions incident on a highly cleaned Mo surface (BRUNNEÉ [*102*] and on a

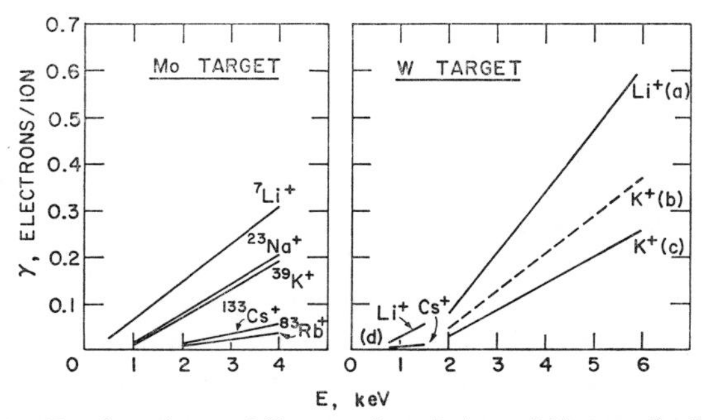

Fig. 14.3.1.1.2.a. The dependence of the secondary-electron yield γ on the kinetic energy of the incident ion for alkali ions bombarding Mo and W targets. The γ values for the Mo target were obtained by BRUNNEÉ [*102*]. The curve *a*, Li^{+}—W, was reported by EREMEEV et al. [*189*]; curve *c*, K^{+}—W, by PETROV [*560*], and curves *d*, Li^{+} and Cs^{+} on W by WATERS [*754*]

degassed W surface (curve (*c*) in Fig. 14.3.1.1.2.a, [*560*]). One notices the interesting result that, with the exception of Li, the yield $\gamma = \gamma_{\text{kin}}$ changes with the square of the incident-ion energy in the energy region studied (1—4 keV). The observation that $\gamma(E) \propto E^2$ is in agreement with theoretical predictions made by von Roos [*612*] (Section 14.4). In contrast, for noble-gas ions incident on the same surface with the same range of energies, γ was observed to increase linearly with increasing E (Section 14.3.1.1.1). On the other hand, as in the case of noble-gas ion bombardment within a certain energy region, one finds that γ decreases with increasing mass and atomic number of the incident ion. However, for the case of Rb^+ and Cs^+ ions incident on a clean Mo surface, Brunneé observed the opposite behavior ($\gamma_{Rb^+} < \gamma_{Cs^+}$).

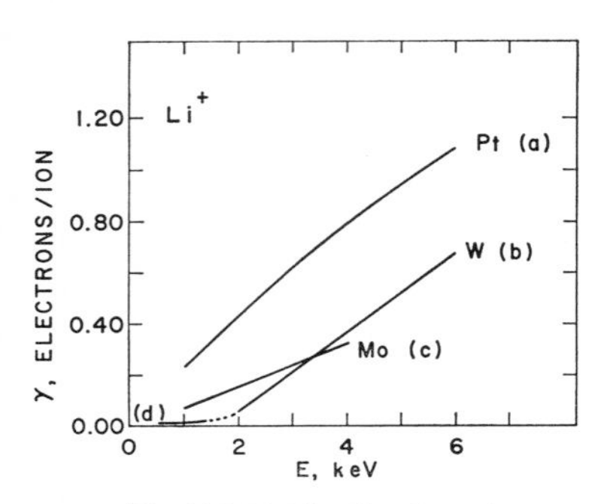

Fig. 14.3.1.1.2.b. The dependence of the secondary-electron yield γ on the incident-ion energy for Li^+ ions striking a Pt target (curve *a*, Ploch [*573*]), a W target (curve *b*, Eremeev et al. [*191*]; curve *d*, Waters [*754*], and a Mo target (curve *c*, Brunneé [*102*])

The $\gamma(E)$ curves shown in Fig. 14.3.1.1.2.b for a Li ion bombardment of gas-free or only slightly gas-covered surfaces (coverage degree θ probably less than 1) illustrate a considerable dependence of the yield on the target material chosen.

14.3.1.1.3. Ions of Various Kinds

In the energy region considered, the emission of secondary electrons from metal surfaces under bombardment with atomic and molecular ions such as H_1^+, H_2^+, N^+, and N_2^+ ions is caused by the mechanism of potential emission as well as kinetic emission. Secondary-electron yields for such systems have been reported recently by Petrov [*559, 560, 561*] (H_2^+—Ta; H_1^+—Mo; N^+, N_2^+ on W), Tel'kovskii [*707, 708*] (H_1^+—Mo, N^+—Mo), and O'Brian et al. [*529*] H_1^+, H_2^+, and D^+ on Ag). Some $\gamma(E)$ curves obtained for these systems are shown in Fig. 14.3.1.1.3.a. Two results seem noteworthy. On the one hand, the $\gamma(E)$ functions vary as $\sqrt{E}$ in the energy region investigated for such systems as H_1^+ on Mo [*707, 708*] and H_2^+ on Ta [*559, 560*], but tend to level off at higher energies. As will be discussed in Section 14.3.1.2, at even higher energies the $\gamma(E)$ curves for the corresponding systems pass through a maximum and subsequently decrease, a behavior which can be considered as an indication of the formation of secondary electrons by the ionization of lattice atoms by ion impact. It may be described by the Born approximation for the cross section of ionic-atomic inelastic collisions. On the other hand, one notices in Fig. 14.3.1.1.3.a that at lower ion energies the lighter ion yields the higher values of γ, while at higher energies the opposite is observed. One may remember that a similar result was obtained in the case of the lighter noble-gas ions.

For the case of atomic and molecular ions of nitrogen incident on a polycrystalline wolfram surface in the energy range $1 \text{ keV} < E < 10 \text{keV}$, PETROV et al. [*561*] reported that the yield $\gamma(E)$ varies linearly with E. Furthermore, they observed that the values of $\gamma(E)$ for the atomic ions were smaller than for the molecular ions. More quantitatively, they

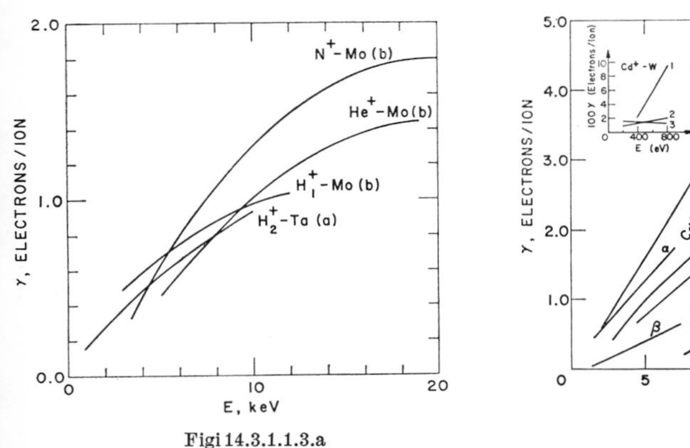

Figi 14.3.1.1.3.a

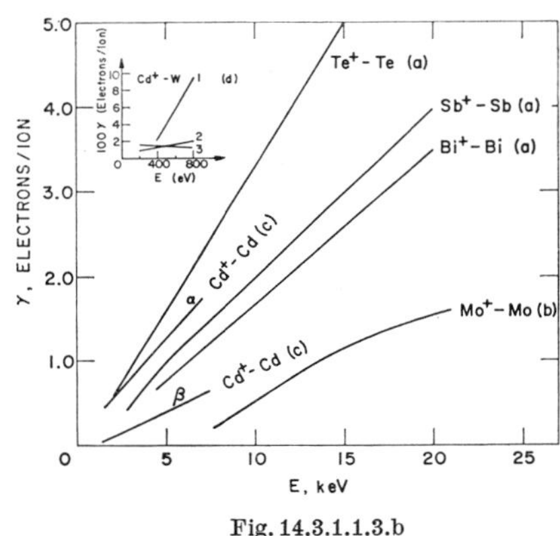

Fig. 14.3.1.1.3.b

Fig. 14.3.1.1.3.a. The dependence of the secondary-electron yield γ on the incident-ion energy E. The values for curve a (H_2^+—Ta) were reported by PETROV [*557, 559*]; the γ values for the curves b were obtained by TEL'KOVSKII [*707, 708*]

Fig. 14.3.1.1.3.b. The dependence of the secondary-electron yield γ on the incident-ion energy E for the case in which the incident ion and the target surface are the same element. The curves a were reported by DUNAEV et al. [168], curve b by TEL'KOVSKII [*708*], and curves c for Cd^+—Cd (α for a heated target, β for an evaporated target) by SCHWARTZ et al. [*645*]. For comparison, insert d shows the curves for a case (Cd^+ ions incident on W) in which the ions are not of the same element as the target. The surface cleanliness increases from curve *1* to curve *3* (PETROV [*558*])

found that $\gamma(N_2^+)$ at an energy E is equal to $2\gamma(N^+)$ at an energy $E/2$. This result indicates that a diatomic molecular ion striking a metal surface dissociates into its atomic constituents, each of which then interacts with the metal surface with half of the energy E of the incident molecular ion.

Fig. 14.3.1.1.3.b shows $\gamma(E)$ curves for the case in which incident ion and target are the same element. The values of γ vary nearly linearly with the incident-ion energy for the energy region studied. A comparison of the values of γ for the systems Cd^+ on Cd and Cd^+ on W, however, shows no significant difference in the values of γ. This indicates that the system in which incident ion and target material are the same play no unique role in secondary-electron emission.

14.3.1.2. High-Energy Range

14.3.1.2.1. Noble-Gas Ions

At higher ion energies, it becomes apparent that the secondary-electron yield and the cross section σ_i for the ionization of an atom by impact with an ion have a similar dependence on the velocity V of the incident ion. For the lighter incident ions both $\gamma(E)$ and $\sigma_i(V)$ reach a maximum and subsequently decrease with increasing V. This behavior is well described by the BORN approximation in this region. Also as V increases the ion impact results in more secondary electrons being formed deeper within the target and the diffusion length of these electrons to the surface is increased. As a result, not all the secondary electrons formed will have a chance to leave the target.

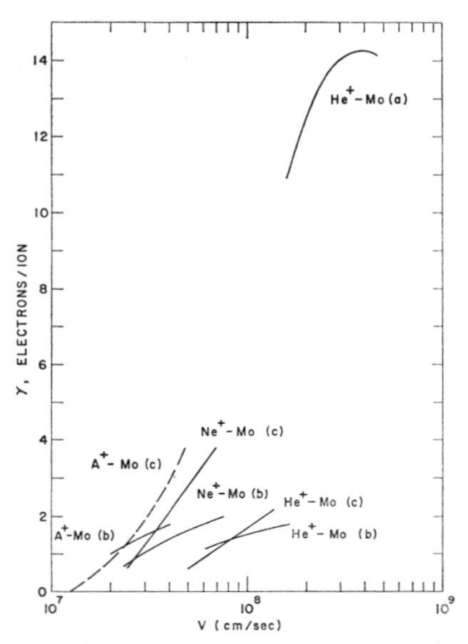

Fig. 14.3.1.2.1.a. The dependence of the secondary-electron yield γ on the velocity V of the incident ions for the case of noble-gas ion bombardment of Mo surfaces. Curve a (He^+—Mo) was reported by HILL et al. [*311*], curves b (He^+, Ne^+, A^+—Mo) by FOGEL et al. [*217*], and curves c (He^+, Ne^+, A^+—Mo) by TEL'KOVSKII [*707, 708*]

Studies of the secondary-electron emission from metal surfaces under bombardment by noble-gas ions in the energy region considered have been reported for instance by HILL et al. [*311*] (He^+—Mo), BARNETT et al. [*49*] (Ne^+, A^+, N^0, A^0—Ni), TEL'KOVSKII [*707, 708*] (He^+, Ne^+, A^+—Mo), COUSINIÉ et al. [*129*] (A^+—Be, Al, Zr, Zn, Mn, Co, Pb, Fe, Ni, Cu, Au, W, Pt, Mo), and FOGEL et al. [*217*] (He^+, Ne^+, A^+, Kr^+—Mo).

Fig. 14.3.1.2.1.a is a set of $\gamma(V)$ curves for the noble-gas-ion bombardment of a gas-covered Mo surface (curve a) and of degassed Mo surfaces (curves b and c). One notices that for the system He^+—Mo the values of γ for the gas-covered Mo surface are considerably higher than the corresponding ones for a degassed Mo surface. Over a broader velocity range, γ is roughly proportional to the velocity until (in the case of He^+—Mo) it starts to level off into a flat maximum. In the velocity region studied, the heavier ions yield higher values of γ as the results of FOGEL et al. [*217*] and TEL'KOVSKII [*707, 708*], for example, have indicated.

14.3.1.2.2. Alkali and Alkaline Earth Metal Ions

Only a few data on secondary-electron emission from metal surfaces under bombardment of alkali and alkaline earth metal ions with higher energies have been reported. Among these are the studies by DUNAEV and FLAKS [*168*] (singly- and multiply-charged Na, Ba, and Ca ions on Ni) and FLAKS [*215*] (singly- and multiply-charged K ions on Pt). The $\gamma(E)$ data for singly-charged Na^+, Ba^+, and Ca^+ ions incident on a gas-covered Ni surface and for K^+ ions incident on a gas-covered Pt surface are illustrated in Fig. 14.3.1.2.2.a. The function $\gamma(E)$ varies almost linearly with the incident-ion energy for the different ions but, in the case of the K^+ ions, starts to level off at higher ion energies. For the energy region investigated, the values of γ for the different ions increase with decreasing mass and atomic number of the incident ion. At considerably higher ion energies, however, the opposite trend must be expected, as has been observed for the lighter ions (Section 14.3.1.2.3).

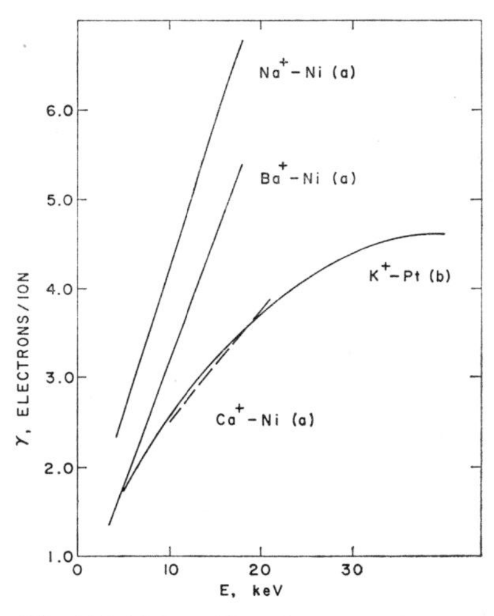

Fig. 14.3.1.2.2.a. The dependence of the secondary-electron yield γ on the kinetic energy E of the incident ion for alkali and alkaline-earth metal ions incident on Ni and Pt targets. The curves (*a*) were reported by DUNAEV et al. [*168*] and the curve (*b*) by FLAKS [*215*]

14.3.1.2.3. Various Kinds of Ions

The secondary-electron emission from metal surfaces under bombardment by high-energy particles has been investigated for light incident ions ($Z < 4$) by numerous authors. Table 14.3.1.2.3.A lists some of the ion/metal-surface systems used to study the function $\gamma(E)$.

Fig. 14.3.1.2.3.a illustrates the dependence of the total secondary-electron yield γ on the velocity of the incident ion for various ions. One notices that the $\gamma(V)$ curves in the lower velocity region increase linearly with increasing ion velocity and then start to level off, pass through a flat maximum, and then drop gradually. The maximum of the $\gamma(V)$ curve seems to shift to higher velocity ranges with increasing mass of the incident particle. These results most strikingly illustrate the close relationship between the cross section for ionization of a lattice atom by ion impact and the formation of secondary electrons.

Another interesting result is that at equal ion velocities the secondary-electron yield is proportional to the mass of the bombarding molecular

Table 14.3.1.2.3.A. *Ion/metal-surface systems ($Z_{ion} < 2$) used to study the function $\gamma(E)$*

Target material	Be	Mg	Al	Fe	St. Steel	Ni	Cu	Zr	Mo	Ag	Pt	Au	Pb	Authors
Incident ion	H_1^+					H_1^+	H_1^+				H_1^+			ALLEN [*5, 6*]
			H_1^+ H_2^+				H_1^+ H_2^+		H_1^+ H_2^+				H_1^+ H_2^+	HILL [*311*]
					H_1^+ H_2^+ H_3^+		H_1^+ H_2^+ H_3^+		H_1^+ H_2^+ H_3^+			H_1^+ H_2^+ H_3^+		MURDOCK [*518*]
		H_1^+	H_1^+ H_2^+	H_1^+		H_1^+ H_2^+						H_1^+ H_2^+	H_1^+	AARSET [*1*]
								H_1^+ H_2^+ H_3^+	H_1^+ H_2^+ H_3^+					TEL'KOVSKII [*708, 709*]
			H_1^+				H_1^+		H_1^+				H_1^+	COUSINIÉ [*129*]
	D_1^+ D_2^+		D_1^+ D_2^+				D_1^+ D_2^+	D_1^+ D_2^+		D_1^+ D_2^+		D_1^+ D_2^+		LEROY [*433*]
									H_1^+ H_2^+					FOGEL [*217, 218*]
							D_1^+							AKISHIN [*4a*]

ion [$\gamma(H^+) : \gamma(H_2^+) : \gamma(H_3^+) = 1:2:3$], as can be seen for the case of a Cu target in Fig. 14.3.1.2.3.a. The close proportionality between the yield and the number of particles comprising the molecular ion leads to the conclusion that the molecular ions break up into their constituent particles at the metal surface, after which their interaction with the metal is of just that of single ions or atoms having the same velocity.

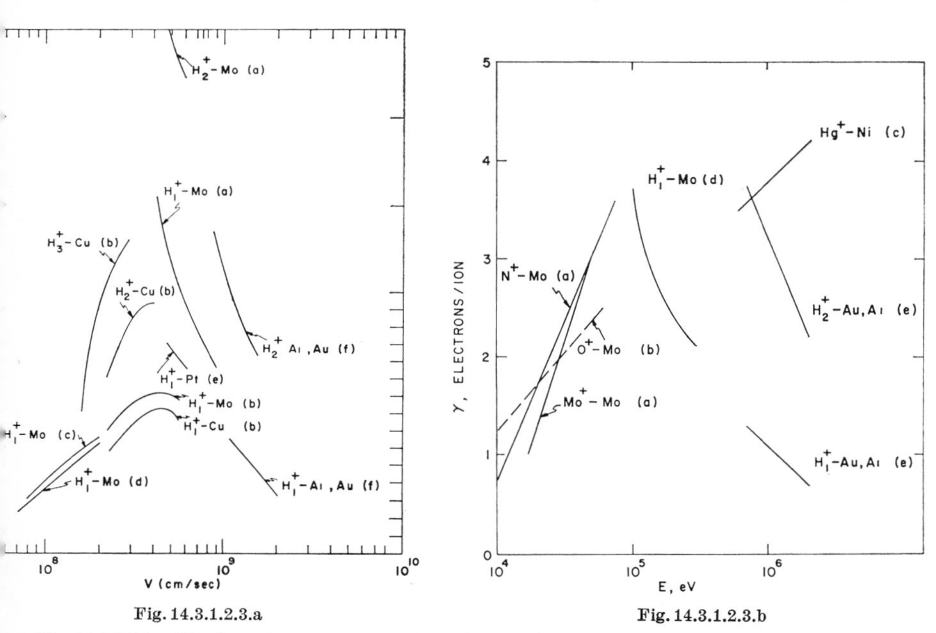

Fig. 14.3.1.2.3.a Fig. 14.3.1.2.3.b

Fig. 14.3.1.2.3.a. The dependence of the secondary-electron yield γ on the velocity V of the incident ions for the case of atomic and molecular hydrogen ions striking various targets. The curves (*a*) were reported by HILL et al. [*311*], curves (*b*) by MURDOCK et al. [*518*], curve (*c*) by FOGEL et al. [*217*], curve (*d*) by TEL'KOVSKII [*708, 709*], curve (*e*) by ALLEN [*5*], and curve (*f*) by AARSET et al. [*1*]

Fig. 14.3.1.2.3.b. The dependence of the secondary-electron yield γ on the kinetic energy E of the incident ions for the case of light as well as heavy ions incident on various targets. The curves (*a*) were reported by TEL'KOVSKII [*708, 709*], curve (*b*) by FOGEL et al. [*217*], curve (*c*) by LINFORD [*439*], curve (*d*) by HILL et al. [*311*], and curves (*e*) by AARSET et al. [*1*]

Fig. 14.3.1.2.3.b also shows the function $\gamma(E)$ for light as well as heavy incident particles. LINFORD [*439*] observed that the $\gamma(E)$ curve for Hg^+ ion bombardment of a gas-covered Ni surface, shown in Fig. 14.3.1.2.3.b, is nearly identical with those for Hg^+ ion bombardment of other gas-covered surfaces such as Cu, Sn, Mo, Al, W, Ag, Mg, Na, and Cd. It is of interest to notice that for the heavier Hg^+ ion the $\gamma(E)$ function still is a rising in an energy region where the corresponding function for the lighter ions (H_1^+, H_2^+) is falling. This result, however, is again understandable in view of the behavior of the cross section for ionization of an atom by heavy-ion impact. (For a discussion of ionization cross sections by ion impact, see the review articles by HASTED [*294*] and MASSEY and BURHOP [*461*].)

14.3.2. Effects of Surface Coverage on the Secondary Electron Yield γ

The dependence of the secondary-electron emission on surface coverage with an adsorbed layer of residual gas or with known gases introduced into the vacuum system has been observed experimentally by numerous investigators [*24, 102, 189, 192, 212, 217, 451a, 452a, 529, 540, 557, 573, 754, 793*]. All authors report a substantial increase in the value of the secondary-electron yield γ when layers of gas became adsorbed on

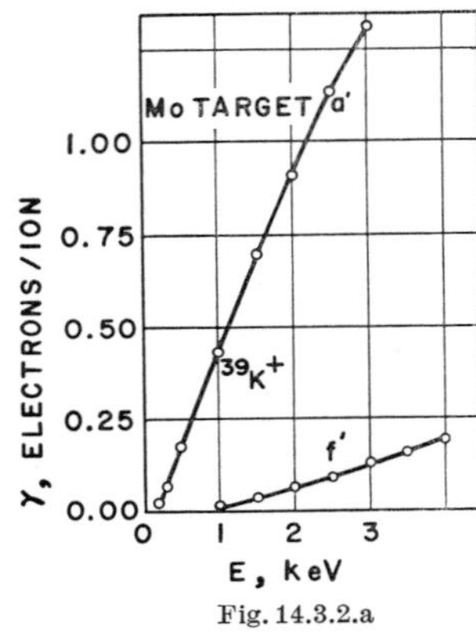

Fig. 14.3.2.a

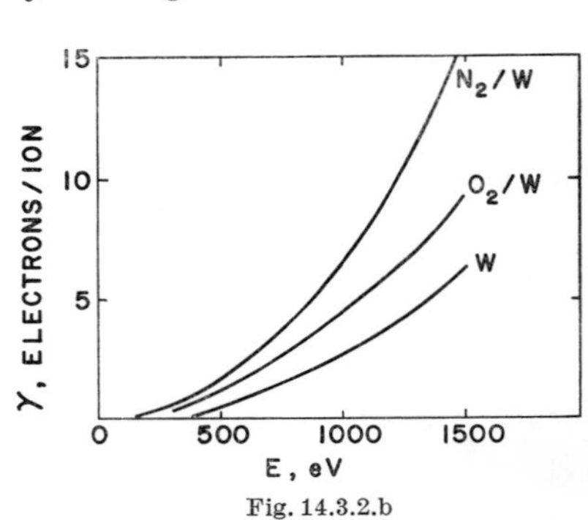

Fig. 14.3.2.b

Fig. 14.3.2.a. The function $\gamma(E)$ for the case of $^{39}K^+$ ion bombardment of a clean (curve f') and a gas-covered (curve a') Mo surface (BRUNNEÉ [*102*])

Fig. 14.3.2.b. Curves of $\gamma(E)$ for Li^+ ion bombardment of a degassed W surface and W surfaces covered with O_2 and N_2 layers (WATERS [*754*])

an initially degassed metal surface. Thus, for example, for the bombardment of a Mo surface with 2-keV $^{39}K^+$ ions (Fig. 14.3.2.a), BRUNNEÉ [*102*] observed that the values of γ for a non-degassed surface (curve a') were more than an order of magnitude higher than the same surface in its initial degassed state (curve f'). EREMEEV et al. [*192*] reported that γ dropped from 100% when 5-keV K^+ ions bombarded a gas-covered W surface (at $T = 293°$ K) to 20% when the surface became degassed by being heated to $T = 1200°$ K. Similarly ARIFOV et al. [*24*], who first bombarded an only slightly degassed Ta surface (at 900° K) and then a more thoroughly degassed one (at 1300° K) with Cs^+ ions of low energy ($E < 1$ keV), observed that the value of γ decreased from 35% to 1%. Most authors also report an increase in the slope of the $\gamma(E)$ curve if the coverage degree of the metal surface is increased (Fig. 14.3.2.a).

Fig. 14.3.2.b shows $\gamma(E)$ curves for Li^+ ion bombardment of a degassed W surface and of W surfaces covered with O_2 and N_2 layers (WATERS [*754*]). One notices the interesting result that over the entire investigated energy range the electron yield for an O_2-covered surface is greater than that for the degassed surface but less than that for a N_2-covered surface.

PAETOW and WALCHER [*540*], who also studied how different gases adsorbed on a wolfram surface influenced the value of γ, observed that γ always increased with increasing coverage degree—regardless of whether electropositive (Cs) or electronegative (O_2) adatoms had been adsorbed. They also found that whenever the coverage degree was large enough ($\theta > 1$), the electron yields γ for cesium-oxide-covered or oxygen-covered wolfram surfaces under bombardment with Cs^+ ions ($E = 500$ eV) had similar values. Furthermore, they observed that the values of γ for Cs^+ bombardment of K- or Cs-covered W surfaces for $\theta \gtrsim 1$ were nearly the same. These results indicate that the work-function change of a metal surface as a result of the gas adsorption is not a dominant factor in the kinetic electron-emission process (PAETOW et al. [*540*]).

FOGEL et al. [*217*] investigated the variation of γ with increasing time of adsorption of the background gas (unknown composition) on a Mo surface. After degassing the Mo surface for 20 min at a temperature of 1800° C at a residual gas pressure of approximately 3×10^{-7} mm Hg, the secondary-electron emission in short time intervals was measured for the case of an A^+ ion bombardment at an energy of 12 keV. They observed that while the heating reduced the coefficient γ by approximately 50%, the initial value of γ was again reached 8 min after the heating stopped. If one compares this time with the time necessary to build up a monolayer at the background pressure of 3×10^{-7} mm Hg, namely the order of several tenths of a second, one may conclude that the surface coverage responsible for reaching a saturation value of γ is several monolayers thick. One can then conclude that a certain number of secondary electrons will be formed even in the surface layers. In addition, the electron emission from certain adsorbed layers may be enhanced since it is known [*390, 682*] that the diffusion length of secondary electrons is orders of magnitude larger in insulators than in metals.

More experimental data, especially for the dependence of γ on a known coverage degree θ, are badly needed.

14.3.3. Dependence of the Electron Yield γ on the Mass of the Incident Ion (Isotope Effect)

One important result about the mass dependence of γ has already been illustrated in the $\gamma(E)$ and $\gamma(V)$ curves of Sections 14.3.1.1. and 14.3.1.2. At lower ion energies (Section 14.3.1.1) and within a certain energy range the yield γ decreases with increasing mass of the bombarding ion, whereas at higher ion energies (Section 14.3.1.2) the reverse behavior has been observed for ions in the lower mass region. For example, the $\gamma(V)$ curves for Ne^+ and A^+ ions bombarding a Mo target (Fig. 14.3.1.1.1.b) cross each other at a velocity of approximately 2.4×10^7 cm/sec, and thereafter the heavier A^+ ions yield more secondary electrons than the lighter Ne ions.

In a more detailed study of the dependence of γ on the ion mass, several authors [*102*, *308*, *573*] investigated the electron emission from metal surfaces under the impact of isotopic ions of the same element.

PLOCH [*573*] investigated the electron emission from non-degassed and also from lightly gas-covered metal surfaces under ion bombardment by lithium and neon isotopes. He found hardly any mass dependence. However more accurate measurements of HIGATSBERGER et al. [*308*], who investigated the secondary-electron emission from non-degassed targets (such as AgMg) under bombardment by isotopic ions of Ne and A, and those of BRUNNEÉ [*102*], who investigated the electron emission from a gas-free Mo surface under bombardment of isotopic ions of different alkali metals, definitely showed that γ depends on the ion mass in the region of low ion energies. Fig. 14.3.3.a shows $\gamma(E)$ curves for $^6Li^+$ and $^7Li^+$ ions bombarding a gas-free Mo surface. One notices that the $^6Li^+$ ions yield more secondary electrons than the $^7Li^+$ ion and that the difference $(\Delta\gamma)_{E=\text{const}}$ between the electron yields of the two isotopes at the same ion energy is considerably larger than the value $(\Delta\gamma)_{V=\text{const}}$ for the same ion velocity. Table 14.3.3.A lists values of $(\Delta\gamma/\gamma)_{E=\text{const}}$ and $(\Delta\gamma/\gamma)_{V=\text{const}}$ which have been determined by BRUNNEÉ from experimental results for different

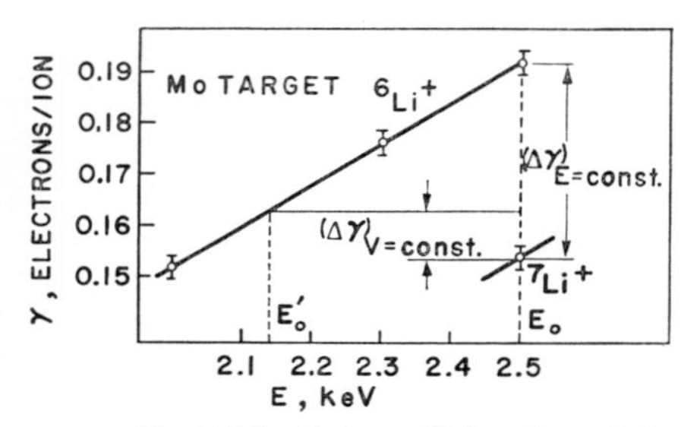

Fig. 14.3.3.a. Isotope effect on the emission of secondary electrons. The $\gamma(E)$ curves for the case of ions of the two lithium isotopes incident on a clean Mo surface were reported by BRUNNEÉ [*102*]. Increments $(\Delta\gamma)_V$ and $(\Delta\gamma)_E$ are the differences in the γ values for the $^6Li^+$ and $^7Li^+$ ions at a constant ion velocity V and a constant energy E_0, respectively

Table 14.3.3.A. *Isotope effect on the secondary-electron emission under alkali-ion bombardment of a gas-free Mo surface at an energy of 2.5 keV* (BRUNNEÉ [*102*])

Isotopes	$(\Delta\gamma/\gamma)_{E=\text{const}}$ experimental	$(\Delta\gamma/\gamma)_{V=\text{const}}$	
		Experimental	Calculated (VON ROOS [*612*])
^6Li, ^7Li	$-(1.4 \pm 0.2)$	$-(0.3 \pm 0.2)$	-0.6
^{39}K, ^{41}K	$-(1.2 \pm 0.2)$	$+(0.4 \pm 0.2)$	$+0.4$
^{85}Rb, ^{87}Rb	$-(1.2 \pm 0.2)$	$+(0.5 \pm 0.2)$	$+1.2$

isotopes of alkali ions bombarding a gas-free Mo surface at an energy of 2.5 keV. One notices the interesting results that the values $\Delta\gamma/\gamma$ for constant velocity is negative for Li ions but is positive for K and Rb ions. The positive sign means that the heavier ions yield more secondary electrons.

Table 14.3.3.A also includes values of $(\Delta\gamma/\gamma)_{V=\text{const}}$, which have been calculated on the basis of the theory given by von Roos [*612*]. He treated the kinetic electron emission under the assumption that a multiple-collision process between incident ions and lattice atoms (considered unbound) results in ionization of the lattice atom. According to his theory, the yield γ depends on the ratio $\mu = m/M$ of the mass m of the incident ion to the mass M of the target atom. For constant incident-ion energy

$$\gamma \propto \frac{(1+\mu)^4}{\mu}. \tag{14.3.3–1}$$

For constant incident-ion velocity

$$\gamma \propto \frac{(1+\mu)^4}{\mu^3}. \tag{14.3.3–2}$$

These results have practical implications for precise measurements of the isotopic ratio R (the ratio of the abundance of a rarer isotope to that of a more abundant one). When isotopic abundances are determined with a high degree of accuracy and secondary-electron multipliers are used as detector components, it is often assumed that the abundance ratio is proportional to the ratio of the secondary-electron currents. This in effect assumes that γ is the same for both isotopes. Ploch [*573*] has suggested that a correction can be made as follows. The correction resulting from the difference in γ for the two isotopes is proportional to

$$\frac{\Delta\gamma}{\gamma} = \frac{\partial\gamma}{\partial m}\Delta m. \tag{14.3.3–3}$$

To evaluate $\partial\gamma/\partial m$, we notice that it is possible to vary E in such a manner that $\gamma(E, m)$ remains constant as m is varied, so that

$$\frac{d\gamma}{dm} = \frac{\partial\gamma}{\partial E}\frac{dE}{dm} + \frac{\partial\gamma}{\partial m} = 0. \tag{14.3.3–4}$$

Hence,

$$\frac{\partial\gamma}{\partial m} = -\frac{dE}{dm}\frac{\partial\gamma}{\partial E}. \tag{14.3.3–5}$$

It was pointed out in section 14.3.1.1 (see Fig. 14.3.3.a) that $\partial\gamma/\partial E$ is much greater than $\partial\gamma/\partial V$ so that

$$\frac{dE}{dm} = \tfrac{1}{2}V^2 + mV\frac{dV}{dm} \approx \tfrac{1}{2}V^2 = \frac{E}{m}. \tag{14.3.3–6}$$

From these last two equations it follows that

$$\frac{\partial \gamma}{\partial m} = -\frac{E}{m}\frac{\partial \gamma}{\partial E}. \tag{14.3.3–7}$$

Then if the subscripts a and r designate quantities associated with the more abundant and the rarer isotope, respectively, Eq. (14.3.3–3) becomes

$$\frac{\gamma_r - \gamma_a}{\gamma_a} = \frac{m_a - m_r}{m_a}\frac{E}{\gamma_a}\frac{d\gamma_a}{dE} = \alpha, \tag{14.3.3–8}$$

where $d\gamma_a/dE$ is to be evaluated from the $\gamma_a(E)$ curve obtained for the more abundant isotope. Then

$$\frac{\gamma_r}{\gamma_a} = 1 + \alpha.$$

Then if I_a and I_r are the currents from the secondary-electron multiplier and n_a and n_r are the numbers of isotopic ions per second, the isotopic ratio obtained experimentally on the assumption that $\gamma_a = \gamma_r$ is related to the true isotopic ratio R by the expression

$$R_{\text{exp}} \equiv \frac{I_r}{I_a} = \frac{\gamma_r n_r}{\gamma_a n_a} = (1+\alpha)R. \tag{14.3.3–9}$$

For example, the value of α already reaches the appreciable amount of 20% for the case of $^6Li^+$ and $^7Li^+$ ion bombardment of a Mo target (at 2.0—2.5 keV), as shown in Fig. 14.3.3.a, and may become even larger at higher energies.

14.3.4. Dependence of the Electron Yield γ on the Charge State of the Incident Particle

Several authors [*71, 168, 215, 216, 615, 645*] who investigated the secondary-electron emission of non-degassed and of degassed but not atomically clean metal surfaces with particles in different charge states observed a definite influence in the value of the electron yield γ. Flaks [*215*] bombarded a gas-covered Pt target with K^+, K^{2+}, and K^{3+} ions over an energy range from 1 to 40 keV and observed a higher yield γ for more highly charged incident ions. Fig. 14.3.4.a shows the $\gamma(E)$ curves for the differently charged potassium ions as observed by Flaks. One notices that the difference $\Delta\gamma = \gamma(K^{n+}) - \gamma(K^{(n-1)+})$ is nearly independent of the kinetic energy of the ion. At an energy of 20 keV, the value of γ for doubly-charged potassium ions is 15% above that for singly-charged ones; and the value for triply-charged ions is nearly 40% above that for singly-charged ones. Also for other systems, such as singly- and multiply-charged Zn, Hg, Tl, and Pb ions striking a non-degassed Pt target, γ increases with increasing ionic charge of the incident ion and the $\gamma(E)$ curves are similar to those obtained with incident potassium ions. FLAKS related the yield changes $\Delta\gamma$ to the potential energy $E_{\text{pot}} = eI^{(n+)} - ne\varphi$ released during the neutralization of an ion of particular charge ne on the target surface. Here $eI^{(n+)}$ is the ionization energy to change the charge state from $(n-1)+$ to $n+$, and φ is the work function of the metal surface.

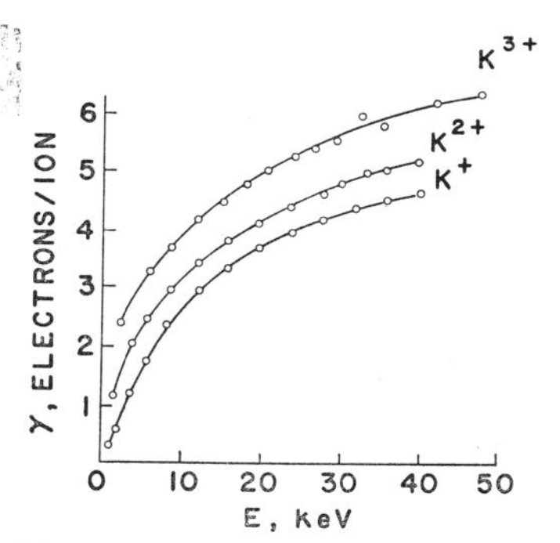

Fig. 14.3.4.a. Secondary-electron yields $\gamma(E)$ for the case of singly-, doubly-, and triply-charged potassium ions incident on a gas-covered Pt surface (FLAKS [*215*])

Table 14.3.4.A lists some values for the ratios $\Delta_r = \Delta\gamma(3+\to 2+)/\Delta\gamma(2+\to 1+)$ and $E_r = E_{\text{pot}}(3+\to 2+)/E_{\text{pot}}(2+\to 1+)$ for different ion-metal systems, according to FLAKS [*215, 216*]. One notices that the values of Δ_r are usually larger than the E_r values and those for the heavier ions are substantially larger than those for the lighter ions. The bad agreement is not surprising if one considers that for the energy region investigated the contribution of secondary electrons emitted by the ion neutralization process (potential emission mechanism) is small relative to the contribution of those emitted by the kinetic emission mechanism.

SCHWARTZ and COPELAND [*645*] investigated the secondary-electron emission from a mercury surface under bombardment of singly- and multiply-charged mercury ions (Hg^{+}, Hg^{2+}, Hg^{3+}). In agreement with the observations of FLAKS [*215, 216*], he found that γ increases with increasing charge of the ion and that the shapes of the $\gamma(E)$ curves for the differently charged He ions were similar over the entire investigated energy range from 4 to 10 keV. However, results which have been obtained at higher incident ion energies and differ from those of FLAKS and of SCHWARTZ et al. have been reported by TEL'KOVSKII [*707*] and by DUNAEV and FLAKS [*168*]. TEL'KOVSKII investigated the secondary-electron emission from degassed molybdenum and zirconium targets under bombardment of singly- and multiply-charged argon ions (A^{+}, A^{2+}, A^{3+}) in the ion-velocity range from 1×10^7 to 7×10^7 cm/sec for a Zr target, and up to 7.8×10^7 cm/sec for a Mo target. The maximum

Table 14.3.4.A. *Ratios of yield changes Δ_r and of potential energies E_r for the transitions $3+ \to 2+$ and $2+ \to 1+$, calculated according to* FLAKS [*215, 216*] *for different ions on gas-covered Pt surfaces*

Incident ion	Δ_r	E_r
K	2.68	2.5
Zn	3.48	2.8
Hg	2.35	2.4
Tl	3.65	2.4
Pb	7.45	3.0

speed corresponds to an energy of 80 keV for the Zr target and 120 keV for the molybdenum target. For both targets, TEL'KOVSKII observed that the yields for the Mo target were higher than those for the Zr target and that the secondary-electron yield γ was virtually identical for A^+, A^{2+}, and A^{3+} ions having the same velocity. This result is not surprising if one considers that for the energy region investigated the contribution of secondary electrons emitted in the ion neutralization process (potential emission mechanism) becomes small relative to the contribution of those electrons ejected by the kinetic emission mechanism. DUNAEV and FLAKS, who bombarded a gas-covered nickel surface (among others) with singly- and multiply-charged Na, Ca, and Ba ions, also observed no appreciable difference between the values of γ for Ca^+ and Ca^{2+} or for Ba^+ and Ba^{2+} ions for identical kinetic energy, but found that the yield for Na^{2+} ions was approximately twice that of Na^+ ions. However, as pointed out by FLAKS [*215*, *216*], the reliability of these values is questionable.

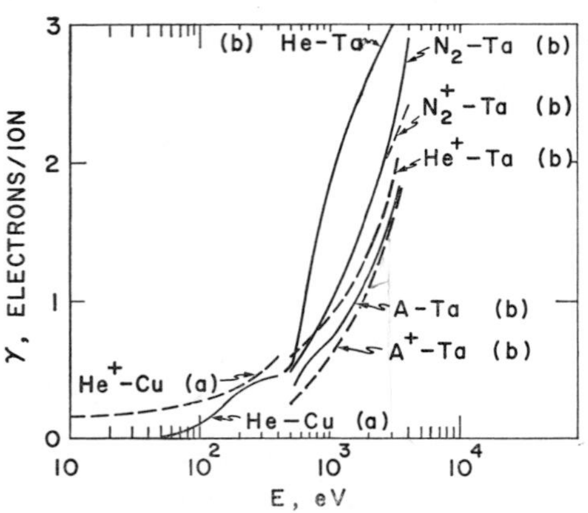

Fig. 14.3.4.b. Secondary-electron yields $\gamma(E)$ for the case in which neutral atoms or molecules as well as ions of the same element strike a metal surface. Curves *a* were reported by ROSTAGNI [*615*] and curves *b* by BERRY [*71*]

The emission of secondary electrons from metal surfaces under the impact of fast neutral particles has also been studied by several authors, including ROSTAGNI [*615*], BERRY [*71*], BARNETT et al. [*49*], MEDVED et al. [*467b*], COMEAUX and MEDVED [*119a*], and ARIFOV et al. [*30a*]. Fig. 14.3.4.b illustrates some of the results of ROSTAGNI [curves (*a*)] and BERRY [curves (*b*)] for the function $\gamma(E)$ for different kinds of neutral atoms and singly-charged ions of the same element bombarding gas-covered Ta and Cu targets. One notices that even at the higher ion energies, where the kinetic electron-emission mechanism dominates, the yield curves for neutral atoms are rather similar to those for ions and the yields for neutral He and A atoms bombarding a Ta surface are larger than those for the corresponding ions of the same energy. The results for He and N_2 are in qualitative agreement with those of BARNETT et al. [*49*] who found that the yields for nitrogen, neon, and argon atoms bombarding a Ni surface in the energy range from 20 to 250 keV are very similar to the corresponding values for ions, but the yields for helium and hydrogen atoms are several percent larger than those for the corresponding ions of the same energy.

Contrary to these observations at higher incident ion energies COMEAUX and MEDVED [*119a*], MEDVED et al. [*467b*], and ARIFOV et al. [*30a*] reported yield curves $\gamma(E)$ obtained in the bombardment of a clean Mo surface by A^+ ions and neutral argon atoms in the energy range from approximately 200 eV to 2.5 keV (see Fig. 14.1a) which show $\gamma_{A^+}(E) > \gamma_{A^0}(E)$. This result is in agreement with expectation, since for ion energies E larger than the threshold energy E_{th} for the kinetic emission it is $\gamma_{A^+}(E) = \gamma_{kin}(E) + \gamma_{pot}$ and $\gamma_{A^0}(E) = \gamma_{kin}(E)$ (see discussion in Section 14.1 and 14.3.1.). COMEAUX and MEDVED confirmed the observation that for the energy region considered $\gamma_{ion}(E) > \gamma_{atom}(E)$ also for the case of He^+ ions incident on a clean Mo surface.

Finally, it should be mentioned that several authors (for instance, FOGEL et al. [*217*], ZANDBERG [*793*], and DUKELSKII and ZANDBERG [*166*]) observed that negative ions cause substantially more secondary-electron emission than do positive ions of the same element. For instance, for a gas-covered Mo surface under positive and negative hydrogen-ion bombardment at 22 keV, FOGEL et al. observed $\gamma_{H^+} = 1.38$ and $\gamma_{H^-} = 3.52$ electrons/ion. For the case of a gas-covered wolfram surface under bombardment by positive and negative ions with an energy of 780 eV, ZANDBERG observed $\gamma_{Na^+} = 0.25$, $\gamma_{Na^-} = 0.41$; $\gamma_{I^+} = 0.15$, $\gamma_{I^-} = 0.26$; $\gamma_{Cl^+} = 0.49$, $\gamma_{Cl^-} = 0.79$; $\gamma_{Bi^+} = 0.06$ and $\gamma_{Bi^-} = 0.18$ electrons/ion.

14.3.5. Temperature Dependence of the Secondary Electron Yield γ

Section 14.3.2 was a discussion of the dependence of the electron yield γ on the gas coverage of the metal surface, a quantity which in turn depends strongly on the temperature of the surface. From this it is evident that the negative temperature coefficient of γ, which was found by many of the earlier investigators [*5, 188, 189, 439, 491, 518, 536*] who worked with gas-covered surfaces, can be related to an increase in the desorption rate of the adsorbed gas layers with increasing temperature rather than to actual changes of the probabilities for electron formation and escape from a gas-free metal surface.

Recently ARIFOV and RAKHIMOV [*28*] tried to determine the temperature dependence of γ_{kin} for a gas-free Mo surface. The authors, working at a residual gas pressure of approximately 1×10^{-7} mm Hg, outgassed the target at a temperature of 2300—2700° K for a prolonged period of time. The measurements, taken immediately after flashing the target, were carried out in the temperature range from 300° to 1600° K. In order to distinguish between secondary electrons and thermoelectrons, a technique of primary-ion-beam modulation (Section 12.2.1) was used. They found, for example, that when a degassed Mo surface was bombarded with 5-keV potassium ions the value of γ_{kin} remained virtually constant over the temperature range investigated. Similarly, for the case of A^+ and A^0 bombardment of a clean Mo surface, MAHADEVAN et al. [*452a*] reported that the secondary yield is unaffected by the temperature of the target. This result is not surprising since there seems to be no reason to expect the temperature to have a significant effect on the process of secondary-electron formation (which probably takes place by ionization of lattice atoms, as discussed in Section 14.4). Changes of the yield as a result of thermal expansion of the metal can probably be neglected also. A small decrease of γ with increasing temperature, however, might be expected from the decrease of the mean free path of the secondary electrons in the metal—a consequence of the thermal vibration of the lattice atoms about their equilibrium positions (STERNGLASS [*683*]).

14.3.6. Variation of γ with the Angle of Incidence α

Several authors* [*1, 5, 433, 536, 658*] have observed a substantial increase in the secondary-electron yield γ with increasing angle α between the incident ion and the normal to the metal surface for the case of non-degassed and slightly degassed surfaces. Unfortunately, some authors [*1, 5, 433, 658*] did not distinguish between secondary emission of electrons and sputtered negative ions; or they neglected effects due to the reflection of the primary beam at the metal surface.

OLIPHANT [*536*] investigated the dependence of γ on the angle of incidence α for low-energy He^+ ions incident on a Mo surface polished to a mirror finish and on a rough polished Ni surface. For 1-keV ions he found the changes in γ to be proportional to $(a - b \cos \alpha)$, where a and b are constants. However, MASSEY and BURHOP [*461*] pointed out that the results could also be represented by $\gamma \propto \sec \alpha$, a relationship which has proved to represent the data of other authors for the higher ion energy region quite well. The data of all the authors cited closely follow the equation

$$\gamma(\alpha) = c\,\gamma_0 \sec \alpha\,, \tag{14.3.6–1}$$

where γ_0 is the yield at normal incidence and c is a constant.

ALLEN [*5*], who bombarded partially degassed Be, Cu, C, Ni, and Pt targets with high-energy protons, found his $\gamma(\alpha)$ data were well represented by this equation. For 120-keV protons incident on Ni he found

$$\gamma(\alpha) = 3 \sec \alpha = 1.75\,\gamma_0 \sec \alpha \quad \text{for } H_1^+\text{—Ni, 120 keV.} \tag{14.3.6–2}$$

AARSET et al. [*1*] bombarded an Al surface with 1.6-MeV protons incident at different angles and found $\gamma(\alpha)$ to be represented best by

$$\gamma(\alpha) = 0.825 \sec \alpha \approx \gamma_0 \sec \alpha \quad \text{for } H_1^+\text{—Al, 1.6 MeV} \tag{14.3.6–3}$$

---

* Studies of the angular distribution of secondaryelectrons for various angles of incidence have been performed by ABBOTT et al. [2].

Fig. 14.3.6.a illustrates more recent results for the function $\gamma(\alpha)$ obtained by LEROY and PRELEC [*433*] for bombardment of a Zr target with 250-keV D^+ and D_2^+ ions. Their data are best fitted by

$$\gamma(\alpha) = 2 \sec \alpha = 0.91\, \gamma_0 \sec \alpha \qquad \text{for } D^+, \qquad (14.3.6\text{–}4)$$

$$\gamma(\alpha) = 4.9 \sec \alpha = 0.89\, \gamma_0 \sec \alpha \qquad \text{for } D_2^+. \qquad (14.3.6\text{–}5)$$

Also SLODZIAN's [*658*] values of γ for normal incidence and for an angle of incidence of 45° on a duraluminum surface bombarded with 5-keV A^+ ions could be fitted by the expression (14.3.6–1).

On the basis of these results, it appears that Eq. (14.3.6–1) represents the data rather well. This relationship might even be expected since increasing the angle of incidence makes the total path of the incident ion run closer to the upper surface layer so that the secondary electrons have a shorter path to reach the surface. The variation of the yield with angle of incidence would then follow a sec α law.

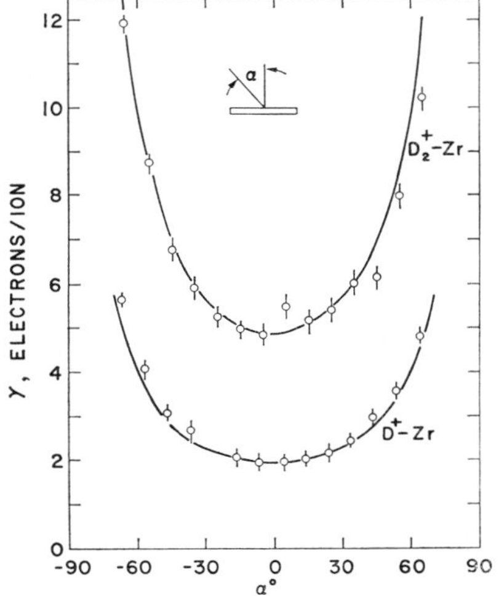

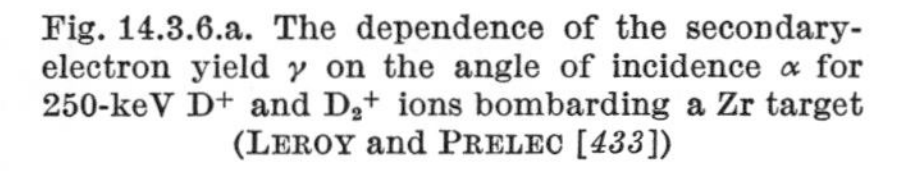

Fig. 14.3.6.a. The dependence of the secondary-electron yield γ on the angle of incidence α for 250-keV D^+ and D_2^+ ions bombarding a Zr target (LEROY and PRELEC [*433*])

14.3.7. Energy Distribution of Secondary Electrons

Most studies of the energy distribution of secondary electrons [*311, 439, 518, 522, 658, 736*] have been conducted with gas-covered surfaces. Additional references to older work are given by MASSEY and BURHOP [*461*] and VEITH [*736*]. The surfaces were more careful degassed in only a few cases [*102, 286, 708, 754*]. For gas-covered surfaces, the energy distribution, particularly the region of the high-energy tail, was commonly found to depend slightly on the nature of the target. For degassed surfaces, on the other hand, no significant changes in the energy distribution for different target materials could be observed, as indicated in the results of TEL'KOVSKII [*708*] and by a comparison of the results of BRUNNEÉ [*102*] (who bombarded Mo with 2.5-keV Li^+) with those of WATERS [*754*] (1-keV Li^+ on W). The majority of the ejected electrons possesses energies less than 5 eV and the energy distribution is nearly independent of the primary-ion energy up to several tens of keV. Table 14.3.7.A contains information on the energy distribution of secondary electrons, obtained by SLODZIAN (*658*] for gas-covered Fe, W, and Cu surfaces bombarded with A^+ ions at 5 keV. One notices that the majority of the secondary electrons have energies less than 5 eV.

Table 14.3.7.A. *Distribution of energies E_e of secondary electrons emitted from different gas-covered targets under A^+ ion bombardment at 5 keV* (SLODZIAN [*658*])

Target	Secondary-electron yield at energy E_e (%)			
	$E_e < 5$ eV	5—10 eV	10—20 eV	20—30 eV
Fe	86	6.5	4.2	1
W	69	20	11.5	1.5
Cu	76	11.3	11.3	1.5

Even for considerably higher primary-ion energies, the majority of the secondary electrons have energies below 30 eV, as has been observed by several authors [*311, 439, 518, 708*]. For instance, LINFORD [*439*] found that only 10% of the secondary electrons emitted from Cu and Sn targets under Hg^+ ion bombardment at 1.32 MeV had energies higher than 10 eV. HILL et al. [*311*] observed for 213-keV protons incident on Mo, and MURDOCK et al. [*518*] found for 137-keV protons incident on Mo and 48-keV protons on Cu, that the majority of the secondary electrons had energies below 30 eV. TEL'KOVSKII, who bombarded various targets (such as Mo, Zr, Ni, Ta, Cu, and C) with hydrogen, helium, nitrogen, and molybdenum ions in the energy region from 10 to 40 keV, found that the energy distribution of the secondary electrons followed a MAXWELL distribution and that the mean energy of the secondaries for all cases lay in the interval from 5 to 8 eV.

Fig. 14.3.7.a (solid curves) illustrates the distribution curves dN/dE_e vs E_e obtained by BRUNNEÉ for a clean Mo target bombarded with 2.5-keV ions of different alkali metals. One notices that the distribution curves for the Li^+, Na^+, and K^+ ions pass through a maximum value in the region $2 < E_e < 3$ eV, but the maximum seems to shift very slightly toward lower energies as the mass increases from Li^+ to K^+. A similar maximum in the distribution curve at ≈ 2 eV has been observed by WATERS [*754*] (Li^+—W, 1 keV) and HAGSTRUM [*285, 286, 287*] (He^+—Mo, 1 keV). (Potential emission is a contributing factor in the last example.) As can be seen in Fig. 14.3.7.a, the number of secondary electrons which have higher energies decreases from Li^+ to Cs^+. One notices also that the slopes of the distribution curves become steeper in the order of incident ions from Li to Cs, a phenomenon which VON ROOS [*612*] related to the number of outer-shell electrons which participate in the secondary-electron formation. These electrons become more numerous as the principal quantum number increases.

The dotted curves shown in Fig. 14.3.7.a have been calculated according to a theory developed by VON ROOS. As will be discussed in more detail in Section 14.4, he considers the formation of secondary electrons as being due to a process in which the incident ions ionize the lattice atoms (assumed to be free, a "gas of lattice atoms"). The calculations were conducted under further simplifying assumptions. For instance, he neglected the probability that some of the secondary

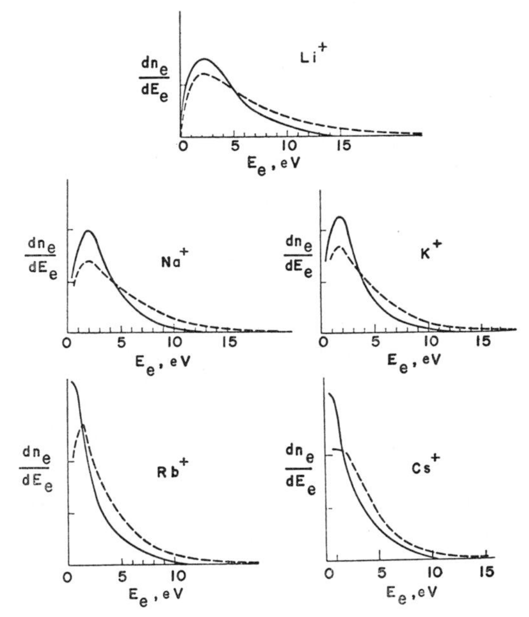

Fig. 14.3.7.a. Energy distribution functions for different 2.5-keV alkalimetal ions striking a clean Mo target. The solid lines represent the experimentally determined data of BRUNNEÉ [*102*], the dotted lines represent the values calculated by VON ROOS [*612*]

electrons formed would be absorbed by the lattice and also the energy losses of the incident ion in displacing lattice atoms from their sites (as discussed in Chapter 8 in connection with sputtering). These approximations limit the applicability of the calculations to the lower ion energies. The major characteristics of the experimental curves, however, can be represented by the theory fairly well despite the approximations made.

Finally it should be mentioned that the experimentally observed existence of electron energies of more than 10 eV indicates that KAPITZA's model [*366*] of the secondary-electron emission as a kind of thermoelectron emission from a hot spot is incorrect.

14.4. Theoretical Aspects of the Kinetic Emission of Secondary Electrons

As already mentioned in Section 14.1, the mechanism of "kinetic" secondary-electron emission from a metal surface becomes important whenever $eI < e\varphi$, where eI is the ionization energy and $e\varphi$ is the work function of the metal and the incident-ion energy exceeds a certain threshold value. Even for systems for which $eI > 2e\varphi$ so that there can be a yield γ_{pot} of secondary electrons from potential emission by an AUGER neutralization process, the yield γ_{kin} of secondary electrons produced by kinetic emission starts to contribute significantly to the total yield $\gamma = \gamma_{pot} + \gamma_{kin}$ when the incident-ion energy exceeds a certain threshold E_{th}. Although the potential-emission mechanism has been successfully treated theoretically by SHEKHTER [*650*], COBAS and LAMB [*116*], and HAGSTRUM [*282*] (Chapter 12), none of the proposed theories of the kinetic-emission mechanism have been generally successful.

KAPITZA [*366*] proposed a thermal-emission theory which considered the secondary electrons as thermal electrons emitted from a microscopic zone in which the target had been heated by the impact of the incident ion. Further developments of the theory have been given by HIPPEL [*316, 317*], MORGULIS [*496*], IZMAILOV [*346*], and SILITTO [*652*]. As KAPITZA himself pointed out, however, the thermal emission implies the doubtful assumption that thermodynamic equilibrium considerations may be employed in treating the very rapid and localized energy exchange between fast ions and lattice atoms. It is therefore not surprising to find a strong disagreement between theoretical predictions and experimental results. For example, the theory predicts that γ should increase with increasing energy of the incident ion, contrary to experimental observations in the high-energy region (Sections 14.3.1.2.1 and 14.3.1.2.3) where for light incident ions on metal surfaces the function $\gamma(E)$ decreases with increasing energy E (see Fig. 14.3.1.2.3.b). In addition, the experimental data do not indicate any noticeable dependence of the function $\gamma(E)$ on thermal constants of the target material (e.g., heat capacity or heat conductivity). Furthermore, the secondary-electron emission would have to have a Maxwellian energy distribution with mean temperatures of several tens of thousands of degrees—an unlikely result indicating also that the assumption of a statistical equilibrium is untenable for the phenomena considered.

FRENKEL [*228*] regarded the emitted secondary electrons as those being stripped off from the bombarding particle in the collision with target atoms. Such a model, however, appears to contradict the adiabatic character of the deformation of the electron shell of the incident ion for the energy region considered. The same objections apply to the attempts to improve and generalize FRENKEL's theory (e.g., AVAKYANTS [*34, 39*]).

IZMAILOV [*347, 348, 349*] based his theory on the hypothesis that the secondary-electron emission is due to an interaction between the free metal electrons and an electromagnetic field caused by the rapid retardation of incident ions by the target (the so-called "radiation theory"). However, such a model is unable to explain the experimentally observed high secondary-electron yields from insulator surfaces where no free electrons exist. Also, for the case of a metal surface, the yield value γ_{kin} is observed to be nearly independent of both the work function ($e\varphi$) of the metal and to depend only slightly on the charge of the incident ion, contrary to the prediction from this model.

As has been pointed out by PLOCH [*573*] and others [*272, 273, 499, 500*], the unfavorable ratio of masses means that the probability for a direct collision between an ion and a free electron in the conduction band is very small. PLOCH assumed therefore that the secondary electrons were formed by a process of ionization of the lattice atom by the incident ion. The experimentally observed close relationship between the energy dependences of the yield γ and the cross section σ_i for ionization of an atom by ion impact may be considered as support of this model. Three theoretical treatments (VON ROOS [*612*], STERNGLASS [*683*], and PARILIS and KISHINEVSKII [*543*]) have been based principally on such a model of secondary-electron production.

The theory formulated by VON ROOS assumes that incident ions make single collisions with the lattice atoms (which are assumed to be free, a "gas of lattice atoms") and are hereby distributed with respect to velocity v and direction r.* To describe the distribution function $N(v, r)$, the BOLTZMANN equation has been applied, with the elastic-collision cross section σ_e as the determining cross section. The formation of secondary electrons is related to σ_i, the cross section for ionization of an atom by ion impact. The distribution function is not influenced by the ionization process since $\sigma_e \gg \sigma_i$. According to VON ROOS, the number n_e of secondary electrons formed per second in the target may be obtained from the expression

$$dn_e = n' d\sigma_i N(x, K, \theta) K \, dx \, dK \, d\Omega \,, \qquad (14.4\text{–}1)$$

where n' is the number of collision centers, $K = mV/\hbar$ is the absolute value of the wave number of the incident ion, θ is the scattering angle with respect to the surface normal (the x axis), and $d\Omega$ is an element of solid angle. Since VON ROOS neglects the probability that some of the secondary electrons formed will be absorbed in the lattice, his theory is restricted to the lower incident-ion energies. The number of secondary electrons having velocities in the range dv per incident ion with velocity

* A critical discussion of this model has been given already in Section 11.4.

V_0 is obtained by von Roos by dividing the expression (14.4–1) by the incident-ion flux $N_i K_0$ and integrating over all coordinates of Eq. (14.4–1). His result is

$$\gamma(v)\,dv = n' \int_0^\infty dx \int_{K_a}^{K_0} dK \int_{F_1}^{F_2} d\sigma_i \int_{4\pi} d\Omega \, \frac{NK}{N_i K_0}\,, \qquad (14.4\text{–}2)$$

where K_a is related to the minimum incident-ion velocity for formation of a secondary electron, and the two functions F_1 and F_2 of the secondary-electron energies are introduced to correct for the fact that only those secondary electrons whose velocity vectors are nearly parallel to the surface normal have a chance to leave the metal. VON ROOS neglects the energy lost by the primary ion in displacing target atoms from their sites.

Calculated energy distributions of the secondary electrons, based on the velocity distribution given by Eq. (14.4–2) and transformed into an energy distribution, are shown as the dashed curves in Fig. 14.3.7.a. The comparison with corresponding experimental results has been made possible by adjusting two parameters. In view of the severe simplifications made in the theory, the agreement between theoretical and experimental values is rather good. It should be mentioned that for the case of a metal surface bombarded by alkali-metal ions at low energies (1—10 keV) VON ROOS's theory predicted the correct energy dependence ($\gamma \propto E^2$) and also the experimentally observed isotope effect. However, the theory fails to reproduce the experimentally observed relationship $\gamma(E) \propto E$ for noble-gas ions incident on metal surfaces for the energy region considered (see Section 14.3.1.1.1.).

While the theory of von Roos is restricted to the lower incident-ion energies, another theoretical treatment of the secondary-electron emission for the higher ion-energy region has been developed by STERNGLASS [*683*]. For the case of high-speed ions, important simplifying assumptions can be made. For instance, it can be assumed that the escaping electrons have originated in a target region in which the incident ions still have nearly their original energy, because the mean free path λ of high-energy ions (e.g., $\lambda = 0.8 \times 10^{-4}$ cm for 500 keV D^+ on Cu) is considerably larger than the mean free path of the secondary electrons (approximately 10^{-7} to 10^{-6} cm [*99*]). As in the theory of VON ROOS, it is assumed that the secondary electrons are formed by excitation and ionization of the lattice atoms by ion impact. Furthermore, it is assumed that only a small fraction of all secondary electrons formed are able to reach the surface with sufficient energy to escape.

STERNGLASS's treatment of the secondary-electron emission consists therefore of two major parts, the derivation of an expression for the number n_e of secondary electrons formed by the incident ion and an expression for the probability $P(x)$ that a secondary electron formed at a depth x below the surface will be able to reach the surface and escape. This means that the yield $d\gamma$ from a thin layer of thickness dx located at a depth x is given by

$$d\gamma = n_e(V, x) P(x) dx\,, \tag{14.4–3}$$

where V is the velocity of the incident ion.

In order to determine the number n_e of low-energy secondary electrons formed in a layer of thickness dx at a depth x, two contributions have to be considered. According to BOHR [85], high-energy ions may lose their energy in the following types of collision processes. In the first process the incident ion causes only a small perturbation of the lattice atoms so that correspondingly small amounts of energy are transferred in each collision. This process accounts for the formation of slow secondary electrons. Another collision process, in effect a "free collision" between the incident ion and an atomic electron, transfers larger amounts of energy to individual electrons. These are the energetic knock-on electrons known as δ rays, which in turn produce secondaries in higher order collisions. As BOHR pointed out, the number of δ rays is very small but the energy needed for their formation is approximately equal to that for the production of slow secondaries at high ion velocities. The number of low-energy secondary electrons consists therefore of two contributions, one for those formed by direct interaction $[n_{1e} = (1/\overline{E}_0)(dE_i/dx)_1]$ and another for those formed by the δ rays $[n_{2e} = f(v, x)(1/\overline{E}_0)(dE_i/dx)_2]$. Here $\overline{E}_0$ is the mean energy loss per secondary formed, $(dE_i/dx)_1$ is the mean energy loss per unit distance in producing slow secondary electrons, $(dE_i/dx)_2$ is the mean energy loss per unit path length in producing δ rays, and $f(v, x)$ is a function representing the fraction of $(dE_i/dx)_2$ expended in formation of secondaries by the δ rays in higher order collisions.

According to BOHR [85] and BETHE [72], at higher ion energies the energy loss is divided equally between the two collision processes, i.e.,

$$\left(\frac{dE_i}{dx}\right)_1 = \left(\frac{dE_i}{dx}\right)_2 = \frac{1}{2}\left(\frac{dE_i}{dx}\right)_{\text{total}}, \tag{14.4–4}$$

where $(dE_i/dx)_{\text{total}}$ is the total energy loss per unit path length. For lower particle velocities, only the outer electron shells of the lattice atom participate in the collision process. For this case, a derivation based on the THOMAS-FERMI model of the atom leads to the approximate expression

$$\left(\frac{dE_i}{dx}\right)_{\text{total}} = 2\pi N e^4 z^2 \left[\frac{4\,Z^{1/3}}{E_R{}^{1/2} E_r}\right], \tag{14.4–5}$$

where N is the number of atoms per unit volume, e is the electron charge, z is the charge of the incident particle, Z is the number of electrons in the outer shell of the lattice atom, E_R is the RYDBERG energy, and

$E_r = (m_e/m)E$, with m_e the mass of the electron, m the mass of the incident ion, and E the energy of the incident ion.

Here it should be mentioned that in a more recent calculation of the number of secondary electrons emitted under high-energy ion bombardment, GHOSH and KHARE [238] replaced the stopping power $(dE_i/dx)_{\text{total}}$ by the cross section for ionization of an atom by ion impact. They also criticized the value $\overline{E}_0 = 25$ eV, which STERNGLASS chose for the mean energy loss per secondary formed, as being smaller than would be expected for loss by electron excitation.

Before the secondary-electron yield can be calculated, it is still necessary to obtain an expression for the probability $P(x)$ that a secondary electron formed at a depth x below the surface will be able to reach the surface and escape. STERNGLASS assumed that the escape mechanism of the secondary electrons consisted of a diffusion process accompanied by an absorption process. Since an exponential law has to be expected for each of these processes, he suggested that $P(x)$ might be expressed as

$$P(x) = P_1 A \exp\left(-\frac{x}{L_s}\right), \tag{14.4–6}$$

where L_s is a characteristic length of the order of the distance traversed between inelastic collisions, and P_1 is the probability that an electron from the interior will arrive at the surface with enough energy to escape. In the case of an isotropic distribution of electron velocities approaching an uniform surface potential barrier, it can be shown that P_1 is given by

$$P_1 = 1 - \sqrt{\frac{e\varphi_D}{E_s + e\varphi_D}}, \tag{14.4–7}$$

where $e\varphi_D$ is the surface dipole energy, estimated by BARDEEN [47] to be about 0.1—0.5 eV for monovalent metals, and E_s is the energy of the secondary electrons as measured relative to the zero energy level ("vacuum level") outside the metal.

The constant A in Eq. (14.4–6) is determined by the distribution of the initial velocities of the secondary electrons and the ratio of the mean free path for adsorption to that for inelastic collisions. STERNGLASS assumes a value of $A \approx \frac{1}{2}$ for the case in which the initial directions are distributed symmetrically about a plane parallel to the surface.

The secondary-electron yield from a thin layer of thickness dx located at a depth x is given by Eq. (14.4–3). When the expressions for $P(x)$ and $n_e(V, x)$ are substituted in this and the result is summed over all layers contributing, one obtains

$$\gamma = \int_0^\infty \frac{1}{2}\,\frac{1}{\overline{E}_0}\left(\frac{dE_i}{dx}\right)_{\text{total}}\left[1 + f(v, x)\,P_1 A \exp\left(-\frac{x}{L_s}\right)\right] dx\,. \tag{14.4–8}$$

Here $f(v, x)$ is a factor that represents the fraction of the mean energy loss $(dE_i/dx)_2$ available for the formation of secondaries in higher order processes at depth x. For the case of medium incident-ion energies (up to the order of 100 keV), BOHR's approximation [Eq. (14.4–5)] for the value $(dE_i/dx)_{\text{total}}$ can be used. Then if the values of $f(v, x)$ and $P_1 A$ given by STERNGLASS are substituted in Eq. (14.4–7), one obtains

$$\gamma = \frac{0.5\,\pi\, e^4 z^2}{\overline{E}_0\, \alpha' \sigma_g} \left(\frac{4\, Z^{1/3}}{E_R{}^{1/2} E_r{}^{1/2}} \right). \tag{14.4–9}$$

Here $\sigma_g = (1.6 \times 10^{-16}\ \text{cm}^2)\, Z^{1/3}$ is the geometrical area of the outermost filled shell as determined by the covalent radii; and $\alpha' = \sqrt{3/\overline{n}_c}\,(\sigma_{cs}/\sigma_g) = 0.23$, where $\overline{n}_c$ is the average number of inelastic collisions the secondary has to undergo before losing its ability to escape and σ_{cs} is the scattering cross section for the secondary electrons.

For the case of a proton bombardment of a metal surface, Eq. (14.4–9) is finally reduced to

$$\gamma = 38 \sqrt{\frac{m}{m_e E}} \tag{14.4–10}$$

where γ is expressed in electrons/ion, m_e and m are the masses of an electron and the incident ion, respectively, and E is the incident-ion energy in eV. The relationship (14.4–10) represents the experimental results rather well in the medium-energy region (up to several hundred keV), but does not describe the more rapid drop of $\gamma(E)$ at higher energies too accurately. This result is not surprising, however, since the $1/\sqrt{E}$ law of BOHR's expression (14.4–5) is a good approximation to the actual $\ln E/E$ law (reference [*72*]) only over a limited energy range.

As mentioned before, GHOSH and KHARE [*238*] modified the treatment of STERNGLASS in replacing the stopping power $(dE_i/dx)_{\text{total}}$ by the cross section for ionization of an atom by ion impact, as given by BETHE. They also used a different method to calculate the number $(n_{1e} + n_{2e})$ of internal secondary electrons produced and the coefficient for absorption of internal secondary electrons. However, for the particular case of proton bombardment of a metal surface, GHOSH and KHARE point out that expression (14.4–10) for γ agrees with the one which can be derived from their theory for the energy region considered. In an additional treatment, GHOSH and KHARE [*238a*] tried to extend the applicability of their theory to lower incident-ion energies (down to approximately 20 keV and over the energy region 20 keV $\leqslant E \leqslant$ 2 MeV), where the cross sections computed directly by BETHE's formula for the ionization cross section are higher than those actually observed. They suggested therefore that the ionization cross section σ_i^{nl} obtained from BETHE's expression for the nl shell should be divided by $(1 + \beta/E)$, where β is a constant. Then the corrected ionization cross section for the nl shell of a lattice atom is

$$(\sigma_i^{nl})_{\text{corr}} = \sigma_i^{nl}/(1 + \beta/E)\,. \qquad (14.4\text{–}11)$$

For the case of proton bombardment of a metal surface, the expression they obtain for the yield is

$$\gamma_{\text{H}^+} = (N/4\alpha)(\sigma_i^{nl})_{\text{corr}}\,, \qquad (14.4\text{–}12)$$

where N is the number of metal atoms per cm^3 and α is the absorption coefficient for internal secondary electrons. For the case of protons incident on an Al target, they calculated the values of γ for the energy region $20\,\text{keV} \leqslant E \leqslant 2\,\text{MeV}$, using $\alpha = 5.28 \times 10^5\,\text{cm}^{-1}$ and $\beta = 74.5\,\text{keV}$.

In their calculation of secondary-electron yields GHOSH and KHARE [*238a*] considered furthermore that incident ions may undergo charge-changing collisions in passing through a metal target. On the basis of data discussed by ALLISON [*8a*], they assumed for instance that a proton beam with energies below 200 keV may be treated as a two-component system consisting of hydrogen atoms and protons; for $E > 200$ keV, it may be considered as a one-component system consisting only of protons. Then by using the electron-capture cross sections for the charge-changing process for protons in a gas as a rough approximation to that for protons in a metal, they made a crude estimate of the equilibrium fractions of hydrogen atoms and ions in a beam traversing a metal and of their relative contributions to the formation of secondary electrons. Their calculated values of γ for the case of protons incident on an aluminium surface over the energy region from 20 keV to 2 MeV deviate less than 10% from the experimental data [ref. *1*, *129*, *311*].

Finally, the theoretical treatment of the secondary-electron emission by PARILIS and KISHNEVSKII [*543*] should be mentioned. They assume that the formation of the ionic residue of a lattice atom struck by an incident ion results in the formation of holes in the filled band. The cross section for ionization of a lattice atom is actually the cross section for formation of an electron-hole pair. They then treat the secondary-electron emission from a metal as the result of a recombination of a conduction electron with a hole formed in the filled bands, accompanied by the transfer of energy to another conduction electron by the AUGER recombination mechanism.

Thus the secondary electrons ejected by the kinetic emission are AUGER electrons as in the case of the potential emission but the actual electron excitation processes in the two cases are different.

According to PARILIS et al., the function $\gamma(V)$ can be expressed as

$$\gamma(V) = N\,\sigma_i^*(V)\,\lambda\,P_e \qquad (14.4\text{–}13)$$

where N is the number of lattice atoms per cm^3; σ_i^* is an effective cross section* for ionization of an ionic residue of a lattice atom; P_e is the probability of emission of an excited electron (AUGER electron) from the metal surface and is approximated by the empirical formula $P_e = 0.016 (eI_n - 2e\varphi)$, where I_n is the depth of the filled band in which a hole is formed and $e\varphi$ is the work function of the metal; and λ is the mean path length of an internal secondary electron for which the probability of absorption varies as $\exp(-x/\lambda)$ if the ionizing collision occurs inside the metal at a depth x. They restricted their treatment to incident-ion/metal-target systems that fulfill the condition $\frac{1}{4} < Z_1/Z_2 < 4$ (Z_1 and Z_2 being the atomic numbers of the incident ion and target atom, respectively) and calculated $\gamma(V)$ for A^+ and Kr^+ ions incident on Mo and W surfaces. For their range of incident-ion velocities (approximately 0.7×10^7 to 2.5×10^7 cm/sec), the parameters they used for Mo were $eI_n = 16$ eV, $\lambda = 13.3$ Å; those for W were $eI_n = 19$ eV, $\lambda = 11$ Å. For the case of A^+ on W, the calculated expressions for $\gamma(V)$ represent the experimental $\gamma(V)$ curves of ARIFOV et al. [*26*] and PETROV [*557*] quite well. For the case of A^+ and Kr^+ on Mo, the experimental $\gamma(V)$ curves observed by ARIFOV et al. [*26*, *30a*], MEDVED et al. [*467b*], and MAGNUSON et al. [*451a*] also agree well with the calculated ones for corresponding velocity ranges. Furthermore, the theory correctly predicts the observed threshold values for the kinetic emission (see, for example, the values of the energy threshold E_{th} indicated in Fig. 14.1–a) and successfully explains the observation that for clean metal surfaces the yield was nearly independent of the charge of the incident ion. It also explains the isotopic effect on the yield.

* PARILIS et al. define the "effective" ionization cross section by

$$\sigma_i^*(V) = \sigma_i(V) - \Delta\,\sigma_i(V),$$

where $\sigma_i(V)$ is the cross section for ionization of an ionic residue of a lattice atom by an incident ion with velocity V and the term $\Delta\,\sigma_i(V)$ accounts for the change of the cross section $\sigma_i(V)$ as a result of the retardation of the incident ion by collisions with lattice atoms. For their quantitative expression for $\Delta\,\sigma_i(V)$, the reader is referred to the original article.

Literature

[1] AARSET, B., R. W. CLOUD, and J. G. TRUMP: Electron Emission from Metals under High-Energy Hydrogen Ion Bombardment. J. appl. Physics **25**, 1365 (1954).

[2] ABBOTT, R. C., and H. W. BERRY: Measurement of the Angular Distribution of Electrons Ejected from Tungsten by Helium Ions. J. appl. Physics **30**, 871 (1959).

[3] AHEARN, A. J.: Mass Spectrographic Studies of Impurities on Surfaces. 6th National Sympos. on Vacuum Technology Transactions, p. 1. Oxford: Pergamon Press 1959.

[4] ALDRICH, L. T.: The Evaporation Products of Barium Oxide from Various Base Metals and of Strontium Oxide from Platinum. J. appl. Physics **22**, 1168 (1951).

[4a] AKISHIN, A. I.: Emission of Secondary Electrons Due to Deuterons with Energies of 1 to 4 Mev. J. techn. Physics (S.S.S.R.) **28**, 776 (1958). Soviet Phys.-Techn. Phys. **3**, 724 (1958).

[5] ALLEN, J. S.: The Emission of secondary Electrons from Metals Bombarded with Protons. Phys. Rev. **55**, 336 (1939).

[6] — Recent Applications of Electron Multiplier Tubes. Proc. Inst. Radio Engrs. April, p. 346 (1950).

[7] ALLEN, J. A.: Evaporated Metal Films. Rev. pure appl. Chem. **4**, 133 (1954).

[8] ALLISON, H. W., and G. E. MOORE: The Use of Radioactive Sr in Some Thermionic Experiments on Thin Films. Phys. Rev. **78**, 354 (1950).

[8a] ALLISON, S. K.: Experimental Results on Charge-Changing Collisions of Hydrogen and Helium Atoms and Ions at Kinetic Energies above 0.2 keV. Rev. mod. Physics **30**, 1137 (1958).

[9] ALMÉN, O. E., and G. BRUCE: Collection and Sputtering Experiments with Noble Gas Ions. Nuclear Instruments and Methods, North-Holland Publishing Co. **11**, 257 (1961).

[10] — — Sputtering Experiments in the High Energy Region. Nuclear Instruments and Methods, North-Holland Publishing Co. **11**, 279 (1961).

[11] ALPERT, D.: Ion Pumping and the Use of Ion Pumps for the Production of Very Low Pressures. Handbuch d. Physik XII, 642 (1958).

[12] ALTERTHUM, H., K. KREBS, and R. ROMPE: Über die selbständige Ionisation von Natrium- und Cäsiumdampf an glühenden Wolfram- und Rheniumoberflächen. Z. Physik **92**, 1 (1934).

[13] AMDUR, I., M. JONES, and H. PEARLMAN: Accommodation Coefficient on Gas-Covered Platinum. J. chem. Physics **12**, 159 (1944).

[14] — Pressure Dependence of Accommodation Coefficients. J. chem. Physics **14**, 339 (1946).

[15] —, and L. GUILDNER: Thermal Accommodation Coefficients on Gas Covered Tungsten, Nickel, and Platinum. J. Amer. chem. Soc. **79**, 311 (1957).

[16] ANDERSON, G. S., and G. K. WEHNER: Atom Ejection Patterns in Single Crystal Sputtering. J. appl. Physics **31**, 2305 (1960).

[17] — — Atom Ejection Patterns in Single Crystal Sputtering. Annual Rep. **2240**, General Mills Electronics Group, Minneapolis, Minn., Nov. 1961.

[17a] ANDERSON, G. S.: Atom Ejection in Low Energy Sputtering of Single Crystals of bcc Metals. J. appl. Physics **34**, 659 (1963).

[18] ANSELM, A. I.: Theory of Surface Ionization on Glowing Metals. J. exp. theoret. Physics (S.S.S.R.) **4**, 678 (1934).

[19] APKER, L.: Surface Phenomena useful in Vacuum Technique. Ind. Engng. Chem. **40**, 846 (1948).

[20] ARCHER, C. T.: The Thermal Conductivity and the Accommodation Coefficient of Carbon Dioxide. Philos. Mag. **19**, 901 (1935).

[21] — Thermal conduction in Hydrogen-Deuterium mixtures. Proc. Roy. Soc. (London) A **165**, 474 (1938).

[22] ARIFOV, U. A., and G. N. SCHUPPE: Positive Surface Ionization of Atoms and Molecules. Akademia Nauk Uzbek. S.S.R. (Fiziko-tekhnicheskii institut) Trudy **2**, No. 1, 19 (1948).

[23] —, and A. KH. AYUKHANOV: Surface Phenomena in the Bombardment of Metals by Positive Ions. Trudy Soveshaniia po Katodnoi Elektronike, Akademia Nauk Ukranian S.S.R. Kiev (June 9, 1951) pp. 99—102 (publ. 1952).

[24] — — Investigation of Secondary Emission from Metals under Bombardment by Positive Ions of Alkali Elements. Izvest. Akad. Nauk S.S.S.R. Ser. fiz. **20**, 1165 (1956) Columbia Techn. Transl., p. 1057.

[25] — —, and S. V. STARODUBTSEV: Ion Scattering Coefficient as a Function of Colliding-Particle Mass Ratio. J. exp. theor. Physics (S.S.S.R.) **33**, 845 (1957); Soviet Phys.-JETP **4**, 845 (1957).

[26] —, and R. R. RAKHIMOV: The Existence of Potential Pull-Out of Electrons from Metal Hit by Inert Gas Ions. Izvest. Akad. Nauk Uzbek S.S.R., Ser. Fiz. Mat. **5**, 5 (1958).

[27] — A. KH. AYUKHANOV, S. M. STARODUBTSEV, and KH. KHADZHIMUKHAMEDOV: On a Method of Investigation of Secondary Ion Processes at High Target Temperatures in the Presence of Thermoelectric Emission. Soviet Phys. Doklady **4**, 86 (1959).

[28] —, and R. R. RAKHIMOV: Variation of Ion-Electron Emission as a Function of Certain Parameters of the Target and the Bombarding Ions. Akad. Nauk S.S.S.R., Bull., Physic. Ser. **24**, 666 (1960).

[29] —, and KH. KH. KHADZHIMUKHAMEDOV: Concerning Neutralization of Fast Positive Ions on Metal Surfaces. Bull. Acad. Sci. S.S.S.R. (English Translation) **24**, 711 (1960).

[30] — A. KH. AYUKHANOV, and D. D. GRUICH: On Scattering of Slow Alkali Metal Ions from the Metal Surfaces. Akad. Nauk Uzbek. S.S.R., Bull. Physic. Ser. **24**, 716 (1960).

[30a] — R. R. RAKHIMOV, and KH. DZHURAKULOV: Secondary Emission in the Bombardment of Molybdenum by Neutral Atoms and Argon Ions. Soviet Phys.-Doklady **7**, 209 (1962).

[31] ARMBRUSTER, M., and J. AUSTIN: The Adsorption of Gases on Smooth Surfaces of Steel. J. Amer. chem. Soc. **66**, 159 (1944).

[32] ARMSTRONG, D., P. E. MADSEN, and E. C. SYKES: Cathodic Bombardment Etching of Metals with Particular Reference to Uranium, Thorium, and their Alloys. AERE M/R 2584 Harwell, Berkshire, June 1958.

[33] ASADA, T., and K. QUASEBARTH: The Extraction of Gold from Cathode Metal by Glow Discharge. Z. Physik. Chem. A **143**, 435 (1929).

[34] AVAKYANTS, G. M.: On a Possible Ionization Mechanism of the Atomic Shell. Doklady Akad. Nauk Uzbek. S.S.R. **6**, 3 (1951).

[35] — On the Theory of Positive Surface Ionization in the Presence of an Outer Electrical Field. Doklady Akad. Nauk Uzbek. S.S.R. **7**, 8 (1952).

[36] Avakyants, G. M.: On the Theory of Surface Ionization. Doklady Akad. Nauk Uzbek. S.S.R. **1**, 17 (1953).
[37] — On the Theory of Surface Ionization. Izvest. Akad. Nauk Uzbek. S.S.R. **3**, 109 (1954).
[38] — Theory of the Phenomena in a Contact Zone. J. exper. theor. Physics (S.S.S.R.) **27**, 333 (1954).
[39] — On the Theory of Contactphenomena. Doklady Akad. Nauk S.S.R. **10**, 10 (1953).

[40] Bader, M., F. C. Witteborn, and T. W. Snouse: Sputtering of metals by Mass-Analyzed N_2^+. NASA TR-105 Washington 1961.
[41] Bailey, T. L.: Experimental Determination of the Electron Affinity of Fluorine. J. chem. Physics **28**, 792 (1958).
[42] Bakulina, I. N., and N. I. Ionov: The Determination of the Energy of Electronic Affinity of Cyanogen from the Surface Ionization of KCN and KCNS Molecules. Doklady Akad. Nauk S.S.S.R. **99**, 1023 (1954).
[43] — — Energy of the Electron Affinity of the Halogen Atoms. Doklady Akad. Nauk S.S.S.R. **105**, 680 (1955).
[44] — — Determination of the Electron Affinity of Sulfur by Surface Ionization. Soviet Phys.-Doklady **2**, 423 (1957).
[45] — — Determination of the Ionization Potential of Uranium by a Surface Ionization Method. Soviet Phys.-JETP **9**, 709 (1959).
[46] Balarin, M., and F. Hilbert: Die Einwirkung energiereicher Ionen auf Metalloberflächen. J. Phys. Chem. Solids. **20**, No. 1/2, 138 (1961).
[47] Bardeen, J.: Theory of the Work Function. The Surface Double Layer. Phys. Rev. **49**, 653 (1936).
[48] — The Image and Van der Waals Forces at a Metallic Surface. Phys. Rev. **58**, 727 (1940).
[49] Barnett, C. F., P. M. Stier, and G. E. Evans: The Relative Secondary Electron Emission for Ions and Atoms. Phys. Rev. **95**, 307 (1954).
[50] Barrer, R. M.: An Analysis by Adsorption of the Surface Structure of Graphite. Proc. Roy. Soc. (London) A **161**, 476 (1937).
[51] Bartz, G., G. Weissenberg, and D. Wiskott: Ein Auflichtelektronenmikroskop. Proc. III Int. Conf. Electron Microscopy London, 395 (1954).
[52] Bassett, G. A., and D. W. Pashley: The Growth, Structure, and Mechanical Properties of Evaporated Metal Films. J. Inst. Metals **87**, 449 (1958/59).
[53] — —, and J. W. Menter: High Resolution Electron Microscopy of Crystals. Discuss. Faraday Soc. **28**, 7 (1959).
[54] Batanov, G. M.: Secondary Emission of No. 46 Glass under the Effect of Positive Ions of Some Gases. Soviet Phys.-Solid State **2**, 1839 (1961).
[55] Baule, B.: Theoretische Behandlung der Erscheinungen in verdünnten Gasen. Ann. Physik **44**, 145 (1914).
[56] Baum, T.: Beiträge zur Erklärung der Erscheinungen bei der Kathodenzerstäubung. Z. Physik **40**, 686 (1927).
[57] Becker, J. A.: Thermionic and Adsorption Characteristics of Cesium on Tungsten and Oxidized Tungsten. Phys. Rev. **28**, 341 (1926).
[58] — Phenomena in Oxide Coated Filaments. Phys. Rev. **34**, 1323 (1929).
[59] —, and R. W. Sears: Phenomena in Oxide Coated Filaments; II. Origin of Enhanced Emission. Phys. Rev. **38**, 2193 (1931).
[60] —, and C. D. Hartman: Field Emission Microscope and Flash Filament Techniques for the Study of Structure and Adsorption on Metal Surfaces. J. physic. Chem. **57**, 153 (1953).

[61] BECKER, J. A.: The Life History of Adsorbed Atoms, Ions and Molecules. Ann. N. Y. Acad. Sci. **58**, 721 (1954).

[62] —, and R. G. BRANDES: On the Absorption of Oxygen on Tungsten as Revealed in the Field Emission Electron Microscope. J. chem. Physics **23**, 1323 (1955).

[63] BEEBE, R. A.: Handbuch der Katalyse, IV, 473. Wien: Springer 1943.

[64] BEECK, O.: Catalysis—A Challenge to the Physicist. Rev. mod. Physics **17**, 61 (1945).

[65] — Hydrogenation Catalysis. Discuss. Faraday Soc. **8**, 118 (1950).

[66] — W. A. COLE and A. WHEELER: Determination of Heats of Adsorption using Metal Films. Discuss. Faraday Soc. **8**, 314 (1950).

[67] —, and A. W. RITCHIE: The Effect of Crystal Parameter in Hydrogenation and Dehydrogenation. Discuss. Faraday Soc. **8**, 159 (1950).

[68] — Catalysis and the Adsorption of Hydrogen on Metal Catalysts. Advances in Catalysis **2**, 151 (1950).

[68a] BEELER, J. R., and D. G. BESCO: Knock-On Cascades and Point-Defect Configurations in Binary Materials. Symposium on Radiation Damage in Solids, Venice, May 1962 (International Atomic Energy Commission, Vienna, 1962), Vol. 1, p. 43.

[68b] BEHRISCH, R.: Festkörperzerstäubung durch Ionenbeschuß. Ergebn. d. Exakt. Naturwiss. **35**, 295 (1964).

[69] BENJAMINS, M., and R. O. JENKINS: The Distribution of Autelectronic Emission from Single Crystal Metal Points I. Tungsten, Molybdenum and Nickel in the Clean State. Proc. Roy. Soc. A **176**, 262 (1940).

[70] — — The Distribution of Autelectronic Emission from Single Crystal Metal Points II. The Adsorption, Migration and Evaporation of Thorium, Barium and Sodium on Tungsten and Molybdenum. Proc. Roy. Soc. A **180**, 225 (1942).

[71] BERRY, H. W.: Secondary Electron Emission by Fast Neutral Molecules and Neutralization of Positive Ions. Phys. Rev. **74**, 848 (1949).

[72] BETHE, H. A.: Zur Theorie des Durchgangs schneller Korpuskularstrahlen durch Materie. Ann. Physik **5**, 325 (1930).

[73] BIERLEIN, T. K., H. W. NEWKIRK, and B. MASTEL: Etching of Refractories and Cermets by Ion Bombardment. Report HW-51028 Hanford Labor., General Electric, Richland, Washington, June 1957.

[74] BLECHSCHMIDT, E.: Kathodenzerstäubungsprobleme. Die Kathodenzerstäubung in Abhängigkeit von den Betriebsbedingungen. Ann. Physik **81**, 999 (1926).

[75] BLEWETT, J. P., and M. B. SAMPSON: Isotopic Constitution of Strontium, Barium, and Indium. Phys. Rev. **49**, 778 (1936).

[76] — Mass Spectrograph Analysis of Bromine. Phys. Rev. **49**, 900 (1936).

[77] —, and E. J. JONES: Filament Sources of Positive Ions. Phys. Rev. **50**, 464 (1936).

[78] BLODGETT, K. B., and I. LANGMUIR: Accommodation Coefficient of Hydrogen; A Sensitive Detector of Surface Films. Phys. Rev. **40**, 78 (1932).

[79] DE BOER, J. H., and J. F. CUSTERS: Lichtabsorptionskurven von adsorbierten Paranitrophenolmolekülen und deren Analyse. Z. phys. Chem. B **25**, 238 (1934).

[80] — — Adsorption by Van der Waals Forces and Surface Structure. Physica **4**, 1017 (1937).

[81] — The Formation of Two Dimensional Nuclei by Ionic Molecules. Kon. Akad. Wetenschap. Proc. Ser. A **49**, 1103 (1946).

[82] DE BOER, J. H.: The Dynamical Character of Adsorption. Oxford: Clarendon Press 1953.

[83] — Adsorption Phenomena. Advances in Catalysis **8**, 110 (1956).

[84] — in "Chemisorption", Editor W. E. Garner, London: Butterworths Scientific Publications, 1957, p. 27.

[85] BOHR, N.: The Penetration of Atomic Particles Through Matter. Kgl. danske Vidensk. Selsk., mat.-fysiske Medd. **18**, 8 (1948).

[86] BOSWORTH, R. C. L.: Studies in Contact Potentials; The Evaporation of Sodium Films. Proc. Roy. Soc. (London) A **162**, 32 (1937).

[87] BOUDART, M.: Heterogeneity of Metal Surfaces. J. Amer. chem. Soc. **74**, 3556 (1952).

[88] BRADLEY, R. C.: Sputtering of Alkali Atoms by Inert Gas Ions of Low Energy. Phys. Rev. **93**, 719 (1954).

[89] — Theoretical and Experimental Investigations of the Atomic Phenomena Occurring on and Near the Surfaces of Solids. Report AFSOR-TN-58-642, AD, 162174, September 1, 1958.

[90] — Secondary Positive Ion Emission from Metal Surfaces. J. appl. Physics **30**, 1 (1959).

[91] — A. ARKING, and D. S. BEERS: Secondary Positive Ion-Emission from Platinum. J. chem. Physics **33**, 764 (1960).

[91a] —, and E. RUEDL: Reflection of Inert Gas Ions from Copper. Bull. Am. Phys. Soc. **5**, 16 (1960).

[92] BRAME, D. R., and T. EVANS: Deformation of Thin Films on Solid Substrates. Philos. Mag. **3**, 871 (1958).

[93] BREMNER, J. G. M.: The Thermal Accommodation Coefficient of Gases I. An Investigation of the Effect of Flashing. Proc. Roy. Soc. (London) A **201**, 305 (1950).

[94] — The Thermal Accommodation Coefficient of Gases II. Determinations at Room and at Liquid Oxygen Temperatures. Proc. Roy. Soc. (London) A **201**, 321 (1950).

[94a] BRINKMAN, J. A.: On the Nature of Radiation Damage in Metals. J. appl. Phys. **25**, 961 (1954).

[95] BRODD, R. J.: Heats of Adsorption from Charge-Transfer Complex Theory. J. Physic. Chem. **62**, 54 (1958).

[96] BROEDER, J. J., L. L. VAN REIJEN, W. M. H. SACHTLER, and G. G. A. SCHUIT: Zur Frage des Bindungscharakters bei Chemisorption von Wasserstoff und Übergangsmetallen. Z. Elektrochem. Ber. Bunsenges.-Physik. Chem. **60**, 838 (1956).

[97] BROWN, R. E., and J. H. LECK: Desorption of gas in cold cathode ionization gauge. Brit. J. appl. Phys. **6**, 161 (1955).

[98] — The Thermal Accommodation of Helium and Neon on Beryllium. Thesis, Univ. of Missouri 1957.

[99] BRUINING, H.: Physics and Application of Secondary Electron Emission. New York: McGraw.-Hill Co. 1954.

[100] BRUNAUER, S., P. H. EMMETT, and E. TELLER: Adsorption of gases in multimolecular layers. J. Amer. chem. Soc. **60**, 309 (1938).

[101] — The Adsorption of Gases and Vapours. Princeton: Princeton University Press, 1943.

[102] BRUNNÉE, C.: Über die Ionenreflexion und Sekundärelektronenemission beim Auftreffen von Alkaliionen auf reine Molybdän-Oberflächen. Z. Physik **147**, 161 (1957).

[103] BUERGER, S. M. J.: The Lineare Structure of Crystals. Z. Kristallographie **89**, 195 (1934).

[104] BULL, T. H., and H. MARSHALL: Life-Time of Potassium Ions on a Tungsten Filament at 1460° K. Nature (London) **167**, 478 (1951).

[104a] BURHOP, E. H. S.: The AUGER Effect and the Radiationless Transitions. Cambridge University Press, New York, 1952.

[105] BURTON, W. K., and N. CABRERA: Crystal Growth and Surface Structure. Discuss. Faraday Soc. **5**, 33 (1949).

[106] — —, and F. C. FRANK: Dislocations in Crystal Growth. Nature **163**, 398 (1949).

[107] CARLSON, A. J., J. T. WILLIAMS, B. A. RODGERS, and E. J. MANTHOS: Etching Metals by Ionic Bombardment. ISC-480 (1954), unclassified.

[107a] CARLSTON, C. E., and G. D. MAGNUSON: Ion Reflection from Metal Surfaces. Bull. Amer. Phys. Soc. **8**, 77 (1963).

[108] CARMICHAEL, J. H., and E. A. TRENDELENBURG: Ion Induced Reemission of Noble Gases from a Nickel Surface. J. appl. Physics **29**, 1570 (1958).

[109] CASSIGNOL, C., and G. RANG: Sur le caractère non linéaire en fonction de l'intensité de la pulverisation cathodique à haute énergie et sa variation en fonction de la température. C. R. hebd. Seances Acad. Sci. **248**, 1988 (1959).

[110] CASTAING, R.: Examen direct des métaux par transmission en microscopic et diffraction electroniqués. Rev. Metallurgie **52**, 669 (1955).

[111] CHANDRASEKHAR, S.: Stochastic Problems in Physics and Astronomy. Rev. mod. Physics **15**, 2 (1943).

[112] CHAUDHRI, R. M., and A. W. KHAN: Emission of Secondary Electrons from Nickel and Molybdenum by Neutral Atoms of Mercury and Potassium. Proc. physic. Soc. (London) **61**, 526 (1948).

[113] CHAVKEN, L. P.: The Measurement of the Pressure of Several Gases with a Thermovapor Manometer. J. techn. Physics (S.S.S.R.) **25**, 726 (1955).

[114] CLAUSING, P.: Über die Strahlformung bei der Molekularströmung. Z. Physik **66**, 471 (1930).

[115] VAN CLEAVE, A. B.: The Adsorption of Nitrogen and Oxygen on Tungsten. Trans. Faraday Soc. **34**, 1174 (1938).

[116] COBAS, A., and W. E. LAMB: On the Extraction of Electrons from a Metal Surface by Ions and Metastable Atoms. Phys. Rev. **65**, 327 (1944).

[117] ČOBIĆ, B., and B. PEROVIĆ: Angular Distribution of Sputtered Particles and Sputtering Rate for High Speed Ions. IV. Symposium on Ionization Phenomena in Gases, p. 260. Amsterdam: North Holland Publishing Co. 1960.

[118] — Investigation of the Characteristics of the Ion Sorption in a Bayard-Alpert Ionization Tube. Bull. Inst. Nuclear Sciences. Boris Kidrich **11**, No. 227, March, 1961.

[119] COCKCROFT, J. D.: On Phenomena occurring in the Condensation of Molecular Streams on Surfaces. Proc. Roy. Soc. (London) A **119**, 293 (1928).

[119a] COMEAUX, A., and D. B. MEDVED: Electron Ejection and Reflection Properties of Neutral Atoms. Proceedings 23rd Physical Electronics Conference, March 1963.

[120] COMPTON, A. H.: The variation of the Specific Heat of Solids with Temperature. Phys. Rev. **6**, 377 (1915).

[121] COMPTON, K. T.: Accommodation Coefficient of Gaseous Ions at Cathodes. Proc. Nat. Acad. Sci. USA. **18**, 705 (1932).

[122] —, and E. S. LAMAR: A Test of the Classical "Momentum Transfer" Theory of Accommodation Coefficients of Ions at Cathodes. Phys. Rev. **44**, 338 (1933).

[123] CONSTABLE, F. H.: Active Centers from the Point of View of Kinetics. Handbuch der Katalyse, Bd. V, p. 141. Wien: Springer 1957.
[124] COPLEY, M. J., and T. E. PHIPPS: The Surface Ionization of Potassium on Tungsten. Phys. Rev. **45**, 344 (1934).
[125] — — The Surface Ionization of Potassium on Tungsten. Phys. Rev. **48**, 960 (1935).
[126] CORRENS, C. W.: Einführung in die Mineralogie. Berlin: Springer 1949.
[127] COULLIETTE, J. H.: The Diffusion of Metastable Atoms in Mercury Vapor. Phys. Rev. **32**, 636 (1928).
[128] COUPER, A., and D. ELEY: Parahydrogen Catalysis by Transition Metals. Nature (London) **164**, 578 (1949).
[129] COUSINIÉ, M. P., N. COLOMBIÉ, C. FERT, and R. SIMON: Electronique. — Variation du coefficient d'émission électronique secondaire de quelques métaux avec l'énergie des ions incidents. C. R. hebd. Seances Acad. Sci. **249**, 387 (1959).
[129a] CRAIG, R. A.: AUGER Electron Ejection from Superconducting Metals. Nuovo Cimento **27**, 257 (1963).
[130] CRAMER, H.: Über die Diffusion durch polierte und geäzte Kupferoberflächen. Ann. Physik (5) **34**, 237 (1939).
[131] CULVER, R. V., and F. C. TOMPKINS: Surface Potentials and Adsorption Process on Metals. Advances in Catalysis **11**, 67 (1959).
[132] CUNNINGHAM, R. E., and A. T. GWATHMEY: The Influence of Foreign Atoms on the Surface Rearrangement Produced by the Catalytic Reaction of Hydrogen and Oxygen on a Single Crystal of Copper. J. Amer. chem. Soc. **76**, 391 (1954).
[133] — P. HAYMANN, C. LECOMTE, W. J. MOORE and J. J. TRILLAT: Etching of Surfaces with 8-keV Argon Ions. J. appl. Physics **31**, 839 (1960).
[134] CUTLER, P. H., and I. I. GIBBONS: Model for the surface potential barrier and the periodic deviations in the Schottky effect. Phys. Rev. **3**, 394 (1958).
[135] D'ANS, A. M., R. DARIOS, and L. MALASPINA: Liberazione di elettroni da superficie metalliche per urto di ioni. Nuovo Cimento **5**, 394 (1948).
[136] DATZ, S., and E. H. TAYLOR: Ionization on Platinum and Tungsten Surfaces. I. The Alkali Metals. J. chem. Physics **25**, 389 (1956).
[137] — — Ionization on Platinum and Tungsten Surfaces. II. The Potassium Halides. J. chem. Physics **25**, 395 (1956).
[138] — — Some Applications of Molecular Beam Techniques to Chemistry. Published in Recent Research in Molecular Beams. New York: Academic Press 1959.
[139] — R. C. MINTURN, and E. H. TAYLOR: Thermal Positive Ion Emission and the Anomalous Flickert-Effect. J. appl. Physics **31**, 880 (1960).
[140] DAVIS, L. D., B. T. FELD, C. W. ZABEL, and J. R. ZACHARIAS: The Hyperfine Structure and Nuclear Moments of the Stable Chlorine Isotopes. Phys. Rev. **76**, 1076 (1949).
[141] DEBYE, P.: Die Van der Waalschen Kohäsionskräfte. Phys. Z. **21**, 178 (1920).
[142] — Molekularkräfte und ihre elektrische Deutung. Phys. Z. **22**, 302 (1921).
[142a] DEDERICHS, P. H., and G. LEIBFRIED: Fokussierende Stoßfolgen in kubisch flächenzentrierten Kristallen. Z. Physik **170**, 320 (1962).
[143] DEKKER, A. J.: Solid State Physics, p. 95. Englewood Cliffs, New Jersey: Prentice-Hall Inc. 1957.
[143a] DEPOORTER, G. L.: Energy Exchange between Cold Gas Molecules and a Hot Tungsten Surface. UCRL-10504, Berkeley, California, Nov. 1962.
[143b] —, and A. W. SEARCY: Energy Exchange between a Hot Tungsten Surface and Cold Gases. J. chem. Physics **39**, 925 (1963).

[144] DEVONSHIRE, A. F.: The Interaction of Atoms and Molecules with Solid Surfaces. VIII. The Exchange of Energy Between a Gas and a Solid. Proc. Roy. Soc. (London) A **158**, 269 (1937).

[145] DICKENS, B. G.: The Effect of Accommodation on Heat Conduction through Gases. Proc. Roy. Soc. (London) A **143**, 517 (1934).

[146] DIENES, G. J., and G. H. VINEYARD: Radiation Effects in Solids. New York: Interscience Publishers Inc. 1957.

[147] DOBREZOW, L. N.: Ionization of Atoms of Alkali Metals on the Surface of Tungsten, Molybdenum and Thoriated Tungsten. J. exp. theor. Physics (S.S.S.R.) **4**, 783 (1934).

[148] — Ionization of Alkali Metal Atoms on Heated Surfaces of Wolfram, Molybdenum and Thoriated Wolfram. J. exp. theor. Physics (S.S.S.R.) **6**, 552 (1936).

[149] — S. V. STARODUBTSEV, and J. I. TIMOKHINA: Surface Ionization of Thin Layers of Calcium and Magnesium Oxides. Doklady Akad. Nauk S.S.S.R. **55**, 303 (1947).

[150] — Electron and Ion Emission of Thoriated Wolfram in Sodium Vapor. J. exp. theor. Physics (S.S.S.R.) **17**, **301** (1947).

[151] —, and P. J. UWAROW: J. techn. Physics (S.S.S.R.) **22**, 175 (1952).

[152] —, and N. M. KARNAUKHOVA: Determination of the Coefficients of Cathodic Sputtering of Metals by Ions of the Same Metal. Doklady Akad. Nauk S.S.S.R. **85**, **4**, 745 (1952).

[153] — Elektronen- und Ionenemission. Berlin: VEB-Verlag Technik 1954.

[154] — On the Theory of Positive Surface Ionization in the Presence of an External Electrical Field. Izvest. Akad. Nauk Uzbek. S.S.R. **3**, 101 (1954).

[155] — On the Problem of the Theory of Surface Ionization. Trans. of the M. I. Kalinin Poltechn. Institute, Leningrad, No. 194, 143—153 (1958).

[156] DORRESTEIN, R.: Anregungsfunktionen metastabiler Zustände in Helium und Neon, gemessen mit Hilfe der von metastabilen Atomen verursachten Elektronenauslösung aus Metallen. Physica **9**, 447 (1942).

[157] DOTY, P. M., and J. E. MAYER: The Electron Affinity of Bromine and a Study of its Decomposition on Hot Tungsten. J. chem. Physics **12**, **323** (1944).

[158] DOWDEN, D. A.: Heterogeneous Catalysis. Part I. Theoretical Basis. J. chem. Soc. (London) **1950**, 242

[159] — Catalytic Activity of Nickel. Ind. Engng. Chem. **44**, 977 (1952).

[160] DRECHSLER, M., and E. W. MÜLLER: Zur Feldelektronenemission und Austrittsarbeit einzelner Kristallflächen. J. Physics **134**, 208 (1953).

[161] — Berechnung Van der Waalsscher Adsorptionsenergien und Platzwechselenergien an Einkristallflächen. Z. Elektrochem. **58**, 327 (1954).

[162] — Vorzugsrichtungen der Oberflächendiffusion auf Einkristallflächen. Z. Elektrochem. **58**, 334 (1954).

[163] — Messung der Oberflächendiffusions-Koeffizienten und Platzwechselenergien von Barium auf Wolfram-Einkristallflächen mit dem Feldelektronenmikroskop. Z. Elektrochem. **58**, 340 (1954).

[164] DU BRIDGE, L. A.: A Further Experimental Test of Fowlers Theory of Photoelectric Emission. Phys. Rev. **39**, 108 (1932).

[165] DUKELSKII, V. M., and N. I. IONOV: Formation of Negative Halogen Ions in the Interaction of Alkali Halides with the Surface of Incandescent Tungsten. J. exp. theor. Physics (S.S.S.R.) **10**, 1248 (1940).

[166] —, and E. YA. ZANDBERG: Secondary Electron mission under the Action of Negative Ions. J. exp. theor. Physics (S.S.S.R.) **19**, **731** (1949).

[167] — —, and N. I. IONOV: Negative Ions of Rubidium and Cesium. Doklady Akad. Nauk (S.S.S.R.) **68**, 31 (1949).

[168] DUNAEV, YU. A., and I. P. FLAKS: Secondary Emission Induced by Bombarding the Metal Target with Multicharge Ions. Doklady Akad. Nauk **91**, No. 1, 43 (1953).

[169] EBERHAGEN, A.: Die Änderung der Austrittsarbeit von Metallen durch eine Gasadsorption. Fortschr. Physik 8, 245 (1960).

[170] EGGLETON, A. E. J., and F. C. TOMPKINS: The Thermal Accommodation Coefficient of Gases and Their Adsorption on Iron. Trans. Faraday Soc. 48, 738 (1952).

[171] — —, and D. W. B. WANFORD: Measurement of the Thermal Accommodation Coefficients of Gases. Proc. Roy. Soc. A **213**, 266 (1952).

[172] —, and F. G. HUDDA: Interaction of Rare Gases with Metal Surfaces. I. A, Kr, and Xe on Tungsten. J. chem. Physics **30**, 493 (1959).

[173] EHRLICH, G.: Molecular Dissociation and Reconstitution on Solids. J. chem. Physics **31**, 111 (1959).

[174] — Molecular Processes at the Gas-Solid Interface, Structure and Properties of Thin Films, p. 423. New York, London: John Wiley & Sons 1959.

[175] — — Surface Migration of Nitrogen on Tungsten. J. chem. Physics **32**, 942 (1960).

[176] —, and F. G. HUDDA: Observation of Adsorption on an Atomic Scale. General Electric Research Lab. Report No. 60-RL-2505 M, August 1960.

[177] — Adsorption of CO on Tungsten and Its Effect on the Work Function. J. chem. Physics **27**, 1206 (1957).

[178] EISINGER, J.: Electrical Properties of Nitrogen Adsorbed on Tungsten. J. chem. Physics **28**, 165 (1958).

[179] — Properties of Hydrogen Chemisorbed on Tungsten. J. chem. Physics **29**, 1154 (1958).

[180] ELEY, D. D.: A Calculation of Heats of Chemisorption. Discuss. Faraday Soc. 8, 34 (1950).

[181] — Catalysis and the Chemical Bond, p. 14. University of Notre Dame Press 1954.

[182] — Electron Transfer and Heterogeneous Catalysis. Z. Elektrochem., Ber. Bunsenges. physik. Chem. **60**, 797 (1956).

[183] ERBACHER, O.: Die elektrochemische Belegung von Metalloberflächen mit einer einatomaren Schicht edlerer Atome. Z. physik. Chem. A **178**, 15 (1936).

[184] — Primäre und sekundäre Adsorption von Metallionen an Metalloberflächen. Z. physik. Chem. A **182**, 243 (1938).

[185] — Unmittelbare Gleichgewichtsverschiebung infolge Metallionenadsorption an Platin. Z. physik. Chem. A **182**, 256 (1938).

[186] EREMEEV, M. A., and M. V. ZUBCHANINOV: Scattering of K Ions by a Ta Surface. J. exp. theor. Physics (S.S.S.R.) **12**, 358 (1942).

[187] — The Collision of Alkali Ions with the Surface of a Metal. Trudy Soveshaniya, Po Katod Elektronika Acad. Nauk Ukrain S.S.R. Kiev, June 4—9 (1951), p. 91.

[188] — Emission of Electrons and Reflection of Ions from Metal Surfaces. Doklady Akad. Nauk S.S.S.R. **79**, 775 (1951).

[189] —, and V. V. SHESTUKHINA: Knockout of Electrons and Reflection of Potassium Ions from Tungsten and Tantalum. J. techn. Physics (S.S.S.R.) **22**, No. 8, 1262 (1952).

[190] — — The Emission of Electrons and the Reflection of Potassium Ions from Tungsten and Tantalum. J. techn. Physics (S.S.S.R.) **22**, 1267 (1952).

[191] — — Knockout of Electrons and Reflection of Lithium Ions from Tungsten and Tantalum. J. techn. Physics (S.S.S.R.) **22**, No. 8, 1268 (1952).

[192] EREMEEV, M. A., and V. G. YUR'EV: Electron Emission and Reflection of Potassium and Lithium Ions from Oxidized Tungsten and Tantalum. J. techn. Physics (S.S.S.R.) **22**, No. 8, 1290 (1952).

[193] ESTERMANN, I., and O. STERN: Beugung von Molekularstrahlen. Z. Physik **61**, 95 (1930).

[194] — Recent Research in Molecular Beams, p. 1. New York: Academic Press 1959.

[195] EUCKEN, A., and A. BERTRAM: Die Ermittlung der Molwärme einiger Gase bei tiefen Temperaturen nach der Wärmeleitfähigkeitsmethode. Z. physik. Chem. B **31**, 361 (1936).

[196] —, and H. KROME: Die Ausgestaltung der Wärmeleitfähigkeitsmethode zur Messung der Molwärme sehr verdünnter Gase durch gleichzeitige Bestimmung des Akkommodationskoeffizienten. Z. physik. Chem. B **45**, 175 (1940).

[197] — Lehrbuch der chemischen Physik, Bd. II/1. Leipzig: 1943.

[198] — Lehrbuch der chemischen Physik. Bd. II/2. Leipzig: 1943.

[199] EVANS, R. C.: The Surface Ionization of Potassium on Molybdenum. Proc. Cambridge philos. Soc. **29**, 522 (1933).

[200] EWALD, H., and H. HINTENBERGER: Methoden und Anwendungen der Massenspektroskopie, S. 43. Weinheim: Verlag Chemie 1952.

[201] FARNSWORTH, H. E., and R. P. WINCH: Photoelectric Work Functions of (100) and (111) Faces of Silver Single Crystals and Their Contact Poten tia Difference. Phys. Rev. **58**, 812 (1940).

[202] — R. E. SCHLIER, T. H. GEORGE, and R. M. BURGER: Ion Bombardment-Cleaning of Germanium and Titanium as Determined by Low-Energy Electron Diffraction. J. appl. Physics **26**, 252 (1955).

[203] —, and R. F. WOODCOCK: Radiation Quenching, Ion Bombardment, and Annealing of Nickel and Platinum for Ethylene Hydrogenation. Ind. Engng. Chem. **49**, 258 (1957).

[204] — — Effects of Radiation Quenching, Ion Bombardment and Annealing on Catalytic Activity of Pure Nickel and Platinum Surfaces.—Hydrogenation of Ethylene, Hydrogen, and Deuterium Exchange. Advances in Catalysis IX, 123 (1957).

[205] — R. E. SCHLIER, T. H. GEORGE, and R. M. BURGER: Application of the Ion Bombardment Cleaning Method to Titanium, Germanium, Silicon, and Nickel as Determined by Low-Energy Electron Diffraction. J. appl. Physics **29**, 1150 (1958).

[206] — Comments on the Paper by J. C. RIVIERE: Contact Potential Difference Measurements by the Kelvin Method. Proc. physic. Soc. **71**, 703 (1958).

[207] FAUST, J. W.: Heat Conduction by Rarefied Gases. Master Thesis, Univ. of Missouri, Columbia, 1949.

[208] — The Accommodation Coefficients of the Inert Gases on Aluminium, Tungsten, Platinum and Nickel and Their Dependence on Surface Conditions. Thesis, Univ. of Missouri, Columbia, 1954.

[209] FEITKNECHT, W.: Über den Angriff von Kristallen durch Kanalstrahlen. Helv. chim. Acta **7**, 825 (1924).

[210] FETZ, H.: Über die Beeinflussung eines Quecksilbervakuumbogens mit einem Steuergitter im Plasma. Ann. Physik **37**, 1 (1940).

[211] — Über die Kathodenzerstäubung bei schiefem Aufprall der Ionen. Z. Physik **119**, 590 (1942).

[212] — A. DIENER, and E. MEYER: Elektronenbefreiung durch energiearme Ionen an gasbedeckten Metalloberflächen. Z. angew. Physik **13**, Heft 6, 292 (1961).

[213] Finkelstein, A. T.: A High Efficiency Ion Source. Rev. Sci. Instruments **11**, 94 (1940).

[214] Fisher, T. F., and C. E. Weber: Cathodic Sputtering for Micro-Diffusion Studies. J. appl. Physics **23**, 181 (1952).

[215] Flaks, I. P.: Secondary Emission Caused by Positive Potassium Ions of Varying Charge. J. techn. Physics (S.S.S.R.) **25**, 2463 (1955).

[215a] Fluit, J. M.: Sputtering of Copper Single-Crystals by 20 keV Noble Gas Ion Bombardment as a Function of Target Temperature and as a Function of the Angle of Incidence. Colloques Internatl. du Centre National de la Recherche Scientifique No. 113, Paris, Dec. 1962, p. 119.

[215b] — P. K. Rol, and J. Kistemaker: Angular-Dependent Sputtering of Copper Single Crystals. J. appl. Physics **34**, 690 (1963).

[215c] — L. Friedman, J. van Eck, C. Snoek, and J. Kistemaker: Photons and Metastable Atoms Produced in Sputtering Experiments (5—20 keV). Proc. Fifth Internatl. Conference on Ionization Phenomena in Gases (1961), North Holland Publishing Company, Amsterdam 1962, p. 131.

[216] Flaks, I. P.: Electron emission owing to Positive ions of Zinc, Mercury, Thallium, and Lead at Varying Charge. J. techn. Physics (S.S.S.R.) **25**, 2467 (1955).

[217] Fogel, Ya. M., R. P. Slabospitskii, and A. B. Rastrepin: Emission of Charged Particles from Metal Surfaces under Bombardment by Positive Ions. J. techn. Physics (S.S.S.R.) **30**, 63 (1960). Soviet Phys.-Techn. Phys. **5**, 58 (1960).

[218] — —, and I. M. Karnaukhov: Mass-Spectrometer Investigation of Secondary Positive and Negative Ion Emission Resulting from the Bombardment of an Mo Surface by Positive Ions. J. techn. Physics (S.S.S.R.) **30**, 824 (1960). Soviet Phys.-Techn. Phys. **5**, 777 (1960).

[219] Found, C. G., and I. Langmuir: Study of a Neon Discharge by Use of Collectors. Phys. Rev. **39**, 237 (1932).

[220] Fowler, R. H., and L. Nordheim: Electron Emission in Intense Electric Fields. Proc. Roy. Soc. (London) A **119**, 173 (1928).

[221] — The Analysis of Photoelectric Sensitivity Curves for Clean Metals at Various Temperatures. Phys. Rev. **38**, 45 (1931).

[222] Francis, G.: Cathode Sputtering Handbuch d. Physik, pp. 22, 154, and 161. Berlin: Springer 1956.

[223] Francis, V. J., and H. G. Jenkins: Electrical Discharges in Gases and their Applications. Rep. Progr. Physics **7**, 230 (1940).

[224] Frank, J., and E. Einsporn: Über die Anregungspotentiale des Quecksilberdampfes. Z. Physik. **2**, 18 (1920).

[225] Frank, F. C.: The Influence of Dislocations on Crystal Growth. Discuss. Faraday Soc. **5**, 48 (1949).

[225a] Frauenfelder, H.: Die Untersuchung von Oberflächenprozessen mit Radioaktivität. Helv. Phys. Acta **23**, 347 (1950).

[226] Fredlung, E.: Über die Wärmeleitung in verdünnten Gasen. Ann. Physik **28**, 319 (1937).

[227] Frenkel, J.: Über die elektrische Oberflächenschicht der Metalle. Z. Physik **51**, 232 (1928).

[228] — Electron Emission from a Cathode Bombarded by Positive Ions. J. exper. theor. Physics (S.S.S.R.) **11**, 706 (1941).

[229] Freundlich, H.: Colloid and Capillary Chemistry. London 1926.

[230] Gazulla, O. R. F., and S. S. Perez: Über die Druckabhängigkeit der Wärmeleitung und die Bildung von Doppelmolekeln im Schwefeldioxyd. Z. physik. Chem. A **193**, 162 (1943).

[231] GEHRKE, E.: Über die Wärmeleitung verdünnter Gase. Ann. Physik **2**, 102 (1900).

[232] GEIGER, H.: Durchgang von α-Strahlen durch Materie. Handbuch der Physik (GEIGER-SCHEEL) XXII, 2, 155 (1933).

[233] GEHRTSEN, CHR.: Der Kanalstrahlenstoß. Physik Z. **31**, 948 (1930).

[234] GIBBS, J. W.: Collected Works. Vol. I. New York: Longmans, Green and Co. 1928.

[235] GIBSON, J. B., A. N. GOLAND, M. MILGRAM, and G. H. VINEYARD: Dynamics of Radiation Damage. Phys. Rev. **120**, 1229 (1960).

[236] GILLAM, E.: The Penetration of Positive Ions of Low Energy into Alloys and Composition Changes Produced in them by Sputtering. J. physic. Chem. Solids **11**, Nos. 1/2, 55 (1959).

[237] GINER, J., and E. LANGE: Elektronenaustrittsspannungen von reinem und sauerstoffbedecktem Au, Pt und Pd auf Grund ihrer Voltaspannungen gegen Ag. Naturwissenschaften **40**, 506 (1953).

[238] GHOSH, S. N., and S. P. KHARE: Secondary Electron Emissions from Metal Surface by High-Energy Ion and Neutral Atom Bombardment. Phys. Rev. **125**, 1254 (1962).

[238a] — — Secondary Electron Emission from Metal Surfaces by H^+, H°, He^+, and He° Bombardment. Phys. Rev. **129**, 1638 (1963).

[239] GLOCKLER, G., and M. CALVIN: The Electron Affinity of Iodine from Space-Charge Effects. J. chem. Physics **3**, 771 (1935).

[240] — — Electron Affinity of Bromine Atoms from Space-Charge Effects. J. chem. Physics **4**, 492 (1936).

[241] GOLDMAN, D. T., and A. SIMON: Theory of Sputtering by High Speed Ions. Phys. Rev. **111**, 383 (1958).

[242] — D. E. HARRISON, and R. R. COVEYOU: A Monte Carlo Calculation of High-Energy Sputtering. Oak Ridge Natl. Lab. Rep. ORNL 2729, 1959.

[243] GOMER, R., R. WORTMAN, and R. LUNDY: Mobility and Adsorption of Hydrogen on Tungsten. J. chem. Physics **26**, 1147 (1957).

[244] —, and J. K. HULM: Absorption and Diffusion of Oxygen on Tungsten. J. chem. Physics **27**, 1363 (1957).

[245] — Adsorption and Diffusion of Inert Gases on Tungsten. J. chem. Physics **29**, 441 (1958).

[246] — Scattering Lengths of Inert Gases for Electrons of Negative Energy. Field Emission through Dielectrics. J. chem. Physics **29**, 443 (1958).

[247] — Field Desorption. J. chem. Physics **31**, 341 (1959).

[248] — Adsorption and Diffusion of Argon on Tungsten. J. physic. Chem. **63**, 468 (1959).

[249] — Field Emission Through Dielectric Layers. Austral. J. Physics **13**, 391 (1960).

[250] GOOD, R. H., and E. W. MÜLLER: Field emission at Higher Temperatures. Handbuch der Physik **21**, 190 (1956).

[250a] GORIS, P.: Improved Sample Bonding and Emission with Tantalum Surface Ionization Filaments. AEC Research and Development Report Chemistry, TID 4500, ED 17, IDO 14590, June 29, 1962, Phillips Petroleum Company, Atomic Energy Division, Idaho Operations Office.

[251] GORSHKOV, V. K.: Integral Mass-Spectrographic Method of Determination of Element Concentrations. Pribory i. Tekh. Ekspt. **2**, 53 (1957).

[252] GOUTTE, R., and C. GUILLAUD: Emission Electronique Secondaire Provoquee par Impact D'Ions Sodium sur des Echantillons Metalliques Contraints. J. Physique Radius **18**, 202 (1957).

[253] GRANOWSKI, W. L.: Der elektrische Strom im Gas. Berlin: Akademie-Verlag 1955.

[254] GRAU, G. G.: Die Akkommodation und Adsorption von Gasen an festen Oberflächen. Z. Elektrochem. angew. physik. Chem. 53 (1949).

[255] — Die Akkommodation und Adsorption von Gasen an festen Oberflächen. Dissertation, Heidelberg 1949.

[256] GREENE, D.: Secondary Electron Emission from Molybdenum Produced by Helium, Neon, Argon, and Hydrogen. Proc. physic. Soc. (London) B **63**, 876 (1950).

[257] GRILLY, E. R., W. J. TAYLOR, and H. L. JOHNSTON: Accommodation Coefficients for Heat Conduction Between Gas and Bright Platinum for Nine Gases Between 80° K (or their Boiling Points) and 380° K. J. chem. Physics **14**, 435 (1946).

[258] GRØNLUND, F., and W. J. MOORE: Sputtering of Silver by Light Ions with Energies from 2 to 12 keV. J. chem. Physics **32**, 1540 (1960).

[259] GROVE, W. R.: On the Electro-Chemical Polarity of Gases. Phil. Trans. Roy. Soc. **142**, 87 (1852).

[260] GÜNTHERSCHULZE, A., and K. MEYER: Kathodenzerstäubung bei sehr geringen Gasdrucken. Z. Physik **62**, 607 (1930).

[261] —, and W. BÄR: Die Elektronenablösung durch den Aufprall der positiven Ionen auf eine MgO-Kathode einer anomalen Glimmentladung. Z. Physik **107**, 730 (1937).

[262] — — Die Elektronenablösung durch den Aufprall der positiven Ionen auf die Kathode einer Glimmentladung. Z. Physik **108**, 780 (1938).

[263] — — Die Elektronenablösung durch den Aufprall der positiven Ionen auf die Kathode einer Glimmentladung. Z. Physik **109**, 121 (1938).

[264] — Neue Untersuchungen über die Kathodenzerstäubung der Glimmentladung. Z. Physik **118**, 145 (1941).

[265] —, and W. TOLLMIEN: Neue Untersuchungen über die Kathodenzerstäubung der Glimmentladung. Z. Physik **119**, 685 (1942).

[266] — Cathodic Sputtering. Vacuum **3**, 360 (1953).

[267] GUNDRY, P. M., and F. C. TOMPKINS: Application of the Charge-Transfer No-Bond Theory to Adsorption Problems. Trans. Faraday Soc. **56**, 846 (1960).

[268] GUREVICH, I. M., and B. M. YAVORSKY: Extraction of Electrons from Metal by Metastable Atoms of Mercury. C. R. Acad. Sci. U.S.S.R. **53**, 789 (1946).

[269] — On the Formation of Metastable Atoms in Pure Optically Excited Mercury Vapors. Doklady Akad. Nauk (S.S.S.R.) Bull. **57**, 665 (1947).

[270] GURNEY, R. W.: The Scattering of Positive Ions from a Platinum Surface. Phys. Rev. **32**, 467 (1928).

[271] — Theory of Electrical Double Layers in Adsorbed Films. Phys. Rev. **47**, 479 (1935).

[272] GURTOVOI, M. E.: Nature of the Kinetic Ejection of Secondary Electrons by Positive Ions. J. exp. theor. Physics. (S.S.S.R.) **10**, 483 (1940).

[273] — Secondary Electron Emission from Thorium-Tungsten in an Ion-Atom Impact in a Mercury Vapor Discharge. J. exp. theor. Physics (S.S.S.R.) **11**, 489 (1941).

[274] GUSEVA, M. I.: The Sputtering Effect of Positive Ions with Energies up to 25 keV in a Small Electromagnetic Separator. Soviet Phys. Solid State **1**, 1410 (1959).

[275] GUSEV, V. M., M. I. GUSEVA, V. P. VLASENKO, and N. P. ELISTRATOV: Investigation of the Interaction of Fast Deuterium Ions with Metals. Bull. Acad. Sci. S.S.S.R. Physical Ser. **24**, 696 (1960).

[276] GUTHRIE, A. N.: Surface Ionization of Barium on Tungsten. Phys. Rev. **49**, 868 (1936).

[277] HAGSTRUM, H. D.: Electron Ejection from Mo by He^+, He^{++}, and He_2^+. Phys. Rev. **89**, 244 (1953).

[278] — Effect of Adsorption of Common Gases on Electron Ejection by Noble Gas Ions. Phys. Rev. **89**, 338 (1953).

[279] — Instrumentation and Experimental Procedure for Studies of Electron Ejection by Ions and Ionization by Electron Impact. Rev. Sci. Instruments **24**, 1122 (1953).

[280] — Reflection of Ions as Ions or as Metastable Atoms at a Metal Surface. Phys. Rev. **93**, 652 (1954).

[281] — Auger Ejection of Electrons from Tungsten by Noble Gas Ions. Phys. Rev. **96**, 325 (1954).

[282] — Theory of Auger Ejection of Electrons from Metals by Ions. Phys. Rev. **96**, 336 (1954).

[283] — Ejection of Electrons from Contaminated Metals by Positive Ions. Phys. Rev. **98**, 561 (1955).

[284] — Metastable Ions of the Noble Gases. Phys. Rev. **104**, 309 (1956).

[285] — Auger Ejection of Electrons from Tungsten by Noble Gas Ions. Phys. Rev. **104**, 317 (1956).

[286] — Auger Ejection of Elctrons from Molybdenum by Noble Gas Ions. Phys. Rev. **104**, 672 (1956).

[287] — Effect of Monolayer Adsorption on the Ejection of Electrons from Metals by Ions. Phys. Rev. **104**, 1516 (1956).

[288] — Detection of Metastable Atoms and Ions. J. appl. Physics **31**, 897 (1960).

[289] — Reflection of Noble Gas Ions at Solid Surfaces. Phys. Rev. **123**, 758 (1961).

[289a] —, and Y. TAKEISHI: Tertiaries in the Distribution of Electrons Ejected from Solids by Ions. Bull. Amer. physic. Soc. **8**, 432 (1963).

[290] HANAU, R.: Cathode Sputtering in the Abnormal Glow Discharge. Phys. Rev. **76**, 153 (1949).

[291] HARKNESS, A. L.: A Crucible Source. ASTM-14-Meeting, New Orleans, June 2, 1958.

[292] HARRISON, D. E.: Extended Theory of Sputtering. J. chem. Physics **32**, 1336 (1960).

[293] HARRISON, D. E., Jr., and D. G. MAGNUSON: Sputtering Thresholds. Phys. Rev. **122**, 1421 (1961).

[293a] HARROWER, G. A.: Energy Spectra of Secondary Electrons from Mo and W for Low Primary Energies. Phys. Rev. **104**, 52 (1956).

[294] HASTED, J. B.: Inelastic Collisions between Atomic Systems. Advances in Electronics and Electron Physics **13**, 1 (1960).

[295] HAYDEN, R. J.: Mass Spectrographic Mass Assignment of Radioactive Isotopes. Phys. Rev. **74**, 650 (1948).

[296] HAYMANN, P.: Action des ions argon de faibles energies sur des surfaces d'uranium. C. R. Acad. Sci. Paris **248**, 2472 (1959).

[297] —, and J. J. TRILLAT: Effects of Low Energy Argon Ions on Uranium Surfaces. C. R. Acad. Sci. Paris **251**, 85 (1960).

[298] — The Action of Argon-Ion Beams on Metal Surfaces. J. Chem. physique **57**, 572 (1960).

[299] HAYWARD, D. O., and R. GOMER: Adsorption of Carbon Dioxide on Tungsten. J. chem. Physics **30**, 1617 (1959).

[300] HEALEA, M., and E. L. CHAFFEE: Secondary Electron Emission from a Hot Nickel Target Due to Bombardment by Hydrogen Ions. Phys. Rev. **49**, 925 (1936).

[301] —, and C. HOUTERMANS: The Relative Secondary Electron Emission Due to He, Ne, and A Ions Bombarding a Hot Nickel Target. Phys. Rev. **58**, 608 (1940).

[302] HEAVENS, O. S.: The Contamination in Evaporated Films by the Material of the Source. Proc. physic. Soc. (London) B **65**, 788 (1952).

[303] — Optical Properties of Thin Solid Films. London: Butterworth Scientific Publications 1955.

[304] HENDRICKS, J. O., T. E. PHIPPS, and M. J. COPLEY: Evidence of Halogen Films on Tungsten in the Surface Ionization of Potassium Halides. J. chem. Physics **5**, 868 (1937).

[305] HENNEBERG, W., and A. RECKNAGEL: Zusammenhänge zwischen Elektronenlinse, Elektronenspiegel und Steuerung. Z. techn. Pysik **16**, 621 (1935).

[306] HERCUS, E. O., and D. M. SUTHERLAND: The Thermal Conductivity of Air by a Parallel Plate Method. Proc. Roy. Soc. (London) A **145**, 610 (1934).

[307] HERRING, C., and M. NICHOLS: Thermionic Emission. Rev. mod. Physics **21**, 185 (1949).

[308] HIGATSBERGER, M. J., H. L. DEMOREST, and A. O. NIER: Secondary Emission from Nichrome V, CuBe, and AgMg Alloy Targets Due to Positive Ion Bombardment. J. appl. Physics **25**, 883 (1954).

[309] HIGUSCHI, I., T. REE, and H. EYRING: Adsorption Kinetics. I. The System of Alkali Atoms on Tungsten. J. Amer. chem. Soc. **77**, 4969 (1955).

[310] — — — Adsorption Kinetics. II. Nature of the Adsorption Bond. J. Amer. chem. Soc. **79**, 1330 (1957).

[311] HILL, A. G., W. W. BUECHNER, J. S. CLARK, and J. B. FISK: The Emission of Secondary Electrons under High Energy Positive Ion Bombardment. Phys. Rev. **55**, 463 (1939).

[312] HILL, T. L.: Thermodynamic Transition from Adsorption to Solution. J. chem. Physics **17**, 507 (1949).

[313] HIN LEW: The Hyperfine Structure of the $^2P_3/_2$ State of Al^{27}, The Nuclear Quadru-Moment. Phys. Rev. **76**, 1086 (1949) (also listed under ref. [435]).

[314] HINTENBERGER, H., and C. LANG: Thermal Ion Source with Very Small Material Use. Z. Naturforsch. **11**a, 167 (1956).

[315] HIPPEL, A. V.: Kathodenzerstäubungsprobleme. Ann. Physik **80**, 672 (1926).

[316] — Kathodenzerstäubungsprobleme. Zur Theorie der Kathodenzerstäubung. Ann. Physik **81**, 1043 (1926).

[317] — Kathodenzerstäubungsprobleme. Der Einfluß von Material und Zustand der Kathode auf den Zerstäubungsprozeß. Ann. Physik **86**, 1006 (1928).

[318] HOLLAND, L.: Vacuum Deposition of Thin Films. New York: John Wiley & Sons Inc. 1956.

[319] HOLMES, D. K., and G. LEIBFRIED: Range of Radiation Induced Primary Knock-ons in the Hard Core Approximation. J. appl. Phys. **31**, 1046 (1960).

[320] HOLST, G., and E. OOSTERHUIS: Le potentiel explosif d'un gaz. Comtes Rendus **175**, 577 (1922).

[321] HONIG, J. M.: Adsorbent-absorbate interactions and surface heterogeneity in physical adsorption. Ann. N. Y. Acad. Sci. **58**, 741 (1954).

[322] HONIG, R. E.: The Application of Mass Spectrometry to the Study of Surfaces by Sputtering. Adv. in Mass Spectrometry, p. 162. London: Pergamon Press 1958.

[323] HONIG, R. E.: Sputtering of Surfaces by Positive Ion Beams of Low Energy. J. appl. Phys. **29**, 549 (1958).

[324] HOTTENROTH, G.: Untersuchungen über Elektronenspiegel. Ann. Physik **30**, 689 (1937).

[325] HÜCKEL, E.: Adsorption und Capillarkondensation. Leipzig: Akad. Verlagsgesellschaft, 1928.

[326] HUGHES, F. L., H. LEVINSTEIN, and R. KAPLAN: Surface Properties of Etched Tungsten Single Crystals. Phys. Rev. **113**, 1023 (1959).

[327] — — Mean Adsorption Lifetime of Rubidium on Etched Tungsten Single Crystals: Ions. Phys. Rev. **113**, 1029 (1959).

[328] — Mean Adsorption Lifetime of Rubidium on Etched Tungsten Single Crystals: Neutrals. Phys. Rev. **113**, 1036 (1959).

[329] HULL, A. W., and W. F. WINTER: The Volt-Ampere Characteristics of Electron Tubes with Thoriated Tungsten Filaments Containing Low Pressure Inert Gas. Phys. Rev. **21**, 211 (1923).

[330] HUNSMANN, W.: Eine Differentialmethode zur Messung kleiner adsorbierter Gasmengen. Z. Elektrochem. angew. physik. Chem. **44**, 606 (1938).

[330a] HUNTINGTON, H. G.: Creation of Displacements in Radiation Damage. Phys. Rev. **93**, 1414 (1954).

[331] HURLBUT, F. C.: Studies of Molecular Scattering at the Solid Surface. J. appl. Physics **28**, 844 (1957).

[332] — Molecular Scattering at the Solid Surface. Recent Research in Molecular Beams, p. 145. Academic Press 1959.

[333] —, and D. E. BECK: New Studies of Molecular Scattering at the Solid Surface. University of California, Berkely, Report No. KE-150-166, Aug. 11, 1959.

[333a] HUSMANN, O. K.: Experimental Evaluation of Porous Materials for Surface Ionization of Cesium and Potassium. Progress in Astronautics and Aeronautics (Development in Electrostatic Propulsion) **9**, 195 (1963).

[333b] — A Comparison of the Contact Ionization of Cesium on Tungsten with that of Molybdenum, Tantalum, and Rhenium Surfaces. Hughes Research Laboratories, Malibu, California, AIAA preprint No. 63019, AIAA Electric Propulsion Conference Colorado Springs, Colorado, March 11—13, 1963.

[333c] — Diffusion of Cesium and Ionization on Porous Tungsten. Progress in Astronautics and Rocketry (Electrostatic Propulsion) **5**, 505 (1961); Academic Press, New York.

[334] HUTSON, A. R.: Velocity Analysis of Thermionic Emission from Single Crystal Tungsten. Phys. Rev. **98**, 889 (1955).

[335] HUXFORD, W. S.: Townsend Ionization Coefficients in Cs-Ag-Photo-Tubes Filled with Argon. Phys. Rev. **55**, 754 (1939).

[336] INGHRAM, M. G., and W. A. CHUPKA: Surface-Ionization Source Using Multiple Filaments. Rev. sci. Instruments **24**, 518 (1953).

[337] —, and R. J. HAYDEN: A Handbook of Mass Spectroscopy. National Academy of Sciences. National Research Council, Nuclear Science Series, Report No. 14, Washington, 1954.

[338] —, and R. GOMER: Massenspektrometrische Untersuchungen der Feldemission positiver Ionen. Z. Naturforsch. **10**a, 863 (1955).

[339] IONOV, N. I.: Formation of Negative Ions in the Process of Surface Ionization of Alkali Halides on Heated Tungsten. Doklady Akad. Nauk S.S.S.R. **28**, 512 (1940).

[340] — Temperature Dependence of the Formation of Negative Iodine Ions on the Surface of Incandescent Tungsten. J. exp. theor. Physics (S.S.S.R.) **17**, 272 (1947).

[*341*] IONOV, N. I.: Mass Spectrometric Investigation of the Formation of Negative Halogen Ions in the Interaction Between Halides and the Surface of Incandescent Tungsten. J. exp. theor. Physics (S.S.S.R.) **18**, 174 (1948).
[*342*] — Surface Ionization of Potassium Chloride and Cesium Chloride Molecules in the Electric Field. Soviet Phys.-Techn. Phys. **26**, 2134 (1956).
[*343*] —, and M. A. MITTSEV: Determination of the First Ionization Potentials of Neodymium and Praseodymium by the Surface Ionization Method. Soviet Phys.-JETP **11**, 972 (1960).
[*343a*] —, and V. I. KARATAEV: Distribution of Initial Velocities of Electrons, K^+-Ions, Cl-Ions Formed by Surface Ionization of KCl on Tungsten and Tantalum. Soviet Phys.-Techn. Phys. **7**, 454 (1962).
[*344*] IVANOV, R. N., and G. M. KUKAWAZE: Two-filament Ionic Source with Surface Ionization for the Mass-Spectrometer. Pribory i Tekh. Exsperimenta **1**, 106 (1957).
[*345*] IVES, H. E.: Positive Rays Produced in Thermionic Vacuum Tubes Containing Alkali-Metal Vapors. J. Franklin Inst. **201**, 47 (1926).
[*346*] IZMAILOV, S. V.: Thermal Theory of Electron Emission under the Impact of Fast Ions. J. exper. theor. Physics (S.S.S.R.) **9**, 1473 (1939).
[*347*] — The Theory of Ion-Electron Emission from Metals. Soviet Phys. Techn. Phys. **3**, 2031 (1958).
[*348*] — Theory of Ion-Electron Emission from Metals. Soviet Phys.-Solid State **1**, 1415 (1959).
[*349*] — Theory of Secondary Electron Emission from Metals under the Action of Fast Neutral Atoms. Soviet Phys.-Solid State **1**, 1425 (1959).
[*350*] — Theory of Secondary Electron Emission from Metals Due to Fast Ions or Neutral Atoms. Soviet Phys.-Solid State **3**, 1046 (1962).

[*351*] JACKSON, J. M., and N. F. MOTT: A Quantum Mechanical Theory of Energy Exchanges Between Inert Gas Atoms and a Solid Surface. Proc. Cambridge Philos. Soc. **28**, 136 (1932).
[*352*] — — Energy Exchange Between Inert Gas Atoms and a Solid Surface. Proc. Roy. Soc. (London) A **137**, 703 (1932).
[*353*] —, and A. HOWARTH: Exchange of Energy Between Inert Gas Atoms and a Solid Surface. Proc. Roy. Soc. (London) A **142**, 447 (1933).
[*354*] JACQUET, E.: Theorie der Adsorption von Gasen. Fortschr. chem. Physik u. physik. Chemie B **18**, (Heft 17), pp. 1—56 (1925).
[*355*] JAGUDAJEW, M. D., and G. N. SCHUPPE: The Negative Ionization of Iodine on Thoriated Wolfram. Akad. Nauk Uzbek S.S.R. (Bulletin) **3**, 3 (1946).
[*356*] JAHNKE, E., and F. EMDE: Funktionstafeln. 4th Edition, p. 237.
[*357*] JOHNSON, R. P.: Drift of Adsorption TH on W Filaments Heated with D. C. Phys. Rev. **53**, 766 (1938).
[*358*] — Construction of Filament Surfaces. Phys. Rev. **54**, 459 (1938).
[*359*] JONES, H.: Theorie of Electrical and Thermal Conductivity in Metals. Handbuch d. Physik **19**, 227 (1956).

[*360*] KAMINSKY, M.: Investigation of the Time Variation of Alkali Ion Emission from Solid Specimens on Glowing Metal Surfaces. Advances in Mass Spectrometry. Pergamon Press 1959, p. 125.
[*361*] — Interaction Between Ion Beams and Metal-Single Crystal Surfaces in the Rutherford Collision Region. Bull. Amer. Phys. Soc. **8**, 338 (1963).
[*362*] — Atomic Collision Sequences Revealed in the Sputtering of Metal Monocrystals by Ion Bombardment in the Rutherford Collision Region. Bull. Amer. Phys. Soc. **8**, 428 (1963).

[363] Kaminsky, M.: Sputtering Experiments in the Rutherford Collision Region. Phys. Rev. **126**, 1267 (1962).

[363a] — Atomic and Ionic Impact Phenomena on Metal Surfaces. Argonne National Laboratory Physics Division Summary Report, ANL-6720, page 85, June 1963.

[363b] — Atomic Collision Sequences in the Lattices of Metal Monocrystals Bombarded with Ions in the Rutherford Collision Region. Argonne National Laboratory, Physics Division Summary Report, ANL-6719, page 25, March-May 1963.

[364] — Ionenbildung an heißen Metalloberflächen. Physik. Verh. Verbands-Ausgabe **1**—**2**, 60 (1962).

[365] — A Pulsed Molecular Beam Mass Spectrometer for Studies of Atomic and Ionic Impact Phenomena on Metal Surfaces. Sympos. Thermionic Power Conversion, Colorado Springs, 1962. Journ. Adv. Energy Conversion. New York: Pergamon Press **3**, 255 (1963).

[366] Kapitza, P. L.: On the Theory of δ Radiation. Philos. Mag. **45**, 989 (1923).

[367] Kaplan, L. E.: [unpublished, referred to by H. D. Hagstrum, Phys. Rev. **96**, 330 (1954)].

[368] Keesom, W. H.: Die van der Waalsschen Kohäsionskräfte. Physik. Z. **22**, 129 (1921).

[369] — Die van der Waalsschen Kohäsionskräfte. Berichtigung. Physik. Z. **22**, 643 (1921).

[370] — Die Berechnung der molekularen Quadrupolmomente aus der Zustandsgleichung. Physik. Z. **23**, 225 (1922).

[371] Kemball, C.: Catalysis on evaporated metal films. I. The Efficiency of Different Metals for the Reaction between Ammonia and Deuterium. Proc. Roy. Soc. (London) A **214**, 413 (1952).

[372] Kenty, C.: Photoelectric and Metastable Atom Emission of Electrons from Surfaces. Phys. Rev. **38**, 377 (1931).

[373] — Photoelectric Efficiencies in the Extreme Ultraviolet. Phys. Rev. **38**, 2079 (1931).

[374] — Line Broadening and the Imprisonment of Resonance Radiation. Phys. Rev. **40**, 633 (1932).

[375] Keywell, F.: Measurement and Collision-Radiation Damage Theory of High Vacuum Sputtering. Phys. Rev. **97**, 1611 (1955).

[376] Keiss, C. C., C. J. Humphries, and D. D. Laun: Preliminary Description and Analysis of the First Spectrum of Uranium. J. Res. Nat. Bur. Standards **37**, 57 (1946).

[377] Kilian, T. S.: Thermionic Phenomena Caused by Vapors of Rubidium and Potassium. Phys. Rev. **27**, 578 (1926).

[378] Kim Heng Pong, and I. L. Sokolskaya: Bull. Leningrad State Univ. No. **12**, 65 (1952).

[379] Kinchin, G. H., and R. S. Pease: The Displacement of Atoms in Solids by Radiation. Rep. Progr. Physics **18**, 1 (1955).

[380] Kingdon, K. H.: A Method for the Neutralization of Electron Space Charge by Positive Ionization at Very Low Pressures. Phys. Rev. **21**, 408 (1923).

[381] —, and I. Langmuir: The Removal of Thorium from the Surface of a Thoriated Tungsten Filament by Positive Ion Bombardment. Phys. Rev. **22**, 148 (1923).

[382] — A Method for Studying the Ionization of the Less Volatile Metals. Phys. Rev. **23**, 778 (1924).

[383] — Thermionic Effects Caused by Vapors of Alkali Metals. Proc. Roy. Soc. (London) Ser. A **107**, 61 (1925).

[*384*] KINGTON, K. L., and J. G. ASTON: The Heat of Adsorption of Nitrogen on Titanium Dioxide (Rutile) at 77.3° K. J. Amer. chem. Soc. **73**, 1929 (1951).

[*385*] KIRKWOOD, J. G.: Polarisierbarkeiten, Suszeptibilitäten und van der Waalssche Kräfte der Atome mit mehreren Elektronen. Physik. Z. **33**, 57 (1932).

[*385a*] KISTEMAKER, J., and C. SNOEK: Surface Phenomena Related with Sputtering. Colloques Internatl. du Centre National de la Recherche Scientifique No. 113, Paris, Dec. 1962, p. 51.

[*386*] KLEIN, R.: Investigation of the Surface Reaction of Oxygen with Carbon on Tungsten with the Field Emission Microscope. J. chem. Physics **21**, 1177 (1953).

[*387*] KNACKE, O., and I. N. STRANSKI: The Mechanism of Evaporation. Progress in Metal Physics **6**, 181 (1956).

[*388*] KNAUER, F., and O. STERN: Über die Reflexion von Molekularstrahlen. Z. Physik **53**, 779 (1929).

[*389*] — Die Verweilzeit adsorbierter Alkalien an erhitztem Wolfram. Z. Physik **25**, 278 (1948).

[*390*] KNOLL, M., O. HACHENBERG, and J. RANDMER: Zum Mechanismus der Sekundäremission im Inneren von Ionenkristallen. Z. Physik **122**, 137 (1944).

[*391*] KNUDSEN, M.: Die Gesetze der Molekularströmung und der inneren Reibungsströmung der Gase durch Röhren. Ann. Physik **28**, 75 (1909).

[*392*] — Die Molekularströmung der Gase durch Öffnungen und die Effusion. Ann. Physik **28**, 999 (1909).

[*393*] — Die molekulare Wärmeleitung der Gase und der Akkommodationskoeffizient. Ann. Physik **34**, 593 (1911).

[*394*] — Der molekulare Gaswiderstand gegen eine sich bewegende Platte. Ann. Physik **46**, 641 (1915).

[*395*] — Radiometerdruck und Accommodationskoeffizient. Ann. Physik **6**, 129 (1930).

[*396*] — Radiometer Pressure and Coefficient of Akkomodation. Kgl. danske Vidensk. Selsk., mat.-fysiske Medd. Copenhagen **11**, 1 (1930).

[*397*] — Kinetic Theory of Gases. 3rd Edition. London: 1950.

[*398*] KOCH, J.: Die Herstellung und die nähere Untersuchung einer neuen Alkaliionenquelle. Z. Physik **100**, 669 (1936).

[*399*] KOEDAM, M.: Sputtering of a Polycristalline Silver Surface Bombarded with Monoenergetic Argon Ions of Low Energy (40—240 eV). Physica **24**, 692 (1958).

[*400*] — Sputtering of copper single crystals bombarded with monoenergetic ions of low energy (50—350 eV). Physica **25**, 742 (1959).

[*401*] —, and A. HOOGENDOORN: Sputtering of Copper Single Crystals Bombarded with A^+, Kr^+ and Ne^+ Ions with Energies Ranging from 300—2000 eV. Physica **26**, 351 (1960).

[*402*] — Cathode Sputtering by Rare Gas Ions of Low Energy (Bombardment of Polycrystalline and Monocrystalline Material). Thesis. State Univ., Utrecht, March 1961.

[*403*] KOLM, H. H.: Rotating Electrometer for Comparative Work Function Measurements. Rev. sci. Instruments **27**, 1046 (1956).

[*404*] KONOZENKO, I. D.: Effect of the Electric Field on the Surface Ionization of Sodium Atoms on the Surface of Thoriated Tungsten. J. exper. theor. Physics (S.S.S.R.) **9**, 540 (1939).

[*405*] KOPFERMANN, H.: Kernmomente. 2. Aufl. Frankfurt, 1956.

[405a] Kopitzki, K., and H.-E. Stier: Mittlere Geschwindigkeit der bei der Kathodenzerstäubung von Metallen ausgesandten Partikel. Z. Naturforsch. **16** a, 1257 (1961).
[405b] — — Mittlere kinetische Energie der bei der Kathodenzerstäubung von Metallen ausgesandten Partikel. Z. Naturforsch. **17** a, 346 (1962).
[405c] — — Geschwindigkeitsmessungen an den bei der Kathodenzerstäubung von Kupfer-Einkristallen ausgesandten Partikeln zur Untersuchung der Fokussierungsenergien des Kristalls. Physics Letters **4**, 232 (1963).
[406] Kossel, W.: Über Kristallwachstum. Naturwissenschaften **18**, 901 (1930).
[407] — Existenzbereiche von Aufbau- und Abbauvorgängen auf der Kristallkugel. Ann. Physik **33**, 651 (1938).
[407a] Krohn, V. E.: Emission of Negative Ions from Metal Surfaces Bombarded by Positive Cesium Ions. J. appl. Physics **33**, 3523 (1962).
[408] Kruithoff, A. A., and F. M. Penning: Determination of the Townsend Ionization Coefficient α for Pure Argon. Physica **3**, 515 (1936).
[409] Kunsman, C. H.: The Thermionic Emission from Substances Containing Iron and Alkali Metal. Phys. Rev. **25**, 892 (1925).
[410] — The Positive Ion Emission from Mixture Containing Fe, Al and Cs, and the Work Funktion $\varnothing +$ for Cs from this Mixture. Phys. Rev. **27**, 249 (1926).
[411] — The Thermionic Emission from Iron-Alkali Mixtures Used as Catalysts in the Synthesis of Ammonia. J. Franklin Inst. **203**, 635 (1927).
[412] Kwan, T.: Some General Aspects of Chemisorption on Catalysis. Advances in Catalysis **6**, 67 (1954).

[413] Ladage, A.: Elektroneninterferenzen an elektrolytisch polierten Oberflächen nach Kathodenzerstäubung. Z. Physik **144**, 354 (1956).
[414] Laegreid, N., and G. K. Wehner: Sputtering Yields of Metals for Ar^+ and Ne^+ Ions with Energies from 50—600 eV. J. appl. Physics **32**, No. 3, 365 (1961).
[415] Lamar, E. S., and K. T. Compton: A Special Theory of Cathode Sputtering. Science **80**, 541 (1934).
[416] Lamb, W. E., and R. C. Retherford: Fine Structure of the Hydrogen Atom, Part I. Phys. Rev. **79**, 549 (1950).
[417] Landau, L.: Über die relativistische Korrektion der Schrödinger-Gleichung für das Mehrkörperproblem. Physik Z. Sowjetunion **8**, 487 (1935).
[418] Landsberg, P. T.: A Contribution to the Theory of Soft X-Ray Emission Bands of Sodium. Proc. physic. Soc. A **62**, 806 (1949).
[419] Langsberg, E.: Analysis of Low-Energy Sputtering. Phys. Rev. **111**, 91 (1958).
[420] Langmuir, I., and K. H. Kingdon: Contact Potential Measurements with Adsorbed Films. Phys. Rev. **34**, 129 (1929).
[421] —, and C. G. Found: A New Type of Electric Discharge Characterized by Electron Emission from the Walls.
[422] — Cesium Films on Tungsten. J. Amer. chem. Soc. **54**, 1252 (1932).
[423] — Extension of the Phase Rule for Adsorption Under Equilibrium and Non-Equilibrium Conditions. J. chem. Physics **1**, 3 (1933).
[424] Langmuir, D. B.: Contact Potential Measurements on Tungsten Filaments. Phys. Rev. **49**, 428 (1936).
[425] Lazareff, P.: Über den Temperatursprung an der Grenze zwischen Metall und Gas. Ann. Physik **37**, 233 (1912).
[426] Lazarus, D.: Diffusion in Metals. Solid State Physics **10**, 71 (1960).
[426a] Lebedev, S. Ya., Ya. Ya. Stavisskii, and Yu. U. Shut'ko: Surface Ionization of Cesium in Diffusion of Cesium Vapor through Porous Molybdenum. Soviet Phys.-Technical Phys. **6**, 836 (1962).

[426b] Lehmann, Chr., and G. Leibfried: Fokussierende 110-Stoßfolgen in flächenzentrierten Kristallen bei kleinen Winkeln. Z. Physik **162**, 203 (1961).

[427] Leibfried, G.: Correlated Collisions in a Displacement Spike. J. appl. Phys. **30**, 1388 (1959).

[428] — Defects in Dislocations Produced by Focusing Collisions in f.c.c. Lattices. J. appl. Physics **31**, 117 (1960).

[429] Leland, W. T., and R. A. Olson: Production of Secondary Ions from Metallic Surfaces Bombarded by High-Energy Positive Ions. Bull. Amer. physic. Soc. **1**, 395 (1956).

[430] Lenel, F. V.: Über die Adsorptionswärme von Edelgasen und Kohlendioxyd an Ionenkristallen. Z. physik. Chem. B **23**, 379 (1933).

[431] Lennard-Jones, J. E.: Processes of Adsorption and Diffusion on Solid Surfaces. Trans. Faraday Soc. **28**, 333 (1932).

[432] —, and C. Strachan: The Interaction of Atoms and Molecules with Solid Surfaces I. The Activation of Absorbed Atoms for Higher Vibrational States. Proc. Roy. Soc. (London) A **150**, 442 (1935).

[433] Leroy, J., and K. Prelec: Etude de l'Emission Secondaire d'Electrons Au Cours Du Bombardement de Cibles Metalliques Par Des Ions Positifs D^+ et D_2^+. Centre D'Etudes Nucleaires de Saclay, Rapport CEA No. 1445 (1960).

[434] Levine, L. P.: Negative Hydrogen Ion Production by the Bombardment of a Tungsten Surface by Positive Ions in the 1 keV Range. Thesis, Syracuse University 1959.

[435] Lew, H.: The Hyperfine Structure of the $^2P_{3/2}$ State of Al^{27}, The Nuclear Electric Quadrupole Moment. Phys. Rev. **76**, 1086 (1949).

[436] — The Hyperfine Structure and Nuclear Moments of Pr^{141}. Phys. Rev. **91**, 619 (1953).

[437] Lewis, L. G., W. M. Garrison, D. King, and R. J. Hayden: Positive Ionemission from Fission Element Oxides. CP-2122, Aug. 5, 1944.

[438] Lewis, T. J.: Theoretical Interpretation of Field Emission Experiments. Phys. Rev. **101**, 1694 (1956).

[439] Linford, L. H.: The Emission by Swiftly Moving Mercury Ions. Phys. Rev. **47**, 279 (1935).

[440] Little, F. P.: Secondary Effects. Handbuch der Physik XXI, 574 (1956).

[441] London, F.: Über einige Eigenschaften und Anwendungen der Molekularkräfte. Z. Physik B **11**, 222 (1930).

[442] — Zur Theorie und Systematik der Molekularkräfte. Z. Physik **63**, 245 (1930).

[443] Longacre, A.: The Scattering of Lithium Ions by a Polycrystalline Nickel Surface. Phys. Rev. **46**, 407 (1934).

[444] Lorenz, R., and A. Lande: Adsorption und übereinstimmende Zustände. Z. anorg. Chem. **125**, 47 (1922).

[445] Lowzow, W. M., and S. V. Starodubtsev: Normal Secondary Ion-Electron and Electron-Electron Emission of Thin Films of KCl. Akad. Nauk Uzbek S.S.R. (Fizik.-tekhnik. institut), Trudy **3**, 45 (1950).

[445a] — — Investigation of the Dependence of Secondary Ion-Electron and Electron-Electron Emission on the Thickness of the KCl. Akad. Nauk Uzbek. S.S.R. (Fizik.-tekhnik institut), Trudy **3**, 57 (1950).

[446] Lueder, H.: Zerstäubung von Metallen durch Aufprall langsamer Ionen und Messung des Schwellenwertes der Zerstäubung. Z. Physik **97**, 158 (1935).

[447] Lukirsky, P. I., and S. Rijanoff: Abhängigkeit der lichtelektrischen Emission des Kaliums von der Anordnung von atomaren Wasserstoff und Kaliumschichten auf ihrer Oberfläche. Z. Physik **75**, 249 (1932).

[448] LUKIRSKY, P. I.: Über die Austrittsarbeit der Elektronen und die photoelektrischen Eigenschaften der Metalle. Physik Z. Sowjetunion **4**, 212 (1933).

[449] MACFADYEN, K. A., and J. HOLBECHE: An Improved Technique for the Measurement of Contact Potential Differences. J. sci. Instruments **34**, 101 (1957).

[450] MACHU, W.: Moderne Galvanotechnik. Weinheim: Verlag Chemie 1954.

[451] MAGNUS, A.: Theorie der Gasadsorption. Z. physik. Chem. A **142**, 401 (1929).

[451a] MAGNUSON, G. D., and C. E. CARLSTON: Electron Ejection from Metals due to 1- to 10-keV Noble Gas Ion Bombardment. I. Polycrystalline Materials. Phys. Rev. **129**, 2403 (1963).

[451b] — — Electron Ejection from Metals due to 1- to 10-keV Noble Gas Ion Bombardment. II. Single Crystal. Phys. Rev. **129**, 2409 (1963).

[452] — Secondary Electron Ejection. General Dynamics Astronautics, San Diego Report AE 62-0517, May 10, 1962.

[452a] MAHADEVAN, P., J. K. LAYTON, and D. B. MEDVED: Secondary Electron Emission from Clean Surface of Molybdenum Due to Low-Energy Noble Gas Ions. Phys. Rev. **129**, 79 (1963).

[453] MALAMUD, H., and A. D. KRUMBEIN: Measurement of the Effect of Chlorine Treatment on the Work Function of Titanium and Zirconium. J. appl. Phys. **25**, 591 (1954).

[454] MANDELL, W., and J. WEST: On the Temperature Gradient In Gases At Various Pressures. Proc. physic. Soc. (London) **37**, 20 (1925).

[455] MANN, W. B.: The Exchange of Energy Between a Platinum Surface and Gas Molecules. Proc. Roy. Soc. (London) A **146**, 776 (1934).

[456] —, and W. C. NEWELL: The Exchanges of Energy Between a Platinum Surface and Hydrogen and Deuterium Molecules. Proc. Roy. Soc. (London) A **158**, 397 (1937).

[457] MARGENAU, H., and W. G. POLLARD: The Forces Between Neutral Molecules and Metallic Surfaces. Phys. Rev. **60**, 128 (1941).

[458] MARTIN, S. T.: On the Thermionic and Adsorptive Properties of the Surfaces of a Tungsten Single Crystal. Phys. Rev. **56**, 947 (1939).

[459] MASSEY, H. S.: The Theory of the Extraction of Electrons from Metals by Positive Ions and Metastable Atoms. Proc. Cambridge philos. Soc. **26**, 386 (1930).

[460] — The Theory of the Extraction of Electrons From Metals by Metastable Atoms. Proc. Cambridge philos. Soc. **27**, 460 (1931).

[461] —, and E. H. S. BURHOP: Electronic and Ionic Impact Phenomena. Oxford: Clarendon Press 1952.

[462] MATSEN, F. A., A. C. MAKRIDES, and N. HACKERMANN: Charge-Transfer-No-Bond Adsorption. J. chem. Physics **22**, 1800 (1954).

[463] MAXTED, E. B.: The Nature of Chemisorptive Bonds. I. Some Observed Regularities. J. chem. Soc. (London) 1987 (1949).

[463a] MAXTED, E. B., and N. HASSID: Thermal Activation Effect in the Adsorption of Hydrogen on Platinum and Nickel. J. chem. Soc. (London), 1532 (1932).

[464] MAYER, H.: Physik dünner Schichten. Bd. I/II. Stuttgart: Wiss.V erlagsgesellschaft M.B.H. 1955.

[465] MCCALLUM, K. J., and J. E. MAYER: A Direct Experimental Determination of the Electron Affinity of Chlorine. J. chem. Physics **11**, 56 (1943).

[466] MCFEE, J. H., and P. M. MARCUS: Velocity Distributions in Direct and Reflected Atomic Beams. Technical Report 1. Jan. 1960, Carnegie Institute of Technology, Dept. of Physics, Pittsburgh, Pa.

[*467*] McKeown, D.: New Method for Measuring Sputtering in the Region Near Threshold. Rev. sci. Instruments **32**, 133 (1961).

[*467a*] Medved, D. B.: Electron Ejection and Reflection Properties of Alkali and Noble Gas Ions and Atoms. J. appl. Physics **34**, 3142 (1963).

[*467b*] — P. Mahadevan, and J. K. Layton: Potential and Kinetic Electron Ejection from Molybdenum by Argon Ions and Neutral Atoms. Phys. Rev. **129**, 2086 (1963).

[*468*] Messenger, H. A.: The Significance of Certain Critical Potentials of Mercury in Terms of Metastable Atoms and Radiation. Phys. Rev. **28**, 962 (1926).

[*469*] Metlay, M., and G. E. Kimball: Ionization Processes on Tungsten Filaments. I. The Electron Affinity of the Oxygen Atom. J. chem. Physics **16**, 774 (1948).

[*470*] — — Ionization Processes on Tungsten Filaments. II. The Adsorption of Fluorine on Tungsten. J. chem. Physics **16**, 779 (1948).

[*471*] Meyer, E.: Über die Elektronen- und positive Ionenemission von Wolfram-, Molybdän- und Tantalglühfäden in Kaliumdampf. Ann. Physik **4**, 357 (1930).

[*472*] Meyer, K., and A. Güntherschulze: Kathodenzerstäubung in Quecksilberdampf bei sehr geringen Drucken. Z. Physik **71**, 279 (1931).

[*473*] Michaelson, H. B.: Work Functions of the Elements. J. appl. Physics **21**, 536 (1950).

[*474*] Michels, W. C.: Accommodation Coefficients of Helium and Argon against Tungsten. Phys. Rev. **40**, 472 (1932).

[*475*] —, and G. White: Heat Losses from a Tungsten Wire in Helium. Phys. Rev. **47**, 197 (1935).

[*476*] — Accommodation Coefficients of the Noble Gases and the Specific Heat of Tungsten. Phys. Rev. **52**, 1067 (1937).

[*477*] Mignolet, J. C. P.: Studies in Contact Potentials. I. The Adsorption of Some Gases on Evaporated Nickel Films. Discuss. Faraday Soc. **8**, 105 (1950).

[*478*] — Studies in Contact Potentials. II. Vibrating Cells for the Vibrating Condenser Method. Discuss. Faraday Soc. **8**, 326 (1950).

[*479*] — Interaction Energy and the Surface Potential of Some Films. Bull. Soc. chem. Belgique **64**, 126 (1955).

[*480*] — An Empirical Relation Between the Surface Potential of Certain Chemical Films and the Electronegativity of their Constituents. Bull. Soc. chem. Belgique **65**, 837 (1956).

[*481*] Miller, A. R.: The Variation of the Dipole Moment of Adsorbed Particles with the Fraction of the Surface Covered. Proc. Cambridge philos. Soc. **42**, 292 (1946).

[*482*] Minturn, R. E., S. Datz, and E. H. Taylor: Thermal Emission of Alkali Ion Pulses from Clean and Oxygenated Tungsten. J. appl. Physics **31**, 876 (1960).

[*483*] Mitchell, A. C., and M. Zemansky: Resonance Radiation and Excited Atoms. New York: McMillan 1934.

[*484*] Mitchell, J. J., and J. E. Mayer: An Experimental Determination of the Electron Affinity of Chlorine. J. chem. Physics **8**, 282 (1940).

[*485*] Mitropan, I. M., and V. S. Gumeniuk: Emission of Negative Ions from Metallic Surfaces Bombarded with Positive Hydrogen Ions. J. exper. theor. Physics (S.S.S.R.) **32**, 214 (1957) — Soviet Phys.-JETP **5**, 157 (1957).

[*486*] —, and V. S. Gumeniuk: Relation Between Secondary Emission of Negative Ions and the Angle of Entry of Primary Protons into a Metal Target. J.

exper. theor. Physics (S.S.S.R.) **34**, 235 (1958). — Soviet Phys. JETP **7**, 162 (1958).

[*487*] Moesta, H.: Der Einfluß adsorbierter Gase auf physikalische Eigenschaften der Oberfläche fester Körper. Fortschr. chem. Forsch. **3**, 657 (1958).

[*488*] Molchanov, V. A., V. G. Tel'kovskii, and V. M. Chicherov: Anisotropy of Cathodic Sputtering of Single Crystals. Soviet Phys.-Doklady **6**, 22 (1961).

[*488a*] — — — Angular Distribution of Sputtered Particles on Irradiation of a Single Crystal by an Ion Beam. Soviet Phys.-Doklady **6**, 486 (1961).

[*489*] — — Variation of the Cathode Sputtering Coefficient as a Function of the Angle on Incidence of Ions on a Target. Soviet Phys.-Doklady **6**, 137 (1961).

[*490*] Molnar, J. P.: Form of Transient Currents in Townsend Discharges with Metastables. Phys. Rev. **83**, 933 (1951).

[*491*] Moon, P. B.: The Action of Positive Ions of Caesium on a Hot Nickel Surface. Proc. Cambridge philos. Soc. **27**, 570 (1931).

[*492*] —, and M. L. E. Oliphant: The Surface Ionisation of Potassium by Tungsten. Proc. Roy. Soc. (London) A **137**, 463 (1932).

[*493*] Moore, G. E., and H. W. Allison: Adsorption of Strontium and of Barium on Tungsten. J. chem. Physics **23**, 1609 (1955).

[*494*] Moore, W. J., C. D. O'Briain, and A. Lindner: The Interaction of Ionic Beams with Solid Surfaces. Ann. N. Y. Acad. Sci. **67**, 600 (1957).

[*495*] Morgulis, N. D.: Thermische Ionisation von Natrium-Dämpfen an einer glühenden Wolframoberfläche. Physik. Z. Sowjetunion **5**, 221 (1934).

[*496*] — The Ionic Space Charge and its Neutralization by Electrons. J. exper. theor. Physics (S.S.S.R.) **4**, 489 (1934).

[*497*] — The Quantum Theory of Ionization and Neutralization on Metallic Surfaces. J. exper. theor. Physics (S.S.S.R.) **4**, 684 (1934).

[*498*] —, and M. P. Bernhadiner: Neutralization and Ionization of Cesium and Potassium at Thoriated Tungsten. J. exper. theor. Physics (S.S.S.R.) **1**, 998 (1939).

[*499*] — Nature of the Cathode Sputtering and the Kinetic Emission of Secondary Electrons. J. exper. theor. Physics (S.S.S.R.) **9**, 1484 (1939).

[*500*] — Nature of Cathode Sputtering and the Kinetic Emission of Secondary Electrons. J. exper. theor. Physics (S.S.S.R.) **11**, 300 (1941).

[*501*] — The Problem of the Ionization of Atoms and Neutralization of the Ions on the Surface of a Semi-Conducting Cathode. J. exper. theor. Physics (S.S.S.R.) **18**, 568 (1948).

[*502*] —, and V. D. Tishchenko: The Investigation of Cathode Sputtering in the Near Threshold Region. Soviet Phys.-JETP **3**, 52 (1956).

[*503*] —, and V. M. Gavriliuk: Effect of an Absorbed Film of Dipole Molecules on the Electronic Work Function. J. exper. theor. Physics (S.S.S.R.) **30**, 149 (1956). — Soviet Phys. JETP **3**, 159 (1956).

[*504*] —, and V. D. Tishchenko: On the Threshold for Cathode Sputtering of Metals. Bull. Acad. Sci. S.S.S.R. Phys. Ser. **20**, 1082 (1956).

[*504a*] —, and D. A. Gorodetskii: Reflection of Slow Electrons from the Surface of Pure Tungsten and from Tungsten Covered by Thin Films. Soviet Phys.-JETP **3**, 535 (1956).

[*505*] Morosow, G. A.: Surface Ionization of Barium on Tungsten. J. techn. Physics (S.S.S.R.) **17**, 1143 (1947).

[*506*] Morrison, J. L., and J. K. Roberts: A New Method for Studying the Adsorption of Gases at Very Low Pressures and the Properties of Absorbed Films of Oxygen on Tungsten. Proc. Roy. Soc. (London) A **173**, 1 (1939).

[507] MÜLLER, E. W.: Elektronenmikroskopische Beobachtungen von Feldkathoden. Z. Physik **106**, 541 (1937).

[508] — Abreißen adsorbierter Ionen durch hohe elektrische Feldstärken. Naturwissenschaften **29**, 533 (1941).

[509] — Oberflächenwanderung von Wolfram auf dem eigenen Kristallgitter. Z. Physik **126**, 642 (1949).

[510] — Feldemission. Ergebn. exakt. Naturwiss. **27**, 290 (1953).

[511] — Work Function of Tungsten Single Crystal Planes Measured by the Field Emission Microscope. J. appl. Physics **26**, 732 (1955).

[512] — Die Adsorption von Sauerstoff auf Wolfram nach Beobachtungen mit dem Feldelektronenmikroskop. Z. Elektrochem., Ber. Bunsenges. physik. Chem. **59**, 372 (1955).

[513] — Field Desorption. Phys. Rev. **102**, 618 (1956).

[514] — Field Ionization and Field Ion Microscopy. Advances in Electronics and Electron Physics **13**, 83 (1960).

[515] MULLIKEN, R. S.: A New Electroaffinity Scale; Together with Data on Valence States and on Valence Ionization and Electron Affinities. J. chem. Physics **2**, 782 (1934).

[516] — Structures of Complexes Formed by Halogen Molecules with Aromatic and with Oxygenated Solvents. J. Amer. chem. Soc. **72**, 600 (1950).

[517] — Molecular Compounds and their Spectra II. J. Amer. chem. Soc. **74**, 811 (1952).

[518] MURDOCK, J. W., and G. H. MILLER: Secondary Electron Emission Due to Positive Ion Bombardment. Ames Laboratory, Iowa State College, Ames, Iowa, AEC Rep. ISC 652, June 1955.

[519] NASINI, A. G., and G. SAINI: Adsorption und Allgemeines über mono- und mehrmolekulare Schichten. Handbuch der Katalyse Bd. 5, Teil 2, Seite 10 (1957).

[519a] NAZARIN, G. M., and H. SHELTON: Theory of Ion Emission from Porous Media. Progress in Astronautics and Rocketry (Electrostatic Propulsion) **5**, 91 (1961).

[520] NELSON, R. S.: Determination of Preferred Crystal Orientation by a Sputtering. AERE-M 416, Harwell, Berkshire, May 1959.

[521] — The Effect of Temperature on the Propagation of Focused Collision Sequences in Au. AERE-Report M 928, Harwell, 1961.

[521a] —, and M. W. THOMPSON: Atomic Collision Sequences in Crystals of Copper, Silver, and Gold Revealed by Sputtering in Energetic Ion Beams. Proc. Roy. Soc. (London) A **259**, 458 (1961).

[521b] — — The Penetration of Energetic Ions Through the Open Channels in a Crystal Lattice. Report AERE, R-4262, 1963, Harwell, Berkshire, England.

[521c] — — Evidence for Focused Recoil Trajectories from a High-Energy Sputtering Experiment with Cu. Physics Letters **2**, 124 (1962).

[521d] — — Focused Collision Sequences in Aluminium. Philos. Mag. **7**, 1425 (1962).

[521e] — —, and H. MONTGOMERY: The Influence of Thermal Vibration on Focused Collision Sequences. Philos. Mag. **7**, 1385 (1962).

[521f] — The Observation of Collision Sequences in UO_2. Report AERE-R 4283, April 1963, Harwell, Berkshire, England.

[521g] —, M. W. THOMPSON, B. W. FARMERY, and M. J. HALL: The Influence of Thermal Vibration on Focused Collision Sequences. Report AERE-R 4044, March 1962, Harwell, Berkshire, England.

[*521h*] NELSON, R. S.: Focused Collision Sequences in Tungsten and Molybdenum. Philos. Mag. **8**, 693 (1963).

[*522*] NEMENOW, L. M., and A. S. FEDJURKO: Electron Emission Under Action of Positive Ions. J. exper. theor. Physics (S.S.S.R.) **9**, 532 (1939).

[*523*] NICHOLS, M. H.: The Thermionic Constants of Tungsten as a Function of Crystallographic Direction. Phys. Rev. **57**, 297 (1940).

[*524*] NORDHEIM, L. W.: The Effect of the Image Force on the Emission and Reflexion of Electrons by Metals. Proc. Roy. Soc. (London) A **121**, 626 (1928).

[*525*] NOTTINGHAM, W. B.: Photoelectric and Thermionic Emission from Composite Surfaces. Phys. Rev. **41**, 793 (1932).

[*526*] — Thermionic Emission. Handbuch d. Physik **21**, 18 (1956).

[*527*] NOVICK, R., and E. D. COMMINS: Hyperfine Structure of the Metastable State of Singly Ionized Helium-3. Phys. Rev. **111**, 822 (1958).

[*528*] OATLEY, C. W.: The Absorption of Oxygen and Hydrogen on Platinum and the Removal of these Gases by Positive-Ion Bombardment. Proc. physic. Soc. **51**, 318 (1939).

[*529*] O'BRIAIN, C. D., A. LINDNER, and W. J. MOORE: Sputtering of Silver by Hydrogen Ions. J. chem. Physics **29**, 3 (1958).

[*530*] OECHSNER, H.: Diplomarbeit, Physikal. Institut, Universität Würzburg, Juli 1960.

[*530a*] OEN, O. S., D. K. HOLMES, and M. T. ROBINSON: Ranges of Energetic Atoms in Solids. J. appl. Physics **34**, 302 (1963).

[*531*] OGILVIE, G. J., and M. J. RIDGE: The Cathodic Sputtering of Silver. J. physik. Chem. Solids **10**, 217 (1959).

[*532*] — The Surface Structure of Silver Crystals After Argon-Ion Bombardment. J. physik. Chem. Solids **10**, 222, Pergamon Press (1959).

[*533*] — Bombardment of Metals by Inert Gas Ions. Austral. J. Phys. **13**, 402 (1960).

[*534*] OLDEKOP, W., and F. SAUTER: Zur Theorie der Austrittsarbeit aus Metallen. Z. Physik **136**, 534 (1954).

[*535*] OLIPHANT, M. L. E.: The Action of Metastable Atoms of Helium on a Metal Surface. Proc. Roy. Soc. (London) A **124**, 228 (1929).

[*536*] — The Liberation of Electrons from Metal Surfaces by Positive Ions. Part I. Experimental. Proc. Roy. Soc. (London) A **127**, 373 (1930).

[*537*] —, and P. B. MOON: The Liberation of Electrons from Metal Surfaces by Positive Ions. Part II. Theoretical. Proc. Roy. Soc. (London) A **127**, 388 (1930).

[*538*] ORR, W. J. C.: Calculations of the Adsorption Behavior of Argon on Alkali Halide Crystals. Trans. Faraday Soc. **35**, 1247 (1939).

[*539*] PADDEN, T. R., and F. M. CAIN: Cathodic Vacuum Etching. Metal Progress **66**, (1), 108, 164 (1954).

[*540*] PAETOW, H., and W. WALCHER: Über den Einfluß von Adsorptionsschichten auf die Auslösung von Elektronen und die Reflexion von Ionen beim Auftreffen von positiven Caesiumionen auf Wolfram. Z. Physik **110**, 69 (1938).

[*541*] PANETH, H.: The Mechanism of Self-Diffusion in Alkali Metals. Phys. Rev. **80**, 708 (1950).

[*542*] PANIN, B. V.: Secondary Ion Emission from Metals Induced by 10—100 keV Ions. Soviet Phys.-JETP **14**, 1 (1962).

[*543*] PARILIS, E. S., and L. M. KISHINEVSKII: The Theory of Ion-Electron Emission. Soviet Phys.-Solid State **3**, 885 (1960).

[544] PARKER, J. H.: Electron Ejection by Slow Positive Ions Incident on Flashed and Gas-Covered Metallic Surfaces. Phys. Rev. **93**, 1148 (1954).

[545] PASHLEY, D. W.: The Preparation of Smooth Single Crystal Surfaces of Silver by Evaporation Technique. Philos. Mag. **4**, 316 (1959).

[545a] PAUL, W., and M. RAETHER: Das elektrische Massenfilter. Z. Physik **140**, 262 (1955).

[546] PAULING, L.: The Nature of Interatomic Forces in Metals. Phys. Rev. **54**, 899 (1938).

[547] — The Nature of the Chemical Bond. Ithaca: Cornell Univ. Press 1948.

[548] — A Resonating-Valence-Bond Theory of Metals and Intermetallic Compounds. Proc. Roy. Soc. (London) A **196**, 343 (1949).

[549a] PAWLOW, W. I., and S. W. STARODUBZEW: Secondary Emission of Conductors under the Effect of Bombardment by Positive Ions. Part I. Fast Ions. J. exp. theor. Physics (S.S.S.R.) **1**, 409 (1937).

[549b] — — Secondary Emission of Conductors Under the Effect of Bombardment. Part II. Positive Ions. J. exp. theor. Physics (S.S.S.R.) **1**, 424 (1937).

[550] PEASE, R. S.: Sputtering of Solids by Penetrating Ions. Rendiconti S.I.F. Corso **13**, 158 (1960).

[551] PENNING, F. M.: The Action of Positive Ions in an Independent Gas Discharge. Physica **8**, 13 (1928).

[552] — Liberation of Electrons from a Metal Surface by Positive Ions. Proc. Roy. Acad. Sci. (Amsterdam) **31**, 14 (1928).

[553] —, and J. H. A. MOUBIS: Cathode Sputtering in a Magnetic Field. Koninkl. Med. Akad. Wetenshap. Proc. **43**, 41 (1940).

[554] PEREZ, S. S., and O. R. FOZ GAZULLA: Über die Druckabhängigkeit der Wärmeleitung und die Bildung von Doppelmolekeln im Schwefeldioxyd. Z. physik. Chem. A **193**, 162 (1943).

[555] PEROVIĆ, B. D.: Cathode Sputtering of Cu and Pb Single Crystals by High Energy A^+ Ions. Bull. Inst. Nuclear Sci. "Boris Kidrich" Vol. **11**, No. 226, 37 (1961).

[556] PERRY, R. L., and J. J. BARRY: Changes in Photoelectric Probability Factor Resulting from Surface Contamination of Aluminum. Phys. Rev. **98**, 281 (1955).

[556a] PETRICK, E. N., O. K. HUSMANN, and H. W. SZYMANOWSKI: Analytical and Experimental Investigation of Compact Charge Ionization. Report CWR-700-10 June 1, 1960, Curtiss-Wright Corporation, Quehanna, Pennsylvania.

[557] PETROV, N. N.: Secondary Emission from Metallic Surfaces under the Action of Positive Ions. Soviet Phys.-Solid State **2**, 857 (1960).

[558] — Secondary Emission from Incancescent Metal Bombarded by Cesium and Potassium Ions. Soviet Phys.-Solid State **2**, 865 (1960).

[559] — Ejection of Electrons from Metals by Ions. Soviet Phys.-Solid State **2**, 1182 (1960).

[560] — Investigation of Collision of Ions with Metal Surfaces. Izvest. Akad. Nauk S.S.S.R. Ser. Fiz. **24**, No. 6, 673 (1960).

[561] —, and A. A. DOROZHKIN: Emission From Tungsten Induced By Certain Positive Ions. Soviet Phys.-Solid State **3**, 38 (1961).

[562] PHILBERT, G.: L'emission électronique provoquée par l'impact d'ions sur des cibles de molybdène et de carbone. Compt. Rendus. Acad. Sci. (Paris) **237**, 882 (1953).

[563] PHIPPS, T. E., and M. J. COPLEY: The Surface Ionization of Potassium on Tungsten. Phys. Rev. **45**, 344 (1934).

[564] — — Reflection Coefficient of Electrons. Phys. Rev. **46**, 144 (1934).

[565] PHIPPS, T. E., and M. J. COPLEY: The Surface Ionization of Potassium on Tungsten. Phys. Rev. **48**, 960 (1935).

[566] — — Surface Ionization of Potassium Iodide on Tungsten. J. chem. Physics **3**, 594 (1935).

[567] — —, and J. O. HENDRICKS: Evidence for Halogen Films on Tungsten in the Surface Ionization of Potassium Halides. J. chem. Physics **5**, 868 (1937).

[568] —, and A. A. JOHNSON: Differential Method Applied to the Surface Ionization of Sodium Halides on Tungsten. J. chem. Physics **7**, 1039 (1939).

[568a] PIERCE, J. R.: Rectilinear Electron Flow in Beams. J. appl. Physics **11**, 548 (1940).

[568b] PIERCY, G. R., F. BROWN, J. A. DAVIES, and M. MCCARGO: Experimental Evidence for the Increase of Heavy Ion Ranges by Channeling in Crystalline Structure. Phys. Rev. Letters **10**, 399 (1963).

[569] PIEROTTI, R. A., and G. D. HALSEY: The Interaction of Krypton with Metals. J. physic. Chem. **63**, 685 (1959).

[570] PITKIN, E. T., M. A. MACGREGOR, V. SALEMME, and R. BIERGE: Investigation of the Interaction of High Velocity-Ions with Metallic Surface. ARL Tr-60-299 Office of Technical Services, Dept. of Commerce, Washington D.C., 1960.

[571] — Sputtering Due to High Velocity Ion Bombardment. Progr. Astronaut. Rocketry **5**, 195 (1961).

[572] PLESHIVTSEV, N. V.: Sputtering of Copper by Hydrogen Ions with Energies up to 50 keV. Soviet Phys.-JETP **37** (10), 878 (1960).

[573] PLOCH, W.: Massenabhängigkeit der Elektronenauslösung durch isotope Ionen. Z. Physik **130**, 174 (1951).

[574] POHL, R., and P. PRINGSSHEIM: Über die Herstellung von Metallspiegeln durch Destillation im Vakuum. Verh. dtsch. physik. Ges. **14**, 506 (1912).

[575] POLANYI, M.: Theories of the Adsorption of Gases. General Survey and Some Additional Remarks. Trans. Faraday Soc. **28**, 316 (1932).

[576] POTTER, J. G.: Temperature Dependence of the Work Function of Tungsten from Measurement of Contact Potentials by the Kelvin Method. Phys. Rev. **58**, 523 (1940).

[577] POWELL, C. F., and L. BRATA: Emission of Metallic Ions from Oxide Surfaces. II. Mechanism of the Emission. Proc. Roy. Soc. **141**, 463 (1933).

[578] —, I. E. CAMPBELL, and B. W. GONSER: Vapor Plating. New York: John Wiley & Sons. Inc. 1955.

[578a] PROBST, F. M.: Energy Distribution of Electrons Ejected from Tungsten by He^{+}. Phys. Rev. **129**, 7 (1963).

[579] PROSEN, E. J. R., R. G. SACHS, and E. TELLER: Van der Waals Interaction of Metals and Molecules. Phys. Rev. **57**, 1066 (1940).

[580] RAINES, B.: The Accommodation Coefficient of He on Nickel. Phys. Rev. **56**, 691 (1939).

[581] RAMSEY, N.: Molecular Beams. Oxford: Oxford Univ. Press 1956.

[582] RAUH, E. H.: Work Function, Ionization Potential and Emissivity of Uranium. ANL-5534, May 1956.

[583] READ, G. E.: The Reflection of Positive Rays by a Platinum Surface. Phys. Rev. **31**, 629 (1928).

[584] REIMANN, A. L.: Thermionic Emission. New York: John Wiley & Sons, Inc. 1934.

[585] — The Temperature Variation of the Work Function of Clean and of Thoriated Tungsten. Proc. Roy. Soc. (London) **163**, 499 (1937).

[586] Research Staff of General Electric, Ltd.: London, Cathode Disintegration. Philos. Mag. **45**, 98 (1923).

[587] Reynolds, J. H.: The Surface Ionization of Lanthanum. Phys. Rev. **85**, 770 (1952).

[588] Reynolds, F. L., and M. C. Michel: Thermal Ionization at Hot Metal Surfaces. US AEC Report, UCRL-3950 (1957).

[588a] — Ionization on Tungsten Single-Crystal Surfaces. J. chem. Physics **39**, 1107 (1963).

[588b] Reynolds, T. W., and L. W. Kreps: Gas Flow, Emittance, and Ion Current Capabilities of Porous Tungsten. Report NASA-TR-TND-871 (1961).

[589] Rhodin, T. N.: Physical Adsorption of Single-Crystal Zinc Surfaces. J. physic. Chem. **57**, 143 (1953).

[590] — Surface Studies with the Vacuum Microbalance-Instrumentation and Low-Temperature Applications. Advances in Catalysis **5**, 1 (1953).

[591] Riddoch, A., and J. H. Leck: Positive Ion Emission from Metal Surfaces Caused by Ion Bombardment. Proc. physic. Soc. **72**, 467 (1958).

[592] Roberts, J. K.: The Exchange of Energy Between Gas Atoms and Solid Surfaces. Proc. Roy. Soc. (London) A **129**, 146 (1930).

[593] — The Exchange of Energy Between Gas Atoms and Solid Surfaces II. The Temperature Variation of the Accommodation Coefficient of Helium. Proc. Roy. Soc. (London) A **135**, 192 (1932).

[594] — The Exchange of Energy Between Gas Atoms and Solid Surfaces. The Accommodation Coefficient of Neon. Proc. Roy. Soc. (London) A **142**, 518 (1933).

[595] — The Absorption of Hydrogen on Tungsten. Proc. Roy. Soc. (London) A **152**, 445 (1935).

[596] — Some Properties of Absorbed Films of Oxygen on Tungsten. Proc. Roy. Soc. (London) A **152**, 464 (1935).

[597] Robinson, C. F.: Effect of Various Gases on Potassium Ion Emission from Hot Platinum. J. chem. Physics **20**, 1329 (1952).

[597a] Robinson, M. T., D. K. Holmes, and O. S. Oen: Ranges of Energetic Atoms in Solids. II. Lattice Model. Bull. Amer. Phys. Soc. **7**, 171 (1962).

[598] Rodebusch, W. H., and W. A. Nichols: Atomic Oxygen as a Reducing Agent. Phys. Rev. **35**, 649 (1930).

[599] —, and W. F. Henry: Molecular Beams of Salt Vapors. Phys. Rev. **39**, 386 (1932).

[600] Roginski, S. S.: Adsorption und Katalyse an inhomogenen Oberflächen. Akademie Verlag, Berlin 1958.

[601] Rohn, K.: Herstellung dünner Metall- und Salzschichten durch Aufdampfen in Vakuum. Z. Physik **126**, 20 (1949).

[602] Rol, P. K., J. M. Fluit, and J. Kistemaker: Sputtering of Copper by Ion Bombardment in the Energy Range of 5—25 keV. Terzo Comgr. Int. sui fenomeni d'Ionizzazione nei Gas. Soc. It. di Fis 872 (1957).

[603] — — — Sputtering of Copper by Bombardment with Ions of 5—25 keV. Physica **26**, 1000 (1960).

[604] — — — Theoretical Aspects of Cathode Sputtering in the Energy Range of 5—25 keV. Physica **26**, 1009 (1960).

[605] — —, F. P. Viehböck, and M. de Jong: Sputtering of Copper-Monocrystals By Bombardment with 20 keV Ar^+. Proc. of 4th Internatl. Conference on Ionization Phenomena in Gases. Amsterdam: North Holland Publ. Co. 1960, ID 257.

[606] ROLF, P.: The Accommodation Coefficient of Helium on Platinum. Phys. Rev. **65**, 185 (1944).

[607] ROMANOW, A. M., and S. V. STARODUBTSEV: Trans. Physico Techn. Inst. Acad. Sci. Uzbek. S.S.R. **4**, 102 (1952).

[608] — — Absorption and Ionization of Sodium on Hot Tungsten. Soviet Phys.-Techn. Physics **27** (2), 652 (1957).

[609] — Ionization of Lithium on Tungsten. Soviet Phys.-Techn. Physics **27** (2), 1125 (1957).

[610] — — Role of Surface Nonuniformity in the Adsorption and Ionization of Sodium and Lithium on Tungsten. Izvest. Akad. Nauk Uzbek. S.S.R. Fiz. Mat. Ser. 2, 11 (1957).

[611] VON ROOS, O.: Theorie der Reflexion positiver Ionen an Metalloberflächen. Z. Physik **147**, 184 (1957).

[612] — Theorie der kinetischen Emission von Sekundärelektronen, ausgelöst durch positive Ionen. Z. Physik **147**, 210 (1957).

[613] ROSENFELD, S., and W. M. HOSKINS: A Modified Zisman Apparatus for Measuring Contact Potential Differences in Air. Rev. sci. Instruments **16**, 343 (1945).

[614] ROSTAGNI, A.: Researches on Positive and Neutral Rays. II. Liberation of Electrons from Metal Surfaces. Nuovo Cimento **11**, 99 (1934).

[615] — Untersuchungen über langsame Ionen und Neutralstrahlen. Z. Physik **88**, 55 (1934).

[616] ROTHMAN, S. J., R. WEIL, and L. T. LLOYD: Preparation of Diffusion Couples by Cathodic Sputtering. Rev. sci. Instruments **30**, No. 7, 541 (1959).

[617] ROWLEY, H. H., and K. F. BONHOEFFER: Über den Energieaustausch an der Grenzfläche Platin/Wasserstoff. Z. physik. Chem. B **21**, 84 (1933).

[618] —, and W. V. EVANS: Accommodation Coefficient of Hydrogen on Iron. J. Amer. chem. Soc. **57**, 2059 (1935).

[619] RÜCHARDT, E.: Durchgang von Kanalstrahlen durch Materie. Handbuch der Physik (Geiger-Scheel) XXII, 2, 75 (1933).

[620] SACHTLER, W. M H.: Calculations of the Heat of Adsorption of Hydrogen on Platinum. Recueil Trav. chim. Pays-Bas **72**, 897 (1953).

[621] — G. J. H. DORGELO, and W. VAN D. KNAAP: Structure and Constitution of Evaporated Nickel Films Determined by the Electron Microscope and Electron Diffraction. J. Chim. physique **51**, 491 (1954).

[622] — Halbempirische Methode zur Berechnung des Elektronenaustritts-Potentials von Metallen. Z. Elektrochem. **59**, 119 (1955).

[623] —, and G. J. H. DORGELO: The Polarity of the Chemisorptive Bonding. Measure of the Potential of the Surface and of the Conductivity on Evaporated Metallic Films. J. Chim. physique **54**, 27 (1957).

[624] SAMPSON, M. B., and W. BLEAKNEY: A Mass-Spectrograph Study of Ba, Sr, In, Ga, Li, and Na. Phys. Rev. **50**, 456 (1936).

[625] SASAKI, N., and T. YUASA: Chemical Studies on Ion Emission. I. The Effect of Beryllia on the Electrical Emission from Alumina Coated Tungsten Filament. J. chem. Soc. Japan, Pure Chem. Sect. **73**, 273 (1952).

[626] —, and M. ONCHI: Effect of Vapours of Halogen Compounds on the Emission of Positive Ions From Platinum. Nature (London) **174**, 84 (1954).

[627] SAWYER, R. B.: The Reflection of Lithium Ions from Metal Surfaces. Phys. Rev. **35**, 124, 1090 (1930).

[628] SCHAAFS, W.: Messung von Voltaspannungen mit Hilfe der Methode des rotierenden Ankers. Z. angew. Physik. **10**, 424 (1958).

[*629*] SCHÄFER, K., and O. R. F. GAZULLA: Über die Druckabhängigkeit der Wärmeleitfähigkeit realer Gase. Z. physik. Chem. B **52**, 299 (1942).
[*630*] —, and G. G. GRAU: Die thermische Akkomodation und die Geschwindigkeit der Energieübertragung von CS_2 an Platin. Z. Elektrochem. **53**, 203 (1949).
[*631*] — Probleme der Wärmeleitung in Gasen bei niedrigem Druck und der Energieübertragung an festen Oberflächen. Fortschr. chem. Forsch. **1**, 61 (1949/50).
[*632*] — Energieübertragungsmechanismus und Reaktionsgeschwindigkeit an metallischen Oberflächen. Z. Elektrochem. **56**, 398 (1952).
[*633*] —, and K. H. RIGGERT: Eine neue Methode zur Bestimmung partieller thermischer Akkommodationskoeffizienten. Z. Elektrochem. Ber. Bunsenges. physik. Chem. **57**, 751 (1953).
[*634*] —, and M. KLINGENBERG: Die partielle thermische Akkommodation verschiedener Gase nach der Band-Draht-Methode an Platin zwischen 0° und 100° C. Z. Elektrochem. Ber. Bunsenges. physik. Chem. **58**, 828 (1954).
[*635*] —, and K. H. RIGGERT: Ein Verfahren zur Bestimmung partieller thermischer Akkommodationskoeffizienten. J. Colloid. Sci. Supplement **1**, 128 (1954).
[*636*] —, and H. GERSTÄCKER: Absorption, partielle thermische Akkommodation von Gasen an Oberflächen und ihr Zusammenhang mit katalytischen Wirkungen. Z. Elektrochem. Ber. Bunsenges. physik. Chem. **60**, 874 (1956).
[*636a*] SCHEER, M. D., and J. FINE: Kinetics of Cs^+ Desorption from Tungsten. J. chem. Physics **37**, 107 (1962).
[*636b*] — — Kinetics of Desorption. II. Cs^+ and Ba^+ from Rhenium. J. chem. Physics **38**, 307 (1963).
[*637*] SCHIEFER, K.: Dissertation. Universität Würzburg. Germany (1955).
[*638*] SCHMIDT, C. G.: Über Ionenstrahlen. Ann. Physik. **82**, 664 (1927).
[*639*] SCHMIDT, R. W.: Über Wachstums- und Abbauformen von Wolframkristallen, insbesondere über den Gleichstromeffekt. Z. Physik **120**, 69 (1942).
[*640*] SCHOTTKY, W.: Thermodynamik der seltenen Zustände im Dampfraum. Thermische Ionisierung und thermisches Leuchten. Ann. Physik **62**, 113 (1920).
[*641*] —, and G. WAGNER: Theory of Regular Mixed Phases. Z. physik. Chem. B **11**, 163 (1930).
[*642*] SCHUIT, G. G. A., and J. H. DE BOER: The Heats of Adsorption of Hydrogen on Nickel-Silica Catalysts. Recueil Trav. chim., Pays-Bas **72**, 909 (1953).
[*643*] SCHULZ, G. J., and R. E. FOX: Excitation of Metastable Levels in Helium Near Treshold. Phys. Rev. **106**, 1179 (1957).
[*644*] SCHWAB, G. M., H. NOLLER, and J. BLOCK: Kinetic der heterogenen Katalyse. Handbuch der Katalyse Bd. V, p. 162. 1957. Wien: Springer-Verlag.
[*645*] SCHWARTZ, M., and P. L. COPELAND: Secondary Emission by Positive Ion Bombardment. Phys. Rev. **96**, 1467 (1954).
[*646*] SEELIGER, R., and K. SOMMERMEYER: Bemerkung zur Theorie der Kathodenzerstäubung. Z. Physik **93** (1935).
[*647*] — Zur Theorie der Kathodenzerstäubung. Z. Physik **119**, 482 (1942).
[*648*] SEITZ, F.: On the Disordering of Solids by Action of Fast Massive Particles. Discuss. Faraday Soc. **5**, 271 (1949).
[*649*] —, and J. S. KOEHLER: Solid State Physics. Vol. 2. New York: Academic Press 1956, p. 305.
[*650*] SHEKHTER, S. S.: Neutralization of Positive Ions and the Emission of Secondary Electrons. J. exper. theor. Physics (S.S.S.R.) **7**, 750 (1937).

[651] SHELTON, H.: Thermionic Emission From a Planar Tantalum Crystal. Phys. Rev. **107**, 1553 (1957).
[652] SILITTO, R. M.: An Extension of Kapitza's Theory of δ-Radiation. Proc. physic. Soc. **60**, 453 (1958).
[653] SILSBEE, R. H.: Focusing in Collision Problems in Solids. J. appl. Physics **28**, 1246 (1957).
[654] SIMS, C. T.: Investigation of Rhenium. Quarterly Progress Report No. 7 for the Period Dec. 25, 1953 to March 22, 1954. Batelle Memorial Institute, Report AD 33442, March 1954.
[654a] SILVERNAIL, W. L.: The Thermal Accommodation of Helium, Neon, and Argon on Clean Tungsten from 77° to 303° K. Thesis, University of Missouri, 1954.
[655] SKOUPY, F.: Chemie und Physik des Glühlampen-Vakuums. Z. tech. Physik **5**, 563 (1924).
[656] SLATER, J. C., and J. G. KIRKWOOD: The Van der Waals Forces in Gases. Phys. Rev. **37**, 682 (1931).
[657] SLOANE, R. H., and R. PRESS: The Formation of Negative Ions by Positive Ion Impact on Surfaces. Proc. Roy. Soc. (London) A **168**, 284 (1938).
[658] SLODZIAN, G.: Sur l'émission électronique secondaire des metaux bombardes par des ions positifs. Compt. Rendus. Acad. Sci. (Paris) **246**, 3631 (1958).
[659] SMIRNOW, B. G., and G. N. SCHUPPE: The work function of the electrons at some of the faces of a tungsten monocrystal. J. techn. Physics (S.S.S.R.) **22**, 973 (1952).
[660] SMITH, G. F.: Thermionic and Surface Properties of Tungsten Crystals. Phys. Rev. **94**, 295 (1954).
[661] SMITH, S. J., and L. M. BRANSCOMB: Atomic Negative-Ion-Photodetachment Cross Section and Affinity Measurements. Research Natl. Bur. Standards **55**, 165 (1955).
[662] SMOLUCHOWSKIJ, M., VON: Über Wärmeleitung in verdünnten Gasen. Wied. Ann. **64**, 101 (1898).
[663] — Über den Temperatursprung bei Wärmeleitung in Gasen. Sitz. Berichte Akad. Wiss. Wien. **107**, 304 (1898).
[664] — Weitere Studien über den Temperatursprung bei Wärmeleitung in Gasen. Sitz. Berichte Akad. Wiss. Wien **108**, 5 (1899).
[664a] SNOKE, D. R., S. H. FAIRWEATHER, and W. J. SKINNER: Development of Ion Propellant Systems. Thompson-Ramo-Wooldridge Inc., Cleveland, Ohio, Report NP-9677, June 24, 1960.
[665] SODDY, F., and A. J. BERRY: Conduction of Heat through Rarefied Gases. I. Proc. Roy. Soc. (London) A **83**, 254 (1910).
[666] — — Conduction of Heat through Rarefied Gases. II. Proc. Roy. Soc. (London) A **84**, 576 (1911).
[667] SONKIN, S.: The Action of Mercury Metastable Atoms on a Tungsten Surface. Phys. Rev. **43**, 788 (1933).
[668] SOSNOVSKY, H. M. C.: Effect of Crystal Orientation of the Activation Energy for the Catalytic Decomposition of Formic Acid on Silver. J. chem. Physics **23**, 1486 (1955).
[669] — The Catalytic Activity of Silver Crystals of Various Orientations after Bombardment with Positive Ions. J. physic. Chem. Solids **10**, 304 (1959).
[670] SPIVAK, G. W., and E. M. REICHRUDEL: On the Secondary Emission from Collectors in Neon Discharge. Phys. Rev. **42**, 580 (1932).
[671] —, and A. A. ZAITZEV: Energy Exchange Between Neon, Argon, and Mercury Atoms and a Solid Wall. C. R. Acad. Sci. U.S.S.R. **2**, 118 (1935).

[672] SPIVAK, G. W., and E. M. REICHRUDEL: Effect of Metastable Atoms on Electron Temperature in the Positive Column. Physica **3**, 301 (1936).
[673] SPORN, H.: Leuchtzonen vor Glimmentladungskathoden. Z. Physik **112**, 279 (1939).
[674] STANTON, H. E.: On the Yield and Energy Distribution of Secondary Positive Ions from Metal Surfaces. J. appl. Physics **31**, 678 (1960).
[675] STARK, J., and G. WENDT: Über das Eindringen von Kanalstrahlen in feste Körper. Ann. Phys. **38**, 921 (1912).
[676] STARODUBTSEV, S. V.: Dissertation, Leningrad (1938).
[677] — Trans. Physica Techn. Inst. Acad. Sci. Uzbek S.S.R. **1**, 5 (1948).
[678] — Investigation of Adsorption Phenomena by the Method of Modulated Molecular Beams. J. exper. theor. Physics (S.S.S.R.) **19**, 215 (1949).
[679] —, and YU. I. TIMOKHINA: Emission of Positive Ions by Incandescent Oxides of Titanium, Zirconium and Silicon. J. techn. Physics S.S.S.R. **19**, 606 (1949).
[680] — — Anthology Commemorating the 70th Birthday of Academian. A. F. Joffe, 1950, p. 117.
[681] STATESCU, C.: Beiträge zur Kenntnis der Reflection des Lichtes an einer dünnen Metallschicht. Ann. Physik **33**, 1032 (1910).
[681a] STAVISSKII, YA. YA., and S. YA. LEBEDEV: Surface Ionization of Cesium Upon Diffusion through Porous Tungsten. Soviet Phys.-Technical Phys. **5**, 1158 (1960).
[682] STERNGLASS, E. J., and M. M. WACHTEL: Measurement of Low-Energy Electron Absorption in Metals and Insulators. Phys. Rev. **99**, 646 (1955).
[683] — Theory of Secondary Electron Emission by High Speed Ions. Phys. Rev. **108**, 1 (1957).
[683a] STEVENS, CH. M., and A. L. HARKNESS: Suggested Method for Mass Spectrometric Isotopic Analysis of Uranium Using the Single and Multiple Filament Surface Ionization Technique. AEC Technical Publication on Selected Measurement Methods for Plutonium and Uranium in the Nuclear Fuel Cycle (editor R. J. JONES). 1963, Available from Superintendent of Documents. U.S. Government Printing Office, Washington D.C.
[683b] — Isotopic Analysis of Solid Samples by Surface Ionization. Section **13—20** in Chapter **13** on Mass Spectrometry. Book title: The Analysis of Essential Reactor Materials, editor, C.T. Rodden, AEC Technical Publication, 1963.
[684] STEVENSON, D. P.: Heat of Chemisorption of Hydrogen in Metals. J. chem. Physics **23**, 203 (1955).
[685] STRACHAN, C.: The Interaction of Atoms and Molecules with Solid Surfaces IX. The Emission and Absorption of Energy by a Solid. Proc. Roy. Soc. (London) A **158**, 591 (1937).
[686] STRANSKI, I. N.: Growth and Dissolution on Non-Polar Crystals. Z. physik. Chem. (B) **11**, 342 (1931).
[687] — Forms of Equilibrium of Crystals. Discuss. Faraday Soc. **5**, 13 (1949).
[688] STRAUMANIS, M.: Die neuesten Kristallwachstumstheorien und der Versuch. Wiener Chem.-Ztg. **46**, 243 (1943).
[689] STROHMEIER, W.: Diss. Heidelberg, 1948. Über den Zusammenhang zwischen Akkommodationskoeffizienten, Relaxationszeit und Adsorptionsisotherme an festen Oberflächen.
[690] STUART, R. V., and G. K. WEHNER: Sputtering Thresholds and Displacement Energies. Phys. Rev. Letters **4**, 409 (1960).

[691] STUART, R. V., and G. K. WEHNER: Sputtering Yields at Very Low Bombarding Ion Energies. Seventh National Symposium on Vacuum Technology Transactions (Paris). Oxford: Pergamon Press 1960.

[692] — — Annual Report on Sputtering Yields. Annual Report No. 2243, General Mills, Minneapolis 3, Minnesota, November 1961.

[692a] STUHLINGER, E.: Electrical Propulsion Systems for Space Ships with Nuclear Power Source. J. Astronautics, 1955, p. 149.

[693] SUHRMANN, R., and W. M. H. SACHTLER: Lichtelektrische Untersuchungen über die elektronische Wechselwirkung zwischen einer Platinoberfläche und adsorbierten Wasserstoff- bzw. Sauerstoffatomen und Molekeln sowie N_2O-Molekeln. Z. Naturforsch. **9**a, 14 (1954).

[694] —, and K. SCHULZ: Determination of Electronic Interaction between Adsorbed Foreign Molecules and the Surface of Thin Nickel Layers at Low Temperatures by the Aid of Electronic Resistance Measurements. Z. physik. Chem. N. F. **1**, 69 (1954).

[695] — — Electronic Interaction between Chemisorbed Molecules and Adsorbing Surfaces. J. Colloid Science Supplement **1**, 50 (1954).

[696] — Electronic Interaction between Metallic Catalysts and Chemisorbed Molecules. Advances in Catalysts, **7**, 303, 316 (1955).

[697] — Electronic Interaction during Chemisorption on Conducting Surfaces. An Introductory Lecture. Z. Elektrochem. **60**, 804 (1956).

[698] SUTTON, P. P., and J. E. MAYER: Direct Experimental Determination of Electron Affinities; The Electron Affinity of Iodine. J. chem. Physics **3**, 20 (1935).

[699] SZHENOV, YU. K.: Surface Ionization of Calcium, Strontium and Magnesium on Oxidized Tungsten. Soviet Phys.-JETP **29** (**2**), 775 (1956).

[700] — On the Mechanism of Surface Ionization of Atoms of the Alkali Earth Metals. Soviet Phys.-JETP **37** (**10**), 239 (1960).

[701] TAKEISHI, Y.: Auger ejection of electrons from barium oxide by inert gas ions and the cathode fall in the normal glow discharges. J. phys. Soc. Japan **11**, 676 (1956).

[702] — Auger ejection of electrons from nickel by inert gas ions. J. phys. Soc. Japan **13**, 766 (1958).

[703] — Secondary Electrons induced by Ions. Shinku-Kogyo **5**, 153 (1958).

[704] TAYLOR, G. B., G. KISTIAKOWSKI, and J. H. PERRY: Platinum Black Catalysts I. Physical Properties and Catalytic Activity. J. physic. Chem. **34**, 748 (1930).

[705] TAYLOR, J., and I. LANGMUIR: The Evaporation of Atoms, Ions and Electrons from Caesium Films on Tungsten. Phys. Rev. **44**, 423 (1933).

[706] TAYLOR, W. J., and H. L. JOHNSTON: An Improved Hot Wire Cell for Accurate Measurements of Thermal Conductivities of Gases over a Wide Temperature Range. J. chem. Physics **14**, 219 (1946).

[707] TEL'KOVSKII, V. G.: Secondary Electron Emission from Metals Induced by Ions and Neutral Particles. Izvest. Akad. Nauk S.S.S.R. **20**, 334 (1956).

[708] — Secondary-Electron Emission under the Influence of Positive Ions and Neutral Particles. Izvest. Akad. Nauk S.S.S.R. Ser. Fiz. **20**, 1179 (1956).

[709] — Secondary Electron Emission of Metals under the Action of Ions and Neutral Particles. Doklady Akad. Nauk S.S.S.R. **108**, 444 (1956).

[710] TEMKIN, M.: Transition State in Surface Reactions. Acta physicochim. S.S.S.R. **8**, 141 (1938).

[711] — The Arrhenium Equation and the Method of the Active Complex. J. physic. chem. S.S.S.R. **14**, 1054 (1940).

[*712*] TEMKIN, M.: Kinetics of Heterogeneous Catalysis. J. physic. chem. S.S.S.R. **14**, 1153 (1940).

[*713*] — Kinetics of Ammonia Synthesis. J. physic. Chem. S.S.S.R. **14**, 1241 (1940).

[*713a*] — The Reactivity and the Mutual Effect of Adsorbed Atoms and Radicals. Symposium on Problems of Chemical Kinetics, Catalysis and Reactivity. Akad. Nauk S.S.S.R. 1955.

[*714*] THOMAS, L. B., and F. OLMER: The Accommodation Coefficient of Mercury on Platinum and the Heat of Vaporization of Mercury. J. Amer. chem. Soc. **64**, 2190 (1942).

[*715*] — — Accommodation Coefficients of He, Ne, A, H_2, D_2, O_2, CO_2, and Hg on Platinum as a Function of Temperature. J. Amer. chem. Soc. **65**, 1036 (1943).

[*716*] —, and R. E. BROWN: The Accommodation Coefficient of Helium on a Bare Tungsten Surface. J. chem. Physics **18**, 1367 (1950).

[*717*] —, and R. C. GOLIKE: A Comparative Study by the Temperature Jump and Low Pressure Methods and Thermal Conductivities of H, Ne, and Co_2. J. chem. Physics **22**, 300 (1954).

[*718*] —, and E. B. SCHOFIELD: Thermal Accommodation Coefficient of Helium on a Bare Tungsten Surface. J. chem. Physics **23**, 861 (1955).

[*719*] THOMSON, J. J., and G. P. THOMSON: Conduction of Electricity through Gases. Cambridge CUP 1928—1933.

[*720*] THOMPSON, M. W.: A Theory of High-Energy Sputtering based on Focused Collision Sequences. Proc. V. Internatl. Symposium of Ionization of Gases, Munich, 85, Oct. 1961.

[*720a*] — The Ejection of Atoms from Gold Crystals During Proton Irradiation. Philos. Mag. **4**, 139 (1959).

[*721*] —, and R. S. NELSON: Atomic Collision Sequences in Crystals of Copper, Silver and Gold Revealed by Sputtering in Energetic Ion Beams. Proc. Roy. Soc. (London) A **259**, 458 (1961).

[*721a*] — — Evidence for Heated Spikes in Bombarded Gold from the Energy Spectrum of Atoms Ejected by 43 keV A^+ and Xe^+ Ions. Philos. Mag. **7**, 2015 (1962).

[*722*] TIMOKHINA, Y. I., and G. N. SCHUPPE: The Negative Ionization of Chlorine on a Thoriated Wolfram Surface. Trans. Physico-Tech. Inst., Acad. Sci. Uzbek S.S.R. **1**, 120 (1947).

[*723*] — Doklady Akad. Nauk S.S.S.R. **58**, 303 (1947).

[*724*] TOMLINSON, R. H., and A. K. DAS GUPTA: Use of Isotope Dilution in Determination of Geological Age of Minerals. Canad. J. Chem. **31**, 909 (1953).

[*725*] TOWNES, C. H.: Theory of Cathode Sputtering in Low Voltage Gaseous Discharges. Phys. Rev. **65**, 319 (1944).

[*726*] TOYE, T. C.: The Effect of Copper on the Work Function of Liquid Tin. Proc. physic. Soc. **73**, 807 (1959).

[*727*] TRAPNELL, B. M. W.: Adsorption on Evaporated Tungsten Films. II. The Chemisorption of Hydrogen and the Catalytic Parahydrogen Conversion. Proc. Roy. Soc. (London) A **206**, 39 (1951).

[*728*] — Chemisorption. New York, London: Academic Press, Butterworths Scientific Publications 1955.

[*729*] TRILLAT, J. J.: Etude des structures superficielles par diffraction electronique et bombardement ionique combines. Applications. J. chim. physique **53**, 570 (1956).

[*730*] TRISCHKA, J. W., D. T. F. MARPLE, and A. WHITE: The Production of Halogen Negative Ions at the Surface of a Thoriated Tungsten Filament. Phys. Rev. **85**, 137 (1952).

[731] von Ubisch, H.: On the Conduction of Heat in Rarefied Gases and Its Manometric Application. I. Appl. sci. Res. A **2**, 364 (1949—1950).
[732] — On the Conduction of Heat in Rarefied Gases and its Manometric Application. II. Appl. sci. Res. A **2**, 403 (1949—1950).
[733] Vapnik, V. N., L. G. Gurvich, and N. V. Zinov'ev: Contribution to the Theory of Scattering of Ions from Metal Surfaces. Izvest. Akad. Nauk S.S.S.R., Ser. Fiz. **24**, 685 (1960).
[734] Varnerin, L. J., and J. H. Carmichael: Trapping of Helium Ions and the Re-Emission of Trapped Atoms from Molybdenum. J. appl. Physics **28**, 913 (1957).
[735] Varney, R. N.: Liberation of Electrons by Positive-Ion Impact on the Cathode of a Pulsed Townsend Discharge Tube. Phys. Rev. **93**, 1156 (1954).
[736] Veith, W.: Elektronenerregung und Trägerreflexion beim Auftreffen von K^+-Trägern auf Metalle. Ann. Physik **29**, 189 (1937).
[737] Veksler, V. I., and G. N. Schuppe: Conversion of Ions on a Metal Surface. J. techn. Physics S.S.S.R. **23**, 1573 (1953).
[738] —, and M. B. Ben'iaminovich: The Production of Secondary Ions from Tantalum and Nickel Bombarded by Positive Cesium Atoms. J. techn. Physics (S.S.S.R.) **26**, 1671 (1956). — Soviet Phys.-Techn. Phys. **1**, 1626 (1956).
[739] — Energy Distribution of Sputtered and Scattered Ions in the Bombardment of Tantalum and Molybdenum by Positive Cesium Ions. Soviet Phys.-JETP **11**, 235 (1960).
[740] — Interaction of Slow Positive Rubidium and Cesium Ions with the Surface of Molybdenum. Soviet Phys.-JETP **15**, 222 (1962).
[740a] — Energy Spectra of Slow Positive Rubidium and Cesium Ions Scattered by Molybdenum Surface. Soviet Phys.-Solid State **4**, 1043 (1962).
[741] Vier, D. T., and J. E. Mayer: A Direct Experimental Determination of the Electron Affinity of O. J. chem. Physics **12**, 28 (1944).
[742] Vol'kenshtein, F. F.: The Electron Levels of Atoms Adsorbed on a Crystal Surface. J. physic. Chem. (S.S.S.R.) **21**, 1317 (1947).
[743] — Electron Theory of Promoting and Poisoning Ionic Catalysts. J. physic. Chem. (S.S.S.R.) **22**, 311 (1948).
[744] — Some peculiarities of adsorption due to "Thermal disorder" on crystal surfaces. J. physic. Chem. (S.S.S.R.) **23**, 917 (1949).
[745] Vollmer, M.: Kinetik der Phasenbildung. Dresden und Leipzig, 1939.
[746] Voshage, H., and H. Hintenberger: Zur quantitativen massenspektrometrischen Bestimmung extrem kleiner Alkalimengen nach der Methode der vollständigen Verdampfung. Z. Naturforsch. **14**a, 216 (1958).
[747] Wachman, Y. H.: The Thermal Accommodation Coefficient and Adsorption on Tungsten. Thesis, Univ. of Missouri 157 (1957), Columbia, University of Missouri.
[748] Wahba, M., and C. Kemball: Heats of Adsorption of Ammonia and Hydrogen on Metal Films. Trans. Faraday Soc. **49**, 1351 (1953).
[749] Wagner, J. B., and A. T. Gwathmey: The Formation of Powder and its Dependence on Crystal Face During the Catalytic Reaction of Hydrogen and Oxygen on a Single Crystal of Copper. J. Amer. chem. Soc. **76**, 390 (1954).
[750] Walcher, W.: Über einen Massenspektrographen hoher Intensität und die Trennung der Rubidiumisotope. Z. Physik **108**, 376 (1938).
[751] — Über die Verwendungsmöglichkeiten von Glühanoden zur Massenspektroskopischen Isotopentrennung. Z. Physik **121**, 604 (1943).

[751a] WALTHER, V., and H. HINTENBERGER: Untersuchung über die Emission positiver Sekundärionen und die Reflexion von Edelgasionen an Festkörperoberflächen. Z. Naturforsch. **18**a, 843 (1963).
[752] WARD, A. F. H.: The Sorption of Hydrogen on Copper. Part I. Adsorption and the Heat of Adsorption. Proc. Roy. Soc. (London) A **133**, 506 (1931).
[752a] WARD, J. W.: A Versatile Cathodic Etcher. Los Alamos Scientific Laboratory, Los Alamos, New Mexico, Report TID-14534, Jan. 16, 1962.
[753] WATERS, P. M.: Kinetic Ejection of Electrons from Tungsten by Cesium Ions. Phys. Rev. **109**, 1466 (1958).
[754] — Kinetic Ejection of Electrons from Tungsten by Cesium and Lithium Ions. Phys. Rev. **111**, 1053 (1958).
[755] WEBB, H.W.: The Metastable State in Mercury Vapor. Phys. Rev. **24**, 113 (1924).
[756] WEBER, S.: Über die Abhängigkeit des Temperatursprunges von dem Akkommodationskoeffizienten nebst einigen verwandten Wärmeleitungsproblemen in Gasen. Kgl. danske Vidensk. Selsk. mat.-fys. Medd. **16**, 9 (1939).
[757] — Über den Einfluß des Akkommodationskoeffizienten auf die Wärmeleitung und Radiometerkraft in Gasen. Kgl. danske Vidensk. Selsk. mat.-fys. Medd. **19**, 11 (1942).
[758] WEHNER, G. K.: Threshold Energies for Sputtering and the Sound Velocity in Metals. Phys. Rev. **93**, 633 (1954).
[759] —, and G. MEDICUS: Sputtering at Low Ion Velocities. J. appl. Physics **25**, 698 (1954).
[760] — Sputtering by Ion Bombardment. Advances in Electronics and Electron Physics **7**, 239 (1955).
[761] — Sputtering of Metal Single Crystals by Ion Bombardment. J. appl. Physics **26**, 1056 (1955).
[762] — Controlled Sputtering of Metals by Low Energy Hg-Ions. Phys. Rev. **102**, 690 (1956).
[763] — Sputtering Yields for Normally Incident Hg-Ion Bombardment at Low Ion Energy. Phys. Rev. **108**, 35 (1957).
[764] — Low Energy Sputtering Yields in Hg. Phys. Rev. **112**, 1120 (1958).
[765] — Sputtering Yield of Germanium in Rare Gases. J. appl. Physics **30**, 274 (1959).
[766] — Velocities of Sputtered Atoms. Phys. Rev. **114**, 1270 (1959).
[767] — Influence of the Angle of Incidence on Sputtering Yields. J. appl. Physics **30**, 1762 (1959).
[768] — Forces on Ion-Bombarded Electrodes in a Low Pressure Plasma. J. appl. Physics **31**, 1392 (1960).
[769] — Annual Report on Sputtering Yields. Annual Report No. 2136, 1960, General Mills, Research Department, Minneapolis 13, Minnesota.
[770] —, and D. ROSENBERG: Angular Distribution of Sputtered Materials. J. appl. Physics **31**, 177 (1960).
[771] — Surface Bombardment Studies. General Mills Research Department, Minneapolis 13, Minnesota, Annual Report No. 2240, November 1961.
[772] — Annual Report on Sputtering Yields. General Mills Electronics Group, Research Department, Minneapolis 13, Minnesota, Report No. 2243, November 1961.
[773] —, and D. ROSENBERG: Mercury Ion Beam Sputtering of Metals at Energies 4—15 keV. J. appl. Physics **32**, 887 (1961).
[774] WEIERSHAUSEN, W.: Ionization of Silver and Copper in a Triple Filament Source. Advances in Mass Spectrometry. Pergamon Press 1958, p. 120.

[775] WEISS, A., L. HELDT, and W. J. MOORE: Sputtering of Silver by Neutral Beams of Hydrogen and Helium. J. chem. Physics **29**, 7 (1958).

[776] WEISSLER, G. L., and T. N. WILSON: Work Functions of Gas-Coated Tungsten and Silver Surfaces. J. appl. Physics **24**, 472 (1953).

[777] WELANDER, P.: On the Temperature Jump in a Rarefied Gas. Arkiv fysik **7**, 507 (1954).

[778] WERNING, J. R.: Thermal Ionization at Hot Metal Surfaces. Thesis, UCRL-8455 (1958).

[779] WESTRICK, R., and P. ZWIETERING: Pseudomorphism in the Iron Synthetic Ammonis Catalyst. Koningl. Ned. Akad. Wetenschap. Amsterdam, Proc. B **56**, 492 (1953).

[780] WHEELER, A. [see R. GOMER, and C. S. SMITH: Structure and Properties of Solid Surfaces. University of Chicago Press **439**, 455 (1953)].

[781] WHITE, W. C.: Positive-Ion Emission, A Neglected Phenomenon. Proc. Inst. Radio Engr. **38**, 852 (1950).

[782] WILSON, A. H.: The Theory of Metals. Cambridge University Press 1953.

[783] WISKOTT, D.: Zur Theorie des Auflicht-Elektronenmikroskops. I. Geometrische Elektronenoptik in der Umgebung des Objekts. Optik **13**, 463 (1956).

[784] — Zur Theorie des Auflicht-Elektronenmikroskops. II. Wellenmechanische Elektronenoptik in der Umgebung des Objekts. Optik **13**, 481 (1956).

[785] WOODCOCK, K. S.: The emission of negative ions under the bombardment of positive ions. Phys. Rev. **38**, 1696 (1931).

[786] YONTS, O. C.: Sputtering of Copper by Ions of Various Energies. TID-7558, 334 (October 1958).

[787] —, C. E. NORMAND, and D. E. HARRISON: High Energy Sputtering. J. appl. Physics **31**, 447 (1960).

[788] YOUNG, D. M.: The Adsorption of Argon on Actahedral Potassium Chloride. Trans. Faraday Soc. **48**, 548 (1952).

[788a] —, and A. D. CROWELL: Physical Adsorption of Gases, London, Butterworth, 1962.

[789] YURASOVA, V. E.: Modern Theories of Cathode Sputtering and the Microrelief of the Damaged Metal Surfaces. Soviet Phys.-Techn. Physics **3**, 1806 (1958).

[790] —, N. V. PLESHIVTSEV, and I. V. ORFANOV: Directed Emission of Particles from a Copper Single Crystal Sputtered by Bombardment with Ions up to 50 keV Energy. Soviet Phys.-JETP **37** (**10**), 689 (1960).

[791] YUTERHOEVEN, W., and M. C. HARRINGTON: Secondary emission of nickel under positive ion bombardment in the positive column in neon. Phys. Rev. **35**, 124 (1930).

[792] ZACHARJIN, G., and G. W. SPIVAK: Energy exchange of He, Ne, and A Atoms with a Metal Surface. Physik Z. Sowjetunion **10**, 495 (1936).

[793] ZANDBERG, E. YA.: Comparative Study of the Interaction of Analogous Positive and Negative Ions of Equal Energy with the Metal Targets. J. techn. Physics (S.S.S.R.) **25**, 8, 1386 (1955).

[794] — Surface Ionization of Potassium Atoms and Potassium Chloride and Cesium Chloride Molecules and Tungsten Filaments in Electric Fields of up to 2 Mv/cm. Soviet Phys.-Techn. Physics **2**, 2399 (1957).

[795] — The Surface Ionization of NaCl and LiCl Molecules on Tungsten in Electrical Fields up to 1.3×10^6 V/cm. Soviet Phys.-Techn. Physics **3**, 2233 (1958).

[796] ZANDBERG, E., YA., and N. I. IONOV: The Surface Ionization of Lithium Atoms on Polycrystalline Tungsten in Electrical Fields up to 1.3×10^6 V/cm. Soviet Phys.-Techn. Physics **3**, 2243 (1958).
[797] — — Surface Ionization. Soviet Phys.-Uspekhi **67** (2), 255 (1959).
[798] ZEMEL, J.: Surface Ionization Phenomena on Polycrystalline Tungsten. J. chem. Physics **28**, 410 (1958).
[799] ZENER, CL.: Interchange of Translational, Rotational and Vibrational Energy in Molecular Collisions. Phys. Rev. **37**, 556 (1931).
[800] — Elastic Reflection of Atoms from Crystals. Phys. Rev. **40**, 178 (1932).
[801] — The Exchange of Energy Between Monatomic Gases and Solid Surfaces. Phys. Rev. **40**, 335 (1932).
[802] — Note on Accommodation Coefficients. Phys. Rev. **40**, 1016 (1932).
[803] ZETTLEMOYER, A. C., YUNG-FANG YU, and T. T. CHESSIK: Adsorption Studies on Metals. IV. The Physical Adsorption of Argon on Oxide-Coated and Reduced Nickel. J. physic. Chem. **59**, 588 (1955).
[804] — Modern Techniques for Investigating Interactions with Surfaces. Chem. Reviews **59**, 937 (1959).
[805] ZIMM, J. B. H., and J. E. MAYER: Vapor Pressures, Heats of Vaporization, and Entropies of Some Alkali Halides. J. chem. Physics **12**, 362 (1944).
[806] ZISMANN, W. A.: A New Method of Measuring Contact Potential Differences in Metals. Rev. sci. Instruments **3**, 367 (1932).
[806a] ZUCCARO, D., R. C. SPEISER, and J. M. TEEM: Characteristics of Porous Surface Ionizers. Progress in Astronautics and Rocketry (Electrostatic Propulsion) **5**, 107 (1961) (Academic Press, New York).
[807] ZWIETERING, P., and T. J. ROUKENS: The Kinetics of the Chemisorption of Nitrogen on Iron Catalysts. Trans. Faraday. Soc. **50**, 178 (1954).

Author Index

Subject Index

Wiesbadener Graphische Betriebe GmbH